PETER MARSHALL is a S
and raised in the Orkne
University of Warwick, a
the Royal Historical Society. Marshall won the prestigious Wolfson History Prize in 2018 for his book *Heretics and Believers*.

A *WALL STREET JOURNAL* BOOK OF THE YEAR

A *TLS* BOOK OF THE YEAR

LONGLISTED FOR THE HIGHLAND BOOK PRIZE

'A surprising page-turner, full of humour and startling details ... In *Storm's Edge*, Marshall set out to "make the peripheral central", and so he has'
The Times

'An astonishing tour de force ... Marshall is keen to show that, for those who lived there, Orkney remained the still point of the turning world: its own small, green universe. So, at the same time as writing about great national and international events, for which he deploys a staggeringly large *dramatis personae*, he deftly weaves in vignettes of a life that remained unique to Orcadians'
The Spectator

'Engrossing and near-faultless ... Orkney already boasts a roll call of distinguished writers. The list has just got longer'
Literary Review

'A brilliantly sweeping and gloriously detailed history of Orkney that is also a history of Britain, and ultimately of the world. Memories of the sheer pleasure it gave me are warming me still'
TLS, Books of the Year, Tom Holland

‘A book that succeeds brilliantly in making the most familiar episodes of history – the coming of Protestantism to Britain, say, or the settling of America – seem strange. It is typical of this humane, uncanny and compulsively fascinating book that it should begin with the murder of one of Mr Marshall’s own ancestors’

Wall Street Journal, Books of the Year

‘Peter Marshall’s new, very readable history of the archipelago is a wonderful corrective to our tendency to see Scottish history through a lowland lens … I have, I am ashamed to say, never been to Orkney. But reading Marshall’s book might just tempt me to make the journey’

The Herald

‘A remarkable and wonderful book – extraordinary in both scale and erudition. Peter Marshall, Professor of History at Warwick University, is himself an Orcadian. His writing is as personal and engaged as it is meticulously researched … warmly personal, full of humour … this remarkable book teaches us that the big truths are often easier to grasp when they are played out on a smaller canvas’ *Church Times*

‘This is a wonderful history, full of passion and affection for the author’s home … compelling, hugely readable and deeply revealing … Marshall places Orkney in its proper context as a remote place which has nevertheless had a surprisingly large impact upon the history of both Scotland and Britain. It helps, of course, that he writes beautifully’ *Scottish Field*

‘A fascinating account of Orkney’s soul and secrets … Marshall negotiates the local and international, the mundane and the epochal, the representative and the anomalous with enviable skill’

Scotland on Sunday

‘Peter Marshall places his native islands … at the centres of three centuries of history. For anyone wanting to understand Orcadian history and culture from a “we” rather than a “they” perspective, this book is a fascinating read’

Country Life

Storm's Edge

Life, Death and Magic in the Islands of Orkney

PETER MARSHALL

WILLIAM
COLLINS

William Collins
An imprint of HarperCollins*Publishers*
1 London Bridge Street
London SE1 9GF

WilliamCollinsBooks.com

HarperCollins*Publishers*
Macken House, 39/40 Mayor Street Upper
Dublin 1, D01 C9W8, Ireland

First published in Great Britain in 2024 by William Collins
This William Collins paperback edition published in 2025

1

Map and family tree by Martin Brown

A catalogue record for this book is
available from the British Library

ISBN 978-0-00-839442-4

Set in Minion Pro
Printed and bound in the UK using 100%
renewable electricity at CPI Group (UK) Ltd

This book contains FSC™ certified paper and other controlled
sources to ensure responsible forest management.

For more information visit: www.harpercollins.co.uk/green

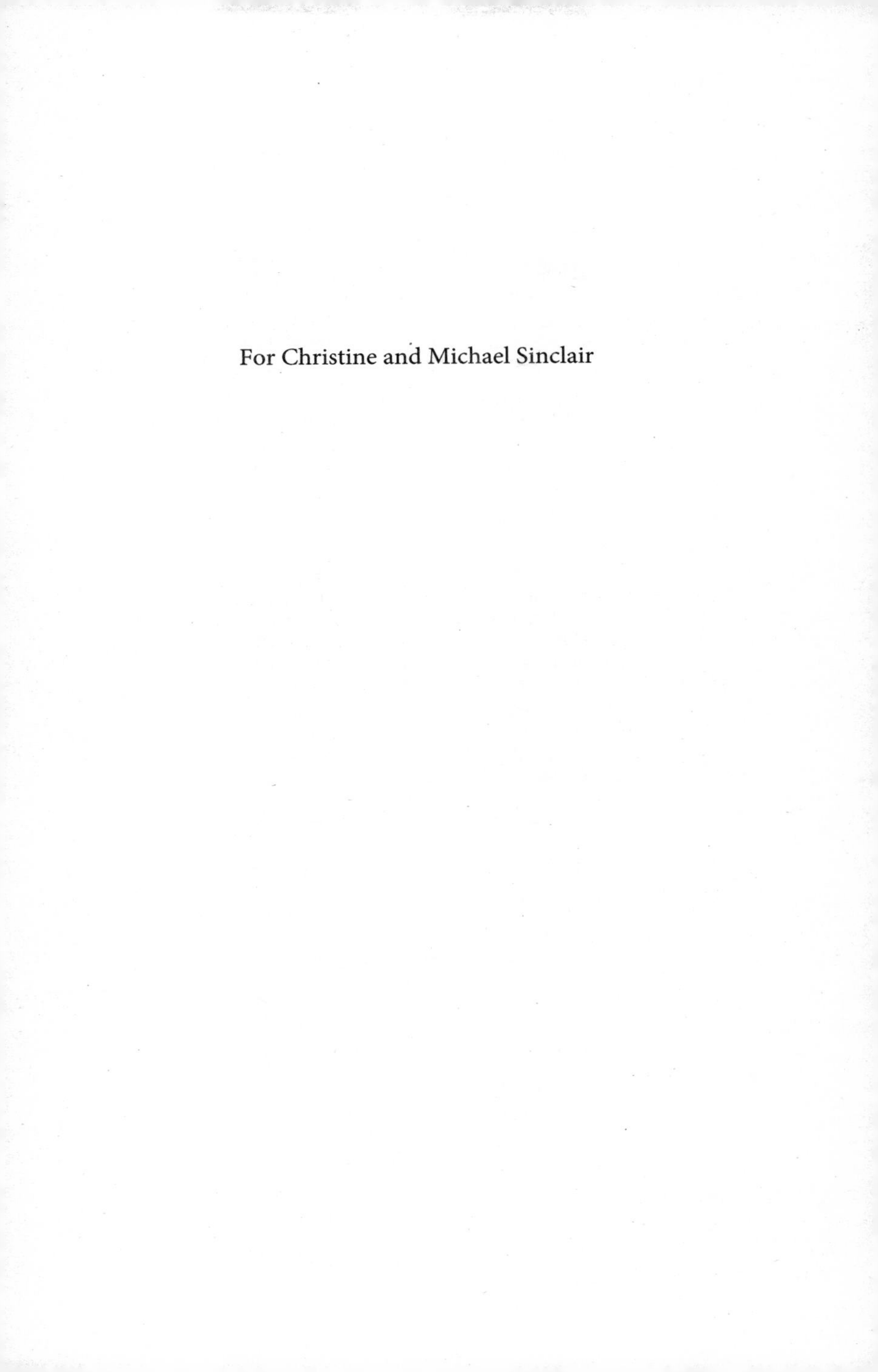

For Christine and Michael Sinclair

Contents

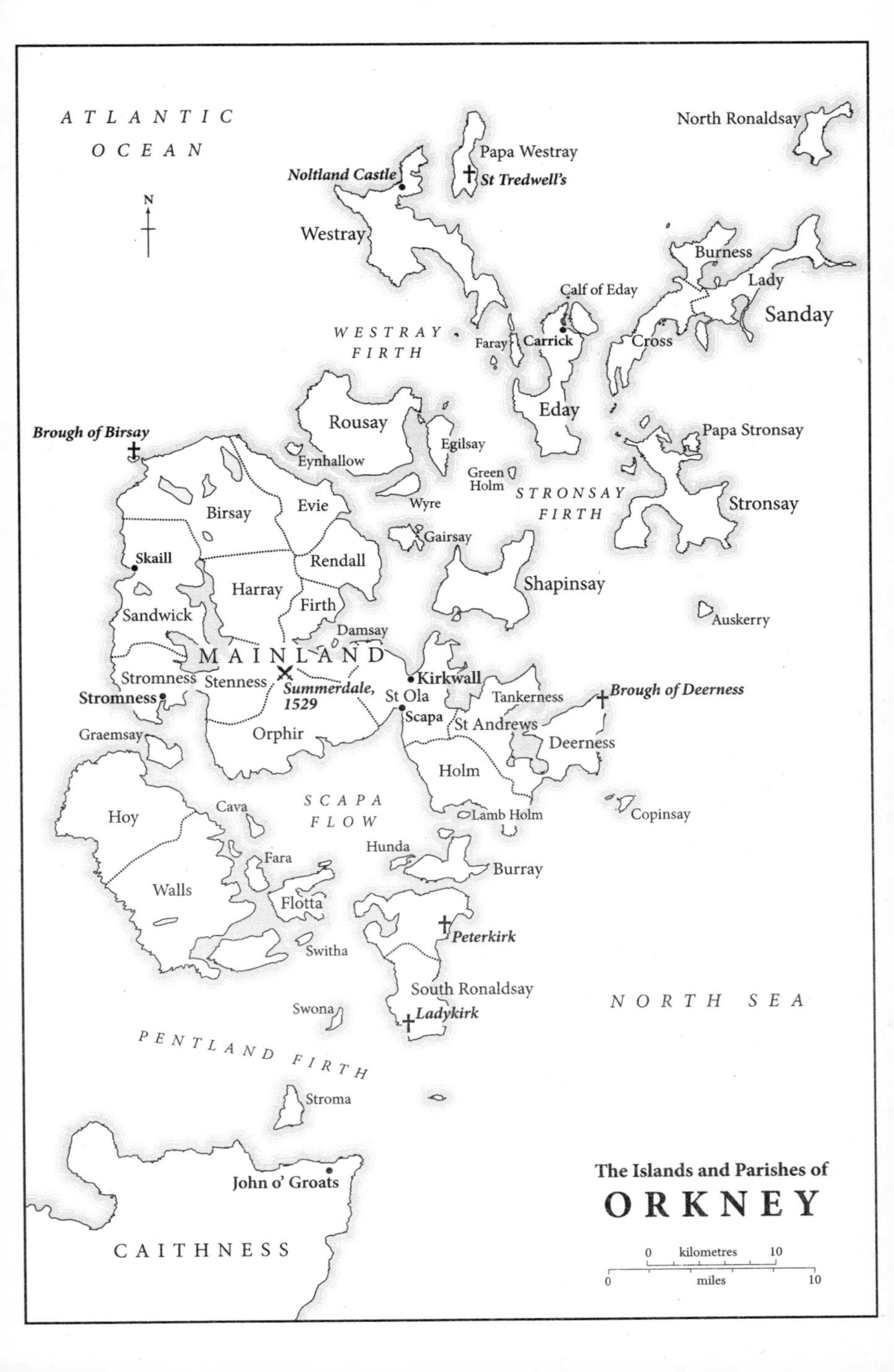
ATLANTIC OCEAN
N
North Ronaldsay
Papa Westray
St Tredwell's
Noltland Castle
Westray
Burness
Lady
Sanday
Calf of Eday
Carrick
Cross
WESTRAY FIRTH
Faray
Eday
Rousay
Egilsay
Papa Stronsay
Brough of Birsay
Eynhallow
Green Holm
Wyre
STRONSAY FIRTH
Stronsay
Birsay
Evie
Gairsay
Skaill
Rendall
Harray
Shapinsay
Firth
Auskerry
Sandwick
Damsay
MAINLAND
Stromness
Stenness
Summerdale, 1529
Kirkwall
St Ola
Tankerness
Brough of Deerness
Stromness
Scapa
St Andrews
Graemsay
Orphir
Deerness
Holm
Hoy
Cava
SCAPA FLOW
Lamb Holm
Copinsay
Hunda
Fara
Burray
Walls
Flotta
Peterkirk
Switha
South Ronaldsay
NORTH SEA
Swona
Ladykirk
PENTLAND FIRTH
Stroma
John o' Groats
CAITHNESS
The Islands and Parishes of
ORKNEY
0 kilometres 10
0 miles 10

Preface

I decided to write this book after I discovered one of my ancestors was murdered by a witch. Her name was Marion Paulson, and she was the wife of James Fotheringhame, who in the early seventeenth century was a farmer in the district of Burness, the northern limb of a thin, three-legged isle called Sanday, which amounts to something under twenty square miles, and lies just above the fifty-ninth parallel in a cluster of windswept islands off the north-east corner of Scotland.

The witch was called Anie Tailzeour, also known as 'Rwna Rowa', or Red Runa. She was, in fairness, provoked. The indictment at her 1624 trial tells us Marion had 'tane ane luik meal furth of your pock'; that is, she had pilfered a handful of ground oats or barley from Anie's bag. In her anger, Anie pronounced a curse on anyone daring to eat her meal: 'whereupon, the said Marion swelled, and now is dead, by your witchcraft and devilry'. This was just one of many accusations against Anie Tailzeour, some going back years. Later that summer, she was burned at the stake, on a bare hillside outside the little port and market town of Kirkwall, the place I was born and grew up.

I cannot demonstrate a direct family connection to Marion and James, but there seems almost certain to be one. Sanday is a small place and Fotheringhame a distinctive local name. My maternal grandfather, Andrew Fotheringhame, a teacher born in 1890, left the island in the 1920s. His own great-grandparents were certainly resident in Sanday in the latter part of the eighteenth century, and the last of my cousins there died in 2009.

Prior to starting this book, I had never been particularly curious about my own family's history. Indeed, professional historians – sad to say – can sometimes be a little snooty about genealogical research and its alleged lack of concern with social and political contexts. Yet the tragically entwined fates of two early seventeenth-century women conjured a moment of unexpected clarity. The dividing walls between our prosaic modern world and an enchanted pre-modern one can be surprisingly thin and brittle. And the cultural and emotional distances between historians and their subjects are only ever as wide as we want to think they are. At least ten members of the assize, or jury, that convicted Anie Tailzeour in 1624 had the same surnames as people with whom I went to school.[1]

There are many things that connect us to our ancestors, but, after DNA, the most important is surely place. The physical environment changes constantly, but we can still walk the same routes, climb the same hills, enjoy the same vistas, as people who lived hundreds of years before us. A sense of connection to lost lives in shared spaces is inescapable in the landscape chronicled by this book: a compact archipelago of around seventy variously sized islands, its limits and possibilities defined by the imminence of the sea, and refined by the whims of the weather.

A word about naming. People will often talk and write about 'the Orkneys', but Orcadians (even long-exiled ones like myself) have a vigorous prejudice against this usage, despite some historical precedent for it. I suspect the reason is the splintered, scattering impression created by 'Orkneys'. 'Orkney', by contrast, conveys a better sense of the unity of a place that, for the entirety of the period covered by this book, was often referred to as 'the country of Orkney'. For similar reasons, I follow the Orcadian habit of locating people, places and events 'in' a particular island, rather than 'on' it – a rooted rather than precarious presence. Orkney folk will sometimes speak about matters 'on the Mainland', though by this they invariably mean the main island of Orkney, not the northern, Scottish half of the next-door island of Great Britain. 'On Orkney' is, of course, a geographical and grammatical nonsense.

These matters seem more concrete from the top of Wideford Hill, a gentle yet elevated climb just to the west of Kirkwall. On a clear day, the summit offers breath-taking views of the Orkney Mainland, east and west, and of the islands, north and south, along an unbroken 360-degree arc. One stands here at the point of imagined convergence of multiple radial lines, from a dispersed array of bays, beaches, lochs, lighthouses, hills, headlands and busy modern wind-turbines.

The effect, paradoxically, is one of both expanse and containment. Scotland's north coast is visible, just, on the southern rim of the horizon, but most of what you can see seems a world in and of itself, domed by the sky and belted by the sea. Here, on a fresh, sunny Monday morning in February 2023, I get talking to a man walking three friendly dogs. It turns out his grandfather was taught by my grandfather at Burness School in Sanday, just after the First World War. The marriage of random encounter and common connection is typical of Orkney, present and past. Kinship and lineage matter here; who you are is mapped and measured by who your people were.

Mention of the Great War dispels any temptation to imagine ourselves gazing upon an enclosed and impenetrable world. This book is a history of Orkney – a most book-worthy place – but it aspires to be more than a 'local' history. The story I want to tell is framed and filled by contact and conjunction. It begins in 1540, with the arrival in the islands of an illustrious royal personage, and ends in 1814, with the visit to Orkney of a scarcely less eminent national luminary.

During the two and three-quarter centuries between those dates, Britain as we know it came into existence. Scholars refer to the era as 'early modern', but that insipid label does little justice to a sequence of dramatic – and often traumatic – processes, which profoundly reshaped patterns of society and redirected the course of countless individual lives.

Across the British Isles, in the sixteenth, seventeenth and eighteenth centuries, ordinary people were submerged by successive waves of uninvited change. They experienced sweeping Reformation of religion; bloody civil conflict, rebellion and dynastic upheaval; the political union of once-sparring nations; the forging of empire, and the pursuit of warfare on a continental and global scale; an at first violent suppression, and then condescending dismissal, of folklore, magic and witches; a fitful flourishing of intellectual 'Enlightenment'; benefits, and costs, of commercial, agricultural and industrial 'revolution'.

Storm's Edge offers a fresh, oblique and, I hope, illuminating perspective on all these momentous processes. It does so from the vantage of a place usually omitted entirely from histories of the making of modern Britain, and often accorded only a passing mention in general histories of Scotland. When it appears at all, Orkney, more often than not, is glimpsed as part of a distant duo, '– and Shetland'. There will be recurrent references to Shetland in the pages that follow, but the more northerly archipelago was, and is, a distinct community with its own character and concerns.

What the 'Northern Isles' have in common is their relatively late and fortuitous admission to the kingdom of Scotland, after centuries as a semi-independent Norwegian lordship. In the islands' homegrown (and often excellent) historical writing, as well as in wider traditions of scholarship, 'Viking' Orkney has attracted considerably more attention than the later, post-medieval centuries – if not quite as much as the Neolithic era, when Orkney, with its extraordinary collection of imposing stone monoliths, beautifully constructed settlements, and elaborate multi-occupancy tombs, was a beacon of civilisation for Britain.

Long after the twilight of Maeshowe, Skara Brae and the Ness of Brodgar, and with the passing of a 'golden age' of jarls and sagas, Orkney's social, cultural and political position seemed to slide towards the margin, the periphery and the edge. The islands' most influential chronicler in the early part of the twentieth century, the novelist and antiquarian Joseph Storer Clouston, believed that with the abolition of the old earldom at the beginning of the seventeenth century (events described in Chapter 4), 'the history of the country of Orkney comes to an end and the annals of a remote Scottish county begin'.[2] As we will see, it is a rather more complicated, and more interesting, story than that.

It is certainly true that the idea of remoteness – with either negative or positive connotations – has defined perceptions of Orkney from the middle ages to the present day. It is a notion islanders themselves have sometimes adopted and amplified, though usually for sensible and strategic purposes of their own.

To call something, or somewhere, 'remote' or 'peripheral' is to draw up rankings of value, and to make what may be unexamined assumptions about what is truly central and essential. 'What was it like to grow up somewhere so remote?' It's a question I have quite often been asked, but never really known how to answer. Everyone lives at the centre of their own social and moral universe.

In historical writing, the relationship between 'centre' and 'periphery' has long been a productive theme, in a variety of periods and geographical settings. Emphasis, however, generally falls on the ways in which the periphery presented a 'problem' for central government, rather than on priorities of people who actually lived there.

The principal aim of this book is to make the peripheral central, to reverse the direction of the telescope and reorient the map. It invites readers to consider what might change in our perception of the past, when people, places and predicaments we instinctively regard as peripheral or remote become the central point of reference, the heart and hub of experience. One possibility is that it might encourage us to reflect on how all history is really 'local' history – that people's most acute ambitions and anxieties are caused by particular conjunctions of presence and time, and not necessarily dictated by the themes picked out later as the defining issues of their age.

In short, *Storm's Edge* is intended as a subversive new way of telling what a famous children's book (by Henrietta Marshall; no relation) called *Our Island Story*. The obvious is worth stating here: 'Britain' – still less England, Shakespeare's 'sceptred isle' – is not an island, but an agglomeration of several thousand adjoining islands, sandwiched between the North Atlantic and the North Sea. *Storm's Edge* is a kind of alternative history of Britain, but at heart it is a story about islands, and what they can teach us about crisis, conflict and the contours of community.

Individual islands, and clusters of islands like Orkney, are revelatory places. They are of course real locations, but they are also storied sites of metaphor and the imagination, where strange things occur, ordinary rules don't always apply and magic has opportunities to flourish. A cottage industry of modern scholarship devotes itself to the 'phenomenology' of islands – the challenges they present to notions of nationality and identity, and the distilled and condensed ways they endow physical space with cultural meaning.[3]

Islands are indeed paradoxical places. As communities, they are prescribed by the clearest, most non-negotiable of boundaries. But they can also be surprisingly open and permeable, the encircling sea as much a conduit as a barrier. In relationship to mainlands, they are by definition marginal, if not isolated and secluded. Yet by virtue of participation in trade and proximity to sea-lanes, islanders often enjoy greater contact with the outside world than inhabitants of inland regions.

Politically, islands can be stubbornly resilient and profoundly vulnerable, hard to invade but easy to dominate. The identities of islanders are resolutely local, formed and fixed by the demands of the environment and the social cement of collective experience. But, at the same time, islanders are constantly challenged to think about themselves in relation to the pull of other places. 'No man is an island', wrote the English poet John Donne, yet no island is ever really an island either. They are always both apart and a part.

A brief overview of content and themes is in order. *Storm's Edge* opens with a short Prologue, describing the visit to Orkney of James V, king of Scots, in the summer of 1540, and reflecting on its national, international and local significance. The first chapter provides a fuller introduction to Orkney around the end of the middle ages, and explores its shifting location – politically, legally, culturally and linguistically – between the kingdoms of Scotland and Norway. There follows, in the next chapter, an assessment of the religious culture of late medieval Orkney, looking at how, within the folds of a theoretically universal Church, belief and meaning were shaped by patterns of land and seascape, and by the habits and traditions of Orcadian people.

Chapter 3 discusses how these habits, beliefs and meanings changed in Orkney, as they did across Britain, with the advent of Protestant Reformation – an abrupt plummeting of stone to water, whose ripples and ramifications will be felt through the remainder of the narrative. Another aspect of what has been called Orkney's 'Scottification' is addressed in Chapter 4: the rise and demise of the 'tyrant' Stewart earls, ruling the islands in high style at a time of national consolidation in Scotland, and tentative union of the 'British' crowns.

In Chapter 5, I draw on previously unutilised evidence to offer a new account of Orkney's role in a trans-European tragedy: the prosecution and execution of ordinary people – mainly women like Anie Tailzeour – for the impossible crime of witchcraft. I seek to establish what was distinctive about Orkney's entanglement in this most disturbing of historical episodes, and to move beyond instinctive moral abhorrence to sense the fears and motivations of those caught up in witchcraft cases, and reflect on what the accusations say about wider webs of meaning in their world.

Britain's dynastic and constitutional travails supply rich material for Chapters 6 and 7: a roller-coaster of revolution, civil war, commonwealth, restoration, revolution (again), unification and rebellion. Though often on the battered edge of these storms, Orkney people were at times makers of the political weather. The islands, I argue, had real practical and symbolic weight in contests over the identity and governance of Scotland and Great Britain.

Chapter 8 puts Orkney's ordinary inhabitants once more under the spotlight, investigating what happened to their distinctive language, culture and beliefs, as the political arguments tracked through the preceding chapters raged, subsided and resurfaced. It evaluates the efforts of the Kirk and its ministers in Orkney to create a pious, well-ordered, Christian society: an island laboratory in an age of experiment with enforced social engineering taking place right across Europe. Chapter 9 returns to the challenge of 'locating' Orkney – mapping the 'progress' of the islands by the close of the eighteenth century, and their significance for narratives of nationhood, and for practices of prosperity, imperialism and war. The Epilogue, following the novelist Sir Walter Scott north in the summer of 1814, invites us to look at Orkney through an outsider's condescending eyes, while encouraging reflection on how celebrations of modernity and nationality remain haunted by preoccupations of the past, and hindered by the peculiarities of place.

Orkney is not, and never has been, particularly large or populous, but a great deal can be said about its history and culture; more, perhaps, than I realised when I began. My approach in this book is that of the mosaicist, placing a myriad of small pieces in combination so that patterns form and emerge. Changing the metaphor slightly, the larger story is woven from intertwining threads of many smaller ones. These often involve humble and undistinguished men and women, whose lives I tend to find more intriguing, as well as more elusive, than those of the great and the good.

Some individuals, and various threads of topic and incident, appear once and then re-emerge later in the course of the book. For purposes of orientation, I have inserted cross-references to earlier passages, though readers can cheerfully ignore these textual waymarkers if they choose. Though *Storm's Edge* is based on extensive archival research, I have benefited enormously from the work of other writers on Orkney, living and departed. Those debts are occasionally acknowledged in the text, but more often – so as not unduly to impede the narrative – they are recorded in the endnotes. I have modernised, and anglicised, quotations from primary sources in Scots, while aiming to retain a flavour of rhythm and diction. A glossary at the end offers help with possibly unfamiliar terms and concepts.

Underpinning all that follows is the hopeful conviction that a history of small places need not be a history of small matters and unimportant themes. Paying attention to the experience of Orkney in the 'early modern' era represents more than locating and inserting a missing piece of the jigsaw. It is an invitation to think in new ways about stories of nation and nationhood, islands and mainlands, tradition and innovation.

After the confessions of this Preface, I attempt (not always successfully) to keep personal experience and family history out of the ensuing pages. It would be wrong, though, to pretend that historical writing can ever avoid being in some measure autobiographical. The relationship between outsiders and insiders, and the question of when and how people pass between these categories, is a fundamental dynamic of Orkney history that has always felt personally resonant. Relevant, too, is the challenge of how to hold, balance or express multiple and overlapping identities – Orcadian, Scandinavian, Scot, Briton, European, Christian, human. I am not alone in having a stake in these dilemmas. We should never look to the past for easy answers to modern predicaments, but history helps formulate the right questions to ask.

Prologue

A Journey and Two Maps

The lords of Scotland were gathering in Edinburgh, as the king prepared to take to the water. A fleet lay impatiently at anchor in the harbour at Leith – twelve ships, according to a report from one of the spies scurrying, in the first weeks of May 1540, to send news to Henry VIII of England. Another estimated as many as sixteen vessels, 'furnished with all the best ordnance, harness and habiliments of Scotland'. 'Ordnance' was artillery, a technology in which James V, king of Scots, took an enthusiastically boyish interest. Guns were brought down from the fastness of Edinburgh Castle; 'in all Scotland', Henry's chief adviser Thomas Cromwell learned, 'was not left ten pieces of ordnance besides that which the king doth take'.

'Harness and habiliments' meant armour and livery coats, clamped to the sweaty backs of men waiting to embark. They had marched from across the realm to Edinburgh, behind an assortment of adamantine earls: Huntly, Arran, Argyll, Marischal, Cassillis, Erroll and Atholl. Five hundred had arrived from Fife and Forfar, rallied by the resolute, and periodically pious, cardinal of Scotland, David Beaton.

A happy event delayed the departure. On 22 May, the queen, Mary of Guise, gave birth at St Andrews – 'ane son and prince, fair and lifelike to succeed to us', as James wrote in a brief triumphal letter to Henry VIII. On 12 June, James dictated his will, 'knowing the uncertain adventures that may fall to all manner of men'; most likely, the fleet sailed the following day. English onlookers wondered nervously where the king of Scots was planning to go, and why.

Relations between Scotland and its southern neighbour were, as often, strained. James was Henry VIII's nephew. But ties of kinship had done nothing to stop English archers and billmen from slaughtering a Scots army, allies to England's enemy France, on the bare hillside at Flodden in Northumberland in September 1513. Among the dead was the charismatic James IV, leaving an infant not yet two years old to succeed him. James V attained his majority in 1528, when religion was already starting to harden old enmities. In 1533, having failed to persuade the pope to dissolve his marriage to Catherine of Aragon, Henry VIII broke with the Catholic Church. The programme of religious reform that followed seemed to some enlightened and liberating, but to others violent and destructive.

Scots and English alike were drawn to compelling new ideas about salvation emanating from Luther's Wittenberg. But it suited James V to fashion himself, in contrast to Henry, as a loyal son of the papacy. In the late 1530s, English friars who had been evicted from their priories, as well as leaders of a failed rebellion, the Pilgrimage of Grace, found refuge in Scotland. Henry was angered by this succour for his enemies, now arrayed in international conspiracy against him. Chief among them was Henry's cousin, Reginald Pole, a Catholic cardinal, in brooding Italian exile. Pole wrote to David Beaton in 1539 to congratulate him on his own elevation to the Sacred College. He hoped Beaton, and his pious royal master, would keep Scotland undefiled from the 'contagion' of its neighbour, which, 'like a painted adulteress', had abandoned its lawful spouse.[1]

In the early summer of 1540, London put ships in a state of readiness, and sent artillery to places the Scots might be tempted to land. The French ambassador at Henry's court relayed the English paranoia: James intended to travel to the continent to ally with the king of France, or he planned to sail to Ireland and 'make himself lord of those who refuse obedience'. Suspicions were confirmed when a border captain described meeting a garrulous Scotsman who had been at James's court during Lent. He had seen eight emissaries with letters for James from all the great men of Ireland, promising to 'take him for their king and lord'.[2]

All this was smoke and sea-mist. James V's real intention, Ambassador Marillac revealed to his master, Francis I of France, was to 'visit some of his islands which are on the coast of Ireland' – as geographically accurate a description of the Hebrides as might be expected of an urbane diplomat from the Auvergne. The Highlands and Islands were a perennial problem for the Stewart monarchs, rulers whose authority was conventionally cast in terms of personal fealty not territorial domination: kings 'of Scots', not 'of Scotland'.

In the middle ages, real power in the west lay in the hands of the MacDonald 'Lords of the Isles'. Their lordship had been brought to an end by James IV in 1493, but the region remained restive. 'Daunting the isles' – military expeditions to intimidate the independent-minded clans – was an old practice of the monarchy. James V's father and grandfather led campaigns to the West Highlands, and his own plans to do so in 1531 were averted only by the unexpected submission of a rebellious chieftain, Alexander MacDonald of Islay. Another uprising in 1539, led by Donald Gorme MacDonald of Sleat in Skye, aimed to restore the abolished Lordship of the Isles. James thus travelled to the Hebrides in 1540 to demand submission from assorted MacDonalds, MacLeans and MacLeods.[3]

The king sailed on the *Salamander*, an imposing warship acquired three years earlier as a wedding present from Francis I. It was accompanied by the *Great Unicorn*, also a gift from Francis, the *Little Unicorn*, the *Lyon* and the *Mary Willoughby* – the latter two, prizes captured from the English. The *Salamander* was furnished with hangings and tapestries, gold plate for the king's meals and musicians for his entertainment. More practical equipment included four clocks and a compass. Another specially commissioned instrument – 'ane whistle of gold to the King's Grace' – was of questionable nautical value, but allowed James to act the part of a stout sea-captain.[4]

The journey was pure political theatre – a stately circumnavigation of regnal territories, affirming their boundaries in the sanctified presence of the monarch. It finished at the royal castle of Dumbarton, at the mouth of the Clyde, from where the king travelled overland back to Edinburgh, arriving around 6 July. In the meantime, his travels had taken him around the far corners of his dominions. It meant that in the summer of 1540, for the first time ever, a king of Scots made landfall on the islands to the north.[5]

'Beyond all the isles of Scotland lies Orkney, some part to the north-north-west seas [Atlantic], and some part to the Almain seas [North

Sea]'. The renowned Scottish scholar Hector Boece's *Scotorum Historia* (History of the Scots) was printed in 1527 and translated at the king's command in the early 1530s. It supplied James with some idea of what to expect at this point of furthest distance on his travels.

Arriving in Orkney meant crossing the Pentland Firth – the point at which the North Sea meets the Atlantic, and a place of hidden, treacherous whirlpools, and one of the world's most powerful tidal currents. Distances to Orkney from the northern coast of mainland Scotland are short – under ten miles at the nearest point – but getting there always involved crossing a line and traversing a barrier, moving 'beyond'. When James IV's servant Alexander McCullough was dispatched north in 1500, his warrant offered protection 'from the day of the passing of him forth of our realm to the said lands of Orkney until his return'. In 1521, John Major, principal of Glasgow University, wrote in his *Historia Majoris Britanniae* (History of Greater Britain) that, 'outside Britain', the king of Scots possessed islands, 'such as, to the north, the Orkneys, which the Greeks and Latins ever spoke of with a sort of horror'. Ancient authors certainly wrote with wonder about 'Ultima Thule', a mysterious place at the furthest boundary of the inhabited world. Usually, it was thought to lie beyond Orkney, but frequent linkage of the places strengthened an association of the islands with the very edge of civilisation.

For James V, as for all visitors since, there was no better time to come. In mid-June, the sun seems scarcely to set; night's darkness is transmuted into a few hours of hazy twilight known as the 'grimleens'. Orkney's air is laced with light breezes, rather than the fierce winds encountered at other seasons. On the best days of summer, the islands sparkle in the sea, low on a squinting eyeline of the horizon, thin layers of subtly changing greens and blues under a vast dome of encompassing sky.

Major's chronicle took a matter-of-fact approach, observing that the Orkney Islands 'produce in plenty oats and barley, but not wheat, and in pasture and cattle they abound. Orkney butter, seasoned with salt, is sold very cheap in Scotland.' Boece was keener to enumerate 'wonders'. Visitors would find greater abundance of birdlife, 'wild fowl and tame', than anywhere in Britain. James knew of this already: royal accounts make regular payments to the master falconer 'to send in Orkney for hawks to the King's Grace'. In December 1535, James gifted six birds from the islands to his uncle, Henry VIII. Orkney horses, according to Boece, were small of stature but unusually hardy, while 'to speak of fish, there is more abundance thereof than any uncouth [foreign] folk may believe'. There was profusion of grains, but Orkney was quite 'naked of wood'. Its people were 'given to excessive drinking', and brewers of 'the starkest ale of Albion'.[6]

In Orkney, according to John Lesley, a Catholic prelate writing a little later in the century, James was 'honourably received' by the bishop, Robert Maxwell. As a self-proclaimed paragon of Catholic piety, the king would have attended mass in Kirkwall's cathedral. The mighty red-and-yellow-sandstone kirk of St Magnus towered over the front-gabled houses of the little trading port. James himself, in a charter issued four years earlier, lauded the cathedral as 'one of the greatest fabrics in our kingdom and an ornament of the place in which its stands'.

The king was in Orkney by 19 June, when he wrote a letter to the provost and baillies of Aberdeen, commending Andrew Buk, skipper of a vessel which accompanied the expedition across the Pentland Firth. We do not know how long James remained in Kirkwall, though long enough for the queen to send a ship there with dispatches for her husband. Bishop Lesley supplied no further details of the visit than to note the reprovisioning there of the fleet, but George Buchanan, in his 1582 *History of Scotland*, wrote how in 1540 the king 'first sailed to the Orkneys, where he quieted the disorders, by apprehending, and imprisoning some of the nobility, and placed garrisons in two castles' – the nature of these 'disorders' is something to which we will return.[7]

Back in Edinburgh in late July, James informed Henry VIII that he had judged it wise 'to visit our isles, north and south, for the ordering of them in justice and good policy'. Local tradition, bolstered by the authority of a plaque in Parliament Close, just off Kirkwall's main shopping street, maintains the Scottish parliament was convened during the visit. That seems improbable, though James probably held court, and listened to local petitions and grievances. Among his first acts on returning was to write to the municipal authorities of Bremen, Schiedam and Rotterdam, complaining of outrages committed by Friesian and Flemish fishermen against his subjects in the Northern Isles.[8]

Memories of the visit lived long in the islands. In Kirkwall, a gilded oak bed James was reputed to have slept in was shown to visitors up to the middle of the eighteenth century. In the parish of Sandwick, in the West Mainland of Orkney, the Kirkness family, of the farm of Stove, maintained into the nineteenth century a conviction they were a cut above their neighbours – in fact, knights. The story was that old John Kirkness, at his daughter's urging, offered work as a gooseherd to a handsome red-haired vagrant, a youth with a tangible air of authority. On taking his leave, the stranger ordered John to kneel, tapped him on the shoulder with his stick and declared his descendants would be known as 'the Belted Knights of Stove'. Their guest, the Kirknesses believed, was none other than James V.

The motif – much favoured by Shakespeare – of a ruler passing incognito among his subjects, either for pleasure or to develop an understanding of their needs, is a familiar one in folklore. Several such tales became in later centuries linked to James V, though the possibility that in June 1540 the king absconded from his courtiers to labour on a West Mainland farm must be thought unlikely.[9]

The visit's most enduring legacy was a cartographic one. In 1583, a map with a remarkable history was printed in Paris by the French geographer royal, Nicholas de Nicolay. It bore the title *Vraye & exacte description Hydrographique des costes maritimes d'Escosse & des Isles Orchades Hebrides avec partie d'Angleterre & d'Irlande servant a la navigation* (A True and Exact Hydrographical Description of the Sea Coast of Scotland, and of the Orkney Isles and Hebrides, with Part of England and Ireland, Used for Navigation). Nicolay based his work on a chart he acquired in England in 1546, along with a text written in Scots. He published the text and map together as *La Navigation du Roy d'Escosse Jacques Cinquiesme du nom, autour de son Royaume* (The Navigation of the King of Scotland, James V, around His Kingdom).[10]

Works of this kind are known as 'rutters'. From the French *routier*, a rutter gives instructions for sailing coastal waters, and surmounting hazardous rocks, winds and tides, by navigating a course from headland to headland. It was achieved with compass, hourglass, lead and line, and a 'traverse board' – a mnemonic device with holes and pegs, to record speed and direction during the long hours of a crew's watch. At James

V's command, the experienced pilot Alexander Lindsay prepared a rutter for the 1540 voyage. He may also have drawn the chart that came into existence around the same time. Lindsay's reckonings allow us to track the fleet along the east coast of Scotland and across the Pentland Firth, as it skirted a low-water site of danger off Duncansby Head, 'le Bar' (the Bore), and a perilous confluence of tides at the north end of Stroma, the teardrop-shaped island parked between the Caithness coast and Orkney proper. If Lindsay's advice was followed, the fleet laid anchor in the safe haven of Scapa Flow, and the king came ashore just to the south of Kirkwall.

The subsequent fate of rutter and chart is a map in miniature of Scotland's history in the turbulent middle years of the sixteenth century. After 1540, relations between James V and Henry VIII worsened. Encouraged by the Francophile Cardinal Beaton, James rebuffed persuasions to follow his uncle's example and throw off the authority of Rome. When James failed to show up for a peace conference in York in 1541, a humiliated Henry made preparations for war. For the Scots, ignominious defeat ensued, an army lost at Solway Moss, Cumberland, in November 1542. Within weeks, James was dead – supposedly of a broken heart, though cholera seems a likelier culprit. In an echo of Flodden, a hapless infant was left on the Scottish throne; this time a girl, Mary, born barely a week before her father's death, the son produced on the eve of the 1540 expedition having failed to see his first birthday.

The years that followed witnessed what was later christened the 'rough wooing' – a campaign of violence and intimidation to force the Scots into acceptance of Mary's betrothal to Henry's son, Edward. Around now, a copy of Lindsay's rutter came into the possession of the English border commander John Dudley, later duke of Northumberland, perhaps handed over by one of the Scots notables taken at Solway Moss.

In 1546, Dudley was in Paris, with an English peace delegation. Here, he made the acquaintance of Nicolay, and clearly took to the young cartographer, whom he invited to accompany him back to London. In England, Nicolay recalled, Dudley allowed him to make a transcription of Lindsay's rutter, 'together with the marine chart, somewhat roughly made'. Back in France in 1547, Nicolay, with the help of a learned

Scotsman, translated the uncouth work into respectable French, and presented fair copies of the rutter and chart to the new king, Henry II.

They were soon pressed into military service. In May 1546, a band of Protestant zealots murdered Cardinal Beaton and seized control of St Andrews Castle. There, with expectations of English support, they held off for months the besieging forces of the regent, James Hamilton, earl of Arran. Equipped with Lindsay's rutter and map, and accompanied by Nicolay himself, a French fleet set sail and at the end of July 1547 bombarded the castle and took it by storm. The defenders, who included the firebrand preacher John Knox, were consigned to back-breaking servitude on the rowing benches of French galleys.

Cartography, the science of making maps, has often been an instrument of subterfuge, strategy and war. And maps have always been more than technical attempts to miniaturise physical topography. Consciously or not, they are interpretations of the meaning of territorial space, and they make political, cultural and ideological claims.

The Lindsay–Nicolay map is an impressively detailed rendering of Scotland's coastline, criss-crossed by 'windrose lines' – a technique that, in the centuries before accurate calculation of longitude, helped gauge distance and direction between destinations. Four and a half centuries on, one marvels at its precocious accuracy. It is a reassuringly familiar representation of the nation.

Yet maps, like photographs, are notable for what they include and what they omit; for what they place at the centre and what is consigned to the margins. In Nicolay's improved version of Lindsay's original, we behold a kingdom of Scotland bounded and centred, a unity and an entirety. Scotland's importance, as much as its relative location, is signalled by the marginal and incomplete presence of 'Angliae pars' and 'Irlandie pars' to the south, and by vistas of apparent absence to the west, north and east.

Within this political ordering of space, the islands of Orkney, 'Orchades Insulae', are allotted a place on the perimeter – the king's other island territory of Shetland, a hundred miles further north, drops entirely from view. Yet despite, or because of, this marginality, Orkney's placement has symbolic importance. The islands – no fewer than twenty-eight of them individually named – sit atop the kingdom of Scotland like a capstone, or a hat's feather. They are part of the greater whole, yet differentiated from it. In a frayed and fragmented affiliation to the fabric of the nation, threads of royal power are revealed. James's widow, Mary of Guise, who in 1554 replaced Arran as regent for her infant daughter,

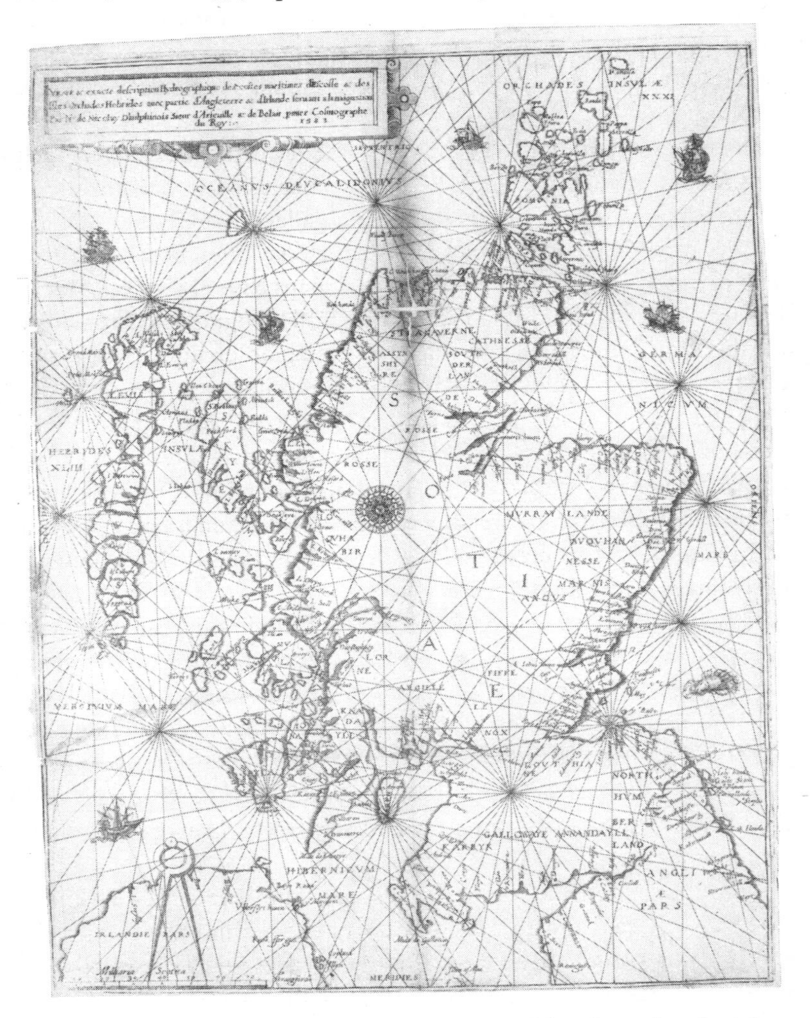

Figure 0.1: Nicholas de Nicolay's 1583 map of Scotland, based on a lost chart drawn up for James V's voyage around the coast of the country in 1540.

was, according to Bishop Lesley, 'well obeyed over all the parts of Scotland, as also in Orkney and in the Isles'.[11]

Without islands there can be no 'mainlands'. In Nicolay's map, the Orkney Islands, through their placement on the boundary, help that larger entity find shape and definition. The uttermost of the islands, North Ronaldsay (Ronalsa), lies just inside the frame – an encompassing circuit of parallel printed lines, directing the viewer's gaze inexorably inwards.

Orkney, then, completed Scotland – and did so in more senses than one. Along with Shetland, it was the last major territorial acquisition of the Scottish crown. In 1468, James V's grandfather, James III, had married Margaret, daughter of King Christian I of Denmark. Christian's territories included Norway, of which the earldom of Orkney was an overseas tributary. Christian could come up with only a fraction of the 60,000 florins fixed as the dowry, and so for the remainder agreed to pledge to his son-in-law 'all and sundry of our islands of the Orkneys', until 'whole and full satisfaction and payment is effectually made by us'; Shetland was pledged separately, for a smaller sum, in 1469.[12]

It is an open question whether either side viewed this mortgaging – the ungainly Scots word is 'impignoration' – as a permanent arrangement. There is uncertainty, too, about whether it necessarily involved transfer from Christian to James, from Denmark to Scotland, of that elusive constitutional quality, 'sovereignty'.[13] In 1540, Orkney had been 'Scottish' for not much over two generations, and Scots observers usually regarded it as a place apart. For James, the Kirkwall visit was more than a convenient stopover. It was intended to demonstrate lordship in the furthest, and newest, reaches of his kingdom.

In Venice, the year before the king's voyage, a very different map was published, with Latin inscriptions and separate explanatory notes in Italian and German editions. Rejoicing in the title *Carta Marina et Descriptio Septentrionalium Terrarum* (Maritime Map and Description of the Northern Land), it depicted Scandinavia and the Baltic, the northern coastlines of Germany, the Netherlands and Britain, and the islands and archipelagos of the North Atlantic. In its surviving hand-tinted copies, the *Carta Marina* is a thing of rare beauty, endowed with vibrant colours, intriguing captions, meticulous indexing and exquisite illustrations (see Plate 3). It is also, in still more obvious ways than the Lindsay–Nicolay chart, a work of brazen propaganda.

The map's creator was a Swedish Catholic churchman: Olaus Magnus, younger brother of Johannes Magnus, archbishop of Uppsala. Both brothers opted for Italian exile after King Gustav Vasa hitched his country to the Lutheran cause. The *Carta Marina* nonetheless resonates with Swedish patriotism. From the map's centre, the kingdom dominates the world of 'Scandia', seeming around twice the size of neighbouring Norway. One would scarcely guess that Denmark – from which Sweden secured precarious independence in 1523 – was the pre-eminent Scandinavian power of the day. In Magnus's map, a geographically shrunken Denmark seems almost to cower at the feet of its northern neighbour.

The legitimacy of Danish control over Norway – and of the Protestant Reformation the Danish authorities were in the process of imposing there – is also called into question. Next to the image of a Norwegian king and coat of arms, we find the caption 'nemo accipiat coronam tuam'. It is a quotation from the Book of Revelation (3:11), the Lord urging the persecuted to hold fast, 'that no man take thy crown'.

Orkney, in the *Carta Marina*, is no kind of peripheral place. It occupies a prominent position in the map's lower left quartile, at the centre of a nexus of locations: Shetland and Norway to the east; Scotland to the south; the Hebrides to the west; and, to the north, the Faroes, Iceland and the island of 'Tile' (Thule) – a place unknown to modern geography, but vivid in the imagination of classical and medieval authors.

While not quite reaching the imposing magnitude of Iceland, Magnus's Orkney is roughly the size of mainland Scotland, and of the (diminished) kingdom of Denmark. The islands, of which a caption declares there to be thirty-three, are not the scattering of arbitrary and angular shapes familiar from modern maps. Instead, we see compact, interlocking pieces, forming a virtually perfect circle. A half-dozen beautifully drawn ships are almost like bridges, across which one might walk from one island to the next.

Defiantly unacknowledged is the fact of Scottish – or indeed Danish – possession. Next to the identifying label 'Orcades' sits the royal crest of Norway: a crowned, axe-bearing lion rampant, on a red field. Another

Figure 0.2: The Islands of Orkney, from Olaus Magnus's Carta Marina *of 1539.*

crown, on the largest of the islands, bears the elegiac legend 'olim regnum' – this was once a kingdom. A glorious past is further evoked in the assertion 'regum antique sepulture', kings of old were buried here. This probably refers to Hakon IV, one of the greatest rulers of medieval Norway, who died in Orkney in 1263, and was buried in St Magnus Cathedral before being taken to his homeland.

Differentiation from Scotland is yet more explicit in the explanatory notes. These contain a section on Orkney, but not on mainland Scotland, which was not regarded as part of the 'northern lands'. In Orkney, Magnus remarks, 'they speak the Norwegian language, as a token that they belong to that kingdom, as do many other surrounding islands' – he likely had Shetland and the Faroes in mind.[14]

The *Carta Marina* includes only one Orkney place name, labelling the archipelago's main island 'Pomona'. The designation was widely recorded in medieval sources, as well as in Boece's *Scotorum Historia*, but resulted from a longstanding scholarly error. The late third- or early fourth-century Roman writer Solinus, in his description of the imaginary island of Thule, reported it to lie at a distance of five days' sailing from Orkney, adding 'sed Thyle larga et diutina pomona copiosa est' – that Thule is fertile and productive of long-lasting crops. Pomona – from *pomus*, fruit tree, fruit or apple – was the Roman goddess of abundance, and medieval chroniclers, sceptical about Thule's fertility, chose to read the word as a proper noun, and to understand Solinus as saying that Thule was large, and Pomona fruitful and abundant. Orcadians rarely if ever referred to the island in this way, something of which James V was aware. In his 1536 charter, the king observed how Kirkwall, held previously by the Danes, was on 'the greatest of all the islands of Orkney, called by us and the inhabitants thereof the Mainland [*continentem*] and then called by them Pomonia'.[15]

Lindsay and Nicolay's map allowed itself a single decorative sea-monster and a few illustrative ships, but is otherwise restrained and functional. The *Carta Marina* abounds with wonders and curiosities – in the vicinity of Orkney, the fabulous sea-creatures Balena, Orcha and Ziphius, with horns, spouts, beaks and fangs. In the unfathomable depths of the waters around Norway, Magnus wrote in his annotations, are many 'strange and terrible beasts'.

Magnus's notes also make sense – if one can call it that – of an at first puzzling feature. From the east coast of 'Pomona', out over the North Sea, extends a large stylised tree branch. Three aquatic birds bob cheerfully on the waters beneath: 'some ducks grown from the fruit of a tree'.

Olaus Magnus returned to this remarkable phenomenon in his 1555 *Historia de Gentibus Septentrionalibus* (Description of the Northern Peoples), a comprehensive work of geography, ethnography and natural history. It tells readers that 'in Orkney ducks are produced when the fruit from a certain tree falls into the sea. After a short while, when they can take wing, they fly off to join other ducks, either domestic or wild.' The passage drew on a widespread medieval belief in the unique genesis of the black-and-white barnacle goose – so-called because it was thought to hatch spontaneously from the similarly striped goose (or gooseneck) barnacle, a crustacean attaching to ships' hulls and pieces of driftwood that might readily be imagined as tree branches fallen into the water. A bonus was that the goose technically counted as a species of fish, and so could be eaten in good conscience during Lent.

Magnus learned of the wonder from the writings of the Italian scholar Aeneas Sylvius Piccolomini, the later Pope Pius II. In 1435, Piccolomini was on Vatican business in Scotland, and wrote an account of his travels.

He was told of a tree growing on the banks of rivers whose fruit resembled a goose. The fruit rotted if it dropped on land, but on falling into water sprouted feathers and wings, swam and took flight. Not unreasonably, Piccolomini asked to be shown this wondrous tree, but was told it was not now to be found in Scotland, but only in the Orkney Islands. 'Miracles', the future pope reflected wryly, 'always seem to flee further into the distance.'[16]

Two maps, created at a time of momentous political and religious realignment, have the islands of Orkney as their point of connection and contention. One is rooted in a trans-Scandinavian (and mythical) inheritance, while the other looks to the era of the emerging nation state. It would be easy – perhaps too easy – to see in them visions of the past and the future. The Orkney of the *Carta Marina*, with its stylised clump of interchangeable islets, is undoubtedly less 'realistic' than the Lindsay–Nicolay map. Its suggestion of copious areas of forestation will raise the eyebrows of anyone who has ever visited Orkney and looked to find shelter from the wind.

Yet if the *Carta Marina* was wrong about the trees, it was right about other things. The map shows a dense network of churches and chapels in the islands. Their placement along shoreline and hillside links the harsh topography of a maritime environment to cycles of birth, bustle and burial, and unseen worlds of petition, blessing and grace. The *Carta Marina* also speaks some truths the Lindsay–Nicolay map does not want to hear. From the latter, we would know nothing of Orkney's connections with the kingdom of Norway, nor that in 1540 most of its inhabitants spoke 'Norwegian'; that is, the form of Old Norse usually referred to as 'Norn'.

Modern maps of Britain follow Nicolay in placing Orkney in the corner and on the edge, but that is an assigned rather than natural position. A map of the North Atlantic fixing Orkney at the centre would look very different. A 'marginal' location would be revealed as a point of confluence and connection, roughly equidistant between Bergen and Edinburgh, Oslo and Dublin, and Reykjavik and London. Because to ask 'where is Orkney?' is to raise important questions about the construction of Scotland and Britain, and the relation of both to rival centres of culture and power.

1

Between Norway and Scotland

A Description of the Orkney Isles

There was no church on the island, so one Christmas the inhabitants, young and old, decided to get in a boat and cross to a neighbouring isle to celebrate the Nativity of the Saviour. Suddenly, a violent storm blew up, and the rage of the waters was such that the boat was overwhelmed and all on board drowned. At this, strange to report, the animals on the island – cows, sheep, calves, pigs, dogs, rabbits and every other living thing – threw themselves into the sea with amazing ferocity and were destroyed. From that day, nobody has ever lived there.

The story was recorded towards the end of the sixteenth century, and was, its author insisted, 'no fable, but an entirely true report'. He called the island Southay, meaning South Isle, but its actual name is Switha, perhaps deriving from Old Norse *sviv-øy*, an island that seems to be floating or swinging. Switha is a droplet of grassland, a mere hundred acres, edged by low cliffs. It lies along the southern approach to Scapa Flow, at rough equidistance from several larger neighbours: Walls to the west, Flotta to the north, South Ronaldsay to the east. To which of these the islanders were heading, our source does not say.

In time, the animals returned. From at least the mid-seventeenth century, cattle, and latterly sheep, have grazed on the island. But their owners come only as occasional visitors. Whether Switha was ever continuously inhabited is uncertain – some ambiguous remnants of a once enclosed area lie along the southern shore. But in prehistoric times people arrived for purposes of ritual and memory. Two small standing stones, one in the northern part of the island and one in the south, are wedged carefully in position, and a couple of cairns may contain Neolithic chambered tombs.

Switha is one of the least of the many isles of Orkney – in the view of a seventeenth-century writer 'it rather merits, taken strictly, the name of holm than of island'.[1] The Yuletide tragedy might be a baseless legend, or a dim folk memory of some real historical episode. Either way, the tale of the lost people of Switha has things to teach us about life and death in Orkney at the close of the middle ages. It opens a window on separate but interdependent communities in a varied island world; it reveals the connectedness of human islanders to other living creatures; it registers a deep-rooted need to mark and sacralise turning points of the year. Most of all, it conjures for us the ever-presence of the sea – at once beckoning and threatening, confining and liberating; a source of life, and, sometimes, a sentence of death.

The story appears in a Latin *Descriptio Insularum Orchadiarum* (Description of the Orkney Isles). It is a fascinating and mysterious document. The original manuscript has long since disappeared, though transcriptions were made in the seventeenth and eighteenth centuries. In these, the title continues 'by me Jo. Ben. who lived there in the year 1529'. This date is certainly wrong, as the text describes events which took place as late as 1582. There is, however, no mention of important happenings from the first years of the seventeenth century, and one possibility is that the first copyist accidentally transposed digits, and the date should properly read 1592. The identity of 'Jo. Ben.' cannot be established with certainty, though a plausible theory points to John Bonar, a clergyman who worked as a schoolmaster in Kirkwall and perhaps grew up in Orkney. A John Bonar was for a short time (1593–4) noted to be serving as a minister in North Ronaldsay.[2]

It is in North Ronaldsay – the most northerly of the islands – that the *Descriptio* opens, a place which 'often causes shipwreck to English and other sailors'. Jo. Ben. proceeds southwards to two other islands he knew personally: the long, low island of Sanday, skirted with smooth white beaches, and half-cultivated Stronsay, with its little satellite, Papa Stronsay ('there is one farmer living there'). Then to the other isles of the northern constellation – Shapinsay, Westray ('most fertile of all the Orcadian islands'), Papa Westray, Faray and Eday – and an inner cluster of islands off the upper shores of the Orkney Mainland: Egilsay, Rousay, Wyre, Eynhallow and Gairsay.

Jo. Ben. had a theory for how the Mainland became Pomona. Slipping from Latin into Scots, he tells us it signifies 'the middle of the apple, because it lies betwixt the North and South Isles [of Orkney]' ('apple' is one meaning of *pomum* in classical Latin). The explanation is fanciful. But there is something astute in this metaphor of a place with its own core and layers. The central point where the Mainland itself narrowed was known as St Ola, 'most blooming and pleasant' of all Orkney parishes, and the location of Kirkwall, 'a most excellent city'. In the low-lying East Mainland were the agrarian parishes of St Andrews and Deerness, and to their south the conjoined parishes of Holm and Paplay ('men cross the sea here when they are travelling to the southern parts').

West from Kirkwall, Jo. Ben. describes the parishes of Firth, Stenness, Orphir, Sandwick and Stromness ('a safe harbour; here is the best departure point for a fleet'). In the north of the West Mainland, we find the moorland parishes of Rendall and Evie, with a narrow strip of fertile coastline 'where monstrous whales come in', the Atlantic-facing Birsay and the uniquely land-locked parish of Harray. Then to the South Isles, set in an irregular crescent around the great natural harbour of Scapa Flow: South Ronaldsay, Flotta, Cava, (another) Faray or Fara, Swona, Burray, Walls, Hoy, Graemsay and – finally – Switha.

Whatever its date of composition, or the identity of its author, the *Descriptio* provides a revealing account of Orkney in the sixteenth century. Other surviving surveys are dry listings of lands and rents, or were written by ill-informed outsiders relying on earlier sources. Jo. Ben. was a first-hand witness and – at least in part – an Orkney insider. He was also – though he would have recognised none of the words – an enthusiastic folklorist, anthropologist and ethnographer, fascinated by the customs, ecology and stories of people among whom he had spent time, yet to whom he did not quite belong.

How many people lived in Orkney is impossible to say for certain. Jo. Ben. estimated a potential fighting force of around 5,000 in the Mainland, and as many again in the isles. On the assumption that adult males constituted around 30 per cent of the population, this suggests a figure above 30,000, which seems too high. Scotland's first attempt at a national demographic census was in 1755, when Orkney's population was stated to be 23,381 in a total of 1.265 million. Estimates of Scotland's population in 1500 tend to hover around 700,000, implying 13,000 or so people in Orkney if its overall share had remained proportional. Demographic historians agree, however, that the population rose sharply over the course of the sixteenth century, and the number of inhabitants in Orkney at the time Jo. Ben. was writing may have been roughly the same, at around 18,000, as when I grew up there in the 1970s.[3]

If so, the population was differently distributed. Kirkwall was much smaller, and the occupancy of the isles, north and south, considerably greater. Several islands noted by Jo. Ben. to be cultivated and inhabited – Eynhallow, Damsay, Fara (south) and Faray (north), Copinsay, Cava, Swona – experienced final depopulation in the nineteenth and twentieth centuries.

Figure 1.1: An Orkney township, Rackwick in Hoy, photographed at the start of the twentieth century. Orcadian farmers lived in close proximity, their shared fields encircled by the community's hill dyke.

In both Mainland and isles, the majority lived in 'toons'. These townships were not villages, but defined blocks of arable land, of varying size, lying within the parish boundaries. Each township contained a scattering of farmsteads, whose occupants toiled alongside each other in large fields subdivided into individually held strips or 'rigs'. Periodically, these were reassigned to ensure everyone received a share of the best, and worst, soil. The land itself, for taxation purposes, was divided into 'urislands' (ouncelands) – districts notionally owing an annual ounce of silver for the land tax known as 'skat'. An urisland was further divided into eighteen 'pennylands'. The pennyland was a basic unit of land valuation which equated roughly to a small farm. Most townships were smaller than a single urisland, though some larger ones encompassed several.[4]

Townships peppered the landscape in fertile, lower-lying parts of the Mainland – Deerness, Harray and Sandwick – as well as some of the larger and better-cultivated islands: South Ronaldsay, Sanday and the southern half of Westray. Elsewhere, the populated land tended to run in a wide or narrow belt along the bends of the shore. These patterns of settlement, based on viable agricultural soil, had changed little since the Iron Age.

The defining feature of almost every township was its 'hill-dyke' – a usually turf-built construction, six or so feet in height, whose maintenance was a solemn collective responsibility of the community. The dyke looped around the cultivated fields or townland on the inner side, marking it off from the rising moorland beyond. Its function, during seasons of sowing, growing and reaping, was to keep animals out of the crops. Cattle grazed at will on the common land of the 'hill', before being let back into the townland once the harvest was brought in. But – as we shall see – the dyke was an important symbolic as well as practical barrier. Orcadian lives were lived in the imagined as well as physical spaces between the push of the hill and the pull of the sea.

Patterns of agrarian life in sixteenth-century Orkney were found elsewhere in Scotland, including the runrig system for shared use of arable land. Yet some things struck Jo. Ben. as unusual. The agricultural cycle began late – 'they till not until the spring of the year, and as they till so they sow'. Jo. Ben., like many later commentators, mocked the flimsiness of the distinctive Orkney plough: 'the culter and the sock [cutting blade and ploughshare] be not two pounds in weight'. Cereal crops, grown in generous quantities, were oats and a strain of barley, unique to the islands, from the variety known as 'bere'. It seemed noteworthy that crops 'are handled only by men; their women neither shake the straw, nor yet winnow the corn', though Jo. Ben. offered no theory on why this was so. The men also kept watch on the phases of the moon, making sure that cattle slaughtered for winter sustenance around Martinmas (in mid-November) were killed while the moon was waxing; Orcadians believed 'they grow in the barrel'.

Jo. Ben. has surprisingly little to say about fishing, but the sea's centrality to the islands' subsistence economy is a recurring theme of his text. Virtually the only manure used was 'seaware', the nutrient-rich marine algae found everywhere along the ebb-tide shore. Marine fauna supplemented the produce of the land. At Orkney's northernmost point, the rocky islet of Seal Skerry off the North Ronaldsay coast, large numbers of the eponymous aquatic mammals came ashore to bask on the rocks. Jo. Ben. describes local farmers setting on the seals with stout clubs – he had once seen sixty killed in this way. In Eday, fierce battles were waged against 'huge monsters'. These, presumably, were pods of beached whales, as the islanders' prize was to chop the carcasses in pieces, and boil them into oil.

The *Descriptio* generalises outrageously – Orcadian men are 'much given to drinking and debauchery', and to fighting among themselves. Yet it often asserts the individuality of islands and Mainland parishes. Cava 'has very good cheese', Swona 'a safe station for foreign ships'. Birches grow on Hoy, but not on other islands, 'which are without trees'. In Firth, 'oysters are caught in abundance'. Deerness once had a gold-mine – an unlikely sounding proposition, though Jo. Ben. documents a fatal accident there in 1506. The women of Kirkwall were much given to wantonness, 'I think on account of the abundance of fish'. On Faray, 'the pleasant isle', good for grazing cattle, 'children sing to the beasts'. The pastoral idyll contrasts with the sparseness of life in North Ronaldsay: 'they lack fuel for fire, except for dried seaweed ... the best fuel comes from cattle dung, spread on a wall and dried in the sun'.

There are scattered suggestions of community rivalries, even antagonisms. For reasons Jo. Ben. does not think to explain, one island's inhabitants were referred to as 'the liars of Walls'. Harray people, renowned for their indolence, were jeeringly called 'the sheeps of Harray'. In Sanday, visiting the graveyard of the church of the Holy Cross, Jo. Ben. saw a collection of large skulls. He asked an old man for an explanation, and was told that the isle had once been subject to the people of Stronsay, who came annually to demand tribute. Eventually tiring of this tyranny, the Sanday folk prepared an ambush, and ruthlessly slaughtered the Stronsay men, their wives, children and servants.

Jo. Ben. heard various time-salted tales of this kind, some, like the sad story of Switha, serving to account for an otherwise inexplicable desolation in places where people ought to have been. The little tidal island of Helliar Holm, off the southern coast of Shapinsay, had once housed two brothers, one godly, the other impious. A quarrel arose, and the ungodly brother denounced his sibling to the bishop for sexual relations with a kinswoman by marriage. When the bishop responded by expelling them from the island, their wives, on bended knee, cursed the place, 'which is why no one has lived there to this day'.

Several passages reveal residues of the distant past in the daily environment of Orcadians. In Stenness, by the side of a large loch, was a circle of tall, wide stones, each 'the height of a spear'. Nearby was a tomb, which Jo. Ben. himself had seen, and from which, he had been told, were retrieved the bones of a man no less than fourteen feet in length, with a sum of money beneath his head. This likely refers to the great chambered cairn of Maeshowe, certainly broken into several times before its excavation in the nineteenth century.

Giants were associated with several mysterious structures. In a section on Hoy, Jo. Ben. described the monument that has come to be known as the Dwarfie Stane: a huge slab of red sandstone, deposited by a retreating ice-flow in an isolated, steep-sided glacial valley. It is a unique British example of a Neolithic or early Bronze Age rock-hewn tomb, with a laboriously hollowed-out inner chamber about three feet high; Jo. Ben. rightly describes it as 'worthy of wonder'. He adds that it was carved as a bed-chamber by a giant and his pregnant wife; how they managed to fit is not clear. An entrance blocking-stone, which now sits beside the monument, was brought over by a jealous rival giant to trap the couple within, though they escaped by making a hole in the roof.

Twenty miles to the north-east, in the small, flat island of Wyre, there also 'once lived a lofty giant ... the outline of his house is still there'. Atop a low hillock on the north side of the island are the remains of a defensive structure. Locals have always called it Cubbie Roo's Castle, and in later centuries tales of this legendary colossus circulated throughout the North Isles of Orkney. He was most likely an imposing twelfth-century Norse chieftain, Kolbein Hruga (Kolbein the Heap), whose 'fine stone fort' is mentioned in the *Orkneyinga Saga*, or saga of the Orkney earls, compiled in Iceland at the start of the thirteenth century.

Another wonder, 'a story told by the old people', concerned a large stone in the church of St Mary, at the southern tip of South Ronaldsay. It bore the impressed image of two human feet, 'a thing no workman could have made'. The enigmatic 'Ladykirk Stone' is still there, housed in an eighteenth-century church occupying the earlier site; modern scholars consider it a Pictish artefact for some sort of coronation ritual.[5] The tale told to Jo. Ben., however, concerned an exiled Frenchman, whose ship foundered in a violent storm. He survived by leaping onto the back of a sea-monster, vowing that, should he be carried safe to land, he would build a church to honour the Virgin Mary. His prayer was answered: the beast bore him to shore, and at once was transformed into stone.

The story is an intriguing blend of medieval Catholic piety and older folk belief about sea-creatures and storms. It makes sense of a mysterious but revered artefact, and ascribes significance to its presence in a specific local site. Orkney's popular culture was a patchwork of such narratives and memories, stitched together with threads not always easy for a stranger to detect.

Jo. Ben. was no hardened sceptic, but neither was he naively credulous – 'if it deserves to be believed' was his cautious (or sarcastic) verdict on the Dwarfie Stane and the giants. If Jo. Ben. was indeed John Bonar, Protestant minister of the first generation after the Reformation, the disapproval of old beliefs is to be expected. 'Here you may discern their fictitious and fantastical traditions', he writes about Eynhallow – the small 'holy island', once perhaps the site of a monastery, anchored in the narrow but treacherous tidal sound between the Mainland and the fast-rising ground of Rousay to the north. It was believed here that if corn was cut after sunset, blood would drip from its stalks. We will return in a later chapter to Jo. Ben.'s disparaging opinions about pilgrimage to old chapels, belief in ghosts and spirits, and the islanders' preoccupation with *napaeae*, the fairies.

The *Descriptio* takes us closer than any other source to the beliefs and customs of late medieval Orkney. But it is not in the end a view from inside. The people about whom Jo. Ben. wrote are always 'they', never 'us'. He thought he had come to know them well, but did not entirely trust them: 'they are cunning and very subtle'. Whether he always understood them in a literal sense is a moot point. A throwaway line grabs the modern reader's attention. 'They use their own way of speaking, as when we say *Good day, Goodman*, they say *goand da, boundae*, etc.'[6]

That 'etc.' is perhaps the most frustrating abbreviation in the history of modern linguistics. It abruptly cuts off the earliest in a meagre handful of recorded speech fragments from a lost native language of the British Isles. Norn was once the dominant tongue of Orkney, Shetland and some parts of northern mainland Scotland. The word means simply 'Norse' or 'Norwegian', deriving from the adjective *norrœn* (northern). Norn's roots lay in western Norway, in the dialects spoken by the first settlers to migrate from those parts across the maritime zone of island settlement known to Norwegians as 'vest i havet', west over sea.

Over time, Norn diverged from speech-forms used in Norway, as happened in other western outposts of the Scandinavian world, Iceland and the Faroe Islands. The salutation *goand da* corresponds to the Icelandic or Faroese *góðan dag* (with a similar pattern of masculine accusative ending on the adjective).[7] *Boundae* is a farmer (*bóndi* in modern Icelandic and *bonde* in Norwegian), but here equates to countryman or resident, and could be translated well enough as 'fellow'. Jo. Ben.'s phrase evokes memories of the cheerful greeting ubiquitous in the Orkney of my youth, still widely used, and heard nowhere else in Britain: 'Whit like the day, beuy?'

Icelandic and Faroese are thriving living languages, but Norn has long since disappeared from Europe's linguistic map. It has not quite disappeared without trace, but those traces – as with a possible linkage of *boundae*–boy–beuy – are often elusive and ambiguous. The 'decline' of Norn is hard to track with certainty. Yet it is a crucial element in the story of how a community on the western frontier of the Scandinavian world came to be absorbed – gradually, awkwardly and never completely – into first a Scottish and then a British cultural and political orbit.

Jo. Ben. was a Scot – a 'guidman' not a *boundae*, and a Protestant not a Catholic. From the late sixteenth century, people of his kind were ascendant in the islands, and worked hard to transform Orkney in their own image. But in this they were never entirely successful, and in the process of seeking change they would be changed themselves. Before Jo. Ben. set foot in Orkney, and long after he left, the islands were a place of dialogue and encounter – a market and a crossroads as much as a margin or an edge.

The Seal of the Community of Orkney

Scots had been making their mark in Orkney long before the political transfer of 1468. It could hardly have been otherwise, given the islands' close proximity to the Scottish mainland. Yet that mainland itself was only gradually prised away from Norwegian power and influence over the course of the middle ages. Why else would one of the most northerly counties of Scotland come to be known as Sutherland, the southern land? The habit of referring to Orkney and Shetland as 'the Northern Isles' predates the union with Scotland. They were the 'Norðreyjar' to differentiate them from the 'Suðreyjar' – the 'southern islands' of the Hebrides, likewise long under the sway of Norway.

In the earldom of Orkney's medieval heyday – that is, during the tenth to thirteenth centuries – the islands were not some isolated outpost of Norwegian rule but a central hub in a maritime empire extending further to the west and south. Royal expeditions heading for Ireland, the Isle of Man, Scotland or England regularly dropped anchor here to provision and raise troops. Orkney jarls and other island chieftains cheerfully raided on their own account in the Hebrides and down the eastern seaboard of Scotland.

Things changed after the Battle of Largs, fought on the Firth of Clyde in 1263, the last confrontation between the forces of Scotland and Norway. The defeated Hakon IV (as we have seen, p. 20) died in Kirkwall on his way home. The Treaty of Perth (1266), signed by his successor

Magnus VI and the Scots king Alexander III, ceded to Scotland at least titular control of the Isle of Man and Hebrides, while confirming Norwegian possession of Orkney and Shetland. In return for ownership of the Suðreyjar, the king of Scots agreed to pay a lump sum and a tribute of 100 merks ('the annual of Norway'), to be deposited yearly at a suitable site in the closest part of Norwegian territory: the cathedral of St Magnus in Kirkwall. Henceforth, as the historian Ian Peter Grohse has suggested, Orkney moved from a position in the interior of the Norwegian realm to one on its outer frontier, at the same time becoming a place of diplomatic rather than of martial encounter between the kingdoms of Norway and Scotland.[8]

Medieval Orkney, like Iceland, Greenland and the Faroe Islands, was a *skattland* of the kingdom of Norway – a tributary territory, owing 'skat' or tax as an acknowledgement of subordinate status. Kings of Norway appointed 'sysselmen' (bailiffs or sheriffs) to collect revenues, and to keep an eye on earls of Orkney whose hereditary claims the crown theoretically denied but was obliged in practice to accept. There was also a chief judicial officer, known as the 'lawman', to administer laws of Norway which were extended to all skattlands in the later thirteenth century. Norwegian law was amended to take account of local customs and traditions, though no island lawbook survives to reveal exactly how the adjustments worked out in the case of Orkney.

We know how decisions were authenticated. The lawman held an official seal, granted to the people of the islands by the Norwegian crown, and bearing the legend 'sigillum: comunitas: orcadie', the seal of the community of Orkney. It was used to validate decrees of the lawthing, an assembly where juries of worthies known as 'lawrightmen' helped the lawman reach judgements. The seal's appearance can be reconstructed from impressions on surviving wax fragments from fifteenth-century letters.

It displayed the royal arms of Norway – the same axe-bearing lion that Olaus Magnus placed on his *Carta Marina* to signal Orkney's connection to the Norwegian crown. The bearers of the arms wear close-fitting embroidered clothing, quite unlike that found on other medieval Norwegian seals – they may be wearing sealskin boots. They represent the elite of Orkney society, the substantial farmers serving as lawright-

Figure 1.2: A reconstruction of the late medieval Sigillum Communitatis Orcadie, *Seal of the Community of Orkney, which authenticated pronouncements of the lawthing. It expresses the strong sense of identity among Orkney's leading inhabitants.*

men, sometimes known also as 'roithmen' (councillors), or simply 'good men' (*godir men*) or 'best men'. The emblem acknowledges the unique identity of a provincial community, as well as links of fealty between the Norwegian crown and Orkney's leading inhabitants.[9]

Those links were becoming strained around the turn of the fifteenth century. From 1397, the three Scandinavian kingdoms of Norway, Denmark and Sweden became united under a single ruler, after an agreement at Kalmar on the Swedish Baltic coast. The Union of Kalmar initiated an era of unstable dynastic politics, during which the monarchy's centre of gravity left Norway and settled in Denmark, and the crown's gaze shifted eastwards, to confront the challenge posed by the powerful trading block known as the Hanse or Hanseatic League.

Orcadians, meanwhile, were increasingly tempted to look to the south for power and patronage. In the middle of the thirteenth century, the old succession of Norse jarls died out, and the earldom passed into the hands of Scots families. Scotland's gravitational pull increased after 1379, when Henry Sinclair was appointed earl by Hakon VI. Sinclair was lord of Roslin in Midlothian, a Scots aristocrat of Anglo-Norman descent. To secure his position, he constructed a formidable castle in Kirkwall, close to the cathedral. His son, Henry II, succeeded as earl in 1400, followed (after a prolonged minority) by his grandson William in 1434. William was invested with the title of earl of Caithness by James II in 1455. This was a return to the situation pertaining before 1350, when Orkney earls ruled the northernmost part of mainland Scotland. It demanded a dexterous dual allegiance, with fealty due to different monarchs for neighbouring territories.

The situation has been seen as 'completely anomalous' in a world of emergent nation states, and an accelerant of Orkney's inexorable pull into a Scottish political–cultural ambit.[10] This may be to overstate the importance of national identities in the highly personalised world of feudal lordship. The arrangement does not seem to have caused any agonies of conscience for William Sinclair, and it may have suited the kings of Denmark–Norway to have someone in authority in Orkney with close links to the Scottish court. The Pentland Firth – on rough days, especially – could feel like the coldest and hardest of barriers. But it was never a regulated national boundary in anything like the modern sense. In the terminology favoured by modern scholars, the islands constituted a 'frontier zone' or 'borderland', rather than a rigid border.[11]

First and foremost, however, the islands were their own place, the location not just of the bundle of political and financial rights constituting the earldom, but also for a parallel public institution, the 'community of Orkney'. The lawthing was more than a court, functioning also as an assembly to represent the views and interests of the people, or at least of the 'good men'. Formally, the lawman was appointed by the king, but, by the early fifteenth century, nominations were made locally. William Thorgilsson, lawman in 1420, was a native islander: we know of his farm and attached private chapel.

Thorgilsson's successors, John Kirkness and Harry Randall, no longer used the customary Norse patronymic. They followed the fashion, already widespread in England and Lowland Scotland, of a hereditary family surname. Almost certainly they were native Orcadians too, choosing the names of locations where they lived and held land. Other surnames, appearing around this time and still common in Orkney, likewise derive from island place names: Baikie, Clouston, Corrigall, Delday, Drever, Flaws, Flett, Isbister, Linklater, Norquoy, Scollay, Yorston.[12]

An emphatic expression of the 'community of Orkney' appears in two letters, stamped with the official seal, and dispatched from Kirkwall in 1425 in the name of 'the whole country of Orkney'. The main document was written in Norwegian, a shorter one in Latin. They were sent to Philippa, queen of Denmark and Norway, regent in Copenhagen while her husband went on pilgrimage to Jerusalem.

The letters complained of extortions and tyrannies at the hands of David Menzies, a Scot appointed by the absentee Earl Henry II (who had died in 1420) as his representative in the islands and guardian to his son, William. Menzies's oppressions included confiscating cargoes, levying illegal fines and making arbitrary arrests, including of the lawman, William Thorgilsson. Menzies took possession of Orkney's seal and lawbook, and used them to forge public documents. The petitioners made a heartfelt plea that no governor 'presume to introduce any new laws, customs or novel constitutions'. All should be bound by the old laws of Norway, along with Orkney's own 'ancient constitutions and customs'.

It is tempting to detect in the complaints of 1425 ethnic tensions between indigenous Norse-Orcadians and immigrant Scots. Menzies had sent 'foreigners from Kattanaes [Caithness]' to ransack the lawman's homestead and chapel. The people of South Ronaldsay complained to him that 'wild Scots' (*willeschotta*) were seizing their goods and doing such great injury that they would rather die than suffer it any longer. Menzies replied callously 'that they should not die all of them on one day, but that they would die every day as long as he had power'.

The wild Scots were in all likelihood islesmen from the Hebrides, raiding in Orkney just as Orcadian Vikings had once pirated in the Western Isles. Letters of 1461 from the burgesses and bishop of Orkney to King Christian I of Denmark complained again about such depredations – followers of the MacDonald Lord of the Isles, sworn enemy to Earl William, had fallen on the islands, 'burning our buildings, carrying away our goods, and destroying all your loyal inhabitants'. The pattern continued into the following century. Surveys of church income in the 1560s noted a baleful condition of lands in Westray, 'wasted by the Lewis men', 'devastated by the highland men this last year'. A generation later, Jo. Ben. was regaled with stories of epic battles between the Lewis men and Westray farmers. Menzies may not have directly encouraged Hebridean raiding, but his accusers were in no doubt that he 'introduced foreigners who heavily oppressed the commons of the country'.[13]

Whether these were hostilities between people of rival *national* identities – Scots or Norwegian – is much less certain. Several of Menzies's victims had surnames suggesting Scots origins: John Logie, William Irving, John Fife, Alexander of Sutherland. Scots, mainly from the Lowland regions of the Lothians, Fife and Angus, were settling in Orkney from at least the fourteenth century, and some had clearly been adopted into the ruling elite of what the letter to Copenhagen called 'landit i orknø', the country of Orkney. One cathedral priest mentioned in the 1425 complaint had a Norse Christian name and Scots surname: Nils Muir. Another of Menzies's opponents with apparent Scots ancestry, 'Jams af Krage', James of Craigie, was son-in-law to Earl Henry Sinclair, and described in a 1422 testimonial as a liegeman (*handgenginn*) of the Norwegian king, one who had taken a personal pledge of loyalty.

Historians of Orkney have written – usually wistfully or regretfully – of the late medieval 'Scottification' of Orcadian society. Yet, as more than one scholar has suggested, a parallel 'Orknification' was likely taking place, as incomers and their descendants became drawn into the structures and attitudes of local landholding society. Some of them – Irvings, Craigies, Cromarties, Sinclairs – went on to become leading 'udallers' under the king; that is, they were landowners holding their estates outright under the Norse system of 'udal' (or *odal*) law, without owing any rents or services other than the universal obligation to pay skat. My own maternal ancestors – Fotheringhames – bore a Scots surname with connections to the village of Fotheringhay in Northamptonshire. But they were securely settled in Orkney before 1446, when Richard Fodrungame, lawrightman, set his seal to a document affirming William Sinclair's right to the earldom. Since immigration to Orkney began in the period when surnames were first adopted, it is even possible that some families with distinctly 'Orkney' names were originally of Scots rather than Norse ancestry.

The quarrels of 1425 were, then, not so much between ethnic Scots and indigenous Scandinavians as between self-identifying members of the 'community of Orkney' and people regarded as heavy-handed intruders. Menzies's most implacable opponent was Thomas Sinclair, a cousin of the young claimant to the earldom, William Sinclair. Thomas was meticulously described in the complaint to the queen of Norway as 'inborin man her i landit', a native born man in the country.[14] He and others of recent Scottish descent could scarcely portray themselves as Norwegians, but they could reasonably claim to be Orcadians, members of a community happy to declare loyalty to the crown of Norway, but at the same time eager to affirm its own interests, customs and identity – a place at the centre, rather than the edges, of the map.

The Common Tongue

The complaint against David Menzies is today chiefly remembered for something that can hardly have been anticipated at the time – it is the last extant text from Orkney in the Norwegian, or Norn, language. The first surviving document in Scots dates from just a few years later: a 1433 record of the gift of a house in Kirkwall. A generation before Orkney's transfer from Norwegian to Scots control, the balance of linguistic pre-eminence was beginning to tilt.

An accelerant was the association of the Scots language with three centres of power and influence. One was the earldom itself. The Sinclair earls were Scots-speaking, as were most retainers and followers at their court. The second was the bishopric, about which more will be said in the following chapter. A Scottish clergyman, Thomas Tulloch, was nominated bishop of Orkney by the pope in 1418. Tulloch was a Scot at ease in the Scandinavian world. In June 1420, he wrote to King Erik of Denmark, Sweden and Norway from the Danish island of Laaland to acknowledge a grant of governorship of Orkney. The letter was drafted in Norwegian, and witnessed by the bishops of Laaland and Bergen. Tulloch wrote again in Norwegian in July 1422, accepting a commission to take charge of 'the castle and fortress of Kirkwall, situated in Orkney in Norway'. Four years later, Tulloch was in Bergen, for talks leading to the reaffirmation of the 1266 Treaty of Perth. He was there, alongside other Norwegian bishops, as a member of the *Riksråd*, the royal council of state, though his background doubtless made him a useful asset in diplomatic discussions with the Scots.[15] In Orkney itself, however, documents produced by Tulloch's clerks and notaries after the mid-1420s were exclusively written in Latin and Scots.

A third avenue of linguistic transmission went through the cathedral town of Kirkwall. Its trade with Scottish burghs was growing in the late middle ages, and merchants and artisans from the south were settling there. In 1358, at a time of political upheaval in Orkney, King David II of Scotland responded to a probable request from Norway by prohibiting travel to the islands, but exceptions were made for those going there on pilgrimage or engaging in commerce. Orkney's Scottish trade expanded at the expense of trade with Norway, a land hit badly by the Black Death in the mid-fourteenth century. For reasons both economic and political, Norway's attention was increasingly turning away from its western colonies – contact with the settlement in Greenland was finally lost near the beginning of the fifteenth century.

Kirkwall exerted cultural and economic influence over its hinterlands in the Mainland and the isles. It was Orkney's market place, and principal seat of law and governance. James V issued a charter of burgh status for Kirkwall in 1536, confirming a grant made by his grandfather in 1486. According to this, the town was 'daily enlarged by building and repairing houses, markets and streets'. Orkney, moreover, was being 'brought to civility' by the 'virtue and industry in foreign trade and navigation' of Kirkwall's inhabitants.

Prosperous Kirkwall merchants entered the property market and established Scots-speaking heirs in the rural parishes. There are some telling contrasts of names and identities in a deed of February 1483. John

Mason, burgess of Kirkwall, purchased a half pennyland in Wasbister, in the parish of Holm. The sellers were a married couple, Magnus Andrew Quhitquoysson and Janet Magnus Cuthamysdochtyr – that is, Magnus the son of Andrew of Quatquoy, and Janet the daughter of Magnus of Cuthamy (Quoythome). Not all Orcadians had yet abandoned Norse patronymics.[16]

The surrender of Orkney by the Danish king in 1468 boosted further the status of Scots speech in Orkney. It was the language of social prestige and political authority, and consolidated its hold as the approved medium of law, governance and written documentation – for precisely that reason, in the historical record Norn is often inaudible.

At the time of James V's visit, however, Scots was not yet for many ordinary folk the daily language of word and thought. In September 1534, a meeting was convened at Holy Cross Kirk on the south side of Sanday. Its purpose was to confirm that Michael Merriman, while bed-bound 'in the article of death', had agreed to sell to Magnus Baikie a parcel of land in the Mainland parish of Birsay. The curate of Sanday, Edward Blair, took a record. He was an immigrant Scot, 'priest of the diocese of Aberdeen'. Perhaps for that reason his account noted that the nine witnesses delivered their testimony 'in the common tongue' – in other words, in Norn.

Another priest and notary public, James Scuill 'of the diocese of St Andrews', was on court duty in Kirkwall in February 1543. The case concerned the 'multure' – the proportion of their grain – that farmers in the East Mainland parish of St Andrews needed to pay for use of a mill belonging to James Irving of Sebay, a wealthy landowner and former lawman. Scuill likewise noted that the assembled parishioners made affirmations 'in a loud voice ... in the common tongue' – someone may have had to translate for him afterwards. That could have been James Irving himself, for in court he likewise 'broke forth ... in the common tongue' to reply to his neighbours. The Irvings were of Scots lineage, but resident in Orkney from at least the mid-fourteenth century. Some Norn-speakers at the meeting in Sanday also had Scots surnames: Sinclair, Muir, Spence. Many medieval immigrants to Orkney acclimatised and assimilated, intermarrying with established families and learning to speak the Norse language. Notable folk like James Irving might be confidently bilingual, but country dwellers in both Mainland

and isles perhaps spoke only Norn, or felt more comfortable speaking it when important matters were at stake.[17]

If Norn was disappearing from written records, it remained imprinted on the descriptive veil of the landscape – the vast majority of place and farm names throughout Orkney were of Norse origin. Nearly all the isles themselves had the Old Norse suffix *-ay*, meaning 'island', though an earlier name for the largest of them, 'Hrossey' (horse island) had long since been dropped in favour of the functionally descriptive *Meginland* (Mainland).

Jo. Ben. assumed that Faray meant 'fair isle'. In fact, the more prosaic derivation was *vaer-øy*, sheep or ram island. Scots immigrants inadvertently rechristened the places they encountered around them. On the Brough of Deerness, a high rocky promontory connected to the eastern end of the Orkney Mainland by only a thin sliver of beach, lay the remains of a small church. It was a site, according to Jo. Ben., 'quod nominatur [which is called] *the Bairnes of Brughe*'. Perhaps it was already known by that name, or perhaps Jo. Ben. simply misheard what an informant was telling him. Either way, there were no allegorical children, but rather an old *bœn-hus*, house of prayer or chapel.

The 'scotticising' of impeccably Norse place names was something noted many years ago by the linguist Hugh Marwick. The most egregious example was Orkney's capital itself – originally 'Kirkjuvágr', the bay of the church. Locals came to pronounce it 'Kirkwa' (and occasionally still do). But by the sixteenth century it had routinely become 'Kirkwall' in written texts – scribes hearing 'wa' as the Scots dialect form of 'wall'.[18]

Blown to Bergen

Ties to Scotland were tightening in the sixteenth century, but Orkney's bonds with Norway were far from wholly loosed. Commerce was a continuing source of connection, especially with the trading port of Bergen, on Norway's western fjord coast. In the early sixteenth century, it was Scandinavia's largest city. When Westray farmers complained to earldom officials about sea-eroded land and abandoned fields in 1492, they made the wry observation that the soil was 'blawn til Bergen' – a place proverbially far off and yet a familiar near neighbour. There was mutual demand for commodities between the west coast of Norway and the islands to the west. Norway was chronically short of grain, which Orkney supplied in the form of bere, malt and oatmeal. Orkney's standing deficiency was in timber, for building work of all kinds and especially the construction of boats and ships. Along with tar, wood was Norway's key export to the British Isles, and to the Northern Isles in particular.

In the summer months, vessels from Orkney and Shetland thronged the little harbours of Sunnhordland, the coastal region south of Bergen. The cargo of four Orkney vessels loading there in 1567 comprised 9 beams of fir, 240 planks, 8,300 *baandstaker* (young trees, for making barrel hoops), 100 *knapholt* (smaller pieces of split oak, for construction or furniture work) and 12 whole boats. The latter were made in Norway, then dismantled for shipment in component parts – a precocious example of the arts of prefabrication and self-assembly for which Scandinavia remains internationally renowned.

Scots played important roles in this North Sea traffic, but Orcadians (rather more so than Shetlanders) were particularly to the fore. Many based themselves permanently in Norway. In 1488, John Reid, 'burgess of Bergen', witnessed a charter transferring a house and land in Kirkwall. Reid was probably a Shetlander, but Katherine Leask, the woman selling the property, asserted that her right to it, conferred on her parents by her grandfather at their wedding, could be confirmed by many people who attended the festivities, and were now living 'in Norway, Shetland and Orkney'.

Bergen's *Borgerbok*, or register of burgesses, began systematically recording the origins of new citizens only in the seventeenth century, but a handful of earlier entrants – like Joenn Pariss in 1558 and Anders Joensen in 1574 – were specifically noted as 'orknøisk' rather than 'Skot'. In the two decades between 1612 and 1632, no fewer than fifty-six Orkney-born merchants were registered as burgesses of the city – not many fewer than the total of sixty-nine coming from the entirety of mainland Scotland. Burgesses in Bergen might readily go native. James, son of Henry Rendall of Ellibister in the parish of Rendall, reverted to a patronymic – he called himself James Hendrykssen when in 1580 he authorised his brother in Orkney to secure his share of the family inheritance.

Orcadians in Bergen – often fluent, or at least competent, in Norn – likely found communicating with Norwegians easier than their notional compatriots from St Andrews or Aberdeen did. But they no longer spoke quite the same language as their hosts. Magnus Maat (Mowat?) and his wife, Dilis Røncke (Rensgar?), lived just south of Bergen in the village of Møllendal, where they had a sideline selling ale. Orcadian women, like those throughout Scotland, did not take their husband's name upon marriage. Magnus and Dilis conversed easily enough with their Norwegian neighbours, but also fell out with them, and in 1594 Dilis was accused of witchcraft. Witnesses in the case noted Magnus and Dilis 'used Orcadian speech' in speaking ill of a local miller, and that in parting on bad terms from the miller's wife, Dilis 'knuerrett paa sit Orkenøesch maall' – she growled in her Orkney tongue. The couple evidently spoke the local dialect, or some lingua franca comprehensible in the Bergen region, but when angered switched to their native Norn. How well their neighbours understood it is not entirely clear, but though Magnus and Dilis were integrated into the community, it seems they were still foreigners of a sort.[19]

Scots and Orcadians were not the most significant foreign presence in late medieval and sixteenth-century Bergen. The city hosted a *kontor* or office of the Hanse, the international trading organisation which dominated seaborne commerce in the Baltic and North Sea. German merchants and artisans with extensive privileges clustered in the warren of wharf-side warehouses known as the Tyskebryggen (the German dock). The Germans resented the competitive intrusion represented by the Scots, and in 1523 their resentment spiralled into violence. On the night of 8 November, German gangs attacked the homes of at least a dozen Scots residents in what became known as 'the Scottish incident'. Property was destroyed, and one man reportedly killed. Others were rounded up and forced to leave the city.

Several Scottish victims were in fact immigrants from the Northern Isles. They included an Orcadian known locally as Lille Jon (Little John) Thomessøn. He was a leading merchant and councillor, and spearheaded a subsequent campaign for redress against the Hanseatic *kontor*. The process spun out over the rest of the century, before an order for compensation was eventually issued to the benefit of Jon's heirs. But

Scots and Orcadian influence in the city was not permanently dented by the riot. In 1543, Lille Jon took office as one of Bergen's two *borgermestere* (provosts or mayors), and he remained a person of substance in the city until his death around 1558.[20]

Ethnic resentments continued to fester in Bergen. Herman Müntzer, the German pastor of St Martin's Church, was reported in a sermon of the late 1560s to have inveighed against 'Rotten, Schotten und Holländer' – rats, Scots and Dutchmen. 'Where they are found, nothing can thrive.' He was quoted with approval by an anonymous German satirical text composed in Bergen in 1584, *Die Nordtsche Sau* (The Norwegian Sow). This added to the list of scoundrels (*hundsfotter*; literally, 'dog-feet'), 'Shetlanders, Orcadians and Faroese; in truth these are people of no use to the country'.

Hostility towards islanders, however, was not general among the city's Norwegian population; several victims in 1523 were married to local women. The Lutheran clergyman Absalon Pederssøn Beyer, whose writings are key sources for mid-sixteenth-century Bergen, thought well of Orcadians. 'Orkney is good corn-land, and the people there are brave warriors.' He remembered Lille Jon Thomessøn as 'God-fearing, pious and righteous'. Indeed, in a chronicle of Norway he devoted a lengthy passage to the history of Orkney, as one of the skattlands 'which belong to Norway's crown'.

Beyer's matter-of-fact assertion of sovereignty was very much the official view, reflected in a competitive advantage that traders from Orkney and Shetland enjoyed over mainland Scots. Up to 1580, their ships were exempt from tolls levied on other vessels in Bergen, on the grounds that the owners and crew were still Norwegian subjects. The point was underscored in letters of complaint drafted by victims of the 1523 riot. They had been forced out of the country, to England and Holland, and not allowed to return to their places of origin – to Scotland, or to Orkney and Shetland, 'which lie under the crown of Norway'. Lille Jon's estimate of his extensive losses was accompanied by a patriotic declaration that he was a man 'born in Orkney, subject to the crown of Norway'.[21] Whether Thomessøn fervently believed this, or simply wrote what the authorities wanted to hear, Orcadians understood the advantages as well as difficulties of living between two kingdoms.

In Pawn to the Crown of Scotland

Norwegians found it hard to reconcile themselves to the loss of empire. In the marriage contract of 1468, Christian I claimed to be pawning the Orkney Islands 'with consent and assent of the prelates, magnates and greater nobles of our realm of Norway'. But whether he sought or received any such assent is extremely doubtful. Leading Norwegians resented the actions of a foreign-born ruler who, in his coronation charter of 1449, promised not to alienate any of the kingdom's castles or fiefs unless there was urgent necessity, and 'only according to the advice of Norway's council of the realm'.

The treaty of 1468 was explicitly conditional and temporary: James III was to hold Orkney only until 'whole and full satisfaction and payment is effectually made by us, our heirs and successors, kings of Norway', though arrangements for repayment of the dowry were left conveniently vague. For his part, James gave little indication of supposing he merely held the islands in trust. In 1470, he granted a series of Scottish estates to Earl William Sinclair, and confirmed him in possession of the earldom of Caithness. In return, Sinclair surrendered to the king all right and claim to the earldom of Orkney.

Whether William Sinclair was the legally constituted earl at the time of the transfer is in fact unclear. He had for some years been closer to the Scottish than to the Danish king, and it is possible Christian I may have deposed him, or tried to depose him, in 1461–2 for failure to appear and do homage. In any event, an act of parliament in 1472 declared James III to have 'annexed and united the earldom of Orkney and the lordship of Shetland to the crown, not to be given away in time to come to na person or persons except anerly [only] til ane of the king's sons of lawful bed'.[22] Were the islands to be redeemed, the sitting earl of Orkney would be none other than the king of Scots – a potential political headache for any future Danish king.

Christian I pledged Orkney and Shetland in his capacity as king of Norway, not of Denmark, but many Norwegians felt that, in so doing, he had damaged the honour of the realm in pursuit of his own ambitions. In 1482, the Norwegian *Riksråd* demanded of the new king, Christian's son Hans, that he get the islands back, and a formal commitment to this effect was included in the customary *Valghåndfestning*, the charter of promises, issued by Hans at his coronation the following year. Similar pledges were made by Hans's successors Christian II in 1513 and Frederick I in 1524.

Frederick's charter appeared just as the Union of Kalmar collapsed. Sweden had broken away from Danish rule, and a brief civil war in Denmark itself saw Frederick oust his despotic nephew Christian II – though Christian spent the next decade scheming to recover the throne. Unlike his predecessors, Frederick issued a separate *Valghåndfestning* for his Norwegian subjects. It frankly and apologetically conceded that Christian I had pawned Orkney 'without the consent and will of the Norwegian Riksråd'.

There was some playing to the gallery in these promises of restoration – Danish kings flattered the nostalgic patriotism of the people of Norway just as they were strengthening their grip on the once-independent kingdom. The issue was also useful for raising revenue – Christian II responded to the complaint of the *Riksråd* by saying that his ability to redeem the islands depended upon Norwegian willingness to pay higher taxes.[23] But successive kings did not regard the loss of Orkney and Shetland as settled, and were usually eager to remind the Scottish government of this. In around 1486, Hans wrote to James III about an alarming rumour that native islanders were being cleared out of Orkney and Shetland to make room for incomers who would be governed by Scots law and language. There were hints of a military response. James replied in conciliatory terms, denying any expulsions or attempts to alter laws and language, though he warned Hans not to revoke pacts his father had agreed to.

In Scottish circles, the suggestion started to be put around that Christian I formally renounced all rights to the islands when his grandson, the future James IV, was born in 1472. The claim appeared in a continuation of Hector Boece's chronicle, compiled in the early 1530s, and a Scots envoy in Denmark reported the existence there in the mid-1520s of a document he described as 'the discharge of Orkney and Shetland'. No such document survives; if it ever existed, it seems extraordinary that neither the Scottish nor Danish governments made much effort to preserve a copy. There was certainly no such recognition from Christian I's successors.

Remarkably, there was a suggestion from the Scottish government itself that Orkney and Shetland might be returned to Danish–Norwegian control. In 1514, in the aftermath of Flodden and desperately short of money and men, the Scots regent, the duke of Albany, sent an envoy to Denmark. He was to negotiate for the raising of 6,000 well-armed soldiers, and was authorised to offer in return 'the lands pledged by the King of Denmark to the King of Scotland for the marriage of his aunt'. In July 1514, in anticipation of a happy reversal, Christian II drafted a letter for dispatch to the inhabitants of Orkney:

> Dear friends, you know that you verily belong by right under the crown of Norway, although you are at present in pawn to the crown of Scotland. But, at the very first opportunity, we will redeem you, so that hereafter you will belong, as you should by right, to the crown of Norway and to us, Norway's king.

It was starting to sound a done deal, but, as the immediate crisis waned, the Scots dragged their feet. In June 1515, Albany assured Christian he would attend to the redemption of the islands 'as soon as the country [Scotland] is quiet and factions less violent' – an open-ended promise in early sixteenth-century conditions. In 1524, with Scotland once more at war with England, Scottish ambassadors were again instructed by Albany to dangle the prospect in front of Frederick I: if he were to send gold to redeem Orkney, then 'we can pay the wages of the soldiers, and the said province can be returned into his power'.[24]

As before, the proposal went nowhere, but neither did it drop entirely off the table. Frederick's son Christian III, in his coronation oath of 1536, swore he would redeem Orkney and Shetland. It was a longstanding convention, but Christian was a determined and decisive king. On the same day, he officially established Lutheranism as the national religion of Denmark, and announced that Norway would cease to be a kingdom in its own right, and ever afterwards remain a province of Denmark. This did not mean, however, that he had no interest in pursuing Norway's historic rights, which were now in fact his own rights as sovereign.

In 1538, Christian went so far as to appoint a lawman for Shetland – a significant infringement of the king of Scots' prerogatives. The nominee, a Shetlander living in Norway called Gervald Williamson, went brazenly to the Scottish court, armed with letters of commendation from Christian, to seek guarantees of safe conduct for his travels. In July 1539, James V, seemingly unaware of the nature of the planned Danish incursion, graciously provided these, and wrote to Christian to say he had done so.

In the event, Williamson seems not to have managed to oust Shetland's resident lawman, Neils Thomasson. But the mortgaged islands were much on the king of Denmark's mind. In the second half of 1539, a memorandum from the royal chancery reported that Norwegians would willingly be taxed for the purpose of redeeming Orkney and Shetland.[25] This, it will be recalled (pp. 19–20), was the very moment at which Olaus Magnus's *Carta Marina* was representing Orkney to a European audience as a tranche of rightful Norwegian territory.

The voyage James V decided to undertake to Orkney in the summer of 1540 was thus no frivolous pleasure cruise. It was imperative to demonstrate to people living there, and anyone else who might be watching, that the islands to the north were the legitimate progeny of the Scottish crown, and not an assortment of sullen foster children, waiting to return to the embrace of their true Nordic parent.

The War of the Sinclairs

There were other motivations behind James's journey. In the years preceding the royal visit, events within Orkney itself had proved unpredictable, volatile and violent, and demanded the attention of outside forces. As with so much of the history of early sixteenth-century Scotland, the difficulties can be traced to the disastrous military defeat at Flodden, in the early autumn of 1513.

Prior to that momentous day, Orkney's internal politics had been relatively stable. In 1489, James IV granted the 'tack' – the lease of the earldom and royal lands in Orkney – to Lord Henry Sinclair, a grandson of the last Sinclair earl. With the grant came custody of Kirkwall

Castle, and rights to hold courts and exercise justice. Lord Henry was an efficient administrator, who made a tidy profit on the tack by encouraging tenants to reoccupy lands that had fallen out of tillage, and by setting realistic new rents payable in kind – often in barrels of butter. He managed the earldom estates alongside so-called 'conquest' lands. Despite the martial label, these were lands acquired peaceably through purchase by the last earl – widely scattered farms and fragments of estates, often intermingled with the lands held in tack, but not pertaining to the earldom proper. The conquest lands were the basis of restored Sinclair wealth and power in Orkney, and also in Shetland, where Lord Henry could count on support from the dominant power on the ground, his uncle, Sir David Sinclair of Sumburgh. David was an illegitimate son of Earl William, with strong links to the crown of Norway.

Lord Henry's tack was renewed in 1501, and in the years following he entrusted local management to his brother, Sir William Sinclair of Warsetter in Sanday, an estate created from amalgamation of conquest properties. Lord Henry was now more often at court, where he cut a figure of cultural as well as political heft – the poet and clergyman Gavin Douglas dedicated to him his translation into Scots of Virgil's *Aeneid*, having undertaken it at Lord Henry's behest. It was a mark of royal favour that, in March 1513, James IV appointed Henry 'Master of all our Machines and Artilleries', devices of which James was inordinately proud. It meant Lord Henry was in the front-line at Flodden, trying to direct the Scottish cannon fire. He was among the first of dozens of Scots dignitaries to fall.[26]

A calamity for Scotland, Flodden also heralded a period of turbulence in Orkney, and unsettled the islands' precarious relationship with authority to the south. The tack passed to Lord Henry's widow, Margaret Hepburn, a sister of the earl of Bothwell. The tacksman's judicial and administrative functions, however, were assumed by his brother, Sir William Sinclair of Warsetter.

Sir William elbowed aside the claims to authority of his youthful nephew, the new Lord Sinclair, also William. Both before and after Flodden, Sir William, 'ane noble and potent man', appears in the role of 'justice', presiding alongside the lawman at meetings of the lawthing. Sir

William's seal was borrowed to authenticate a court ruling in July 1522, but this is his last appearance in the records, and he probably died soon after.[27] The young Lord Sinclair now expected to assume a place of rightful pre-eminence in Orkney, but his hopes were thwarted by the ambitions of Sinclair of Warsetter's sons – not so much those of his legitimate heir, Magnus, as of a bold pair of bastard half-brothers: Edward Sinclair of Strom and the charismatic, erratic and violent James Sinclair of Brecks.

Even before Sir William's death, Margaret Hepburn faced difficulties collecting rents and taxes in Orkney, and accounting for them to the exchequer. She was in arrears in 1515, 1519–20 and 1522. In 1525, she was granted a remittance of £80 Scots on the total due 'because in the preceding year the said lordship [of Orkney] was completely laid waste by the sons of Sir William Sinclair and their adherents, and the lordship of Shetland was completely harried and wasted by the English'.[28]

In 1526, Margaret sought to loosen the grip of her wayward nephews, granting her son William powers of jurisdiction in Orkney and appointing him custodian of Kirkwall Castle. By early 1528, William had taken up residence, with intentions of 'good rule and ministration of justice to the inhabitants'. For what happened next, we are reliant on William's embittered letter of complaint, sent to the king and council, sometime in the second half of the following year.

It is a one-sided, self-pitying account, though likely sound enough in its essentials.[29] Just before Easter 1528, Lord Sinclair and his entourage were attacked outside the castle in Kirkwall by James and Edward Sinclair with a host of well-armed supporters. They killed three of James's own 'brother bairns' (nephews) serving in Lord William's retinue, as well as seven other gentleman attendants. The killings took place after these men had surrendered, 'crying mightily for God's sake on their knees'. Lord Sinclair had no choice but to yield the castle to James and Edward.

Their hold on power complete, the Sinclair brothers let their cousin go. He sought refuge with another cousin, John Sinclair, earl of Caithness, a grandson of the last Sinclair earl of Orkney. Together they plotted the next moves. In May 1529, Lord William formally complained to the king, and received royal letters commanding the rebels, under pain of outlawry, to surrender the castle and submit to justice. If they did not,

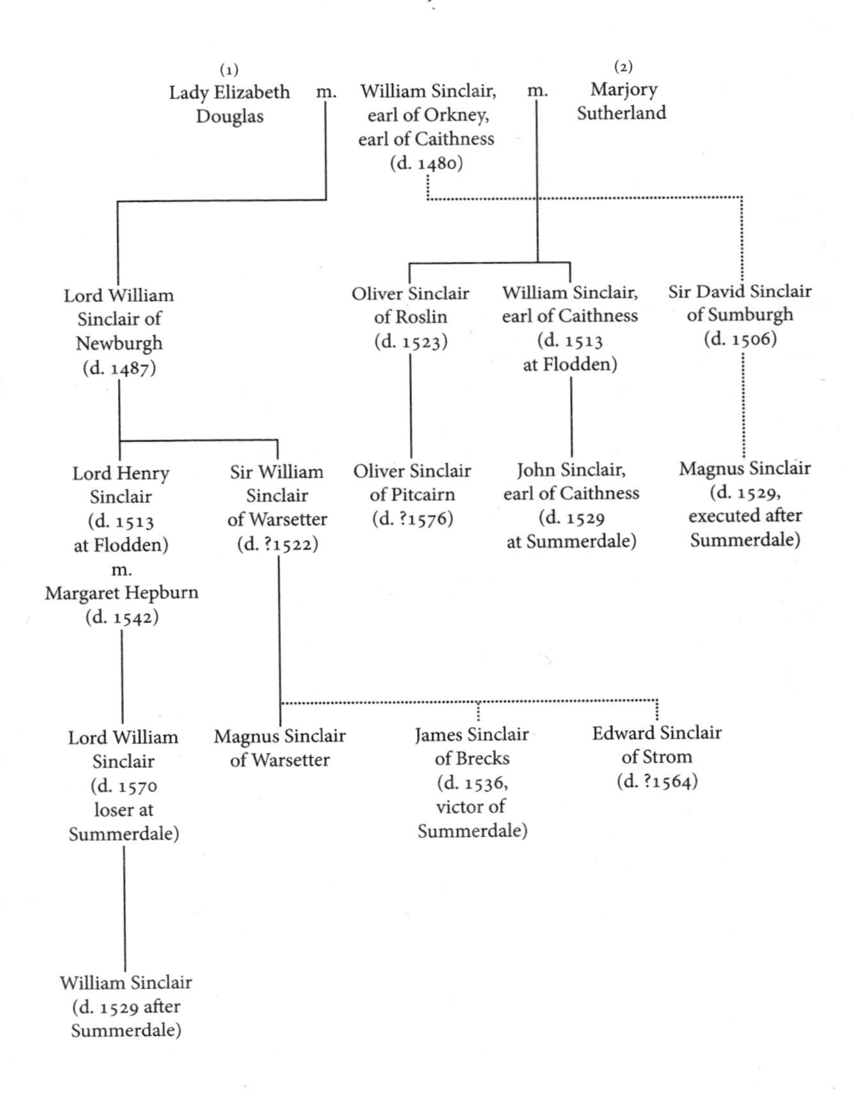

Figure 1.3: Relationships between warring Sinclairs in the early sixteenth century, with broken lines indicating illegitimate birth. The winner of the Battle of Summerdale, James Sinclair of Brecks, and the loser, Lord William Sinclair, were first cousins.

the earl of Caithness was authorised to assist Lord Sinclair in imposing the mandate by force.

A messenger went to Orkney to deliver the judgement, but James Sinclair simply locked him up. Defiance was probably expected, as the invasion force was ready to depart before the ultimatum was dispatched. John Sinclair, earl of Caithness, was perhaps motivated by familial affection, or a sense of outrage at the illegalities in the north. More likely, he spied a chance to fish in troubled waters. Perhaps, when the Orkney tack came up for renewal, Earl John would seem a more promising candidate than the ineffectual Lord William.

In early June 1529, Lord Sinclair and the earl of Caithness crossed the Pentland Firth. Their ships passed to the east of the deserted isle of Switha, and the force made landfall in the Mainland parish of Orphir, at one of the sheltered bays looping around the northern rim of Scapa Flow. Curiously, the invaders did not march directly to Kirkwall – perhaps the route was heavily defended, or they received intelligence that the main enemy force was already elsewhere. Instead, they headed due north, marching up the valley which led them past the little Loch of Kirbister and into the arc of heather-clad hills straddling the parish boundary between Orphir and Stenness. There, on 7 June, the Sinclair brothers were waiting for them, at a saddle in the hills known as Summerdale – the valley of the *sunn-mœrr*, the southern border, of Stenness.

The brothers launched an immediate assault, killing thirty of the Caithness men, including Earl John. Another hundred were slain as they tried to escape to the boats. Lord Sinclair's account is hazy about his own role in the debacle. He was present at the battle, but seems to have avoided injury and capture; no doubt he fled with the panic-stricken rest.

James Sinclair sealed his victory with exceptional ruthlessness. On the day of the battle, a boat was forced ashore containing twenty-two of the defeated army's 'carriage men' – non-combatant transport workers. All were killed on the beach, by executioners who 'tirfitt [stripped] them naked and let the sea have them away'. Remnants of the earl's army claimed immunity in 'divers kirks of the country', including the cathedral of St Magnus, Our Lady of Grace in Harray and the church of St Lawrence – one or other of two parish kirks dedicated to the saint, in Rendall, or in the island of Burray.

This may have given their pursuers pause, as it was a further three weeks before they acted. Yet, in the end, the victors were undeterred by pious custom, or by Church laws that granted fugitives forty days of sanctuary before making their way into exile. The asylum-seekers were dragged from kirks and kirkyards and immediately put to death, 'in high contempt of God and Holy Kirk'. James Sinclair also commanded his brother Edward to execute, without legal process, Magnus Sinclair, an illegitimate son of the late Sir David of Sumburgh. The Sinclairs' civil war was a homicidally incestuous business. Three other men were summarily killed with Magnus, as were seven Shetlanders who came to Orkney to serve Lord Sinclair, 'beheaded and slain in a barn'.

In July 1529, James Sinclair sailed north and consolidated his coup by eliminating resistance in Shetland. He seized and beheaded the lawman, along with several supporters. Others were hanged by James Sinclair 'at his own hand, without any authority, commission or power of Your Grace'. Three young boys were slain for no other reason than 'malice he had to their masters'. In all, Lord Sinclair reckoned, over 300 were put to death. The victims included Lord Sinclair's own son, William, taken prisoner by James, presumably in the wake of Summerdale. Lord Sinclair secured royal letters demanding his release, but James disregarded them, and young William died in his custody.

Lord Sinclair's letter painted for the king a heart-rending picture of the 'oppression and extortion' wreaked by James Sinclair, his brother and accomplices. In addition to those slain, opponents of James were banished from Orkney and 'with their wives and bairns are now begging their meat in England and Norway'. By stressing to the king how royal subjects had been driven from Orkney out of Scotland itself, Lord Sinclair perhaps hoped to appeal to an affronted sense of royal dignity – it was 'great shame that such things should be heard within this realm unpunished'.

Summerdale's aftermath may have been bloodier than the day itself, but tales about the clash in the Orphir hills grew in the telling. In his account of the 'cruel battle', written in the 1560s, Bishop John Lesley claimed that, along with the earl, 500 Caithness men, intending 'to have passed and occupied that country', were slain or drowned in the sea that day by 'the Orkney men'. Jo. Ben. knew for a fact that 'all the Caithnessians were overthrown and slain, so that not one of them survived'.

The bodies of the slain were supposedly buried without distinction in pits; for centuries afterwards, local people believed mounds near the battle site to be invaders' unmarked graves. In the eighteenth century, there were reports of peat-diggers coming across bodies uncannily preserved in the mossy ground.[30] The most memorable traditions about Summerdale were recorded by the Victorian antiquarian James T. Calder, who printed in his 1861 *History of Caithness* what was communicated to him by an Orkney resident. For Calder, the accounts provided 'a striking picture of the superstition and savage barbarity of the people of the north at the period in question'. It is impossible to say if some embryo of these stories dates from the sixteenth century itself. Yet the tales testify to the powerful mystique which Summerdale – still regularly but wrongly called the last pitched battle on Orkney soil – came to assume in the islanders' perceptions of their own history.

The earl of Caithness's army, Calder reported, was accompanied by a witch. On the march from the Orphir beachhead she walked at the head of the advancing troops, unwinding two balls of thread, one blue and one red. The red ball finished before the other, and the witch explained to the earl that this meant the side to suffer earliest loss of blood would be the one to experience defeat. Earl John cynically gave orders for his men to kill without hesitation the first local they encountered. This turned out to be a boy herding cattle; but when the pitiless deed was done, some soldiers recognised the corpse as that of a Caithness native, whose family had been exiled to Orkney a few years earlier.

Fate's implacable face was set against the invaders. The place the two armies met was an area of smooth grass, where, prior to the morning of the battle, no stones littered the ground. But as James Sinclair's troops took up position, the Orkneymen found such abundance of rocks that they were able to lay aside their weapons and direct a withering barrage at the Caithnessians, causing their ranks to break. The earl sought refuge at a nearby farm called Oback. There, in a muddy lane between house and outbuildings, he ran straight into a party of Orcadians, was immediately slain and afterwards beheaded.

Yet the cruelty of fate proved, if not quite evenhanded, at least indiscriminately capricious. Only a single Orkneyman was said to have died that day, returning home in triumph clad in garb looted from the field. His mother mistook him for an escaped enemy soldier and struck a fatal blow with a stone wrapped in the foot of a stocking belonging to her son.[31] Such stories employ timeless themes of hubris and nemesis, lending grandeur and pathos to a bloody island quarrel. Their infusions of destiny and magic owe little to any Christian notions of morality or justice.

Calder lamented Summerdale as the Flodden of his native Caithness, but Orcadian authors of the nineteenth and early twentieth centuries saw a valiant assertion of the rights of freeborn udallers in the face of Scots oppression; a moment which delayed, even if it could not prevent, the imposition of feudalism in the islands. *The Fight at Summerdale*, a stirring 1913 adventure novel by the patriotic Orcadian writer John Gunn, is narrated by a hero with the implausibly pure Nordic name of Rolf Ericsson. Decades later, from exile in a French abbey, Rolf looks back on the 'breaking of Scots promises' and imposition of 'hated Scots law'. Among the book's coloured plates, a picture of the battle evokes a sartorial clash of historical cultures: tartan-clad highlanders versus Orcadian peasants in an Edwardian version of medieval Viking attire (see Plate 7).[32]

All this, of course, was fanciful. The Summerdale War revolved around raw power not principle, and was a fratricidal struggle between two branches of a family with common Scottish roots, rather than an ethnic confrontation between colonising Scots and indigenous Orcadians. Yet contemporary commentators portrayed the conflict as pitting a local community – 'the Orkneymen', 'the country' – against invading forces from outside. For the Scottish court, godparent of the invasion, it was an uncomfortable message about a place which, within living memory, had been a quasi-independent earldom owing formal allegiance to a neighbouring kingdom. Lord Sinclair laid it on thick in his letter to James V. In the afterglow of rebellion, James Sinclair of Brecks was disporting himself as though 'he were ane king in those parts, and like as there were no law, king nor justice in this realm'.

James V did not rise to the bait. In 1528–9, the first year of his adult reign, he had other things on his mind – principally, the ambitions of the earl of Angus, who effectively had held him prisoner during his minority. Lord Sinclair's concerns were a distraction, and perhaps a dereliction of the duty he owed to his king at this critical juncture. After Summerdale, it was Margaret Hepburn, Lady Sinclair, rather than her humiliated son, who continued to make returns for Orkney and Shetland to the exchequer. Indeed, the financial potential of the islands may have been of most interest to the king: a valuation in 1532, as part of negotiations around the jointure (widow's settlement) for James's prospective French bride, suggested the earldom lands were worth considerably more than the sums annually received for them.

In the years after the bloody summer of 1529, James Sinclair ruled the islands with the free hand of a Viking jarl. He styled himself – without authority – 'justice of Orkney', and his justice was of the pre-emptive kind. In 1533, Sinclair seized in North Ronaldsay the cargo of a ship from King's Lynn, ignoring an imperious summons to appear before the English privy council. Another despoiled English merchant, Thomas Miller, had no better luck complaining about Sinclair to the Scottish authorities in 1535.

Orkney had always been vulnerable to the depredations of pirates, and the unwelcome attentions of English crews bound for the fishing grounds of Iceland. James V himself complained in 1535 that folk in the North Isles were taken for 'slaves, servants and prisoners' by passing English ships. There was payback for the islanders in the experience of a suspected pirate vessel passing through Orkney in 1533, captained by one Richard Saw: 'as he was going to land by boat with six of his company [they] were taken by one Sinclair, a gentleman of those parts and there detained as prisoners'. When Sinclair sent his own boats out to seize the ship, the pilot judged discretion the better part of valour, abandoned Saw and sailed to the south, later to tell his story from inside the Liberty of Westminster – an ecclesiastical sanctuary even James Sinclair's hand could not reach.[33]

In the end, the king decided that if he could not punish Sinclair of Brecks he might as well reward him. In June 1535, James V knighted Sinclair, declared him legitimate and made him tacksman for Orkney and Shetland. Not long after, the king issued a pardon to thirty-one instigators of the Summerdale Rebellion. Edward Sinclair of Strom headed the list, followed by Magnus Sinclair of Warsetter, John Sinclair of Tollop, William Sinclair of House and Oliver Sinclair of Havera in Shetland, Magnus Sinclair, Lawrence Sinclair and James Sinclair.

The pardon covered other Scots name-bearers, too – assorted Cromarties, Craigies, Harcuses and Sclaters. But some participants had traditional surnames derived from places in the landscape: Edward Berston, Magnus Midhouse, John Paplay, John Rendall. It is further confirmation that Orkney's governing elite comprised men of both immigrant Scots and indigenous Norse descent. They were indemnified for their part in the death of John, earl of Caithness, and any other 'slaughters, mutilations, oppressions, robberies, forethought felonies, treasons, crimes, transgressions and offences' – all of which was no deal-breaker for a king who considered his best option was to work with the grain of homegrown authority.

The summer of 1535 saw a further gesture of favour to James Sinclair, and to his wife, Barbara Stewart. She was a daughter of Andrew Stewart, Lord Avondale, another noble victim of Flodden, and, more to the point, a sister of Lord Henry Stewart, current husband of the king's mother, Margaret Tudor. By royal charter, James V granted the Sinclairs the whole islands of Sanday and Stronsay, with their associated 'holms' of Papa Stronsay, Auskerry and 'Rymstay' – the latter no holm, but the substantial island of North Ronaldsay.

The grant, in return for an annual payment of 200 merks, was 'in feu-farm'. This was a striking innovation, the first case in Orkney of the feudal form of landholding prevalent elsewhere in Scotland which assigned perpetual tenure of an estate in return for an annual fee. Such estates had to be passed on entire, with the heir paying an enhanced fee in the year of entry. In these ways, 'feued' lands differed from udal ones, which were held in absolute ownership, without superior, and were subject to partible inheritance among the proprietor's children.

In its wording, the charter was technically illegal, and probably betrays ignorance of the situation in Orkney on the part of the Scots lawyers drafting it – though the later story that James Sinclair secured a generous grant by passing off the rich islands of Sanday and Stronsay as uninhabited holms is almost certainly groundless. The king might indeed choose to feu his own earldom estates, but the islands in question contained extensive bishopric lands, as well as the farms of numerous udal landowners. The beneficiary started nonetheless to style himself 'Sir James Sinclair of Sanday'.[34] It is ironic that the Orcadian leader later celebrated for resisting feudalism in fact implemented its first legal toehold in the islands.

James Sinclair did not serve long as a royal governor in Orkney. Even allowing for exaggerations on the part of Lord Sinclair, James's actions after Summerdale suggest a volatile lack of restraint, and in 1536 he committed another shocking act of violence – against himself. A report came to the ears of Adam Abell, a friar at Jedburgh in the Borders, who in the 1530s was revising his moralistic chronicle of world and Scottish history. The first known owners of the manuscript after Abell's death were the Sinclairs of Roslin – another branch of the family descended from the last Sinclair earl of Orkney – who may have been the friar's patrons, as well as a channel of information on affairs in the Northern Isles.

Abell's chronicle confirms accusations to be found in Lord Sinclair's complaint – the breaking of sanctuary, the slaying of unarmed prisoners. In Abell's judgement, James Sinclair was the Devil's servant rather than God's; though pardoned and rewarded by the king, 'he deserved rather to have been hanged for his wicked life'.

But 'behold here the false smiling of Dame Fortune, and how perilous it is to them that she smiles on'. In the summer of 1536, the arrival of a writ from the king seemingly overturned the balance of James's mind. Abell does not say what the letter contained – presumably an expression of displeasure or an announcement of pending investigation, though we can probably discount the later tradition that it revealed a royal discovery of James's supposed deceit about the holms. The writ was delivered into the courtroom where James was presiding 'with great gloriation and arrogance'. Since this was at midsummer ('about the Feast of the Nativity of St John'), the occasion was almost certainly the annual meeting of the lawthing in Kirkwall.

The king's message caused James to break into fits of mad dancing: he 'cast gamounts' – stylised leaps or springs – with none able to restrain him. The next day, James rose early, placing the writ and his ring on the bed by his sleeping wife. Near the house, as at all Orkney farmsteads, stood a stack of drying peats. James began to dance wildly about the stack, scattering the peats. He then jumped repeatedly into 'ane foul deep dub' (a pool of stagnant water) before casting off his bonnet and clothes and running to a high cliff. There he threw himself into the water. 'Others', Abell added, 'say that the evil spirit led him by the oxter [upper arm] in the sea and drowned him.'

James Sinclair of Brecks was neither the first nor the last strong man to lord it over the islands, but his ghastly fate ensured he would be remembered. People in seventeenth-century Orkney knew the location of James's suicide: the Gloup of Linksness, close to the Sinclairs' house in Tankerness.[35] A gloup, from a Norse word meaning 'chasm' or 'ravine', is a rent in the ground formed by the collapsing roof of an elongated sea-cave, slowly but surely carved into the face of a cliff by the waves. Orkney's frayed coastline is pitted with such gloups – perilous vertical channels from the land into the sea, and, like other anomalies of the landscape, potential repositories of memory and fable.

Lawman and Sheriff

Since Sinclair 'wilfully slew himself', his property was forfeit to the crown, and in 1537 his widow and twelve associates were 'put to the horn' (declared outlaws) for failing to answer charges of forcibly extracting from tenants rents due to Lady Margaret Sinclair in the years following Summerdale.

Lady Sinclair's tack was renewed in July 1536. But the process of 'horning' – so-called because a public declaration of outlawry was accompanied with three blasts of a horn – was not pursued, and in April 1539 Barbara Stewart received from the king a grant of her late husband's forfeited estates. In the first week of September, the surviving leaders of the opposing Summerdale armies, Lord William Sinclair and Edward Sinclair of Strom, were summoned to Falkland Palace, the royal hunting estate in Fife north of Edinburgh, to set aside the 'deadly feud and enmity between them'. With hands on the gospels, they were invited to make a bond of 'manrent', a formal declaration of friendship and alliance. All quarrels and rancour would be forgotten 'for the tenderness of blood standing betwixt them, and for good service to be done by them to the king's highness in time coming'.

Among the witnesses to the agreement was yet another kinsman of the recently warring parties – Oliver Sinclair of Pitcairn, a younger son of the laird of Roslin. Oliver was that perennial appurtenance of Renaissance courts, a royal favourite, who had already been rewarded with wardship of the earldom of Caithness, still in the hands of a minor after Earl John's violent death at Summerdale. It was probably Oliver who steered the king towards conciliation with the Orkney rebels. Almost certainly he accompanied James on the voyage to Kirkwall in 1540. Lord William Sinclair was likely there, too, but it was Oliver who had the king's ear.[36]

In April 1541, Oliver Sinclair was himself installed as tacksman of the king's Orkney and Shetland estates, and was soon returning greater sums to the crown than either Lady Sinclair or Sinclair of Brecks. At the same time, Oliver and his heirs were constituted 'justices, sheriffs, admirals and baillies of all and sundry the foresaid lands'. He had no intention, however, of taking up permanent residence in Orkney, and appointed Edward Sinclair of Strom, the power on the ground, as 'his very lawful and undoubted baillie'.

The royal visit nonetheless heralded a sea-change, or the beginnings of one, in Orkney's relationship with Scotland. Oliver Sinclair was the first office-holder to be designated with the Scots title of 'sheriff', rather than the Norwegian position of lawman. The lawman's principal court, the lawthing, became the sheriff court. and the leading local citizens in attendance were henceforth to be known as 'suitors of court', rather than 'roithmen'.

The change came hard on the heels of another step towards incorporation, itself a sequel to the summer expedition. In December 1540, parliament passed an 'Act of Annexation', confirming the king in possession of 'such lands and lordships as are now presently in his hands that are not annexed of before'. This meant principally the ancient Lordship of the Isles, but the act also declared 'the lands and lordship of Orkney and Shetland and the isles pertaining thereto' to be part of the inalienable inheritance of the crown.[37]

It looked like a concerted policy of royal centralisation, but there was a bit less to it than met the eye. The years after 1540 saw no immediate or sustained drive towards Orkney's 'Scottification'. Though the nomenclature changed, the accustomed patterns of justice continued more or less unaltered. The lawman/sheriff or his deputy convened local courts as they travelled the isles on circuit – referred to in the early sixteenth century as 'ogang'. There was an annual 'head court' in Kirkwall in the summer (the lawthing), in the autumn (the All Hallows court) and at New Year, the 'Hirdmanstein' or 'Hird' court, its name derived from a word for the personal armed retinue of the jarl. The old titles for the courts stayed in use for at least a couple of generations – another instance of Orkney's instinctive cultural bilingualism.

A more pressing question was that of which law the courts actually administered. For some time, the idea had been gathering pace in Scotland that the identity of the kingdom was defined by its law, and that only Scots law should pertain in it. A measure was drafted for James IV's parliament of 1504 that 'all our sovereign Lord's lieges, both within Orkney, Shetland and the Isles … be ruled by our sovereign Lord's own laws and the common laws of the realm, and by none other laws'. Before the act passed, however, the reference to Orkney and Shetland was abruptly removed, in belated recognition that the status of the 'impignorated' territories (p. 18) differed from the forcibly abolished western Lordship of the Isles.[38]

This was more than a matter of abstract legal theory, but something that affected local people and their ability to make law work for them. In a case heard at Kirkwall in 1547, Agatha, a daughter of Sir William Sinclair of Warsetter, successfully asserted legitimate birth to claim the 'sister's part' of her father's estate that she was entitled to under udal law. In attendance were 'all and sundry the inhabitants of the country of Orkney', who expressed approval of the judgement 'as the custom of the country of Orkney'. Transfer of the lands was to proceed 'according to the force, form and tenor of the civil laws of Norway and Orkney'.

Cases mentioning sisters' shares often appear in the Orkney judicial records. They sometimes involved male heirs exercising their legal option to buy out such shares in 'the bu', the chief farm of the inheritance. But just as often they show women acting as agents of their own interests, asserting or relinquishing their property rights under law – even when, as with Janet Irving in 1574, the declaration was made with 'my hand at the pen, led by the notary at my command, because I cannot write'.

Another feature of Orcadian udal law was the practice of 'upgestrie' – voluntary transfer of property in return for guarantees of lifelong support. It was valued by widows who might not trust entirely to the warmth of filial affection. In 1527, Margaret Sinclair made over to her son John all her lands and goods in Orkney and Shetland, on condition he 'uphold me honestly all the days of my life'.[39]

Orkney land law differed in an additional way that was of particular significance to an island community. Under Scots law, ownership of the foreshore and the seabed was vested in the crown, but this was not so in the udal system. Sixteenth-century Orkney charters described property boundaries as running 'from the highest stone in the hill to the lowest stone in the ebb'; from a named point in the landscape 'to the lowest of the sand and sea'.[40] They conferred proprietary rights to invaluable resources: the shellfish, seaweed, driftwood – and occasional bounty like seals or whales – presented for collection or capture along the sinuous length of the low-tide shore.

The ebb was sometimes a fractious place. In the East Mainland district of Toab, John Irving of Sebay resented incursions into his property which local farmers regarded as their customary right. The lawman and roithmen ruled in 1509, and again in 1519, that 'the neighbours of Toab' must not graze their animals on the land of Sebay, nor gather any 'ware' (seaweed) or 'wrack' (anything washed up by the sea) from its shores. The land pertained to Irving and his heirs, 'from the lowest of the sea and ebb to the highest and fairest of the said ground'.[41] Tensions between claims to outright ownership and expectations of communal enjoyment were not easy to resolve, especially in an environment where boundaries of earth and ocean were never entirely fixed.

Time of Ancient Memory

Orkney's new political arrangements barely had time to settle before they were again thrown into disarray. War with England broke out in 1541, after James V – urged on by Oliver Sinclair – failed to attend a summit with Henry VIII. Orkney was on the edge of the ensuing conflict, but it was nearly at the heart of the action.

The king of England's awareness of the Orkney Islands was, in all probability, both vivid and hazy. From childhood, Henry VIII was steeped in the mythology of the British King Arthur, a figure invoked by his father to glamorise the accession of the usurping, semi-Welsh Tudors. In the 1530s, Henry VIII drew on Arthurian ideology to justify his break with the pope – Arthur, like Henry, was an 'emperor', refusing demands for tribute from external powers both political and spiritual.

Orkney was a tenacious if ambiguous presence in the canon of medieval Arthurian texts: a distant, exotic land of uncertain location. Sir Gawain was supposed to be a son of King Lot of Orkney, who was married to Arthur's sister, Morgause. Sometimes Orkney was included in lists of Arthur's conquests, and sometimes noted to be a site of particular resistance to him. Henry VIII might have fantasised about reconstituting the empire of Arthur, but in early 1535 he was offered the chance to acquire Orkney without drawing his sword. Christian III wanted Henry to withdraw support from the rebellious Hanseatic city of Lübeck. His ambassador suggested to Thomas Cromwell that, though the king of Scots currently held Orkney as security for Queen Margaret's dowry, if Henry would pay an equivalent sum, the islands could be handed to him – Iceland was mentioned as an additional inducement. Christian perhaps reckoned the lost province might as well be mortgaged to the English as the Scots, and could be redeemed again at a later date.[42]

The proposal was not seriously pursued, but Henry's interest in the islands had been piqued. In September 1542, he wrote to the duke of Norfolk and other commanders at York. 'The isles of Shetland and Orkney', Henry confidently pronounced, 'be great nourishes [nurses] to Scotland, both for bestial and also for plenty of corn.' If they were to be attacked and destroyed, Scotland would suffer a crippling blow. Henry ordered his generals to investigate further, and draw up plans for a naval expedition. A fortnight later, the king wrote to the commanders again. A land invasion of Scotland was now on the cards, but, as a back-up, ships were to sent to Orkney and Shetland 'to devastate and destroy all the corn and cattle of the same'.

A few weeks later, Norfolk and his fellow commissioners responded, having 'by all this time travailed to know and perceive the state of those isles'. Their report to the privy council was a stream of cold water on the king's enthusiasm for a northern offensive. Shetland, they had learned, was so far distant, and the year already so far gone, that no sensible sailor would venture there – Englishmen headed for the Iceland fisheries 'dare not tarry longer upon those coasts than St James's Tide [the end of July]'. Getting there involved passing through a strait called 'Pentley Firth, which is reckoned the most dangerous place of all Christendom'.

Orkney was not quite so far off, but 'also very dangerous and full of rocks'. Its value to the enemy war effort had been exaggerated: a place 'of so little fertility as the same is but a small relief to the realm of Scotland'. If the king commanded it, they would make the arrangements, but the commissioners made clear they thought the proposed expedition a bad idea: 'the same would not acquit the tenth part of the charges, besides the danger and loss of the ships'.[43]

In the event, the decisive blow was delivered in the north of England, after James V authorised a retaliatory strike across the border. A report current at the time, and widely believed afterwards, maintained that James gave overall command of the army to the inexperienced Oliver Sinclair of Pitcairn, sheriff of Orkney, over the head of Lord Maxwell, veteran warden of the West March. Henry VIII heard that Oliver's wife, Katherine Bellenden, maintained a mistress for the king at Tantallon Castle in East Lothian, where her husband was governor, and that James went there when word reached him of the battle. It was news of a disaster almost as huge as Flodden, but without any of the tragic honour. On 24 November 1542, the army became trapped in boggy ground at Solway Moss and ignominiously surrendered. Not many were slain, but dozens of notables, including Oliver Sinclair, were taken prisoner by the English. A few weeks later, the king himself was dead – caused, the Imperial ambassador in England believed, by 'grief, regret and rage'.[44]

Sinclair was released from English captivity in 1543, but the shame of Solway Moss, and loss of his royal patron, undermined him both nationally and locally. James's widow, Mary of Guise, took away the Orkney tack and granted it to the earl of Huntly. The post of sheriff, and custodianship of Kirkwall Castle, were afterwards bestowed on one of the queen's retainers, a Monsieur Bonot, about whom little is known, but who seems to have at least intermittently visited Orkney.

Oliver Sinclair was entangled in vexatious litigation with Lord William Sinclair, who never gave up on asserting his own inherited rights in Orkney. Mary of Guise also claimed that Oliver owed her 'great sums' in arrears of the rent. In November 1543, Katherine Bellenden wrote to the queen dowager, woman to woman, to deny any fiscal wrongdoing, and to petition fruitlessly for a renewal of Oliver's tack. 'We think it great lack', she concluded plaintively, 'to gang fra our native rooms, which my husband and his surname has brukit [enjoyed] there three or four hundred year'.

This was shameless special pleading on the part of Katherine and her husband – Lowland Scots, by birth, upbringing and residence. The former Sinclair earldom of Orkney originated rather less than two centuries earlier, not four. Yet ancient right, and a desire not to be regarded as interlopers in Orkney, clearly mattered to them. Katherine's was not quite as jaw-dropping a claim as that made a few years earlier by her husband's cousin, Lord Henry Sinclair. In 1507, he petitioned King Hans for assurances of favour and protection for all Orcadians and Shetlanders visiting Norway on business. The earldom lands, he reminded Hans, 'were alienated heritably to my progenitors, even before the birth of Christ'.[45]

In the mid-sixteenth century, the precise place of Orkney on the political map of northern Europe was far from finally settled. A seemingly endless conflict between Scotland and England encouraged Christian III once more to try his hand. After the death of Henry VIII, the English government continued his policy of 'rough wooing', but plans to compel a betrothal between the child monarchs Mary Queen of Scots and Edward VI came to nothing in early 1548, when Mary was betrothed to the French dauphin, Francis, and shortly afterwards went to live in France. It was not the only possible outcome. In 1546, Francis I had mooted a marriage between Mary and Christian's son Magnus. It would be a way, he slyly suggested, for Denmark to regain the lost islands of Orkney.

The issue flared up again in late 1548, when Regent Arran sent an ambassador to Denmark seeking ships and munitions. Christian informed the ambassador of his intention to recover Orkney, and dispatched an emissary to the Scottish court to amplify the claim. He again wrote to Arran in September 1549, reminding him the islands were 'part of the kingdom of Norway from time of ancient memory', and demanded a meeting in Orkney the following year, at which the redemption sum could be handed over by his representatives.

Christian III had the money and evidently meant business. In 1549, he levied a special Orkney tax in Norway, while writing about the justice of his claim to the new king of France, Henry II, and a string of leading French nobles. Henry replied sympathetically, though he thought the matter might reasonably be delayed until Mary Queen of Scots came of age. This delaying strategy was suggested to the French king by Arran, who feared that Christian might be tempted to take Orkney by force.

In the event, the crisis passed, due to a combination of Arran's skilled prevarications and Christian's ultimate unwillingness to draw the sabre he was loudly rattling. Arran explained to Christian that, on account of the war with England, he had been unable to search the archives for relevant documentation. An ambassador further informed Christian in 1550 that, for now, the islands could not be handed over, as James V had gifted them to his wife, and Queen Mary was presently underage and so could not rescind the arrangement. Christian III reluctantly agreed to a postponement, but conceded no jot of his legal right.[46]

Nearly a hundred years after the islands were acquired by the Scottish crown, they remained a place pulled between Norway and Scotland – in political and legal status, and in cultural and linguistic identity. Possession of Orkney was a touchstone of prestige and authority – both for Scottish monarchs seeking to mould a unified kingdom out of an inheritance of fractious lordships, and for Danish ones wielding ancestral claims as an instrument of international diplomacy and looking to strengthen their tenuous hold over the conquered kingdom of Norway.

For people who actually lived there, however, matters of national affiliation and international politics were usually not the most pressing ones. The 'community of Orkney' did not perceive the islands as a place on the precarious edge, but as a stage for homegrown and heartfelt performances about inheritance, land and wealth, and status, power and law. The arena of Orkney was also, for all members of the resident cast, the scene of final exit. Islanders prioritised the prosperity of their bodies, but wondered about the welfare of their souls. We must turn, then, to the question of faith, and how it reflected and shaped uniquely island identities in the broader North Atlantic world.

2

The Islands of St Magnus

Nidaros and St Andrews

The tiny island of Lamb Holm, nestled between Burray and the south of the Orkney Mainland at the eastern approach to Scapa Flow, houses one of the treasures of modern Orkney. It is a small church, known as the 'Italian Chapel' – a converted pair of conjoined Nissen huts, decorated from whatever materials they had to hand by pious and homesick Italian prisoners of war. After surrendering in North Africa in 1942, they were incarcerated in a camp on the island, as labourers in the construction of a string of concrete-block barriers, linking the islands between the Mainland and South Ronaldsay, and closing the entrance of Scapa Flow to predatory German U-boats.

In the barrelled interior of the little chapel, the end-wall behind the small concrete altar is filled by an exquisitely painted image of the Christ child and Madonna, his hand clutching an olive branch and her eyes cast demurely down. To either side are arched windows, the glass painted to seem stained, with images of the illustrious Italian saints Francis of Assisi and Catherine of Siena. It is a site of immense charm and serenity. Bus-borne visitors flock to the chapel along the occasionally wave-lashed causeway atop the 'Churchill Barriers'. Orcadians have long since taken the place to their hearts.

This beguiling fragment of Mediterranean religion, splendid in isolation on a thumbnail of desolate grassland, makes an impression by its very incongruity; embraced and celebrated precisely because it does not fundamentally belong. Although now a place of warm ecumenical encounter, it is also a reminder of how Catholicism had come to appear in Orkney by the start of the twentieth century: a recent and alien presence; at best intriguingly exotic, at worst sinister and threatening.

Four hundred years before the building of the Italian Chapel, however, Catholic faith and practice in Orkney were indigenous and ingrained. Decrees of the lawthing, rentals and charters were routinely headed with crosses, or devout invocations of the Saviour and his mother: 'Jesu, Maria'. They were sometimes dated in meticulous reference to faraway Rome: 'in the first year of the pontificate of our most holy lord, Pope Julius the Second'; 'the thirteenth year of the pontificate of the most holy father and lord in Christ, Lord Paul the Third, by divine providence, pope'.[1] Orcadian time flowed with the rhythms of the wider Catholic world.

Now and again, the pope in Rome heard about the distant Orkney Islands. In April 1541, memories of his northern trip still fresh, James V

Figure 2.1: Interior of the 'Italian Chapel', built by prisoners of war in the 1940s on the tiny island of Lamb Holm. No Orkney kirk would have looked quite like this four hundred years before, but all would have contained imagery of Catholic devotion.

wrote to Paul III. The king's host for the visit, Bishop Robert Maxwell, had died before Christmas last, and James was writing to recommend a successor. The bishopric, he helpfully informed His Holiness, comprised 'some islands just under the pole, lying not far from Norway, Denmark and Germany'. And here, he added ominously, 'Catholic faith and law are little observed'.

There is no real reason to believe that assessment to be true – or much truer than for anywhere else. It was intended to persuade the pope to accept the king's episcopal candidate, as popes nearly always did. In any case, James V's priorities were not the spiritual welfare of Orcadians. His letter went on to request 800 merks be subtracted annually from bishopric revenues, as a pension for nine-year-old John Stewart. The boy was one of more than half a dozen of the king's illegitimate sons, all of whom the previous pope had graciously dispensed from the bar of illegitimacy, enabling John to take holy orders when he came of age.

James's candidate for Orkney was Robert Reid, abbot of the Cistercian monastery at Kinloss in Moray. There were to be dispensations for him, too. Reid would retain his abbacy, and various other streams of revenue. In the summer, Paul III wrote to him to confirm that on all these matters 'Our most dear son in Christ, James, illustrious king of Scots, has given Us counsel' – counsel the busy pope was happy enough to accept.[2]

The bishopric of Orkney was not the obscure backwater painted in the king's letter. In the cathedral of St Magnus, it boasted one of Scotland's most impressive churches, and extensive estates in Orkney and Shetland made it an attractive proposition for a high-flyer such as Reid, an able administrator with impressive scholarly interests. In the time of the Norse earldom, bishops were important power-brokers. As representatives of the Norwegian king, they served as counterweights to the ambitions of independent-minded earls. It was a role kings of Scots expected them to carry on performing.

The medieval bishopric of Orkney, including an archdeaconry of Shetland, was part of the archdiocese of Nidaros, a name for the city later known as Trondheim. If Bergen was Norway's commercial capital, Trondheim, further up the western coast, was its spiritual one, seat of the country's sole archbishop. Trondheim's magnificent cathedral housed the shrine of St Olaf, an eleventh-century king, aggressive in pursuit of

Norway's conversion to Christianity, who soon after his death in civil war was hailed as a martyr and national patron.

We again find Orkney not on the edge but at the centre. In the middle ages, the archdiocese of Nidaros was a seaborne ecclesiastical empire, with ten bishoprics under its jurisdiction. Four were in Norway itself. Moving westwards, Orkney came next, with the bishopric of the Faroes to Orkney's north-west, and that of Suðreyjar or Sodor – the 'southern islands' of Man and the Hebrides – to the south-west. There were another two, north and south, in Iceland, and a far-flung outpost at Garðar on the southern tip of Greenland, where the medieval world's westernmost bishop worked to save the souls of intrepid Norse colonists.

The empire was already contracting when the Northern Isles were pawned. The Isle of Man was taken by the English at the end of the fourteenth century, and contact with the ill-fated Greenland colony was lost early in the fifteenth. The marriage treaty of 1468 said nothing about the bishopric, but James III was eager for the spiritual order to align with the temporal. In 1472, Sixtus IV was persuaded to elevate St Andrews to an archbishopric – a boost to Scotland's international stature. He ordered the dioceses of Sodor and the Isles, and of Orkney and Shetland, to be transferred to the new province.[3]

In spite of this, kings of Denmark–Norway continued to assert regal rights over the Orcadian Church. In 1491, Hans wrote in Danish to Sir David Sinclair of Sumburgh, 'our beloved man and servant' – Sinclair was James IV's governor of Shetland, but also served the Danish sovereign as *Lensherre* (feudal lord) of Bergen Castle. Hans granted Sinclair 'the rights and revenues of Ourselves and of the Crown over all the servants of the Church in Orkney'; that is, authority to collect skat from tenants on the bishopric estates. Hans's father, Christian I, had made a similar grant to Bishop Thomas Tulloch a few decades earlier. An urgent note in a contemporary hand appears on Sinclair's original copy: 'this letter in Northin of the king of Norway is of the scats of Orkney … Put this in Inglis.'

James IV was not amused. He promptly issued his own grant to Sir David Sinclair, 'our familiar knight', asserting that all rents and scats from Orkney 'kirklands', once belonging to 'our most noble grandsire, the king of Denmark … now pertains to us by reason of the renunciation and over-giving of our said grandsire of his right and claim of the lordship of Orkney'.

Denmark, of course, recognised no such 'over-giving'. When the Shetland archdeaconry fell vacant in 1501, a priest called Magnus Herwood petitioned Hans, paid a fee and was appointed to the post. James IV made his own appointment, and ordered Lord Henry Sinclair and Sir David Sinclair to oust the interloper. No one was to solicit ecclesiastical promotion from Denmark 'under the pain of treason'.[4]

In Trondheim itself, church authorities simply refused to accept the loss of the Orkney bishopric. In 1500, Archbishop Gaute Ivarsson lodged a formal complaint with Rome. This led nowhere, but Norwegian churchmen continued to lobby their Danish sovereign. Christian II's letter to the people of Orkney (p. 47), drawn up at his coronation as king of Norway in 1514, ordered them to carry on paying accustomed fees to the archbishop in Trondheim. 'Acknowledging the Archbishop of Trondheim' was something Regent Albany was expected to accept, as he brought up the status of Orkney with Christian in 1515.

The papacy, meanwhile, seemed to have forgotten what it decreed in 1472; registers of papal appointments continued to list Orkney as a 'suffragan' diocese of Nidaros. In April 1520, Leo X – not yet as worried as he ought to have been about the troublesome German friar Martin Luther – wrote to the bishop of Orkney, Edward Stewart. The subject was close to Leo's heart: the magnificent basilica of St Peter, and the plenary indulgence offering purchasers remission from time in purgatory that he had issued to raise money for its construction. It was this indulgence that aroused the ire of Luther in 1517, and triggered the Reformation. Giovannangelo Arcimboldi, Italian collector of the indulgence across Scandinavian lands, had informed Leo that monies paid for it were being held back in Orkney. Stewart was to ensure these were forwarded to the archbishop of Nidaros, for sending on to Germany.

The fate of the lost bishopric preyed on the mind of a talented and ambitious priest from Norway's far north. Olav Engelbrektsson became dean of Nidaros in 1515, and archbishop in 1523. In 1521, a new bishop of Skálholt in southern Iceland, Ögmundur Pálsson, was consecrated in Trondheim. The following year, at the request of the cathedral chapter, Pálsson interrupted his home journey to make an ecclesiastical visitation of Shetland – a provocative assertion of Nidaros's rights. At around the same time, Engelbrektsson approached a contact in Rome, a German priest called Zutpheldus Wardenburg, with experience of representing Danish interests at the papal court. His brief was to discover how, 'contrary to right and reason', the bishopric of Orkney became subject to the see of St Andrews.

In March 1525, after long searching, Wardenburg found a copy of Sixtus IV's 1472 bull in the register of the papal financial office. He forwarded it to Engelbrektsson, with an evaluation of the case. Wardenburg believed Sixtus had been deceived by the fraudulent claim that Orkney was immediately subject to Rome, and not a dependency of Nidaros. Since there was copious evidence to prove the contrary, Wardenburg was confident the ruling could be overturned.[5] This was an optimistic assessment, and more urgent matters were already pressing themselves on the archbishop's attention. As president of the Norwegian *Riksråd*, Engelbrektsson struggled to steer a path between Christian II's and Frederick I's rival claims to the throne, and to stem the tide of a Reformation soon to be officially blessed by Frederick's successor, Christian III. A valiant but unsuccessful insurgent against the crown's religious policy, Engelbrektsson fled into exile in 1537 and died the following year. Leaders of the Lutheran church which was then imposed on Norway had little interest in the islands to the west.

The Kirk of Orkney

Engelbrektsson was a rebel, but Orkney's late fifteenth- and early sixteenth-century bishops could generally be trusted to uphold royal authority. In 1490, as a counterweight to Lord Henry Sinclair, James IV gave extensive civil and criminal jurisdiction to Bishop Andrew Pictoris. The bishopric became, in Scots legal terminology, a 'regality'. James confirmed the bishop's rights to skats on church lands, which probably prompted Hans's grant of them to Sir David Sinclair, possibly at the initiative of the latter's nephew, Lord Henry. As the bishop's estates lay scattered throughout the islands, the regality represented, as the Orkney historian Willie Thomson suggested, a parallel jurisdiction to that of the tacksman. James IV confirmed the grant in 1501, and warned Lord Sinclair not to interfere.

In addition to presiding in regality courts, Orkney's bishops claimed jurisdiction over matters of faith and morality – slander, sexual lapse and doctrinal deviation. No records of any such 'court of audience' survive, but a folk memory of its judgements is preserved in Jo. Ben.'s story of the warring brothers of Helliar Holm (p. 29). Presidency of such courts in Scotland was usually delegated to a deputy known simply as 'the official', and 'officials of Orkney' appear regularly in early sixteenth-century charters. It is likely, too, that archdeacons of Orkney and Shetland convened courts in their respective jurisdictions.[6]

Across medieval Europe, Latin was the language of record in church courts, but witnesses were questioned and gave evidence in the vernacular. Did the slanderers and fornicators of late medieval Orkney make their excuses in Norn, or in Scots? The Church in Orkney, even before 1468, was a vehicle for Scots language and influence. The clergy comprised a very large proportion of those able to read, and still more of those who could write. Thomas Tulloch, a native of the diocese of Brechin, was appointed bishop by Martin V at the close of the papal schism – a time when popes in Rome (recognised by Norway) and in Avignon (recognised by Scotland) had named rival candidates to the diocese. During Tulloch's long episcopate (1418–61), Scots replaced Norn as the language of written administration.

Orkney's subsequent bishops were all Scots, with the intriguing exception of the first selected after 1468, Andrew Pictoris ('Andrew the painter') – a stray German from the diocese of Meissen in Saxony. Pictoris arrived in Scotland in the entourage of Margaret of Denmark and made a favourable impression on James III. He was appointed to Orkney in 1477, when Bishop William Tulloch was transferred to Moray. The following year, Rome granted Pictoris a dispensation from visiting the papal court in person, 'since the church of Orkney is situated beyond Scotland … and in going from there to Italy one has to navigate many dangers of seas and islands'.

Bishops of Orkney took relatives and chaplains with them – William Tulloch was a kinsman of his predecessor, Bishop Thomas, as most probably were Andrew Tulloch, archdeacon of Orkney in the 1440s, and three successive Tulloch archdeacons of Shetland. The Black Death's devastating impact a century earlier in both western Norway and Orkney had left many positions vacant, and clergymen from Scotland increasingly came north to fill them. The Norwegian historian Anton Wilhelm Brøgger (1884–1951) wrote of the diocese: 'We can say with fair certainty that not only was there not a single Norse ecclesiastic after 1450, but not even a native of the Orkneys or Shetlands.'[7]

That judgement has been repeated by later scholars, but is too emphatic. A scribe called Henry Murray described himself in 1548 as 'priest of the diocese of Orkney and born therein'. The name suggests Scots ancestry, but there is a stronger Orcadian flavour to Sir William Flett, documented in 1447, and Sir Donald Mansoun, or Magnusson (1547). Sir Hew of Rendall was surely Orkney-born. He appears in a 1503 rental holding land in Rousay bestowed by Earl William Sinclair as reward for going on a mission to Norway – a task for which he may have been thought qualified by his ability to speak Norn.

The courtesy title 'Sir' – in Scots, 'Schir' – was reserved for ordinary priests without a university degree. Many Schirs pop up in the records of the Orkney civil courts. Some were recent incomers, but others had deep island roots. One well-established udaller family, the Halcros of South Ronaldsay, had a particular affinity for the Church. In the mid-sixteenth century, Malcolm Halcro was archdeacon of Shetland as well as provost of the cathedral in Kirkwall. His brother Hugh was a cathedral canon, and parson of South Ronaldsay. The cathedral precentor was Nicholas Halcro, parson of Orphir and vicar of Stenness. In 1556, he was succeeded in those offices by Magnus Halcro.

The medieval Church in western Europe claimed to be an international, universal entity, as the name 'Catholic' implied. But it was also a conglomerate of local franchises, bound up with homegrown identities and interests. A surviving charter, transferring ownership of a few fields in Deerness, is dated to April 1488, 'after the compt [reckoning] of the Kirk of Orkney'. The resonant phrase conferred an aura of sacred approval on an everyday monetary transaction.[8]

Priests and Progenies

Family business and the business of the Church were entangled in a society where churchmen came from local landholding families and the Church itself was a major local landowner. Sir Nicholas Halcro appeared in a South Ronaldsay parish court in 1508, in the 'umboth' of his father, David; that is, to represent his father's interests. The case concerned a division of Halcro lands in Our Lady parish from those of the 'Trinity Stouk' of St Magnus Cathedral. A 'stouk' – from Old Norse *stúka*, a sleeve – is a chapel built out from the side of a church; St Magnus had several, with supporting landed endowments. The term is an apt metaphor for the arms of the church reaching into the economic and agricultural life of the community. In the mid-1540s, the Halcro brothers busily bought up parcels of land in South Ronaldsay, using their talents and connections as priests to shore up the family estate.

Church business could be family business in another sense. The opposing party in the South Ronaldsay courtroom in 1508 was the parson of Trinity Stouk, and archdeacon of Shetland, Henry Phankouth. The curious Germanic name – *pfannkuch* means 'pancake', or 'pancake-maker' – is accounted for by Henry's being an illegitimate son of the former bishop, Andrew Pictoris. Clerical dynasticism was an ingrained local habit. Cathedral precentor Magnus Halcro was a bastard son of the cathedral provost Malcolm Halcro – one of at least four natural sons (by various mothers) for whom Malcolm received royal letters of legitimation in 1545. A couple of years later, Malcolm signed a contract for his daughter to marry the son of a significant Aberdeenshire landowner, Patrick Mowat of Balquholly.

Malcolm Halcro's children were previously acknowledged in a charter of 1544, whose chief beneficiary, the eldest boy, Hugh, was coyly referred to as a beloved *consanguineus* (close relative or cousin). Phankouth was the bishop of Orkney's 'cousin' in royal letters of legitimation from 1497. Sir Nicol Cragie, vicar of Holm, bestowed a house and lands on Gilbert Cragie, 'my natural son and cousin', in 1565. One suspects a similar relationship in the grant of land from Sir Henry Pearson, parson of Stronsay, to 'his cousin', David Pearson. A 1548 document vividly describes David's widow, Katherine, taking possession of the property – the farm of Weland on the north shore of Shapinsay – 'after the use, consuetude [custom] and rite of the country'. In presence of onlookers, she was presented with 'stane and muyld' (stone and earth) as a symbol of rightful title. The hearth fire was extinguished, then 'kindled again in her and her bairns' name'.[9]

The bairns of priests don't seem to have carried much social stigma in late medieval Orkney, nor their fathers for siring them. It is possible, the Shetland historian Brian Smith suggests, that Edward Sinclair of Strom's first wife was a natural daughter of Archdeacon Phankouth. When Oliver Sinclair of Havera, another veteran of the Summerdale revolt, agreed to an exchange of lands with Malcolm Halcro in 1546, it was in recognition of 'great labours that he has done for my weal and mine [the welfare of me and my family] in times bepast'. The non-celibates Malcolm Halcro and Henry Pearson were courteously if conventionally described as 'venerable and discreet men' in a charter issued by the East Mainlander Thomas Louttit in 1531. Having no seal of his own, he borrowed theirs to authenticate the document – a helpful service clergymen regularly performed for the Orkney land market.

It is conceivable James V was thinking of the sexual morality of the Orkney priesthood when he told the pope that 'Catholic faith and law are little observed'. But clerical 'concubinage' was common across later medieval Scotland, especially in the clan-based society of the Highlands. In any case, when it came to carnal restraint, James was hardly one to talk.[10]

No rumours attached themselves to James's man, Bishop Reid, who had the reputation of a (Catholic) reformer. Reid was present in Edinburgh in 1549, at a council convened by Archbishop John Hamilton of St Andrews to counter the looming threat of Protestantism. Ignoring the inconvenient fact that Hamilton himself maintained a mistress, the council passed stringent measures, demanding that priests should avoid the company of their offspring, not seek promotions for them in the church, nor marry their daughters to landowners – a veritable litany of the practices of Master Malcolm Halcro.

One council decree, surely suggested by Reid, targeted Orkney specifically. In making wills and testaments, drawing up inventories and appointing executors, islanders were to follow 'the same law and custom as do other people in all the other provinces and dioceses of Scotland'. It seems like a clear instance of 'Scottification', but the motive was probably financial rather than political. The informality of the existing arrangements meant inventories were not exhibited to the bishop's commissary for confirmation, nor associated fees paid.[11] The Church purported to represent society as a whole, but it was a corporate body alert to its rights as well as responsibilities.

Learning, Morals and Right Government

In the late medieval Catholic Church, a tendency to see reform in institutional terms was hardly unique to Robert Reid. But the instinct was hardwired in the bishop of Orkney. His major local achievement, in 1544, was to reorganise the governing cathedral chapter, with a new constitution laying out duties and assigning income for the church's clergy. There was no existing document to replace, 'on account of the wetness of the country, whereby everything is easily destroyed'. The new text conspicuously overlooked the fact that James III's charter to Kirkwall of 1486, reissued in 1536, vested ownership of the cathedral building not in the Church but in 'the provost, baillies, council and community of our said burgh'.

Reid contended that the existing staff of six canons and six chaplains was inadequate for the round of services such a great kirk demanded. Henceforth, there were to be fourteen canons or prebendaries, assisted by thirteen chaplains and six choristers. Half of the new canons were to hold senior administrative posts, the others had mainly liturgical duties, and responsibility for St Magnus's various altars and chapels. The prebendary of Holy Cross, for example, was to maintain the cathedral clock, and to supervise bell-ringing. St Peter's chaplain, 'an erudite grammarian', was to be master of the Grammar School; the chaplain of St Augustine had charge of the 'Sang School', where choristers were trained. All canons were obliged to be resident for at least part of the year.

The centre of Kirkwall, meanwhile, was to take on the character of a genuine cathedral close. Suitable plots were assigned to each office-holder, who was required to build a manse commensurate with his income. In authority over the cathedral staff was the provost, a doctor or at least bachelor of theology, as well as 'a man of good fame [i.e. reputation], conversation [conduct] and name'. A nominee was already lined up: Malcolm Halcro. He at least met the first part of the stipulation – Halcro was 'Magister', a learned man, rather than humble 'Schir'.

Reid's constitution shines oblique but penetrating light on the spiritual and ceremonial face of an exceptional building in the autumn of its medieval glory. The services in St Magnus would have seemed impressive anywhere, but they surely awed the otherwise down-to-earth burghers of Kirkwall, let alone curious countryfolk and occasional visitors from the isles. Worship commenced with a solemn procession of priests and candle-bearing choristers, preceded by a sacristan in a smart surplice, 'with a white staff like a beadle'. Choral singing was accompanied by the subchanter, a man required to be 'a skilled player upon the organ'.

'Ornaments, vestments and precious jewels' were in the charge of the treasurer. The inventories he was ordered to produce have not survived, so we will never know quite how numerous or lavish the accoutrements were. The vestments of the high altar, and those of the altars of Our Lady and the Holy Blood, were to be washed four times a year – by contemporary standards, a remarkable level of fastidiousness.

The treasurer also had responsibility for 'lights of the church', a forest of sacred illumination. At times of service, candles burned on top or in front of all the altars in the church. There were many such altars, if now challenging to identify and situate. In addition to the high altar – to which Sir David Sinclair of Sumburgh bequeathed 'my red coat of velvet' in 1506 – documents mention altars dedicated to Our Lady, the Holy Blood, St Nicholas, St Andrew, St Barbara, St Christopher, St Catherine, St Lawrence and, of course, St Magnus.

We know that the chaplainry of Our Lady, founded by the formidable Sir William Sinclair of Warsetter, had a physical location in the church. If other endowments described as prebends and chaplainries also possessed their own altar, then we can add St Olaf, St Columba, St Duthac, St Ninian, St John the Evangelist, the Holy Trinity, the Holy Cross, St Peter and St Augustine to the list. It seems unlikely there was room in the cathedral for nineteen separate altars: most probably, there was some combining of dedications. Masses of the Holy Blood, celebrated every Thursday, took place at the high altar, which may also have been called the altar of St Magnus.

Pious contemplation of the blood of Jesus, shed on the cross, grew in popularity across Catholic Europe in the fifteenth century. In Scotland, its status was formalised by the inclusion of texts for a mass of the Holy Blood in the Aberdeen Breviary of 1510 – a work compiled by Bishop William Elphinstone to provide Scotland with a distinctive set of liturgical forms, and the country's first ever properly printed book. The Holy Blood mass in Kirkwall began decades before this. Orkney – or at least its cathedral – was no spiritual backwater, untouched by currents of renewal in the wider Church.

Another fashionable aspect of 'Christocentric' piety was devotion to the Five Wounds of Jesus, whose distinctive emblem appears on a sixteenth-century choir stall in the cathedral. In 1554, the chapter ordered five masses of the Five Wounds to be celebrated annually at the altar of St Andrew.

The altar dedications in St Magnus exhibit a cheerful intermingling of the local, the national and the universal – by no means untypical for the Catholic world of the early sixteenth century. There is the expected devotion to the Virgin Mary, along with a scattering of saints exercising Europe-wide appeal – Christopher, Lawrence and Catherine of Alexandria. Scotland's patron, St Andrew, is represented, and a prebend was dedicated to the sixth-century apostle of Scotland, St Colm or Columba. Its altar was in 'St Columba's aisle', a favoured location for the transaction of secular business.

There was also a prebend of St Duthac, who in addition had a chapel, endowed by Earl William Sinclair, just to the west of Kirkwall at Pickaquoy. The shrine of Duthac – a difficult figure to identify historically – was at Tain in the northern Highlands, and his cult spread across Scotland even before James IV enthusiastically sponsored it. The king visited the shrine annually between 1493 and his death at Flodden. Along with St Andrew and St Ninian – likewise venerated in Kirkwall – the heavenly protector of the Scots during the Flodden campaign was, according to a scoffing English poet, 'Doffin, their demigod of Ross'.[12]

Packed alongside these Scottish patrons in the sacred spaces of the cathedral, and in the hasty or heartfelt prayers of its clergy, were the Scandinavian prince-martyrs, Magnus, earl of Orkney, to whom the cathedral was dedicated, and Olaf, king of Norway. Olaf was the dedicatee of Kirkwall's parish kirk, a five-minute stroll to the north, along what was then the town's only proper street, and for centuries overshadowed by its episcopal big brother. Across the street, in the literal shadow of the cathedral, was a fine manse built for the archdeacon of Orkney, John Tyrie. In 1554, in a further round of reformist housekeeping, Bishop Reid induced Tyrie to hand his manse to the choristers and chaplains as a place where they could prepare for divine service under the watchful eye of the subdean, responsible for the junior clergy's 'betterment in learning, morals and right government'.

'Betterment' was Reid's watchword, both for masonry and men. Under the new constitution, he undertook to pay for annual repairs to the roof and fabric of the cathedral. Office-holders, 'upright and circumspect', were to examine incoming masters of the Grammar School for their fitness. The chancellor was to read weekly extracts from canon law to the prebendaries and chaplains, and to ensure their access to the cathedral's library of books, 'as often as they wish to enter for the study or resolution of questions'. From the inscription 'Orcadensis' on the binding of surviving copies, we know what some of these books were. They included a 1546 edition of the revised version of the Roman Breviary, produced in 1535 at the request of Paul III by the reforming Spanish cardinal Francisco de Quiñones. Kirkwall was up to speed with developments in the wider Catholic world.[13]

Teinds and Yields

It all had to be paid for. The bishopric had sizeable Orkney estates, to which James IV added handsomely with the grant of the island of Burray to Bishop Pictoris in 1495. In 1550, with the consent of the chapter, Reid leased Burray to Barbara Stewart, widow of the infamous James Sinclair of Brecks, and now married to the equally formidable Ruairidh MacLeod, lord of Lewis in the Hebrides. The rent was to be £60 Scots, along with

'24 straw baskets each containing 1,000 little fishes or 500 bigger fishes, and 80 pairs of rabbits if they can be found'.

Among the agreement's signatories was Sir John Reid, prebendary of St Catherine – perhaps a relative of the bishop. Gifts of land, ancient and modern, sustained the cathedral's various stouks. A few were funded by meagre rental income from solitary Kirkwall tenements, but others enjoyed comfortable landed endowments. St Catherine's Stouk, bolstered by benefactions from the Sinclair earls, had lands in Sanday (some of them purchased from hard-up udal farmers), along with estates in Westray, Stronsay, Shapinsay, Birsay, Deerness and the agriculturally rich parish of St Ola.

The Church was an unsentimental landlord. Reid was usually away from Orkney on ecclesiastical or government business – he is known to have visited five times. To manage his secular affairs he brought in as chamberlain Thomas Tulloch, a minor laird from Forres in Moray. Tulloch had a difficult time of it, or made one for himself. In 1551, royal commissioners were appointed to hear complaints against him. 'Murmurs and quarrels' among the bishopric tenants rumbled on into 1557, when Tulloch was absolved of wrongdoing by Mary of Guise's French sheriff, Monsieur Bonot.

Before this, Tulloch was ordered to produce for inspection the weights he had been using for measuring rents in kind from the bishopric estates – bere, malt, oil and butter. Orkney had its own traditional sets of weighing-beams, of a unique local pattern: the 'pundlar', for heavier goods, and the 'bismar' for lighter items. It may be significant that, in 1554, the Scottish parliament appointed Bishop Reid head of a commission for enforcing throughout Scotland long-ignored statutes about standard weights and measures. If Tulloch did try interfering with customary measuring devices, the attempt was rapidly abandoned. But this issue – of profound concern in Orkney – would rear its head more than once in the decades and centuries ahead.[14]

Rents were one of the Church's financial legs. The other was the parishioners' duty to pay an annual amount – notionally, 10 per cent of produce and income – for sustaining their local priest. These payments, rooted in scripture and enforceable by the church courts, were known in England as 'tithes' and in Scotland as 'teinds'. Across Europe – and in

Scotland particularly – such revenues had long been leaching away from the pockets of parish priests, becoming 'appropriated' to other purposes, such as the coffers of a monastery. Teinds were also divided between those of the parsonage, going to the parson or rector with formal control of the parish living, and those of the vicarage, earmarked for the vicar – the word meaning 'replacement' or 'deputy' – who did the actual work.

In Orkney, nearly all parsonages were appropriated to the bishopric. No monasteries survived into the later medieval period, probably because of insufficient funds to support any. There had at one time been more than thirty parishes in Orkney, but before the sixteenth century several were paired to produce a functioning total of twenty-four. The most valuable teinds – those of corn – were assigned to the parsonage, but Orkney's late medieval bishops were generally content to divide these with the vicar.

The diversion of parish revenues to cathedral prebends began in the fourteenth century, but Reid's 1544 constitution carried the process further. Some rectories, and the vicarage teinds of over half the parishes, were attached to specified offices. The provost, for example, received the vicarage of South Ronaldsay – no doubt in recognition of Malcolm Halcro's family holdings in the island, where his brother Hugh was parson. The treasurer got the vicarage of Stronsay; the subdean, the rectory of Hoy.

For each parish, the bishop undertook to appoint a 'perpetual vicar-pensioner', to reside and minister to the people. The 'pension', from vicarage revenues, was set at '10 merks Scots in money, and half a last of victual'. A 'merk' was two-thirds of a pound Scots, and a 'last' was a unit, usually of cereals, comprising twenty-four meils. A 'meil' of grain, on the Orcadian 'bere pundlar', was roughly a hundred pounds weight. It was enough for a priest to live on, but not much more.[15]

Historians of Orkney have been almost universally critical of Bishop Reid's priorities, condemning a willingness to sacrifice the local parish ministry – if not quite literally – on the altars of St Magnus Cathedral. Two dozen impoverished and undereducated curates can have done little to instruct or inspire Orkney's farmers and fisherfolk, and the Church was left in poor shape to weather the gathering storm of Reformation.

Consciously or not, such critiques channel an inherited Protestant understanding that Christian ministry ought to prioritise informed personal expression, rather than routine participation in ritual and ceremony. They may also undervalue the extent to which Reid's cathedral was itself an educational and pastoral institution, sponsoring a grammar school and shining as a beacon of salvation for the community. There was, for example, a long tradition of a daily 'morn mass', celebrated at dawn at the altar of Our Lady, to offer blessing and protection at the start of the working day.

We do not know how many Kirkwall tradesmen and sailors habitually resorted to the cathedral at sunrise, but Reid's reforms had 'buy-in' from the great and good. When the new constitution was signed and sealed in the cathedral on 28 October 1544, the witnesses included Edward Sinclair of Strom, James Craigie of Burgh and John Rendall – three veteran leaders of the 1529 revolt. Also on hand was Patrick Mowat, future father-in-law to the illegitimate daughter of the new provost.

Reid wanted his cathedral to be a centre of preaching: the provost and archdeacon were each to deliver four sermons a year in the cathedral, while the subdean was to preach three times. The statutes required these sermons to be 'in the common tongue' ('in vulgari'). What counted as the 'common tongue' of sixteenth-century Orkney is, of course, a moot point. Fluency in Norn was very likely in the skill set of that wily priest-udaller Provost Malcolm Halcro. But the dominant language on the streets – or street – of Kirkwall was Scots.

Whether in Norn or Scots, these sermons must have been noteworthy occasions. Preaching played no part in cycles of regular worship in late medieval Scotland. It was associated with occasional performances by preaching specialists, the friars, and often attracted large outdoor crowds. It is possible that Dominican or Franciscan friars from Aberdeen sometimes visited early sixteenth-century Orkney, but if so, they left no trace. How eager ordinary Orcadians were to hear sermons, particularly in country and island parishes, is simply unknowable.

Parishes and Purgatory

What we can say is that Orcadian religion was mainly a matter of patterns and practices – though that need not make it less meaningful. Reid's vicars pensioner baptised bairns, solemnised marriages and blessed with holy oil the bodies of the dying. They recited daily, with whatever fluency in Latin they could muster, the words of the mass – a ritualised re-enactment of Jesus' Last Supper and sacrifice on the cross.

Parish kirks where mass was celebrated were generally small and unprepossessing, but they were the only real communal buildings of the districts in which they stood. Baillie courts were held in them, as were ogang sessions (p. 60) of the lawman or sheriff. Special rules applied to kirks and their precincts. An earldom rental of 1503 reveals that a parcel of land in Gorseness, in the parish of Rendall, was confiscated from a udaller named Baddi, 'for bloodshed in kirkyard'. The boundaries of sacred and secular in Orkney, like those between sea and shore, were fluid, but folk were expected to understand their ebb and flow.

Parishioners were required by church law to attend Sunday mass; not for them a matter of active participation, but a sacred spectacle, at which God incarnate – in the form of consecrated bread – was made present to the people through priestly prayers of intercession. Mass was a simpler affair in the kirks of Stenness or Stronsay than in the cathedral of Kirkwall, but the essentials were the same. Once a year, at Easter, congregations were expected to partake in a physical way, by consuming the body of Christ in sacramental communion. They certified their worthiness by making an advance confession of sins. Priests were authorised to offer forgiveness, but absolution for really serious offences – such as homicide or sacrilege – was reserved to the bishop himself, although Reid delegated this responsibility to his 'penitentiary', the subdean.

The question of whether penitent Orcadians named their sins in Scots or in Norn, or had any choice in the matter, is once again tantalisingly unanswerable. A stray fragment, however, suggests that the late medieval Church encouraged, or at least permitted, expressions of spirituality in the old tongue. The only known written text in Orkney Norn, transcribed towards the end of the seventeenth century, is a version of the

Lord's Prayer, beginning 'Fa vor i ir i chimrie, Helleur ir i nam thite' ('Our Father, who art in heaven, hallowed be thy name'), and ending with the familiar, heartfelt petition, 'min delivera vus fro olt ilt, Amen' ('but deliver us from all evil').[16]

Deliverance from evil was the true purpose of much Christian faith and practice. Evil manifested itself in multiple forms – as bodily sickness for humans or animals; as unexplained failures of the harvest or the catch; in the nocturnal machinations of unpredictable and potentially malevolent otherworldly forces. These might be the demonic spirits of orthodox Christian teaching, or one of a variety of creatures with no pedigree in scripture – fairies, trows, sea-dwelling selkies and fin-men, and mound-haunting hogboons (see Chapter 8).

Christianity, at the time of the Reformation, was a thousand years old across most of western Europe. In formerly Viking Scandinavia, the faith was but half that age. In the middle-case of Orkney, a probably incomplete Christianisation of the Picts was erased – probably incompletely – by pagan Norse invaders. Christianity was later gradually restored in a distinctive Scandinavian form. It made for an exceptionally rich and complex heritage of culture and belief.

The dread of something after death united priests and layfolk. The Church taught that the souls of the departed, if not wicked enough to be cast down into hell, would be slowly purified in the painful fires of purgatory. The prospect was somewhat alleviated by assurances that the prayers of the living, above all the sacrifice of the mass, could hasten the journey towards heaven. Protestants would denounce this notion as a fiction and a fraud, but late medieval Orkney shows few signs of scepticism. In deeds and charters, names of the deceased are routinely followed by the pious interjection 'whom God assoil' (absolve).

An almost complete absence of pre-Reformation wills for Orkney makes it hard to say how far its religion resembled what has elsewhere been called 'a cult of the living in the service of the dead'. Yet bequests of land to the stouks of St Magnus invariably involved requests for requiem masses. The cathedral was a humming engine of intercessory prayer, its graves and monuments busy sites of memory and commemoration. Grandest of these was that of Thomas Tulloch (d. 1463), the last bishop to serve under wholly Danish rule. Tulloch's elaborate canopied tomb was a local landmark, where debts were settled and agreements made. Later bishops were buried elsewhere; Reid himself in Dieppe, where he died on a diplomatic mission in 1558.

To be prayed for was to be remembered, and to be remembered was to be prayed for. An inducement for Archdeacon Tyrie to hand over his manse to the cathedral vicars was the requirement on them to 'celebrate an anniversary each year on the day of his death'. In 1545, Precentor Nicholas Halcro gifted a house in Kirkwall to his 'beloved servant' William Tulloch and his wife. In return, Tulloch would pay for a yearly mass and 'dirige' (service for the dead) at the cathedral's St Nicholas altar on the anniversary of Halcro's death. An attraction of anniversary celebrations was their public, theatrical character, reminding onlookers of the significance of the day and encouraging them to contribute their own prayers. In making over her lands to her son in 1527, Margaret Sinclair insisted he undertake 'after my decease to uphold yearly my dirige and soul [mass] for my forebears' souls and mine'.[17]

Pilgrimage, Chapels and Wells

Before the journey along the post-mortem path to paradise got going, there were trails of tribulation to follow in this world – and techniques for finding God's favour in the here and now. One was to travel to sacred sites, which in Orkney, almost by definition, were unlikely to be far away. Orcadians held an assumption common to pre-modern people, and perhaps did so in a particularly acute form: the belief that holiness was more concentrated, divine power more accessible, in some places than in others.

In a sea-encircled world, fresh water was a practical necessity, but also a substance of wonder, enchantment and transformation. Wells with sacred or healing properties were a common destination for those in need of help. They were to be found all over Orkney, often in places like Keldro in Rousay, Keldamurra in Eynhallow or Kelday in Holm, with names deriving from Old Norse *kelda*, spring or fountain.

Wells were places of power long before the coming of Christianity, but the Church cheerfully recruited them into its system of the sacred. There were several 'Lady Wells' – named for the Virgin Mary – in North Ronaldsay, Rousay and South Walls, and at Weyland in the parish of St Ola. The most popular wells attracted visitors from across the islands, and even further. Above the beach of a deep curved bay on the east coast of Stronsay, three mineral springs gushed out near each other among the rocks. They comprised the Well of Kildinguie, perhaps originally *kelda-geo*, well of the inlet. It was said to have drawn pilgrims from Norway and Denmark to drink its remarkable water. The well's curative properties, along with those of the seaweed from a beach a couple of miles to the south-east, generated a local proverb, written down in the eighteenth century. 'The Well of Kildinguie and the dulse of Geo Odin can cure all maladies but Black Death' – even miraculous cures knew their limitations.[18]

Religion, or magic? The distinction makes sense to modern historians, sociologists and theologians, though even they sometimes struggle to define the differences. It probably meant little to illiterate, subsistence-economy farmers, who knew only that their world was charged with invisible forces, some benign and some malevolent – and that it was sensible to petition the former and propitiate the latter. It is equally futile to inquire whether their worldview was 'really' Christian, or just a thinly veneered ancestral paganism. Late medieval Orcadians believed themselves to be Catholic Christians. The only historically meaningful approach is to take them at their word, while seeking to understand what their belief system was for, and how it may have worked in practice.

A few yards inland from the Well of Kildinguie was a small chapel, whose barest outlines can be discerned today. There was a chapel, too, near the most northerly of Orkney's healing springs, the Lady Well at the upper end of North Ronaldsay. Such chapels, often not more than a couple of dozen feet long, were ubiquitous in the landscape of late medieval Orkney. Possible sites of around 150 chapels have been identified, in addition to thirty-six ecclesiastical buildings serving at one time or other as parish kirks. There were at least ten chapels in South Ronaldsay alone.

Chapels sprang up haphazardly during the first century of Norse Christianisation. Wealthy landowners, or groups of humbler farmers, needed places of worship in or close to the township – kirks for mass and sacraments, and sites of burial for the dead. Most chapel remains are close to the shore, and in a remarkable number of cases lie in close proximity to locations of known Iron Age habitation – a testimony to deep continuities in patterns of settlement. Over the course of the twelfth century, the parish system crystallised and payment of teinds was imposed. Some Orkney chapels were promoted to the status of parish kirk, and later rebuilt and extended for congregational use. These tended to belong to the most politically significant landowners, and so were not always centrally or conveniently situated.

In the sixteenth century, a few chapels were still owned by heirs of the founders, and used as places of private worship, but many were already in ruin and decay. That need not have robbed them of their aura of sanctity. Indeed, the fact that old chapels stood outside the times and templates of regular worship may have enhanced a view of them as lightning rods for sacred electricity.

This was particularly true of the minority of chapels sited at a distance from centres of habitation. Some were on miniature holms – there was one, for example, on Corn Holm, a tiny satellite of Copinsay, itself among the smallest of the inhabited isles. Others were squeezed onto slivers of land jutting into lochs. The most renowned of these was St Tredwell's Chapel – a twelfth-century construction, atop a probable Celtic site, which sat on the conical mound of a tiny peninsula, lapped by the waters of a small loch in the northern isle of Papa Westray. Another prime location was tidal islands, such as the Brough of Birsay – places connected to, yet separate from, an adjoining mainland.[19]

Anthropologists will want to talk here about 'liminality' – the idea, from the Latin *limen*, a threshold, that cultural meanings attach themselves to transitional spaces and moments between more fixed and stable states of existence, whether in the calendar year, the individual life cycle or the topography of the lived environment. The ragged coastlines of an irregularly populated archipelago produced many mysteriously liminal places. It is no coincidence the wonder-working Well of Kildinguie lay so close to the ebb, that place of precarious human possession, sometimes land and sometimes sea. According to Jo. Ben., there was a well, 'pure and sparkling, a thing of wonder', near the chapel on the Brough of Deerness – the site of cliff-edge pilgrimage he mistakenly christened 'the Bairnes of Brughe' (p. 41). Jo. Ben. does not tell us what visitors to the site were seeking, merely that having made prayers and offerings they left satisfied, 'affirming they had fulfilled their vows'.

Jo. Ben. is an invaluable, if infuriatingly vague, guide to Orkney's micro-economy of pilgrimage in the period he was writing and in the preceding medieval centuries. He describes the lure of the tiny island of Damsay, in the Bay of Firth, west of Kirkwall. Here, a chapel dedicated to the Virgin Mary was located, in a doubly liminal position, on a thin neck of land between a small loch and the north shore; today the site has been almost entirely reclaimed by the sea. According to Jo. Ben., the chapel was frequently visited by pregnant women – presumably either to give thanks for their condition, or to petition the Virgin for a safe delivery. His odd observation that women living on the island were all sterile, bringing forth no live children even when they conceived, might explain a later tradition – for which there is no solid evidence – that Damsay once housed a nunnery.

There was another chapel dedicated to the Virgin in the heart of the West Mainland, built on a spit of land, formerly the site of an Iron Age broch, protruding into the eastern edge of the large, freshwater Loch of Harray. It is a location which – as I can attest – is very tricky to access today (see Plate 9). About this kirk, Jo. Ben. remarks mysteriously, 'men tell many fabulous tales', adding that people congregate there from various islands. Ben describes it as a 'big church'. Examination of the site suggests it was around thirty-six by fourteen feet, with an apse at the eastern end – hardly a basilica, but towards the generous side of the spectrum of small Orcadian chapels. He also tells us the kirk was commonly called 'the Lady of Grace', the name implying the presence of a venerated statue, and an enhanced openness here on the part of the Virgin to the granting of favours and petitions. This, surely, was the 'Our Lady of Grace' to which supporters of Lord Sinclair beat a demoralised path in June 1529, seeking refuge after the defeat at Summerdale – and from which James Sinclair's henchmen dragged them to their deaths (p. 52).[20]

Perhaps in response to the trauma of Summerdale, a charter was drawn up at Kirkwall on 7 September 1529. Christian Tulloch, a daughter of Thomas Tulloch of Ness in the East Mainland, relinquished to her elder brother all rights to her 'sister part' of their udal inheritance (see p. 61). In return, she received a sum of money and other goods 'given to me in my great mister [distress] and necessity'. She needed the money for a spiritual purpose: 'for so meikle [for as much] as I am to pass in pilgrimage to the Holy Cross of Fana in Norway'.

The parish church of Fana, just outside Bergen, had for centuries housed a miraculous silver crucifix. Christian was perhaps sick or crippled, for Fana was a healing shrine, where grateful pilgrims deposited redundant canes and crutches. It seems unlikely she was the first or last Orcadian to make this journey. More than half a century after the king of Denmark signed the islands away, Orkney's links with Norway had spiritual as well as commercial dimensions. But the clock was ticking. In 1537, less than a decade after Christian travelled from Tankerness to Hordaland, the Danish commandant of Bergen Castle removed the crucifix from the church. Not long after that, a pastor of the now Lutheran parish burned the piles of sticks and wooden limbs people continued to leave at the site, their tokens of piety and gratitude turned to ash on the wind.[21]

St Magnus, Prince of Orkney

Before this rupture, the most powerful bond of spiritual kinship with Norway was a shared pride in the islands' patron saint, whose tale is told in the thirteenth-century *Orkneyinga Saga*, and in several related works of hagiography. Magnus Erlendsson, an early twelfth-century earl of Orkney, displayed a personal piety which shaded into pacifism, and he lost a struggle for mastery with his more ruthless cousin, Haakon Paulsson. In around 1117, he was lured by Haakon to a peace conference on the island of Egilsay and treacherously executed – the sagas say Haakon commanded his cook, Lifolf, to strike the killing blow, after his standard-bearer refused to perform the dishonourable deed.

A brutal political assassination was reimagined in these sources as an inspiring martyrdom, patterned on the passion of Christ. Like Jesus, Magnus forgave his persecutors. He prayed to be 'washed clean by the spilling of his own blood', and when the blow fell 'his soul passed away to Heaven' – martyrs had no need to worry about purgatory. He never, however, forgot who he was. Magnus asked Lifolf to stand in front of him and strike him hard on the skull – 'it's not fitting for a chieftain to be beheaded like a thief'.

At the request of Magnus's mother, Haakon gave permission for the body to be relocated to Birsay, then headquarters of the Orkney diocese. The bishop, William the Old, was initially annoyed by reports of miraculous healings at the tomb. He changed his tune, however, after being struck blind, and praying to Magnus to restore his sight. The next step was to transfer the saint's body to Kirkwall, where Magnus's nephew, Rognvald Kali Kolsson, a claimant to the earldom, had promised to honour his saintly uncle with a stone minster 'more magnificent than any in Orkney'.

Construction began in 1137, and Magnus's relics were soon moved there from an interim resting place in the town's parish kirk of St Olaf. The politics of this – enhanced prestige for the bishopric of Orkney, and heavenly approbation for the rule of Earl Rognvald, himself acclaimed a saint after his death in 1158 – are not really our concern. But the cult of St Magnus – appealing to a pronounced Nordic taste for princely martyrs as heavenly intercessors – spread across the medieval Scandinavian world, taking root in Norway, Denmark, Iceland, the Faroes and Shetland.[22]

In Kirkwall, epicentre of the cult, Magnus's relics were solemnly preserved in a shrine in the cathedral, barring some bones sent to episcopal churches in Iceland and the Faroes, to Aachen in Germany, and from there to Prague. The shrine itself was probably first located in a rounded apse behind the high altar at the far end of the cathedral in Kirkwall. The need for greater space and freer access for pilgrims likely prompted a decision in the thirteenth century to demolish the original east end and extend the choir with an additional three wide bays. The shrine's shape and appearance, and the precise manner of its architectural setting, are unknown, but even before its relocation to the cathedral it was clearly richly ornamented – one of Magnus's miracles, recorded in the *Orkneyinga Saga*, was to punish the sacrilege of two men, an Orcadian and a Caithnessian, who stole gold from it.

It may have resembled the shrine of St Olaf at Nidaros, whose appearance can be reconstructed from contemporary descriptions, and from a distinctive pattern of reliquary, based on the Nidaros model, found in other Scandinavian churches.

Figure 2.2: Reliquary, of a pattern which might have been used to house the bones of St Magnus in the cathedral in Kirkwall. This example, dating from the mid-thirteenth century, comes from the church of St Thomas in Filefjell, on the 'King's Road' linking Western and Eastern Norway.

If so, Magnus's relics were set inside an ornate casket, decorated with gold and silver, and showing scenes from the saint's life. The shape resembled a ridged rectangular house, with dragon heads protruding from the gables, and an open arcade in the lower portion allowing visitors to see, and perhaps even touch, an inner coffin containing the actual bones. The whole was mounted on a pedestal, at about head height.

The popularity of Magnus's shrine in the later middle ages can be gauged from a letter of indulgence, issued by Pope Eugenius IV in May 1441 in response to a request from Bishop Thomas Tulloch. Papal indulgences involved some intricate theological gymnastics. Relief from purgatory was offered in units of equivalence to specified periods of earthly penance, but ordinary people probably saw the numbers as straightforward guarantees of 'time off' – in this case, seven years, plus an additional seven sets of forty days. The usual maximum a bishop could grant on his own account was forty days, so the pope, keeper of the keys to heaven, was flexing his theological muscles. To be eligible for these benefits, pilgrims had to make a pious confession, and visit the cathedral of St Magnus, giving alms for its upkeep, on the feast day or octave (week following the feast) of St John the Baptist (24 June), or those of the martyrdom and the 'translation' of St Magnus to the shrine – 16 April and 13 December respectively.

Churches across Europe sought similar indulgences in the century before the Reformation. The incentive in this case may have been Tulloch's need to pay for building work on the cathedral's west façade. The impression from the pope's letter, however, is not so much that the bishop hoped to revive a flagging pilgrim traffic as to cash in on a thriving one. It described St Magnus as one of the greatest churches of the kingdom of Norway, housing the bones of the martyr along with those of other saints – a nod to St Rognvald, whose relics were almost certainly on display in close proximity to those of his uncle. Every year, so the pope had evidently been told, the shrine was visited by a 'great number' of faithful Christians, from various parts of the kingdom of Norway.[23]

We do not know how pilgrim traffic from Norway may have slowed after 1468, or the extent to which the tomb attracted pious visitors from Scotland. In the Orkney Museum in Kirkwall is a medieval brass mould, found in the cathedral precincts, and used for manufacturing small lead crosses. These were probably produced as souvenirs for visitors, though we might expect pilgrim badges to display an image of the saint. The existence of an otherwise unattested wonder-working crucifix in the cathedral seems unlikely, though not entirely impossible.

Silences in the documentary record are compensated for by marks made on the fabric of the cathedral itself. A recent survey of graffiti in St Magnus has found many examples of medieval carving by worshippers and pilgrims on stone walls and pillars, including stars and daisywheels which may have been signs of ritual protection. Most intriguing, and easiest to overlook, is the evidence of scratched grooves and 'pecking'. It seems the very stone of the building was thought to have powerful sacred properties, and possibly dust was removed for its curative potential, mixed with food or drink.

Surprisingly, Bishop Reid's regulations of 1544 say nothing about custodianship of the shrine. That might mean its importance had declined during the preceding century, though it is possible that shrine income was administered separately, under arrangements well understood at the time but leaving no documented trace. Certainly, pre-Reformation bishops of Orkney, and their largely Scots clergy, were not indifferent to the spiritual sponsorship of the Norse earl-saint. Reid's constitution concluded with warnings of divine retribution against anyone disregarding its instructions: 'we pray that the wrath of Almighty God, the Blessed Virgin Mary, and of all the saints, and particularly of St Magnus our patron, may fall upon him!'

The constitution was authenticated under the seal of the cathedral chapter. It depicted a bareheaded Magnus, standing under a canopied niche with a sword in his right hand, to either side a priest kneeling in veneration. Similar imagery appears on a fourteenth-century statue of the saint from Kirkwall, and it adorned three new cathedral bells, commissioned by Bishop Maxwell in 1528. The *Orkneyinga Saga* does not specify the weapon used to slay Magnus, but a sword became the universal symbol of his martyrdom – perhaps to distinguish him from St Olaf, usually depicted with an axe (see Plates 10 and 11).[24]

Magnus, like Olaf, was a Scandinavian saint, but popular in Scotland too. Gaelic was never spoken in Orkney, but echoes of Highland devotion to the saint can be heard in a beautiful vernacular prayer-poem, *Manus mo ruin* (Magnus of My Love), recited down the centuries and recorded – perhaps embellished – by an enthusiastic folklorist in the late nineteenth century. The poem calls on Magnus, 'saint of power', to protect his people from spectres, giants and oppression, and 'sprinkle dew from the sky upon kine/Give growth to grass, and corn, and sap to plants'. There were several church dedications in Caithness, including a St Magnus Chapel at Spittal near Halkirk, providing accommodation for pilgrims on the road to Orkney. Custodian-priests were appointed there through the fifteenth century and beyond.

The sixteenth century saw attempts to co-opt Magnus as a fully fledged patron of the Scottish Church. James IV, in confirming the bishop of Orkney's regality (p. 72) in 1501, piously announced a special devotion to 'the glorious martyr Saint Magnus'. He was numbered among the saints of Scotland in the Aberdeen Breviary of 1510, which supplied texts for celebrations of his feast days, and incorporated a 'legend' of the saint based on an otherwise lost twelfth-century life. There were similar rubrics for the Office of St Magnus the Martyr in the Nidaros Breviary of 1519, another precociously printed work, through which Archbishop Erik Valkendorf attempted to produce liturgical uniformity in the Norwegian church.[25]

Magnus was credited with a role in a defining event of Scottish national history: the victory of King Robert the Bruce over the English at Bannockburn in 1314. The sixteenth-century historian Hector Boece described how on the day of the battle 'ane knight, with shining armour, showed to the people at Aberdeen how the Scots had gotten ane glorious victory of Inglismen'. The apparition then passed north over the Pentland Firth 'and was holden by the people to be Saint Magnus, Prince of Orkney'. The tale was repeated by the friar-chronicler Adam Abell, who added this 'fair man, clad all in harness', introduced himself as Magnus, and 'vulgar people said that it was Saint Magnus, umquhile [former] earl of Orkney'.

For Boece, the story was a satisfying explanation for the origins of an annual disbursement from the customs revenues of Aberdeen: a payment for bread, wine and wax – wherewithal for saying mass – to the cathedral in Kirkwall. In all probability, the arrangement began before the fourteenth century, and was linked to Kirkwall's status as delivery site for 'the annual of Norway' (p. 33). An embarrassing reminder of tributary obligation to a neighbouring kingdom was thus turned into a tableau of Scotland's national independence, with St Magnus co-opted into the role of supernatural royal herald, carrying news of victory to a place not yet, but destined to become, the furthest boundary of the kingdom.

There was a consolatory connection, too, with Scotland's darkest hour. On the day of the Battle of Flodden, so it was said, Magnus appeared in the little Aberdeenshire village of Auchmedden. There, he pronounced a blessing on the harbour, promising no boat belonging to it should ever be lost by shipwreck. It was known thereafter as 'St Magnus Haven'.

Magnus, then, was a symbol of Scottish as well as Norwegian identity. Yet it was not to authorities in St Andrews or Nidaros that he principally belonged, but to the inhabitants of the islands. At mass on 16 April, they received the exhortation 'Gaude, tellus felix Orcadia, novae lucis refulgens gratia' – 'Rejoice, O Orkney, happy land, shining by the grace of a new light!'[26] With the possible exception of Shetland, Orkney had more Magnuses than any part of the British Isles. In an early cluster of written testaments, from the start of the seventeenth century, Magnus comes just behind William, and ahead of James, John and Thomas, in the ranking-order of male testators' Christian names.

Early sixteenth-century Orcadians, as we have suggested already, probably did not think much in terms of nationhood. Of more immediate importance were neighbourhood and honest dealing, which St Magnus might be called upon to witness. In the 1540s and 50s, William Sinclair of Warsetter, a grandson of the formidable justice, solemnised land deals by swearing to set down the purchase money 'betwixt the sun rising and going to rest, on St Magnus's altar within the cathedral kirk of Orkney'.

Orkney's Bannockburn came in 1529, when William's uncles James and Edward Sinclair repelled the invading army of the earl of Caithness. From Bishop Lesley's account of the battle we learn that 'the Orkney men hold opinion that Saint Magnus their patron was seen that day with them in the field, fighting for their defence'. The peace-loving earl was turned ferocious heavenly warrior – easy perhaps to imagine, as he was so often depicted with a sword. From Lesley, the report passed into successive editions of Raphael Holinshed's *Chronicles of England, Scotland and Ireland*, Shakespeare's go-to resource for the plots of his historical plays. In 1577, the book's English Protestant editor felt moved to add a censorious marginal note – 'the blindness of the Orkney men'.[27]

A more literal blindness lay behind miracles attributed to Magnus in a Latin hymn sung at the service of Lauds (morning prayer) on his feast day in the cathedral. 'Sight comes back once more to sightless eyes; The dumb find speech, the impotent arise.' It was as a source of physical healing that the saints' favour was most eagerly sought. And who to care more about the health of the people of Orkney than a saint who was once their lord and earl? A miracle in the *Shorter Magnus Saga* is arresting in its very ordinariness. It involved the healing at Magnus's tomb of 'a man from Orkney called Thorkeld, who fell down from the top of his barley-stack, badly injuring one side of his body'.

It is no surprise that one of Orkney's most renowned healing wells was the Mans Well in Birsay, near where the saint's body lay after being shipped across from Egilsay. Modern passers-by – including walkers on the recreational pilgrimage route 'The St Magnus Way' – are still (or were in 2018) offered opportunities to drink the curative water, from a mug on a nail in the adjoining fence. The tradition in Birsay was that the saint's body was washed here prior to enshrinement in the bishop's kirk. It attests to belief in the idea – found across pre-modern Europe – that holiness and the power to heal were transferable, almost infectious, qualities.

Magnus's martyred body itself traversed Orkney, in the process seeding the landscape with memories and meaning. Its journey – which the modern St Magnus Way seeks to recreate – led from the site of initial blood-letting in Egilsay, via the miracle-rich sojourn in Birsay, and on to the first temporary resting place in Kirkwall, and the final interment in the cathedral shrine.

Egilsay itself boasted an exceptionally fine round-towered church of St Magnus, built in the late twelfth century to commemorate the martyrdom. It remains, in its ruined condition, an imposing landmark, visible from the Orkney Mainland as well as neighbouring Rousay and Wyre.

Jo. Ben. noted its presence, along with implausible snippets of pious local folklore: that Magnus was born and brought up in Egilsay, that his nurse built a chapel there, designed for domestic residence and fitted out with stone furniture. He omitted to mention that there is a Manse Loch in Egilsay, and that Rousay has a Mansmas Hill, named for the feast day of the saint. A field on its western slope is the 'bonie-hole' – no attractive indentation in the ground, but a *bœnar-hóll*, a prayer-hill. Jo. Ben. did, however, report that the island of Wyre itself, 'some say', was formed out of the boat of St Magnus as he fled to the island from Egilsay.

More credible linkages between the saint and physical features of the landscape were located in the ability of local people to point to current or past sites of Mans or Mansie Stones. These marked places where the saint's body was supposed to have stopped and rested on its processional route through the West Mainland. The archaeologists Sarah Jane Gibbon and James Moore have identified over thirty locations bearing this designation at one time or another, mostly found along ancient pathways and the traditional drove roads down which livestock were moved. Mans

Figure 2.3: The Kirk of St Magnus, Egilsay, one of Scotland's finest examples of Norse ecclesiastical architecture, built in the late twelfth century near the site where the saint was killed. Photograph taken July 2017, during a pilgrimage to mark the 900th anniversary of the martyrdom.

Stones are linked to a wider network across Orkney of Wheelie-Stones or Wheelie-Crosses – places of pause for funeral processions on accustomed routes to the parish kirkyard, where refreshment would be served to the pall-bearers; in Norn, *hvila* means 'to rest'.

Resting the coffin on a point of elevation – a stone, mound or other raised ground, rather than directly on the earth – was most likely a protective ritual. It guarded the corpse from attack by evil spirits, and prevented any escape into the community of the wandering ghost of the deceased. Official Christian doctrine taught that souls and bodies parted at the very moment of death, but popular intuition supposed the separation to be slower and messier, and to require careful management. There was significant overlap, too, between locations of Mansie Stones and places of ancient social and spiritual significance – prehistoric burial mounds and menhirs.[28] New meanings were laid, at times literally, on top of old ones.

Layers

Modern Orkney is a paradise for archaeologists. The uncovering and mapping of layers, the strata of preserved remnants in the earth, lie at the heart of archaeology as a discipline. But layers and layering are appropriate for thinking about Orkney's history in all its aspects. Deep continuities in patterns of agriculture and settlement, and a defined range of possibilities set by the conjunctions of land and sea, meant societal change had to be received and processed in particular ways. In spheres of spirituality, faith and belief, no less than in politics, law and language, Orkney's past was buried close beneath the surface of its present.

Through the first century of Scottish lordship, there was periodic conflict, and occasional bloody violence. Yet, in the main, change was absorbed rather than repelled, and, in the process, the new was inflected with tones and accents of the old. In time, incomers became insiders, and learned to value local ways of doing things. In the century to follow, the scale of encroaching change would become greater, and the challenges of surmounting it considerably starker.

Some thought they could read the runes, and comfortably weather the storms ahead. They included the provost of St Magnus Cathedral, Malcolm Halcro, and his priestly brother Hugh. Under the terms of the charter of 1544 that we have referred to already (p. 75), the Halcro brothers bestowed their considerable accumulated landed properties on 'their beloved cousin Hugh Halcro' – the eldest of Malcolm's bastard sons. The younger ones, with various other kinsmen, were meticulously listed in the order of succession, but the charter insisted all the lands were henceforth to pass to the next heir as a 'whole estate', 'without partition among brothers and sisters'. This was essential 'for the welfare and stability of the house of Halcro'.

The priest-brothers thus renounced inheritance practices enshrined in Norse udal law, and cemented their piecemeal patrimony with the Scots principle of inflexible primogeniture. It was a step of questionable legality, though Malcolm and Hugh correctly judged that they were sufficiently well connected and versed in the law to be able to get away with it. Udal estates, whatever their other attractions, were always vulnerable to erosion and even disintegration. Malcolm and Hugh sensed the times starting to change, and that the adoption of Scottish customs was the way for old Orkney families to survive and thrive.

In other respects, they proved to be less far-sighted. Beyond a token annual rent of a silver penny, only one obligation was placed on the younger Hugh and his descendants: to pay a priest to celebrate mass 'every Sabbath day yearly and perpetually and to pray for the souls of the granters' father and mother, and themselves, and their successors'. This once-and-future ancestral commemoration was to take place – where else? – in the family's private kirk, 'the chapel of Our Lady of Halcro'.[29] Malcolm Halcro died in 1554, secure in the knowledge that for an infinity to come the whispered words of the requiem mass would wend their way from South Ronaldsay to heaven, and move God to take pity on his imperfect soul. It was a comforting final thought, but the provost's infallible instincts had for once missed their mark.

3
Mutation of Religion

Empress of Orkney

'The Gospel Light ... has at last arrived at the extreme limits of the Ocean – in fact, at the Orkneys themselves – and so, with the circle of its journey thus completed, it has no further places to which it can extend.' This was the eye-catching conclusion to a short Latin book, printed in the Swiss city of Basel at the start of 1559. In translation, the title reads *A Thanksgiving from Germany to England on the Restoration of the Light of the Gospel*. Thanks was being given for the death of Mary I, Catholic queen of England – the author was one of several hundred Protestants who had fled abroad after Mary returned England to papal obedience. Those remaining risked the terrible punishment of burning at the stake: almost 300 died that way between 1555 and 1558.

They were not to be forgotten. Among the exiles, an enterprising clergyman named John Foxe gathered materials for what became known as 'Foxe's Book of Martyrs' – an angry, eloquent history of those giving their lives for God's truth, and those who persecuted them. It was to become, translations of the Bible aside, perhaps the most influential Protestant book ever published in England. Yet, at the start of Elizabeth I's reign, Foxe – for he was the author of the *Thanksgiving* – looked north, to the islands of Orkney, for proof of the ultimate triumph of the Reformation.[1]

Elizabeth herself, shining hope of Protestants, unveiled her official title in a proclamation issued on the day of her coronation, 15 January

1559. Like her predecessors, she was sovereign of England, France and Ireland, as well as 'defender of the faith' – the title granted to her father by the pope for bad-mouthing Martin Luther, and one which subsequent monarchs have stubbornly refused to relinquish. There was, however, a conspicuous change from the style of preceding rulers. Elizabeth declared herself not only queen but 'most worthy Empress from the Orkney Isles to the Mountains Pyrenée'.

It was not, or not really, a formal assertion of sovereignty. Such an assertion *was* made in the following year, when another newly crowned monarch, Frederick II of Denmark, reminded the lords of the Scottish council, 'in a friendly way', that Orkney belonged to him. Elizabeth's extravagant claim to an extensive European 'empire' echoed the ideological style of her father, Henry VIII, with its flashy Arthurian overtones. In a play performed before Elizabeth later in the reign, Arthur boasts how 'The Irish king and nation wild we tamed;/The Scots and Picts and Orcade Isles we won'.[2]

Foxe could not have known of Elizabeth's title when he wrote the *Thanksgiving*. But the prominence of Orkney, in both a Protestant declaration of victory and a Tudor assertion of sovereignty, was more than quirky coincidence. It attests to the surprisingly important place occupied by remote localities, and the Orkney Islands in particular, in a geography of the British and European imagination. Small places of the world mattered to the great and powerful precisely because they were distant and peripheral, or were thought to be so: 'even here', was the message they were called on to proclaim. The scale of the achievement was measured by the extremity of its edge.

Heretical Depravity

Did Orcadians know, in the dark winter of 1558–9, that the light of the Gospel was shining brightly among them? From the 1520s, Protestantism was making converts in Scotland, initially among intellectually adventurous clergymen; then, with growing numbers of townsfolk, merchants and lairds. The movement had its leader-in-waiting – the zealous preacher John Knox, who, like Foxe, was an exile from Mary I's England, the country he went to upon release from the French galleys (p. 16). Knox ended up in Calvin's Geneva, famously naming it 'the most perfect school of Christ that ever was in the earth since the days of the apostles'. This was not just a matter of doctrine – there were other places where Christ was preached truly. 'But manners and religion so sincerely reformed, I have not yet seen.' There was implied emphasis on the *yet*, for Geneva was a beacon for what Scotland, including Orkney, might one day become.

Bishop Reid, counsellor to the regent, Mary of Guise, did his best to hold heresy at bay. He participated in reforming councils of the Scottish Kirk, and on at least one occasion helped in the process against a Protestant suspect – Adam Wallace, tried and burned in Edinburgh in 1550. Wallace's 'sacramentarian' views were the most dangerous of heresies – a refusal to believe that bread and wine transformed into Christ's body and blood. This was to deny Jesus' words at the Last Supper – 'for this is my body' – as well as to undermine the power of priests, however unworthy, to act as channels for God's grace. 'Believest thou not', Reid demanded, 'that the bread and wine in the sacrament of the altar ... is the very body of God, flesh, blood, and bone?' Wallace offered the standard Protestant argument that Christ's natural body was in heaven, and could not be in two places at once. 'A horrible heresy', Reid retorted.[3]

The need to keep horrible heresies out of Orkney was in Reid's mind when he issued his constitution in 1544. One task for the new provost was to deputise as 'chief inquisitor of heretical depravity'. There is no evidence Bishop Reid or Provost Halcro were kept busy searching for Protestants in the 1540s. Yet a hint of troubles peeks through the language in which Reid granted a lease of bishopric lands in Rousay to Edward Sinclair of Strom in 1549. It was in recognition of 'faithful kindness and support, especially in the defence of Christian faith and the liberty of Holy Kirk'.

A few years later, Edward Sinclair, ageing victor of Summerdale, would indeed come to the defence of Reid's kirk, and in a quite literal way. In 1557, Scotland, the ally of France, was once again at war with England. Under the command of Sir John Clere, a dozen English ships were sent to protect the Iceland fishing fleet and make trouble in the

Northern Isles – the very strategy Henry VIII's advisers had talked him out of fifteen years earlier (p. 63).

On 11 August, the English landed at Kirkwall and set fire to the town, destroying the little parish church of St Olaf. They came back the following day with artillery, seized control of the cathedral and proceeded to bombard the castle at close range, though the walls held. Returning from their ships for a third day, Clere's men were confronted by a hastily assembled army under Edward Sinclair, estimated by one surviving English officer at 3,000 strong. The invaders were put to flight, and Clere and three of his captains drowned as they scrambled to get back to the fleet. The English government heard that a hundred men of the expedition were slain; it had become 500 in a newsletter sent to Spain in October. Either way, the Battle of Papdale – the last pitched battle on Orkney soil – was a resounding victory for the islanders, and a humiliation for English seamen accustomed to raiding there with impunity. According to the chronicler Robert Lindsay, 'the Orkneymen got great spuilzie [plunder] of the Inglismen ... with many prisoners which paid them great sums of money in ransom'.[4]

We will never know if Sinclair's followers credited their victory to the intercession of St Magnus, as they reportedly did in 1529. But, by the time of the battle, a handful at least of people in Orkney were starting to wonder if saints really had the power to heal and protect, or if God could truly be encountered in a roundel of bread.

One of the junior clergy of St Magnus Cathedral was Sir James Skea, a priest with a traditional Orcadian surname who first appears in 1523 as witness to a charter. Sometime thereafter, he began having spiritual doubts, and by the spring of 1546 had left Orkney, seeking refuge across the Pentland Firth. George, earl of Caithness, was ordered by the lords of session, Scotland's supreme judges, to arrest Skea and deliver him to Bishop Reid. But Skea had moved on to Edinburgh, and around Christmas fled south into England.

Early in 1547, Skea petitioned the duke of Somerset, head of the Protestant government advising the young Edward VI. He declared himself a man 'born in Orkney in the parts of Scotland', and now in England 'for fear of burning for the Word of God'. Skea eagerly offered to betray his allegiance to Scotland, promising 'to show unto Your Grace all the use, fashion and order of the said country'. The proposal was taken up, for in August 1547 and June 1549 he was paid significant sums for unspecified services.

Soon, however, he was in other employ. Who better to find uses for a renegade Orcadian than the king of Denmark? By 1550, Skea was in Christian III's service, just as Christian intensified his efforts to assert Danish sovereignty over Orkney (p. 47). Yet exile, even in Lutheran Denmark, was not to Skea's taste. In September 1550, he begged the Scottish government for permission to return, notwithstanding 'tenacity and pertinacity in holding of opinions concerning the faith'. He was offered a nineteen-year suspension of sentence, conditional on abjuring his heretical views. We do not know if Skea accepted, but Christian, who evidently valued his services, wrote in November to the regent, the earl of Arran, requesting he be allowed to pass safely to 'the Orkney Islands or Scotland' – a carefully chosen phrase.[5]

At least one other Orkney priest was an early adherent of the Reformation. In 1551, Andreas Guielmus (Andrew William or Williams) turned up in Wittenberg. He planned on journeying to England, and requested a recommendation from Philip Melanchthon, former right-hand man to Martin Luther. The letter explains how Guielmus, on account of his evangelical preaching, was expelled from Orkney by his bishop, 'Robertus Rufus' – Robert the Red, or Reid. It is possible – Orcadian-Scots transposed into Latin via Saxon-German – that Guielmus was Andrew Lowson, a stray Scot studying in 1549 at Frankfurt-an-der-Oder, a few days east of Wittenberg.[6] Perhaps, through the networks of news-hungry British exiles, word of the wanderings of a priest from Orkney reached the ears of John Foxe, and prompted his premature pronouncement about the furthest bounds of the ocean.

Two swallows, even in Orkney, do not make a summer. Yet it seems unlikely that Skea and Lowson were the only islanders to take a sympathetic interest in ideas reshaping society and politics across northern Europe. Evangelical preachers taught salvation was a free and unmerited gift of God, not a conditional offer people must strive anxiously to satisfy. Departed souls went straight to heaven or hell; there was no purgatory, for it was not mentioned in the Bible, the sole lexicon of religious truth. The papacy was a human invention – or, as Luther suspected, an instrument of Antichrist. The priesthood was a ministry, not a caste, and its principal responsibility was to preach rather than to hear confessions or say mass. The mass, indeed, was blasphemy and idolatry, as it encouraged adoration of a disc of bread. Protestant communion services were 'the Lord's Supper' – no priestly sacrifice, but a shared commemorative meal. God's grace resided in receptive hearts, not in prescribed rituals, sacred objects or the images and relics of saints. No place, time or season was holier than any other – though the Sabbath, the Lord's day of rest, demanded an especial reverence.

To most early sixteenth-century people, these were jarringly offensive notions – perhaps especially so to those, like most Orcadians, whose lives were shaped by rhythms and rituals of the agricultural year, and vagaries of weather and harvest. Yet others found them inspiring and exhilarating, and felt furious to have been deceived so long by self-interested custodians of the old ways. Meanwhile, those custodians, the Catholic bishops and theologians, regarded heresy as an infectious disease. It did not erupt spontaneously, but was caught and transmitted through contact – with books, places and people.

How Protestant ideas spread to Orkney is a puzzle but not necessarily a mystery. The islands were not – as Foxe seems to have supposed – a terminal frontier of human exploration. They were a place of commerce and connection, on shipping routes from Scotland, England, Scandinavia and other parts. The Orkney clergy, though rooted in their localities, were potentially mobile employees of a transnational institution.

In Scotland, as elsewhere, universities were vectors of the new ideas: idealistic young clerics encountered them at college and brought them back to the parishes. Some at least of the Orkney clergy were graduates – five of six cathedral canons authorising a 1539 lease were 'Maister' rather than 'Schir'. Yet higher education could reinforce orthodoxy as readily as question it; several graduates were pillars of the Orcadian establishment. Malcolm Halcro matriculated at St Andrews in 1512, just a year after his future bishop, Robert Reid. Halcro's legitimised sons Magnus and Ninian were in due course also sent to study there.[7]

James Skea was no graduate, but his flight to Denmark asks whether longstanding links with Scandinavia provided routes of access for reforming ideas. It is possible, though it cannot be convincingly demonstrated. The Orkney-born *borgermester* of Bergen, Lille Jon Thomessøn, was, as we have seen (p. 44), remembered by a Lutheran chronicler as God-fearing and pious, yet this might mean only that he conformed obediently to the change of official religion. Many German Hanse merchants in Bergen were partisans for their compatriot Luther and eager collaborators with the Danish-imposed Reformation; other Bergen citizens were much less enthusiastic for change.[8]

Nonetheless, Orkney merchants trading with Bergen and other Scandinavian and North German ports in the 1540s and 50s would have brought home startling news. They had been to places where the mass was abandoned, purgatory abolished and shrines of saints dismantled. Folk in Kirkwall likely shook their heads and assumed it could never happen there. They were soon to find that they were mistaken, and that the storm, when it broke, came from an unexpected direction.

An Angry Multitude

It took nearly a year after Reid's death in December 1558 for a new bishop of Orkney to be named. Such delays were not unusual, allowing the crown to reap the income from vacant bishoprics. These 'temporalities' might be granted to favoured individuals, and in March 1559 those of Orkney were bestowed on a prominent judge, Sir John Bellenden of Auchnoul, clerk to Scotland's high court of justiciary. Bellenden never visited Orkney, but he was part of a network of sharp-eyed Scots alert to opportunity and profit in the north.

Sir John's aunt was the formidable Katherine Bellenden, wife of Oliver Sinclair of Pitcairn, the former sheriff. Katherine's previous husband had been an Edinburgh merchant, Francis Bothwell, and their son, Adam, was a promising young churchman, about thirty when the vacancy arose. Bellenden–Sinclair influence secured Queen Mary's nomination of Adam as bishop of Orkney on 24 July 1559. The pope, once again, was happy to oblige. Paul IV drew up bulls of appointment, which were brought back from Rome by Adam's brother-in-law, Gilbert Balfour – himself to become a significant figure on the Orkney scene. By October, the legalities were complete, though the new bishop sensibly decided the moment was not ideal to brave the Pentland Firth.[9]

By the time Adam Bothwell headed north, in still unseasonable February, political events in Scotland were accelerating at dizzying speed. Mary of Guise had failed either to intimidate or conciliate the powerful Protestant noblemen, now sworn to each other as 'Lords of the Congregation'. Emboldened by the death of Mary Tudor, they stepped up their campaign. In May 1559, John Knox returned to Scotland under their protection, whipping up urban crowds to smash saints' statues and attack monasteries and friaries.

Mary of Guise, however, with military backing from Catholic France, was confident of crushing the rebels. The Lords of the Congregation pleaded with the new queen of England to intervene. In January 1560, an English fleet blockaded the port of Leith, and in April an army crossed the border. By the start of June, the French were ready to come to terms, a treaty was signed and foreign troops withdrew. Four days later, the queen regent died, the second of three Marys at whom Knox aimed his misogynistic diatribe of 1558, *The First Blast of the Trumpet Against the Monstrous Regiment of Women.*

The third of them, Mary Queen of Scots, was still in France with her husband, recently ascended to the French throne as Francis II. A parliament convening in Edinburgh in August 1560 was not supposed to discuss religion, but the triumphant Protestant lairds showed scant regard for this. Parliament endorsed a Confession of Faith, drafted by Knox and his allies, summarising the Calvinist principles henceforth to shape kirk teaching. Papal authority and the mass were abolished; the sacraments cut from seven to two – baptism and the Lord's Supper. Knox and other ministers were tasked with producing a *Book of Discipline* – a chance to emulate Geneva's 'perfect school of Christ'.

The *Book* recommended the overhaul of Kirk structures on Genevan lines: ministers elected by local congregations, with manners, morals and social welfare placed in the hands of parish-based kirk sessions staffed by lay elders and deacons. Plans to transfer the entire wealth of the old Kirk into the hands of the reforming ministers were, however, a step too far. Many landowners, including Protestant ones, had fingers deep in the existing pie.

Scotland was to be a Protestant nation, but in early 1560, as Adam Bothwell headed for Orkney, much remained uncertain – not least, what role, if any, the new Kirk would find for bishops. Bothwell's journey north was uncomfortably eventful. The boat he was travelling in was seized by the English fleet blockading the Firth of Forth, and Bothwell was held for several weeks at St Andrews before being allowed to proceed.

In the meantime, the lords of the council delegated to Bothwell and the provost of Orkney, Malcolm Halcro's successor Alexander Dick, authority to hear a legal case pitting Orcadians against German merchants of the Bergen Hanse. It involved a ship from Rostock, chartered by the Germans, and in late 1557 seized off the Norwegian coast by Breton pirates. They took it to Kirkwall, where locals colluded in disposing of the valuable but illicit cargo. Among the items were 'books in Latin, Greek, German and Danish, large and small, to the number of three thousand volumes and more'. On the eve of the Reformation, Orkney was suddenly awash with learned books. The councillors wanted Bothwell to deal with the matter, on account of 'great troubles now being in this realm, and of the far distance fra their parts to Orkney'.[10]

Orkney had troubles of its own, as Bothwell discovered on his arrival in April 1560. John Bellenden expected recompense for strings pulled on his cousin's behalf. So, too, did the new bishop's brother-in-law, Gilbert Balfour. Like Knox, Balfour – a hard-nosed mercenary captain – had spent time as a French galley-slave after the murder of Cardinal Beaton (p. 16). But, as far as Knox was concerned, Balfour and his various brothers were 'men without God'. One of those brothers, John, accompanied Gilbert to Orkney. They were soon at odds with Thomas Tulloch, the previous bishop's constable, who accused them of seizing his property in Kirkwall and other 'cruel actions'.

Gilbert Balfour had spent considerable sums securing Bothwell's bulls of confirmation from Rome. His reward, in June 1560, was a massive grant of episcopal lands, in the form of a 'feu'. They were concentrated in Birsay, heartland of the old bishopric, and in the northern isle of Westray. There, Balfour began to construct a fortress. Noltland Castle is to this day a grim, foreboding sort of place, bristling with narrow gun-loops. Inscribed over the forecourt's arched doorway was a chilling passage from the Book of Exodus, 'when I see the blood I will pass over you in the night'. Balfour was a hard man to satisfy. In a letter of December 1560 to another brother-in-law, Archibald Napier, Bothwell complained that Gilbert was quarrelsome 'because I would not give him all that I had'.

Balfour was on hand in September 1560 when Bothwell announced another of what was soon to become a flurry of feu-farm leases of bishopric property. Also present was the new bishop's stepfather, Oliver Sinclair of Pitcairn, provocatively styled in the charter 'sheriff of Orkney', on the basis of appointment by James V twenty years before.

Oliver's reappearance in the islands ignited the smouldering feud between rival branches of the Sinclair family. In July 1560, a contract was drawn up between George, earl of Caithness, son and heir of the nobleman slain at Summerdale, and Magnus Halcro, precentor of the cathedral, and son of the former provost. Magnus promised assistance to

Figure 3.1: Noltland Castle, Westray, an austere and imposing fortress built in the 1560s by Gilbert Balfour, brother-in-law to the bishop of Orkney, Adam Bothwell. Balfour, a ruthless adventurer from Fife, saw Orkney as a place of profit and site of security from his enemies.

the earl should he decide to 'invade the country of Orkney in persecution of his auld enemies'. Bothwell was evidently nervous: another charter, in October 1560, leased to Oliver Sinclair the bishopric lands in Eday, in return for support against 'whatsumever invaders'.[11]

The trouble, when it came, had a different source, explosively combining quarrels over lands, rents and revenues, and the first, deeply divisive, moves towards reform of the Kirk. Bothwell detected the hand of the 'justice clerk', Sir John Bellenden – the bishop's letters to Napier of 1560–1 are peppered with complaints that his cousin, in pursuit of financial reward, was stirring up problems in Orkney and briefing the government against him. Bothwell's loyal chamberlain, James Alexander, assured the bishop's sister, Napier's wife, Janet, that 'na Orknanaze born' were behind the vexation of his master.

That was contrary to appearances, for, as Alexander admitted, everyone knew how 'the Sinclairs made insurrection against his Lordship'. These Sinclairs were Henry and Robert, sons of Edward Sinclair of Strom, victor of Summerdale and Papdale, and loyal ally of the late Bishop Reid. At around Christmas 1560, the Sinclair brothers seized Bothwell's palace in Birsay, a house known as 'Mons Bellus', situated just the Mainland side of the ancient settlement on the Brough of Birsay. Accompanied by a 'great number of commons', they tried to intercept Bothwell as he returned from a visitation of his diocese. Their purpose, the bishop informed Napier, was 'to have either slain me or taken me'.

In 'visiting' – formally inspecting – Orkney's parishes, Bothwell was beginning to implement the acts recently passed by parliament. These ordered that 'no manner of person or persons say mass, nor yet hear mass, nor be present thereat'. Doing so risked confiscation of goods for a first offence, banishment for a second, and death for a third. Hearing mass was a spiritual health hazard. A few months later, John Knox would declare that the saying of a single mass was 'more fearful to him than if ten thousand armed enemies were landed in any part of the realm'.

It was a staggering inversion of everything current and former generations grew up believing to be true. There was no hint of it as late as June 1560, when, in a charter witnessed by Gilbert Balfour and members of the cathedral chapter, Henry Sinclair agreed to transfer some lands to the bishopric in return for a sum to be deposited, in the time-honoured way, 'on the altar of St Magnus'. In another charter of the same month, Bothwell styled himself bishop of Orkney 'by the grace of God and the Apostolic see' – that is, the pope.[12]

Six months later, Henry and his brother Robert, with about twenty other notable signatories, presented 'certain petitions' setting out their grievances. These seem to have involved religious issues and demands for continued celebration of the mass. The Sinclair brothers also reportedly stirred up the commons with talk of how they might hope again to 'live freely, and to know no superiors in time coming'. The bishop's policy of feuing, and the unwelcome intrusion of Gilbert Balfour, threatened the status of small udal landowners. What Bothwell called Henry Sinclair's 'conjuration' (conspiracy) was a serious business. Similar convergences of agrarian discontent and conservative religion had, from the 1520s to the 1540s, caused significant peasant rebellions in Sweden; they had also plunged much of south-west and midland England into uproar in 1549.

Bothwell appealed 'in the Queen's name' to the sheriff-depute – an accredited substitute performing the actual judicial functions of the nominal sheriff. In 1561, this was none other than Edward Sinclair, father of the rebel leaders. The pragmatic old warrior remarked wearily that his sons were 'fools that wist not [did not know] what they did'. For his own part, he did not think the old form of worship should continue – 'he wald on na sort consent the mass were done'. But, at the same time, he was unwilling to take steps to restore the bishop to his palace, and in the interests of 'concord' he advised Bothwell to answer the petitions – something the aggrieved bishop refused to consider until the rebels vacated his residence.

Bothwell was back in possession by 25 March 1561. But in the intervening weeks things had gone from bad to worse. Matters came to a head just after New Year at the Hirdmanstein in Kirkwall, which usually took place inside the cathedral itself. This court of law retained the character of a popular assembly, and 'a great multitude of the commons' were in attendance. Perhaps encouraged by Edward Sinclair's concession about the mass, Bothwell sent representatives to the Hirdmanstein formally to request the sheriff and others to 'be content of mutation [change] of religion'. If he expected dutiful acquiescence, he was to be disappointed. All refused to countenance it – Sinclair, it would seem, included – amid scenes of confusion and disorder. In a letter to Archibald Napier (our only source for these events), Bothwell described how he caused the cathedral doors to be locked, 'and has tholed [allowed] no mass to be said therein since then'. The crowd 'required me sundry times to let them in' to attend mass, but, by his own account, Bothwell stood firm.

Events were taking their toll. Bothwell had experienced 'evil health ... continually since my coming in this country', and at the height of the uproar he again took to his bed, probably in the bishop's palace, a stone's throw from the sealed cathedral door, and the angry crowd milling around outside.

It was Sunday 2 February – the feast of Candlemas, which commemorated the purification of the Virgin Mary, forty days after the birth of Christ. The ritual for the occasion involved processions with lighted candles, afterwards taken home by participants as objects of blessing and protection. A festival of light, it must have resonated with people in the clipped, dark days of Orcadian winter. To Protestants, it was flummery and superstition.

There was a chapel in Bothwell's palace, 'hard at the cheek of the chamber where I was lying sick'. The demonstrators forced their way in, bringing a priest along to say mass. This clergyman proceeded 'to marry certain pairs in the old manner' – a remarkable instance of wedding ceremony functioning as a form of political protest, and one which hints at the possible prominence of women in the agitation.

It was in all respects an alarming incident, with potential to turn more serious still. Histories of the Scottish Reformation have properly paid attention to the mobs of Protestant supporters in Perth, St Andrews, Dundee and elsewhere who demanded the overthrow of the old order and an end to the suffocating rule of bishops. In topsy-turvy Kirkwall, a Catholic mob demanded a halt to heretical changes which their own bishop, appointed by the pope, was trying to impose upon them.

There is much we could wish to know about the events of February 1561. Other than the unnamed priest, who among the Orkney clergy was involved in stirring up the passions of the crowd? What exactly was the role played by Edward Sinclair of Strom and other local notables? Writing about the events only days after they occurred, Bothwell feared false reports of his conduct – or even complicity – might already be making their way to the lords of the council. He could not have stopped the demonstrations, he protested, 'without I would have committed slaughter'. Bothwell begged Napier to use his influence in Edinburgh to help clear his name. It was certain that John Bellenden, 'my small friend', had 'stirred me up all sorts of cummers [troubles] here', using Henry Sinclair and Thomas Tulloch as his instruments. Bothwell, by his own reckoning, was a man of courage and integrity: 'I will not commit me [submit] to ane angry multitude.'[13]

The Reformation in Orkney was not stopped in its tracks in 1561. The pattern across western Europe at this time was of people in power generally getting their way, whatever others felt. The Protestant John Bellenden may or may not have been a puppet-master pulling the strings of Catholic grievance. But it seems clear that the loyalties, rivalries and betrayals of notable folk – what historians call 'the elites' – cut across lines of confessional allegiance and probably prevented religious protest developing in any very coherent way.

In Kirkwall itself, temperatures remained high through 1561. In September, a band, eighty strong, 'issued out of the castle of Kirkwall, and ... searched and sought Henry Sinclair of Strom and Mr William Moodie for their slaughter'. Moodie was a politically active clergyman about whom we will hear more in due course. The would-be murderers failed to find their intended victims, and the matter was referred to the courts with no known sequel. The ringleaders included the bishop's nephew – an ex-friar called Francis Bothwell – and the cathedral precentor, Magnus Halcro, who had seemingly abandoned his allegiance to the earl of Caithness. They were joined by another Edward Sinclair, a brother to the laird of Roslin. This was no ideological confrontation between Protestants and Catholics, but an attempt by Adam Bothwell's supporters to purge his opponents. The vendetta had elements of the longstanding tension between 'bishopric' and 'earldom' portions of the town – a rivalry which, in modern times, finds more or less friendly annual expression in the Ba', an unruly Yuletide ball game, pitting 'uppies' against 'doonies' in a swirling throng through Kirkwall's streets and narrow lanes.

'The Bishop of Orkney begins to reform his diocese, and preaches himself.' That was news the English ambassador, Thomas Randolph, sent to Elizabeth I's chief minister, William Cecil, in March 1561, barely a month after the riotous scenes in Kirkwall. Randolph's information came from a letter written by Bothwell himself to the earl of Arran. The bishop of Orkney was eager to get his own version into the ears of people who mattered.

Arran was a supporter of Mary of Guise, yet, sensing which way the wind was blowing, he jumped ship in the summer of 1559 and threw in with the Lords of the Congregation. Was Adam Bothwell's conversion equally 'political'? It is hard to be certain, though there are hints of genuine spirituality in a letter sent to his sister Janet in January 1561, at the height of his woes in Birsay. Hearing of tensions in her relationship with her husband, Archibald Napier, he wrote to assure her such troubles were sent by God, 'to prove you, to try you if you love Him, and are, as the cross of the faithful uses ever to be, the fatherly chastisement and most special signs of God's undoubted favour and love'.[14] It was a Protestant instinct to see in earthly affliction proof of divine affection, an attitude comforting – and perhaps psychologically necessary – in times of trial and persecution. Bothwell, even as he doled out fraternal counsel, was probably reflecting on his own experience.

Like Reid before him, Bothwell had no intention of residing permanently in the far-flung and fractious islands. By the end of April, he had left Orkney for France, with the intention of seeking remedy from the queen for 'extortion done to him'. He was in Mary Queen of Scots' retinue when she returned to Scotland in August 1561, to encounter political and religious challenges it would prove beyond her capacity to resolve. First among the Scottish nobles to greet Mary on her arrival at Leith was her half-brother, Lord Robert Stewart, one of James V's brood of bastard sons, and a man soon to play a momentous role in the affairs of Orkney.

Bothwell arrived back in the islands in June 1562, having come to terms with Bellenden, who agreed to an annual pension of 400 merks from the bishopric revenues. The *Book of Discipline* had planned to abolish dioceses and divide Scotland into ecclesiastical 'provinces' of about 100 parishes each; this would have seen a large chunk of Caithness thrown in with Orkney and Shetland. But the tidy arrangement was scuppered by a shortage of money, as well as by the unexpected and not wholly unmixed blessing that three Catholic bishops – of Orkney,

Caithness and Galloway – proved willing to conform to the Reformation. Governance of the Church was now vested in a 'General Assembly' of ministers and leading laymen, which met for the first time in December 1560. In 1563, making a virtue of necessity, the General Assembly formally appointed the three bishops as 'superintendents' in their dioceses, with authority to 'plant kirks'. For the next few years, Bothwell planted and tended, at least intermittently – he seems to have spent a month or two in the islands in summer or early autumn.[15]

Historians of the Reformation in Britain once conceived of it as an 'event', taking place in England in 1533–4, and in Scotland in 1559–60. They now tend to describe it as a 'process', beginning somewhere around those dates, but lasting for many years afterwards. Bothwell's preferred term, as we have seen, was 'mutation'. In sixteenth-century Scots, the word signalled a distinct change, particularly in matters of state or government. Its range of modern meanings – across genetic science, mathematics, linguistics and virology – is more complex. Mutations can be sudden and unpredictable, but they are also likely to be recurring, and their evolutionary development is determined by inheritance and by patterns of adaptation to the host environment. Orkney's mutation of religion, in the early 1560s, had barely begun to run its course.

Idolatry Cast Down

After the tumults of 1561, there is not much evidence for how reforms were received in Kirkwall or across the rest of the islands. The conclusion drawn by some modern writers is that the Reformation in Orkney cannot have been too traumatic or disruptive, and that people adjusted readily enough to the new realities. Caution is required here. The absence of evidence of trauma is not necessarily evidence of its absence.

When the archive is silent, historians can turn to the imagination. In *An Orkney Tapestry*, the Orcadian poet George Mackay Brown, himself a convert to Catholicism, imagined the abrupt end of medieval worship in Hoy, at a fictional chapel in the secluded and achingly beautiful Rackwick Valley (see Fig. 1.1), which fans out onto a cliff-ringed, storm-blown bay on the west side of the island.

> A dozen horsemen rode through the hills from Hoy. They dismounted at the chapel of Our Lady. The valley people heard the sounds of blows and smashing and dilapidation inside – it went on all morning. Presently some of the horsemen came out with bulging sacks and staggered with their loads to the edge of the crag and emptied them into the sea below. A young man with a pale face stood at the end of the chapel and told the people that now they could worship God in a pure form; the Pope and his bishops had been cast down from their high Babylonish places; the idolatry of the mass was abolished … When the strangers had ridden off the Rackwick folk peered through the door of their chapel. The strangers had made it starker than any stable.

There is poetic truth to this description, which captures the disorientation and bewilderment island folk must have felt at the nullification of immemorial pieties. It is unlikely, though, that soldiers were sent to purge Orkney's churches in quite this dramatic fashion. The required reforms were probably spelled out by Bothwell in the course of his visitations. After the bishop and his attendants had gone, it was down to local clergymen to implement the changes, and do their best to explain to congregations why they, and the priests themselves, had been in error their entire lives. The sense of puzzlement across Scotland at this time was neatly captured by the Catholic exile Ninian Winzet. He wondered 'how that might be, that Christian men professing, teaching and preaching Christ and his Word so many years, in ane month's space or thereby, should be changed so profoundly in so many high matters'.[16]

The role of priests – Protestants preferred the term 'ministers' – and the reformation of church buildings were linked inextricably. Stone altars were sites where priestly sacrifice was offered to God, and the abolition of the mass made them redundant. Instead, communion was to be celebrated at a long wooden table, placed for the occasion in the main body of the church, and around which congregants would gather to consume wine as well as bread – before the Reformation, only priests communicated 'in both kinds'.

The pulpit, previously an optional or neglected item, was now the centre of attention, as preaching came to assume a central role in worship. Statues of saints, paintings on walls, and screens dividing the nave from the chancel where the altar stood were all symptoms of false religiosity and could not be permitted. Scottish reformers generally made the pragmatic decision to use existing places of worship, rather than build new ones. This produced both an experience of continuity and a heightened awareness of the immensity of change. In Orkney, the fact that most churches were simple rectangular constructions, unencumbered with aisles, ambulatories, apses and transepts, probably made the business of repurposing the buildings, if not the minds of congregations, relatively straightforward.

Within churches, services were said in the vernacular, not Latin. That vernacular was not, of course, Norn, nor even Scots. From the first, English was the written language of the Scottish Reformation. Reformers under the Guise regency, and for a time afterwards, used the Prayer Book produced in England by Archbishop Thomas Cranmer in 1552, itself a more radical revision of an earlier prototype. Even this, however, was germinated in too much common soil with the mass for the tastes of Knox, who preferred the form of service used by English exiles he worshipped with in Geneva. In 1562, the General Assembly prescribed a version of the Genevan liturgy, known as 'the Book of Common Order', for administration of baptism and communion, and in 1564 for all other forms of public worship. To some extent, and in sharp contrast to the situation in England, the Book's compilers trusted ministers to do their own thing – to speak 'these words following, or like in effect', to pray 'as the Spirit of God shall move his heart'. But they also provided a fixed structure for Sunday worship: prayers, scripture readings and singing of psalms, as prelude to preaching of the minister's sermon.

Celebrations of the Lord's Supper were intended to take place at least monthly, but this was hardly ever achievable. Across Scotland, the medieval habit of infrequent communion persisted, if anything strengthened by Calvinism's heavy emphasis on intrinsic human failure. Public worship commenced with an unflinchingly bleak anthropological assessment: 'O Eternal God and most merciful Father, we confess and acknowledge here before Thy Divine Majesty that we are miserable sinners, conceived and born in sin and iniquity, so that in us there is no goodness.'

If not goodness, then perhaps patience and capacity for endurance. The mass, even on high feast days, was a tithe of time demanding no more than an hour. For many medieval Orcadians, the route to church and back took longer than the service itself, the cultural significance of that journey enshrined in the habit of naming paths and tracks, and the places through which they passed, 'kirkgate' or 'messigate'. The distances grew no shorter with the advent of Protestantism, but an hour was now the minimum duration of the sermon alone, and reformers expected people to return for a second service on Sunday afternoons, when youths would be catechised, while parents listened with profit to the questions and answers.[17]

None of this was the work of a day, and there were probably few overnight transformations of the sort envisaged by George Mackay Brown. For a while, at least, old habits of thought and speech manifested themselves. The instinctive prayer for the soul of a deceased person – 'whom God assoil' – appears in charters through to the mid-1560s. Some Orkney clergy continued to style themselves 'Schir', and to superscribe letters with the invocation 'Jesus Maria'.

Most intractable of all was the reckoning of time by the Catholic calendar of saints' and holy days. Late sixteenth- and early seventeenth-century charters, leases, rentals and court records allude promiscuously to Candlemas, Allhallowmas, St Colm's Day, St James's Day, 'Our Lady Day of Lentren' (the Feast of the Annunciation) and Festronisevin (the eve of the Lenten fast, Shrove Tuesday). Orkney was not unique in this reflex, which pertained across post-Reformation Scotland. The old names were woven deeply into the fabric of social and economic life – as nomenclature for the terms operating in law courts and universities, and as accustomed times for payment of rents and holding of fairs and markets. Kirkwall's fairs took place on Palm Sunday, at Lammas in August and Martinmas in November. The Feast of St Martin was a traditional time for slaughtering cattle. Lammas was 'Loaf mass' day, when bread from the harvest was brought to church for blessing – a celebration associating agricultural fertility with the Catholic doctrine of the Eucharist.

Officially, the Reformed Kirk regarded as 'holy' no day but the Sabbath, not even Christmas and Easter. The Book of Common Order nonetheless carried a prefatory calendar incorporating traditional designations for various times of the year. Astonishingly, it included an entry for 15 August of 'Assum. ma' – the Assumption of the Blessed Virgin Mary, a Catholic doctrine repellent to Calvinist beliefs and instincts. Similarly, old names and dedications of churches were often retained in writing, speech and community memory – a habit that seems to have been particularly strong in Orkney, with its profusion of Peterkirks and Ladykirks.[18]

We know little about the processes and timetables by which those kirks were Protestantised, though they must sometimes have run slowly. One remarkable shard of evidence relates to the kirk of St Lawrence in Burray. In the autumn of 1643, a full eighty years after the official implementation of the Reformation, there was a visitation of Burray church, then in the process of being repaired. Afterwards, the clergyman responsible for the island, the minister of neighbouring South Ronaldsay, was quizzed by fellow preachers 'anent [concerning] St Peter's image, which was ordained at the visitation to be burned'. He assured them it was done.

It is a frustrating fragment from an undoubtedly richer story, the details destined to remain lost, like a snatch of conversation overheard on the bus. But the implication is of a medieval statue that managed for decades to remain either inside the kirk of Burray or somewhere in its precincts – a statue, moreover, of St Peter, a saint whose papal associations were particularly offensive to Protestant sensibilities. Either the authorities were unaware of its existence, or its presence was tacitly tolerated, until new waves of fervour rose to scorn such compacts with the popish past.

It was harder to enforce new orthodoxies in the smaller islands like Burray, which lacked resident ministers and depended on sporadic visits from clergymen based in another part of a conjoined parish. The inhabitants of North Ronaldsay, according to Jo. Ben., were 'completely ignorant of divine ways of speech, for they are seldom or never taught'. Papa Westray was equally unclergied, and reliant on visits from the Westray minister whenever, as an eighteenth-century pastor wearily observed, 'the weather permits him to pass the ferry'. The island had its own parish church – as it happens, the only one of Orkney's medieval parish kirks still standing in unruined condition today. Papay, as the locals call it, also had a place of still greater sanctity – the little pilgrimage chapel of St Tredwell (p. 88), jutting into the loch of the same name atop an Iron Age crannog, a dwelling constructed on an artificial island.

Figure 3.2: St Tredwell's Chapel, Papa Westray, one of Orkney's most renowned sites of pilgrimage, frequented before and after the Reformation. The twelfth-century chapel ruins sit on top of an earlier Celtic site in a place of 'liminal' sanctity, surrounded by the waters of a loch which were themselves believed to have curative powers.

A tradition about the sixteenth-century inhabitants of Papa Westray was recorded in the early nineteenth: 'it was with difficulty that the first Presbyterian minister of the parish could restrain them, of a Sunday morning, from paying their devotions at this ruin, previous to their attendance on public worship in the reformed church'.[19] We will hear more in due course about the reluctance of Orcadians to wean themselves from the habit of pilgrimage to sacred places.

Silence around the dismantling of Catholicism in Orkney is almost deafening at the place where we might expect the loudest reverberations, the cathedral of St Magnus in Kirkwall. Following Bishop Bothwell's suppression of the mass in 1561, the various altars were certainly all removed, but exactly when, or on whose authority, is unclear. The process was complete before May 1565, when Francis Bothwell, the bishop's nephew, became prebendary of St Lawrence. According to the deed registering the appointment, installation took place 'where the altar of the said prebend formerly was'. A presence was defined by an absence, and the responsibilities of the present by memories of the recent past.

A still more startling continuity surrounds Francis Bothwell's appointment. The document is authenticated by reference to the reign of Pope Pius IV, a full five years after parliament solemnly decreed the bishop of

Rome to have 'no jurisdiction nor authority'. The actual date is left blank – 'Indictione … Pii pape … anno …' – the phrasing suggesting not so much defiant Catholicism as bureaucratic inertia on the part of a scribe unsure how to write out the accustomed form in any other way.[20] That scribe, William Peirson, prebendary of Holy Cross, conformed to the new regime. But no reformer of Knoxian convictions would have permitted himself such a lapse.

St Magnus's transformation in the early 1560s was on a scale matched by no other church in Orkney, and few in the entire kingdom. Along with the multiple altars, a plethora of liturgical furnishings and sacred ritual objects were rendered, in Knox's phrase, 'monuments of superstition and idolatry'. They included frontals and coverings for the altars, and the ornamented candlesticks placed on them; the embroidered vestments worn by canons and their assistants; processional crosses and banners; chalices and silver plate for celebrating mass; and a fixed tabernacle or hanging 'pyx' for housing the consecrated and reserved bread of the Eucharist. It is impossible to say what happened to these items, along with the 'precious jewels' referred to in Bishop Reid's constitution of 1544. Were they publicly destroyed, sold for profit or quietly squirrelled into private hands? No traces now remain.

The same is not quite true of saints' images in the kirk, of which there were probably at least as many as there were altars. Two small statues of St Magnus and St Olaf were somehow spared from destruction (see Plates 10 and 11). The saints, equipped with sword and axe, were carved in relief within stone niche frames. The original placement is unknown, but they evidently remained in or around the cathedral. In the mid-nineteenth century, the antiquarian Sir Henry Dryden came across them 'in the room over the S. chapel', and they have subsequently been moved to the Orkney Museum.[21] A white stone statue of a third Scandinavian saint, the cathedral's founder Rognvald, stood until recent times in an exterior niche halfway up the imposing round tower which Reid added to the Bishop's Palace (see Plate 12). It is an odd place to encounter a statue of a saint – the adjacent niche, more predictably, carries heraldic arms, now badly worn but presumably those of Reid himself.

The statue seems likely to have been housed originally in the cathedral: it is older than the tower and bears traces of paint. Whether it was moved across with formality and fanfare on the tower's completion, or was relocated there at the Reformation, is unknowable. Either way, it was permitted to remain in its new home. Observers in later centuries were puzzled as to the subject's identity, which has been established by noting at his feet the presence of what seems to be a stringed lyre – Earl Rognvald was renowned as a musician. But Kirkwall folk in the 1560s knew exactly who was perched in what has long been known as the 'Moosie Toor', looking down on comings and goings before the cathedral west door.[22] It is unlikely to be coincidence that Orkney's only surviving medieval statues are stately images of the three Norse ruler-saints, Magnus, Olaf and Rognvald. There were aspects of faith and identity that the denizens of Kirkwall's cathedral close seemed distinctly reluctant to relinquish.

The 'purification' of St Magnus was not, it appears, undertaken in any spirit of iconoclastic zeal. Cathedrals, according to Knox's *Book of Discipline*, were sites of idolatry, and should be 'utterly suppressed in all bounds and places of this realm'. A grudging exception was made for grand ecclesiastical buildings that 'presently are parish kirks or schools'. After the destruction of St Olaf's Church by the English, St Magnus became the place of congregational worship in Kirkwall, which protected it from would-be suppressers. Painted interior walls were whitewashed, but the core fabric escaped significant damage. Indeed, some items of beauty, as long as they served no directly devotional function, were permitted to remain in place – at least until Victorian 'improvers' decided to dispense with them. These included a set of ornately carved choir stalls from the time of Bishop Maxwell, and the screen separating the choir from the main body of the church, which in the early nineteenth century still displayed 'some rich wooden sculpture', though the crucifix resting on top was undoubtedly taken down.

What could not be permitted to remain was the cathedral's original raison d'être: the shrine of St Magnus. Elsewhere in the British Isles, the suppression of saints' shrines provided occasions for shrill triumphalism on the part of Protestant reformers, and for anguished lament from bereaved conservatives. For St Magnus, after 400 years as the epicentre of Orcadian religious culture, there is only stillness and silence. No scrap of documentary evidence refers to the closure of the shrine. It is a baffling – almost an eloquent – blank in the historical record.

Jo. Ben., a generation after the event, recounted various memories and traditions about St Magnus, but said nothing about the shrine in Kirkwall. A 1633 description of Orkney by the laird of Egilsay, Robert Monteith, praised Kirkwall's 'fair cathedral church, dedicated to St Magnus'. Yet Monteith identified the parish church of Egilsay as the place 'wherein, they say, this saint lies interred' – a striking instance of collective false memory, and selective cultural amnesia.[23] Elsewhere in Orkney, visits to holy places persisted long after such practices were formally proscribed, but prohibitions were easier to enforce at the centre of religious and political power; there is no evidence from after the Reformation of clandestine resort to the erstwhile shrine in Kirkwall.

What, then, became of the bones? For centuries, no one had any idea, or admitted to having one. But, in time, a hard carapace of silence around the fate of the shrine started to crack, and it became possible to at least guess at what may have happened in the 1560s. In the mid-eighteenth century, a workman attaching a monumental tablet to a pillar at the east end of the cathedral's choir struck unexpectedly into a hollow recess. Inside were a quantity of bones, along with some fragments of an oak coffin. A young clergyman, George Low, reflected the consensus of local opinion: these were 'in all probability the relics of St Magnus, part of whose bones are said to have been deposited here, and perhaps never disturbed till now'. In the nineteenth century, the bones were examined several times and doubts about the attribution grew – there was an indentation on the skull, but it appeared to be the mark of some old injury, not a killing blow. There were also disparities with a record of some bones of Magnus gifted to the Metropolitan Cathedral in Prague in the fourteenth century.

Many years later, with another unexpected discovery, things fell finally into place. In March 1919, the clerk of works overseeing a restoration programme in the cathedral noticed some dressed facing stones that appeared to be loose. This was on the south side of the choir, on a rectangular pier mirroring the one where the earlier bones were found. There was a similar hollow cavity, with a wooden casket inside. It contained a damaged skull and an array of other bones.

Forensic examination suggested a death consistent with that of Magnus described in the saga accounts. The man died from a vertical

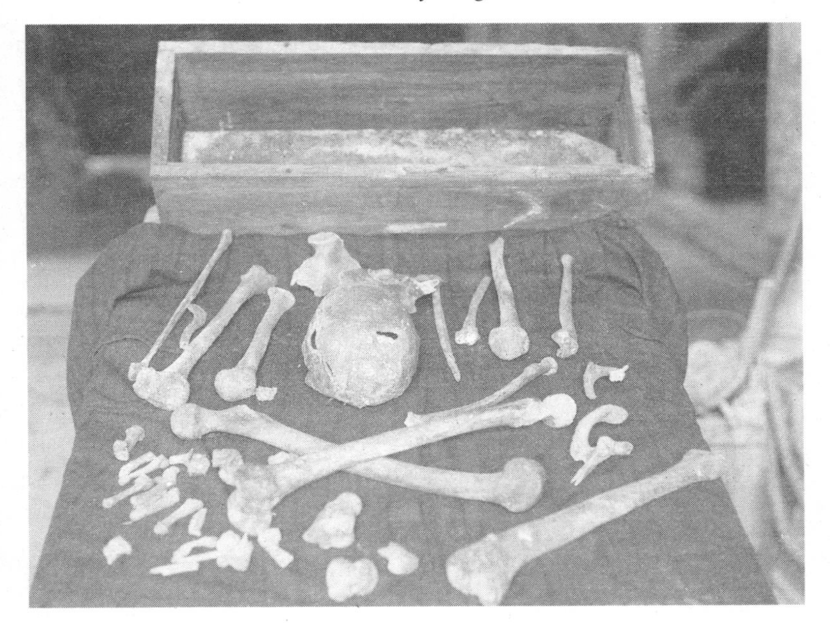

Figure 3.3: Bones from a wooden box found in a pillar of St Magnus Cathedral during renovation work in March 1919; almost certainly, the relics of St Magnus that were housed in the medieval shrine. After examination, the bones were replaced in the pillar where they remain to this day.

blow, not a decapitation, though the killing weapon was most likely an axe, rather than the sword of medieval iconography. If these were the bones of St Magnus, then those from the pillar opposite were surely those of St Rognvald.[24] The relics of the saints, uncle and nephew, had not been rooted up and destroyed, like the contents of countless shrines throughout Europe in the course of the sixteenth century. They were moved a few feet, and reverently hidden from view.

There are no certainties here, but a working hypothesis can be cautiously advanced. The reinterment does not seem hasty or impulsive, the clandestine act of some dissident individual. The placement is careful and considered; it would have taken time and appropriate tools, and created a fair amount of noise. The shrine's gold and jewels were clearly otherwise disposed of, but its true treasures, the relics, were retained –

almost certainly in the original wooden coffins which formed an inner core of the public display.

The elevation of the cavities is interesting; both apertures are approximately nine feet from the floor. It is possible an attempt was made to preserve something of the arrangement of the dismantled shrine.[25] The height is consistent with the reliquaries having formerly rested on a supporting beam, either above the high altar or the altar of a shrine chapel. In either case, they would have been just to the east of the pillars to which the caskets were moved. If an existing juxtaposition was kept in the process of removal, then Rognvald sat on the north side of the beam and Magnus on the south – a holier location in the sacramental geography of medieval religion.

In short, the relics of the shrine were not cast out; they were again 'translated' – sealed and absorbed into the stone musculature of their ecclesiastical host. Once more, we are dealing with 'layers', complex cultural archaeologies of emotion and belief.

It is hard to imagine that this ritual, or at least respectful, reburial of the relics could have taken place without the authorisation of the cathedral chapter, and perhaps even the knowledge and consent of Bishop Bothwell. Maybe – it can be put no more strongly – the issue of the shrine was discussed and resolved at a meeting of the chapter which we know to have taken place in the cathedral on 12 April 1561, ten weeks after the Candlemas riot. The sole documentary trace of the meeting is a record of the canons' consent to the feuing of two properties in Kirkwall, but there must have been much else to discuss. In attendance were the provost, Alexander Dick, the chancellor, James Annand, the treasurer, Francis Bothwell, the precentor Magnus Halcro, the subchanter Magnus Strang, and the prebendaries of St Magnus and Holy Cross, Robert Cheyne and William Peirson.

These men, who oversaw the reform of the cathedral, were no cabal of eager Protestants. Peirson we have met already, inserting the pope's name into an official document. The two Magnuses, Halcro and Strang, were stalwarts of the Reid regime. They held onto their titles (and incomes) but neither was willing to reinvent himself and serve as a Protestant parish minister. Did they, one wonders, feel special stirrings of devotion to the saint after whom they were named? James Annand, a former minor canon recently promoted to the chancellorship, proved more flexible. He would soon take on pastoral responsibilities in the North Isles – where, if tradition is to be trusted, he struggled to dissuade Papa Westray folk from frequenting the chapel of St Tredwell.

The provost, Alexander Dick, eventually decided, in the early 1570s, to serve as minister of South Ronaldsay and Burray. In the meantime, he may have been wrestling with his conscience. In 1567, Dick was indicted in the justiciary court in Edinburgh for saying mass, and for 'adulteration of religion'. Francis Bothwell was the newcomer in the group, and his uncle Adam's man. But he was no Knoxian paladin either. The bishop's critics were soon muttering 'he retained in his own company Sir Francis Bothwell, a papist'.[26] Orkney's journey along the winding road of Reformation began at a leisurely speed, steered by cautious drivers with one eye on the rear-view mirror.

The Unicorn and the Rock

'Sheep wandering without a pastor'. That was the assessment of Orkney's benighted people made by the General Assembly in Edinburgh, at a session held – with conspicuous disregard for the day's old associations – on 25 December 1567. 'Adam, called Bishop of Orkney', was charged *in absentia* with neglecting his 'country', visiting kirks there only between Lammas and Hallowmas (August to November), and behaving himself unspiritually as a judge in the court of session.

Bothwell's greatest offence, however, was his performance of a pastoral act that was also profoundly political. On 15 May 1567, Bothwell officiated at the queen's marriage to James Hepburn, earl of Bothwell, astutely summed up by the English ambassador as 'a vainglorious, rash and hazardous young man'. It was the least of it that, only a week earlier and in murky circumstances, Hepburn had been divorced from his previous wife. He had also kidnapped the queen, and very probably raped her. He was, moreover, widely suspected of orchestrating the murder of Mary's previous husband, the feckless Henry Stewart, Lord Darnley. Darnley's strangled body was found in the garden of his Edinburgh house, blown up by gunpowder on 10 February 1567.

Hepburn had Orkney connections: his mother was a daughter of Lord Henry Sinclair, grandson of the last earl. Three days before the wedding, Mary created him duke of Orkney – first holder, other than in Arthurian myth, of this ostentatious title. The ceremony proceeded according to Protestant rites, but the General Assembly's verdict on Bishop Bothwell was that 'he transgressed the act of the Kirk in marrying the divorced adulterer'.[27]

Mary's marital bliss, such as it was, soon vanished. A confederation of nobles rose in revolt, imprisoned the queen and forced her to abdicate. In July 1567, her one-year-old son was crowned as James VI, in a Protestant ceremony at which Bishop Bothwell – ever eager to prove his usefulness – anointed the infant monarch.

The duke of Orkney, meanwhile, fled to the north, with the intention of using his new fiefdom as a base from which to defy the regime. Like the English admiral Sir John Clere, he found Kirkwall Castle too hard a nut to crack (p. 104). The current sheriff was Bishop Bothwell's brother-in-law, Gilbert Balfour, a former ally of the new duke, and likewise suspected of involvement in Darnley's murder. But Balfour, like Bothwell, was eager to end up on the winning side. He trained the castle guns on Hepburn's ships, and ensured his Westray fastness of Noltland was ready to repel assault. The duke stayed two days, and sailed on to Shetland, to seek assistance from his mother's Sinclair kinsfolk.

Four government ships were in hot pursuit, under the command of William Kirkcaldy of Grange, a relative of Adam Bothwell. The bishop himself joined the expedition as a local guide. The fleet surprised Hepburn and his men ashore, their vessels at anchor in Bressay Sound, near to where the port of Lerwick would later develop. Hepburn managed a rushed embarkation, and a daring escape to the north, scraping out of the sound over some barely submerged rocks. Kirkcaldy's own ship, the *Unicorn*, struck the same obstacle, was holed and sank. Last to leap into a crowded lifeboat was the armour-clad bishop – a feat that would be recounted about him, with mixtures of admiration and mockery – for years to come.

Hepburn's ships, and Kirkcaldy's remaining three, clashed once more off the coast of Unst, the most northerly point in Scotland, before a storm allowed him to escape again and cross the North Sea to Bergen. By chance, resident in the town was a former mistress, Anna Throndsen, who promptly revealed his identity and sued for breach of promise and unpaid debts. In September 1567, Frederick II instructed the governor of Bergen to arrest the duke of Orkney and send him to Copenhagen. The Scottish government pressed for Hepburn's extradition, but he was never released, dying insane in his cell in Dragsholm Castle in 1578.

In the first months of his confinement, the duke penned a lengthy confession for the Danish king, and added an enticing offer. He had authority from the queen and council, he claimed, 'to surrender the islands of Orkney and Shetland, free, quiet and without hindrance, to the crown of Denmark and Norway'. Frederick had merely to draw up the letters of transfer, 'with such rigid conditions as to his said Majesty ... may appear most binding', and Hepburn would ensure they were signed by Mary and the councillors of Scotland.

It was a reckless plunge into perilous waters. Just over three years earlier, a Scots ambassador bound for Frederick's court had been warned something might be proposed to him 'touching Orkney and Shetland'. The ambassador should 'touch no wise that string', but quickly change the subject. For the government in Edinburgh, this was a boat best left unrocked. Perhaps coincidentally, or perhaps in response to rumours of Hepburn's intriguing, the question was raised at the parliament of December 1567 'whether Orkney and Shetland shall be subject to the common law of this realm'. It was an opportunity, a century after their acquisition, to bind the islands more closely to the kingdom of Scotland. But the matter was resolved in favour of the status quo: 'they ought to be subject to their own laws'.[28]

Frederick probably did not quite believe Hepburn's offer, or thought the nominal duke of Orkney had little ability to deliver it. But the prisoner was worth holding on to, just in case. It was encouraging to know that to some of the nobles of Scotland the territorial integrity of the kingdom mattered less than the prospect of returning to power.

Betwixt God and the Devil

The charges against Adam Bothwell in December 1567 resulted in his suspension from the ministry. His restoration, in July 1568, was conditional on preaching a sermon confessing 'the offence in marrying the Queen'. In March 1570, Bothwell was again attacked in the General Assembly. Ignorance and vice had increased in his bishopric, and churches were ruinous and deprived of ministers, 'leaving the flock destitute'. Harsh appraisal was made of Bothwell's conduct and credentials as that curious Scottish entity, a Protestant bishop: he 'styleth himself with Roman titles, as Reverend Father in God, which pertaineth to no ministers of Christ Jesus'.

Bothwell claimed never to have delighted in such titles, and he acknowledged himself 'a worm of the earth, not worthy any reverence'. In other respects, he offered a robust defence. Admittedly he had, since 1567, remained south, 'by reason of his infirmity and sickness'. But, before then, he had tirelessly promoted Protestantism. 'When idolatry and superstition was suppressed, he suppressed the same also in his bounds, preached the word, administered the sacraments, planted ministers in Orkney and Zetland, disponed [distributed] benefices, and gave stipends out of his rents to ministers.' In implicit rebuke to those with comfortable inland responsibilities, he stressed how, as a commissioner of the General Assembly, he 'visited all the kirks of Orkney and Shetland twice, to the hazard of his life, in dangerous storms on the seas'.

Modern historians have praised Bothwell for an 'exemplary' record in ensuring reformed services for Orkney parishes, redeploying parasitic cathedral canons as a pastoral workforce in places from which they drew their rents.[29] It is too sanguine a view. In the decades after 1560, the newly reformed Kirk struggled to meet its responsibilities, its resources stretched to breaking point and beyond. It did not help that, in Orkney, as elsewhere in Scotland, the old Kirk's assets were not available to fund the activities of the new. In February 1562, the privy council decreed a compromise on church finances: existing office-holders would retain for their lifetimes two-thirds of the 'fruits' of their benefice, the remaining third to be divided between the crown and ministers' salaries. It was,

Knox bitterly remarked, 'twa parts freely given to the Devil, and the third must be divided betwixt God and the Devil'.

There was thus little financial incentive for the cathedral clergy to sign up for new and unfamiliar duties. Among those in office before 1561, the precentor, Magnus Halcro, and the subchanter, Magnus Strang, held aloof, as did the subdean, Peter Houston, and the prebendaries of St Duthac and Our Lady of Woodwick, John Houston and John Anderson. Alexander Dick, the provost, and William Peirson, prebendary of Holy Cross, hesitated, but conformed in 1574. The chancellor, James Annand, was more enthusiastic, or pliable, as was the holder of St Catherine's Stouk, James Maxwell, prepared to shoulder duties as vicar of Stronsay.

In all, perhaps half of the thirty-four clergymen in office in Orkney in 1560 continued working in the Kirk in some capacity. They included many of the former vicars pensioner (p. 82), who had few other options open to them. Gavin Watt in Deerness, David Anderson in Evie and Thomas Rattray in South Ronaldsay agreed to serve – their stipends from the respective 'thirds' meticulously calculated from sales of butter, oil, meat, wax, peats and salted fish.

What such men couldn't do was write and deliver sermons. Preaching was the defining act of the Protestant minister; indeed, the essential means by which salvation was achieved, in the sense of awakening individual Christians to an awareness of God's irresistible call to them. A list drawn up in 1567–70 named twenty-two clergymen serving in Orkney parishes. Of these, only seven were ministers, licensed to preach and administer the sacraments of baptism and communion. Two others were 'exhorters', a kind of probationary status, allowing simple explanation of scriptural passages, but not administration of sacraments. The rest were 'readers' – restricted, as the name implied, to reciting forms of prayer set out in the Book of Common Order. God's Word was spread scrapingly thin in Reformation-era Orkney. Thomas Stevenson was minister of Orphir, Stenness and Firth; Jerome Tulloch for Stromness, Sandwick and Graemsay. James Annand, with just two readers to assist him, had responsibility for the whole of Sanday and Westray, as well as Papa Westray and North Ronaldsay.

Alongside the rentals drawn up to calculate thirds in the mid-1560s, an assessment was made of 'kirks of Orkney that has need of ministers to serve the people of God'. The list was dismayingly comprehensive. Nothing had been done to address the problem of conjoined parishes – like Birsay and Harray – with kirks several miles apart from each other. Some proposals were heartfelt, but hopelessly unrealistic: 'Burray, with the holms thereof, has ane parish kirk and needs in special ane minister'; 'North Ronaldsay, distant from any isle four miles, needs ane minister'. So, too, 'the isle of Papa Stronsay, inhabited, three miles wide'. Neither Burray nor North Ronaldsay would see a resident minister before the mid-nineteenth century. The tiny island of Papa Stronsay would never entertain one – though, in a development to horrify the compilers of the rentals, it is now home to a community of traditionalist Catholic monks.[30]

The exception to a pattern of patchy provision was Kirkwall, where a string of ministers preached from the pulpit in St Magnus. The first, serving for two decades, was Gilbert Foulzie, a former monk from Aberdeenshire, whom Adam Bothwell in 1561 appointed as archdeacon of Orkney. When the cathedral ceased hosting a community of resident priests, Foulzie combined his manse with that of the subchanter to create a sizeable family dwelling. The space was needed, for Foulzie was among the first Orkney ministers to marry. To Protestant reformers, celibacy was both an unattainable goal and an artificial barrier between layfolk and clergy.

For priests to have wives was a theological statement, but perhaps not so much of a social revolution. The attitudes of Orkney clergy towards celibacy had always tended towards the casual. There was a deep-rooted instinct for dynasticism, the habit of acquiring estates and arranging the marriages of 'natural' children. The Reformation, quite literally, legitimised the process.

In 1574, over the gateway to his dwelling – later known as Tankerness House and now the home of the Orkney Museum – Foulzie placed a plaque showing his arms and initials, and those of his wife, Elizabeth Kinnaird, with an accompanying inscription: 'Nisi Dominus Custodi, Erit Frustra Semen Nostrum' – without the protection of the Lord, our seed shall live in vain. A son, James, predeceased him; but, with the Lord's protection, three daughters lived to adulthood. Ursula married Edward Sinclair of Essenquoy, a grandson of Edward Sinclair of Strom. The veteran warrior himself died about 1564, and was honourably buried in the south choir aisle of the cathedral. His tombstone was embellished with the image of a two-handed sword, of the kind he wielded against both Caithnessian and English invaders.

Clerical marriage was a doctrine of the Reformation for which the precentor Magnus Halcro, himself the younger son of a priest, showed positive enthusiasm. He had already started building a patrimony. In 1556, Halcro purchased from its udal proprietor a fine house and estate in Rousay, and thereafter styled himself 'Mr Magnus Halcro of Brough'. In 1560, he persuaded the bishop and chapter to grant him a tack of all bishopric lands in the island. It made him a plausible partner for a notable heiress – Margaret Sinclair, only child of the ill-fated victor of Summerdale, Sir James Sinclair of Brecks, and his noble-born wife, Barbara Stewart.

The marriage proved messy, for Margaret was seemingly already wed, to a man named James Tulloch. This resulted in Halcro's excommunication for adultery, but he proved adept at the old Orcadian habit of selective appeal to outside authority. Halcro secured an absolution from John Douglas, archbishop of St Andrews, who was himself an anomaly, the first new bishop appointed since Mary's deposition. Douglas was a kinsman of the earl of Morton, regent for the infant James VI, and in 1571 Morton arranged his promotion of behalf of the crown. The General Assembly argued that such appointments needed the approval of the Kirk, and that there should be no church benefices without cure of souls. A stand-off was resolved by the 1572 Concordat of Leith, which awkwardly integrated episcopacy into the Reformed Kirk, by laying down dual accountability to the crown and the General Assembly, and insisting that all bishops double as parish ministers. Nonetheless, Magnus Halcro's absolution was laid to Douglas's charge in 1573, when he, like Adam Bothwell before him, was brought before the General Assembly to face accusations of misconduct.[31]

William Moodie was another Catholic priest who reinvented himself as a Protestant minister on the way to becoming a respectable laird. A graduate of St Andrews, notary public and chamberlain of the crown estates, we last saw him as an intended victim of assassination (p. 114). He was also, in 1557, one of the alleged despoilers of the hijacked German ship (p. 109). Magnus Halcro had tried to kill him in 1561, but there was no lasting ill feeling, as in 1565 Halcro signed over to Moodie parts of his benefice. In 1570, Moodie served as minister in the Halcro heartlands of South Ronaldsay, and by 1574 had oversight of the southern isles of Flotta, Hoy and Walls, where his family put down roots at the start of the sixteenth century. Moodie married a woman called Katherine Sinclair, in all likelihood a relative of the Oliver Sinclair who in 1565 sold Moodie the West Mainland estate that allowed him to begin designating himself – in proper lairdly fashion – as Mr William Moodie of Breckness.[32]

Men like Magnus Halcro and William Moodie were skilled at surmounting challenges and turning them to worldly advantage. They seemed little more likely to be dislodged by the arrival of Reformation than the cliffs of Orkney were to be vanquished by the furious impotence of the sea. Waves of change, however, were starting to erode and reshape the coastlines of Orcadian society, the old families losing ground to incomers like Balfour and Bellenden. The new lairds were frequent feuers of bishopric and cathedral estates – the wealth of the old Kirk passed inexorably into their hands, rather than finding investment in church buildings or ministers' stipends. They regarded Orkney as a place of refuge and a source of profit, a frontier society where they could make their mark while evading too close a scrutiny of their doings. They did not know it, but they were harbingers of a still more formidable incomer from the south, a dangerous enemy for Halcro and Moodie, and for Bishop Bothwell, too. A quarter of a century after James V accepted the hospitality of Kirkwall, his son came north to claim an inheritance.

4
The Return of the Earls

Rex Scotorum

Birsay village, a tiny cluster of houses peering out at the Atlantic from the north-west extremity of the Orkney Mainland, is an unlikely setting for the sprawling remains of a grand Renaissance palace. Tourists are sometimes surprised to discover it, as they pull off the road leading from the archaeological site at the Brough of Birsay, to acquire supplies at the little shop, and fill up at the solitary petrol pump.

Work on the 'place', as Birsay folk call it, began around 1569. The structure was designed to be defended, built around a central courtyard, with gun-ports and narrow windows in the lower floors. Yet it was no grim fortress like Noltland in Westray. Robert Monteith of Egilsay, in the early seventeenth century, called it 'sumptuous and stately'. Decades later, as the palace slid into decay, its decorated ceilings, painted with lively biblical scenes, still evoked awed admiration. There were gardens for herbs and flowers, archery butts and a bowling green, as well as adjoining rabbit warrens that doubled as a golf-course.

It was a palace fit for a prince, or at least the son of a king. Robert Stewart, born 1533, was the third, perhaps the fourth, of James V's seven acknowledged boy bastards. His young adult life was a parable of the trials and temptations laid before the Scottish nobility by the politics of the Reformation. Influenced by the earl of Moray, his elder half-brother, Robert sided with the Lords of the Congregation in the war against Mary of Guise, and at the 1560 parliament he was judged by Knox to have 'renounced papistry, and openly professed Jesus Christ with us'. But Robert's concern with faith was only ever fitful. He soon attached himself to Queen Mary's second husband, Henry, Lord Darnley – a religious weather-vane, and man of few discernible principles beyond self-regard. As for Robert Stewart himself, the English ambassador reckoned him 'vain and nothing worth, a man full of all evil'.[1]

Lord Robert's interest in Orkney may have stemmed from a position his father secured for him as a child: 'commendator' (lay head) of the abbey of Holyrood in Edinburgh. The abbey's tenants included Oliver Sinclair of Pitcairn and Sir John Bellenden (p. 109), whose business dealings perhaps drew the attention of 'the abbot' – as he was still sometimes called – to the profits to be made in the north.

Upon her return to Scotland in 1561, the queen's instinct was to be generous to her elder half-brother. In December 1564, she bestowed on him 'all and hail the lands of Orkney and Zetland'. It was not a fixed-term tack of the earldom lands, of the kind made to Oliver Sinclair, but a feu, granting Robert permanent, heritable lordship, with the office of sheriff and sweeping rights of patronage over the Kirk. The legality of the grant was questionable, not only because it seemed to make Robert the feudal superior of udal landowners – a contradiction in terms – but because the 1472 act annexing the earldom to the crown insisted it could go only to 'ane of the king's sons of lawful bed'.

These seemed niceties, if they were considered at all. 'My Lord Robert', Ambassador Thomas Randolph wrote to William Cecil in June 1565, 'shall be, as he sayeth himself, Earl of Orkney.' That did not happen yet, and for the next couple of years after the death of Sinclair of Strom power in Orkney tilted towards the Bellendens. Sir John's hard-nosed brother Patrick – laird of Stenness after a feu-grant of church lands from Magnus Halcro – became sheriff in 1565. But in 1566 he was abruptly replaced, over Lord Robert's head, by Gilbert Balfour, then a follower of the earl of Bothwell.

The events in the north were echoes from an explosion in Scottish national politics: the growing estrangement between the queen and her husband. It reached a horrifying pitch in March 1566: the murder, with Darnley's connivance, of Mary's private secretary, David Rizzio. A band of lairds burst into the royal apartments at Holyrood Palace, dragged

Rizzio from behind Mary's skirts and stabbed him repeatedly in an adjacent hallway. Among the assassins was Patrick Bellenden of Stenness. To prevent the queen intervening, according to an account received by the English privy council, Bellenden 'offered a dag [pistol] against her belly with the cock drawn'. Mary was six months pregnant. Lord Robert, one of a small party dining with the Queen, stood helplessly by.

Mary's overthrow in the wake of Darnley's murder, and the duke of Orkney's flight to the islands (p. 128), persuaded Robert to assert his rights there in person, arriving around the end of October 1567. Birsay – perched on the storm-lashed rim of the Mainland, far from the bustle and business of Kirkwall – seems a curious choice of abode. The parish, however, was an ancient seat of episcopal power, and Lord Robert had acquired lands there, as well, probably, as an existing house – the Mons Bellus from which Bishop Bothwell had been briefly evicted in 1560–1 (pp. 111–12). Robert likely demolished this residence, using some of the stone to build his grander replacement. A slab inscribed with the arms

Figure 4.1: Ruins of the Earl's Palace, Birsay – a magnificent place of residence constructed by Robert Stewart, bastard son of James V, between 1569 and 1574.

of Bishop Edward Stewart (*c.* 1520) was later found embedded in the palace walls.

On another slab, above the main gateway, there was an inscription, now long vanished, but still legible at the start of the eighteenth century. It read: 'Robertus Steuartus Filius Jacobi 5ti Rex Scotorum Hoc Aedificium Instruxit' – Robert Stewart son of James V king of Scots constructed this edifice. There was something curious about the phrase. Latin case endings are required to 'agree'. In the genitive – 'of James, king of Scots' – it ought to have read 'Jacobi 5ti Regis Scotorum'. As it was, the sentence's grammar made Robert Stewart himself into the king of Scots. It was, perhaps, an innocent mistake, though one that any educated sixteenth-century person would surely have spotted.[2]

A brazen usurpation, or a Freudian slip in stone, the inscription proved an apt motto and epitaph, for delusions of grandeur, and whispers of treason, were to be the defining marks of Lord Robert's tenure in the north. The Stewart earls of Orkney – Robert, and his son Patrick – presided over a grim, gruelling period in the history of the islands, one which stretched to tearing point the ligaments of the 'community of Orkney', and which coincided, and intersected, with seismic changes in the political landscape of Scotland and Britain. It began with riot and murder, and ended with rebellion, invasion and conquest.

Murder in the Cathedral

Lord Robert's arrival created waves, which soon began to submerge the existing enclaves of power. Patrick Bellenden complained of allies ejected from offices by Robert, an incomer with dubious legal authority: 'ane pretended heritable infeftment'. The followers of Bishop Bothwell likewise resented the intrusion. In March 1568, Robert's servant John Brown was discovered inspecting the cathedral, and forced to leave at gunpoint by five of the bishop's men. Brown crossed the street to the castle to gather a posse of fellow servants. 'In ane anger', they charged the cathedral, and the bishop's men opened fire, killing Brown with a shot to the head. His companions broke through and discharged their guns within St Magnus itself, slaying two defenders.

Robert denied responsibility for the act of sacrilegious violence, countering that Bellenden and supporters, including Magnus Halcro, William Moodie and Gilbert Balfour, were planning to raise soldiers and expel him from Orkney. The incident caused difficulties with Regent Moray, but did not deflect Robert from his goal – 'superiority of this country, as well of the bishopric as the rest'.

Victory over the bishop came in the form of an 'excambion', an exchange of lands, spelled out in a contract of September 1568. Magnus Halcro called it a 'black excambion'. Bothwell granted Robert a feu of all bishopric lands, with the exception of those already feued to Gilbert Balfour, the Bellendens and the inhabitants of Kirkwall. In return, Bothwell would become commendator of Holyrood, though a good portion of the abbey estates were given to Sir John Bellenden in exchange for his lands in Birsay. Robert swore to put aside 'whatsumever anger, hatred or displeasure' he had against Patrick Bellenden, Magnus Halcro, William Moodie and other of 'the said bishop's friends, servants or partakers'.

It was an arrangement denounced by the General Assembly as 'contrary to all laws of God'. Simony – monetary trade in ecclesiastical offices – was one of the charges levied against Bothwell in 1570. The bishop denied resigning any *spiritual* office into Robert's hands, and insisted he might lawfully set his benefice in tack for a 'just duty'. He also claimed to have been forced to agree 'for mere necessity', after Robert 'violently intruded himself'. Yet secretly it may have suited Bothwell – especially after his near-drowning in Shetland (p. 128) – to consolidate his means around Edinburgh. Through to his death in 1593, he remained active as a privy councillor and court of session judge, bishop of Orkney in nothing but name.[3]

The next bastion to breach was the burgh of Kirkwall. In 1569, Robert secured election as provost, in succession to Patrick Bellenden. He promised to defend the town's 'liberties and auld privileges', but burgesses soon found themselves unable to trade without Robert's permission. It was reported in Kirkwall in the following century that Robert seized the municipal charter chest to 'burn and destroy all the said town's papers'. It may well be true: a free royal burgh, alert to its history and legal rights, represented an obstacle to untrammelled authority.

Like dictators of modern times, Robert used the law as an agent of oppression. His favoured instrument was 'escheat' – forfeiture of land upon conviction of a crime. In Scotland, the crown was the usual beneficiary, but Robert claimed to be 'freely infeft with all escheats of Orkney and Shetland'. There were confiscations for murder and theft, as well as for suicide and witchcraft, and the questionable 'crime' of leaving the islands without permission. Later historians of Orkney, primed to detect the imposition of an alien Scottish tyranny, decried trumped-up charges against freeborn udallers, so lands could be let out again on feudal terms. It may not have been a systematic policy, but interference with udal property was becoming increasingly common.[4]

Robert's rule did not go unchallenged. His attempts to wrest control of Noltland Castle from Gilbert Balfour led to complaints before the privy council. Patrick Bellenden and Adam Bothwell also remained stubbornly litigious. The political climate became less congenial for Robert after the assassination in 1570 of his brother, the earl of Moray; the replacement regent, James Douglas, earl of Morton, did not wish him well. In the summer of 1575, hoping to make new alliances, Robert decided to come south, and was promptly detained in Edinburgh Castle. By the end of the year, the privy council had in its possession a 'Complaint anent usurpations by Lord Robert Stewart … upon the poor inhabitants of Orkney and Shetland'.

The unsigned document was compiled by Robert's enemies, Patrick Bellenden, Magnus Halcro and William Moodie among them. But it paints a credible picture of an aggressive outsider's progressive violations of the balances and accommodations underpinning island life: bans on using ferries to the south without licence; night-time arrests and imprisonment without charge; confiscations of property and banishment; collusion with pirates, and bringing into Orkney 'highland men and broken men … auld enemies and oppressors of the people'.

Robert's financial exactions were alien to 'the auld order of the country'. He claimed exclusive rights to shoreline wrack, and appropriated 'common moors and pastures'. He removed the customary right of udal landowners to have local people grind grain at their mills. An old concern resurfaced. Robert was accused of tampering with weights and measures by altering the bismar and pundlar (p. 81). Since rents were paid in kind, this was serious fraud.

Moreover, he had introduced 'new laws and consuetudes [customs], forged of the laws of Norway'. The examples given hardly seem egregious – one law was altered to allow women to retain possession of a 'bu', or head farm, in the absence of male heirs, and another to require compensation when wandering pigs rooted up a neighbour's crops. But the petitioners – playing on anxieties about Orkney as a legal and political borderland – wanted to stress how Robert, 'but ane simple sheriff', 'alleges himself to be as free lord and heritor in Orkney and Shetland as the King of Scotland is in his own realm'. It was the same charge Lord Sinclair had levelled against James Sinclair in 1529 – that of thinking himself a king in the north.

The first item in the Complaint's list of articles was the most serious, a tangled tale of treason. In 1572, fearing punishment by 'some righteous regent', Robert secretly sent the master of his household, Gavin Elphinstone, to Denmark, with a signed commission under his great seal. He was to offer Frederick II 'supremacy and dominion of the countries of Orkney and Shetland ... upon conditions specified' – presumably Robert's appointment as earl. Frederick's written agreement was conveyed to the islands by a Bremen merchant, hidden in a bolt of cloth. The Danish king dispatched a lawman, Laurence Carnes, to Orkney and Robert let him in.

If true, this was Frederick's second unsolicited offer of Orkney and Shetland in four years (p. 129). The allegation was impressively precise, with names and circumstantial details, and other evidence lends it plausibility. Gavin Elphinstone – perhaps a relative of Robert's mother, Euphemia Elphinstone – appears independently in the Danish archives: as an envoy to Frederick II from the earl of Moray in 1568, and as party to a 1570 quarrel among expatriate Scots. He was perhaps recruited by Robert for his diplomatic experience.

There is, moreover, a copy of a reply to Robert's offer, dated 23 January 1574. It was signed by officials, rather than Frederick himself, and is studiously non-committal, expressing a willingness to discuss the status of Orkney while casting doubt on Elphinstone's credentials. The appointment of a lawman, however, suggests serious Danish interest. No other references can be found to a Laurence Carnes, but it sounds like an Orkney name (Carness is just north-east of Kirkwall), and he may have been an Orcadian living in Denmark or Norway. The possibility of a first approach from the Danish side cannot be ruled out.

Robert Stewart, great-grandson of the king who took Orkney and Shetland from Denmark, was willing to hand them back to secure his power there. The historian Peter Anderson suggests a connection with a report received by William Cecil in June 1572. Patrick Bellenden, 'who long hath envied the Lord Robert … for dispossessing him of somewhat he enjoyed in Orkney, being supported by the earl of Caithness who greatly assists him, is now for revenge prepared to essay the same'. Threatened with a replay of the 1529 invasion, Robert was gathering his forces.

No treason trial took place in 1576: either the evidence was judged insufficient, or the political risks too great. Robert was kept prisoner in Edinburgh Castle for two years, and in the royal residence at Linlithgow west of Edinburgh for six months after that. In the meantime, the privy council decreed that inhabitants of Orkney and Shetland be allowed to travel without hindrance on the ferries, to seek redress in courts of law and to pursue 'other their lawful errands'.

The Complaint's authors wanted the regent to come north, or send commissioners 'to take trial and inquisition' of many more grievances waiting to be heard. No inquiry was established for Orkney, where the Complaint originated – or at least the records of one do not survive. In November 1576, however, William Moodie of Breckness and William Henderson, a Scots laird who had come to Orkney as estate manager for Sir John Bellenden, were commissioned to hear the complaints of Shetland, where administration lay in the grasping hands of Robert's half-brother, Laurence Bruce of Cultmalindie.

Parish by parish, Shetlanders turned out and begged the visitors 'for God's sake' to grant them justice. They were robbed and oppressed by false measures and excessive skats; their locally elected lawrightmen displaced by baillies 'of the laird's own inputting'.[5] The depositions are a vivid portrait of local communities roused to action by threats to their traditional way of life. For Orkney, however, as their superiors loudly schemed and squabbled, it is harder for us to hear the voices of ordinary people.

A window into their lives and anxieties is nonetheless briefly opened for us in June 1577, as Lord Robert languished under ward in Edinburgh. The three ships of Captain Martin Frobisher, engaged on his second attempt to discover the North-West Passage through the Arctic, and open trade routes to the Indies, stopped in Orkney to take on fresh water. Dionyse Settle, one of several gentlemen accompanying the expedition, wrote and published an account of the voyage, and of the brief sojourn at islands 'subject and adjacent to Scotland'. When a party from the ships came ashore, 'the people fled from their poor cottages, with shrieks and alarums, to warn their neighbours of enemies'. With 'gentle persuasion', the Englishmen persuaded them they were not pirates, and to return to their homes.

In Settle's description of those homes, we encounter a cultural landscape alien and perplexing to a cultivated Elizabethan gentleman:

> Their houses are very simply builded with pebble stone, without any chimneys, the fire being made in the midst thereof. The goodman, wife, children, and other of their family, eat and sleep on the one side of the house, and their cattle on the other, very beastly and rudely, in respect of civility. They are destitute of wood, their fire is turfs and cow shards. They have corn, barley, and oats, with which they pay their king's rent, to the maintenance of his house. They take great quantity of fish which they dry in the wind and sun. They dress their meat very filthily, and eat it without salt. Their apparel is after the rudest sort of Scotland.

A second account of the voyage was published by George Best, lieutenant of Frobisher's flagship. The fleet came into 'St Magnus Sound', a name not in use locally, but seemingly referring to an east–west route through the islands, favoured by English fishermen of the Icelandic fleet. Most likely, they anchored in Deer Sound and came ashore in Deerness in the East Mainland, since Best reported Kirkwall lying to the west and referred intriguingly to a 'mine of silver' – Jo. Ben. mentioned a mine in sixteenth-century Deerness. Once the locals were made to understand the visitors' intentions, 'after their poor manner they friendly entreated us'. Best discovered 'they have great want of leather, and desire our old shoes, apparel, and old ropes (before money) for their victuals'.

Settle's narrative was a European bestseller, with translations into French, German and Latin. In this account, Orkney's inhabitants seem scarcely less exotic than the Inuit hunters Frobisher's men would encounter on Baffin Island – people with whom they likewise began to trade, then deviously proceeded to kidnap.[6]

The 'otherness' of Orkney was useful for affirming a variety of 'civilised' identities. Robert Sempill, soldier and balladeer, was a survivor of the notorious massacre of Protestants in Paris in 1572. He thought the events of St Bartholomew's Day so infamous that the like would scarce be attempted 'in isles nor in Orkney'. A better poet, the courtier William Fowler, found himself banished to Orkney in around 1587, cast by fate and fortune 'upon the utmost corners of the world ... the borders of this massive round'.

Still more revealing is a sermon on the sacrament of the Lord's Supper, preached in Edinburgh in 1589 by Robert Bruce, minister of St Giles' Kirk. Christ's body, Bruce assured his listeners, was something all had title of ownership to, even though it was located at the Father's right hand, in faraway heaven. He invited his audience to consider this: 'if any of you have a piece of land lying in the furthest part of Orkney; if ye have a good title to it, the distance of the place cannot hurt your title'. The strained metaphor may not have resolved ambiguities about Calvinist Eucharistic teaching. But it captured perfectly the paradoxical place of Orkney in the perception of educated Scotland – a site of proverbial distance and isolation, but also a place where well-to-do Scots could imagine themselves with investments to safeguard.[7]

Earl of Orkney, Knight of Birsay

At the start of 1578, as the earl of Morton's stock fell, Lord Robert emerged from prison, basking in the favour of his young nephew, James VI, and he was back in Orkney by November 1579. A year later, Morton was arrested, and in June 1581 executed, for complicity in Darnley's murder. Robert's star now rose higher still: in a ceremony at Holyrood

Palace, afterwards ratified by parliament, the king dubbed him 'earl of Orkney, lord of Shetland, knight of Birsay'.

In re-establishing the earldom, the act declared it was necessary to have 'the power of a great and noble person' resident in Orkney, given 'how far the said lands lie distant from the remainder parts of this realm ... and that our said sovereign and his predecessors have had seldom resort thereto'. It was an inversion of the fears about quasi-independent lordship invoked in the 1575 Complaint. A strong hand was needed to counter the people's oppression by 'highland and broken men ... and by pirates and strangers'. The act did not say that these were oppressions at which Robert himself was widely believed to have connived.[8]

In Orkney, Robert's strong grip reasserted itself, while his fraternal henchman Laurence Bruce resumed his rule in Shetland. The deaths of two other brothers, meanwhile, left Robert as James V's last surviving son. A younger, confusingly also called Robert, passed away in January 1581. The strangely obscure Lord Adam Stewart died in June 1575. He followed, or perhaps preceded, his brother to Orkney, for he was buried in St Magnus Cathedral – a tombstone emblazoned with the royal arms was erected there by his daughter, 'domina de Halcro'. This 'lady of Halcro' was Barbara Stewart, wife of Henry Halcro, a grandson of Provost Malcolm, and nephew of the inveterate plotter Magnus. A king's granddaughter, even from the wrong side of the sheets, was no bad catch for the udaller Halcro lairds of South Ronaldsay.

Robert had a brood of illegitimate children of his own, as well as three daughters and five sons with his wife, Jean Kennedy. The countess set foot in Orkney as little as possible, returning from one of only two known visits there 'very sick upon the sea'. The boys were educated in the south but groomed for power in the north. The eldest, Henry, was 'Master of Orkney' – the title formally bestowed on heirs-presumptive to a Scottish peerage. It passed to his brother Patrick on Henry's sudden and seemingly not much lamented death around Christmas 1585. By then, his father suspected Henry of plotting against him with the earl of Caithness. Robert's closest relationships, it seems, were with his illegitimate sons James Stewart of Graemsay and William Stewart of Egilsay.[9]

Once back in Orkney, the earl intensified his efforts to reduce the independent udal proprietors. Scores were settled with the widow and heirs of Magnus Halcro. In November 1584, an assize packed with Robert's supporters confiscated the Rousay estate of Brough, on grounds of non-payment of skat. In the East Mainland, Magnus, Gilbert and Edward Irving were stripped of their rights to the lands of Sebay, in favour of their brother William, an adherent of the earl.

Robert was skilled at turning agencies of local justice into tools of inventive oppression. A traditional task of the courts of perambulation and ogang (p. 60) was to reallocate rigs (strips of cultivated soil in fields) within the township and ensure a fair division of agricultural land. Robert's commissioners were ordered to ensure earldom or bishopric lands were not 'any way hurt by the udal men'; they were to be equal 'in yearly rent and other commodities as the udal land in all places'.

Fair dealing was not the aim. Courts of perambulation documented the patches of hill land that farmers had brought into the cultivation of the township. In Orkney, they were called 'quoys' and 'outbreaks'; townships had for centuries grown incrementally like this, as evidenced by the many places mentioned in this book containing 'quoy' as part of the name. Robert, however, began confiscating quoylands – probably relying on a provision in old Norse law that new settlers on common land became tenants of the king. In March 1584, Magnus Sinclair of Toab was stripped of his holdings in the East Mainland for benefiting from 'his father's vice for stealing and gripping the king's lands'.

Rising protests caused Robert to row back. In September 1587, all quoylands seized from 'the gentlemen udallers' were returned into their hands. It was less of an about-face than it appeared. Robert proposed giving back the lands on a feudal rather than udal basis: beneficiaries would hold them from the earl as 'vassals', and forfeit them if they failed in 'dutiful and true service'.[10]

Robert's eagerness to pose as a just ruler was a sign of his weakened position. In August 1586, the French ambassador reported people saying that James VI disliked his uncle, believing 'he only serves his own ends'. The king, emerging into adulthood in the mid-1580s, had a new political priority, and one for which he feared the earl of Orkney might prove a liability. He intended to marry the king of Denmark's daughter.

Frederick II hoped the union would secure the return of Orkney and Shetland. In May 1585, he wrote to Elizabeth of England to let her know he was sending an embassy to Scotland for the recovery of islands 'acknowledged indisputably to form part of our kingdom'. Danish ambassadors arrived at Leith in June, bringing with them the redemption money and a scholar of law, Dr Nicolaus Philosophus, who treated the Scottish court to a long, learned historical disquisition on Denmark's rightful claims.

The Scottish government did what it had done on previous occasions when the question of Orkney arose: prevaricate and obfuscate. James explained that an outbreak of plague in Edinburgh meant he could not consult his records, but he would make a formal reply in due course. Yet Scottish ambassadors who visited Denmark in 1586 and 1587 were unsure what to say if the Danish king 'will have an answer anent [concerning] Orkney before he enter on any other treaty'.

Frederick's death in April 1588, and the accession of the eleven-year-old Christian IV, strengthened James's position. The final round of negotiations opened in 1589 with a Scottish demand for the Danes to give up title to Orkney. There was little prospect of that, but a compromise – or fudge – proved possible. Both claims would remain open, with the king of Scots staying, for now, in peaceful possession of the islands. In August 1589, the bride, Anne, was duly dispatched, but storms drove her fleet into harbour in Norway. In an uncharacteristically gallant gesture, James crossed the North Sea to find her, and on 23 November the couple married in Oslo.[11]

In all this, the earl of Orkney played conspicuously little role. Robert's star fell rapidly through 1587, the year James attained his full majority, and of his mother's execution in England. It was perhaps now, as he reflected on kingly prerogatives, and on his prospects of succeeding to the English throne, that James VI was made aware of those reckless words above the gateway of Robert's palace in Birsay.

In July 1587, parliament decreed Robert had exceeded his authority in bestowing church offices in Orkney; soon after, James confiscated the earldom lands, and granted them to Lord Chancellor John Maitland and Sir Lewis Bellenden, successor to his late father in the office of justice clerk. In December, the king commissioned them, with Bellenden's uncle Patrick, to investigate oppressions by the 'late earl of Orkney'.

It was decided to dislodge him, by force if necessary. Under the command of Patrick Bellenden, three ships were kitted out at Leith. The expedition was furnished with royal letters demanding Robert hand over his palaces and the cathedral and castle in Kirkwall. Robert, meanwhile, sent money south to prepare ships of his own, under pretence of needing to protect fishermen from pirates.[12]

In the event, Bellenden's fleet did not sail in 1588, though a greater one of course did. In the early summer, Philip II of Spain launched his 'invincible' Armada against England, to avenge the death of Mary Queen of Scots and to cauterise English support for Protestant rebels in the Netherlands. James VI assured Elizabeth of his friendship, but did little to restrain anti-English and pro-Catholic nobles sympathetic to the Spanish cause.

It was to these men that Robert turned as his estrangement from the court deepened. Helping to organise ships for Orkney's defence against Bellenden was the High Admiral, Francis Stewart, earl of Bothwell, a nephew to both Lord Robert and the late unlamented duke of Orkney. Another ally was George Gordon, earl of Huntly, with whom Robert entered a bond of alliance in December 1587. The Spanish ambassador to France placed both Huntly and Orkney on a list of nobles friendly to Spain.

In August 1588, the defeated Armada began its demoralised journey home, taking a northern route between Orkney and Shetland. The fleet's lead supply ship, *El Gran Grifon*, was wrecked off Fair Isle. Its crew spent an uncomfortable six weeks on the island, losing fifty men to starvation. Whether the unwilling visitors left any permanent legacy is doubtful, though the possibility permits the conceit of a Fair Isle detective named Jimmy Perez in Ann Cleeves's popular series of Shetland crime novels. Traditions about shipwrecked Spaniards remaining and intermarrying with locals are in fact stronger in Orkney than in either Fair Isle or Shetland. They were championed by the Victorian folklorist Walter Traill Dennison, a collector of stories about the 'Westray Dons'. These were the supposed descendants of a boatload of Spanish sailors whose ship foundered in the Roust – a treacherous stretch of clashing currents on the upper side of North Ronaldsay. The survivors were welcomed in Westray, where their offspring formed a tight knot of intermarrying families, skilled seamen, with fiery temperaments. In the view of Traill Dennison – voicing the conventional racial theorising of his day – 'the union of Spanish blood with the Norse produced a race of men active and daring'.[13]

The best to be said for these traditions is they cannot positively be disproved. Had a boatload of Spaniards recently disembarked in Westray, we might expect Jo. Ben. to have heard about it, if, as seems likely, he was writing in the early 1590s. Ongoing research by the geneticist Professor Jim Wilson has not yet found any DNA evidence to support the idea. Still, the legend of the Westray Dons underlines a phenomenon Jo. Ben. encountered frequently in Orkney: the willingness of islanders to embrace exotic stories about themselves and their origins, stories that rooted them in place, while asserting connection with worlds beyond their shores.

Earl Robert's association with the pro-Spanish faction was opportunistic not ideological, and he did little to assist the Armada through the waters of his earldom. Indeed, at the start of 1590, the English ambassador thought that Francis Stewart, earl of Bothwell, had turned against his uncle Robert, and was planning to seize Orkney, in order to help the Spaniards 'find harbour and refuge in any distress'. In an era of unprecedented maritime warfare, the strategic value of the Northern Isles was considerable.

That value was recognised by William Sempill, an exiled Catholic soldier who advised the Spanish monarchy on Scottish affairs. In 1591, Sempill presented Philip II with a memorandum on 'ways to establish the Catholic religion in Scotland'. He suggested sending 3,000 men to Orkney, a place 'fertile and abundant of all the necessary things for the sustenance of the aforesaid number', which 'could be made in a short time impenetrable'. The islands would provide the bridgehead for an invasion of mainland Scotland, which Sempill expected the earl of Caithness to support.[14]

The plan was not adopted. Nonetheless, at the start of the 1590s, Orkney became a flashpoint for relations between Scotland, England and Spain. Despite his reticence during the 1588 campaign, Robert became entangled in the affair of 'the Spanish barque', a vessel hanging around Scotland's waters, picking up Armada survivors and preying on English shipping. It anchored at Cairston (Stromness) in 1589, and in the following summer made Kirkwall its base for an aggressive campaign of patriotic piracy.

The English ambassador dispatched agents to Orkney, and sent rolling reports to William Cecil. Earl Robert was feasting with the captain and crew, while the Spaniards boasted about the number of Englishmen they had captured and killed. Robert's illegitimate son William joined the action. He was with the Spaniards on 19 July 1590 when they seized four English fishing boats off Fair Isle. The ships were brought in to Kirkwall, and sold to interested parties, though the Spaniards gave one to the earl in exchange for some artillery pieces. There was an element of payback in this: in June 1590, Robert's son Patrick, Master of Orkney, was robbed by English pirates on his way to the Scottish court. They took, he claimed, money, jewels and other things to the value of £36,000 Scots – an eye-watering sum.

We can only guess at what the burghers of Protestant Kirkwall thought of Captain Alvarez de Merida and his Spanish papist crewmen strutting around their little town. They were evidently not shunned. At the request of the Master of Orkney, Gilbert Foulzie, retired minister of St Magnus, redeemed a fishing boat from the Spaniards for a payment of £50 sterling (about £500 Scots).[15] It was not the last time an Orkney clergyman would do Patrick's unquestioned bidding.

Black Patie

Robert Stewart died in February 1593, in the episcopal residence in Kirkwall that by the later sixteenth century had come to be known as the Palace of the Yards (i.e. courtyards). His will left gifts to a gaggle of largely Scots servants and to three of his mistresses. He bequeathed nothing to his wife, then suing him for divorce on grounds of adultery. Power, in practice, had already passed to Patrick. A royal charter, ratified by parliament in June 1592, confirmed Robert's possession of the earldom lands, though only 'in liferent'; they were by the same charter granted 'in fee' to Patrick and his heirs. King James thought well of his cousin Patrick, in 1591 naming him lord of Shetland. Even before his father's death, he was being referred to as earl of Orkney, though it would take time for the honour to be formally confirmed.

Patrick Stewart possessed a level of self-confidence unperturbed by uncertainty over titles, or losses at the hands of pirates. In 1594, he sent an unofficial embassy to the court of Maurice of Nassau, stadholder of the Netherlands. It presented Patrick as a candidate for marriage with Maurice's sister, Emilia – a potential union scarcely less glittering than that of James VI with the king of Denmark's sister.

In the event, Emilia declared herself unwilling to 'dwell so far from her brother and kindred'. William Cecil – who kept a weather eye on the affairs of Orkney – learned that influential voices at the Dutch court had vetoed the proposal on the grounds of Patrick having 'no assurance of the Orcades because they have been claimed by the kings of Denmark'. Copenhagen's assertions were taken seriously in Europe.[16]

In the islands themselves, Patrick's methods proved just as aggressive as those of his father. In 1592, parliament received a petition from disgruntled udal landowners, among them William Irving of Sebay and, in Shetland, Earl Robert's half-brother Laurence Bruce. Patrick seemingly had little use for his uncle, who became, and remained, an inveterate opponent. Patrick, the udallers claimed, planned to confiscate their lands on grounds of 'nonentry' – the failure of an heir, under Scots law, to secure investiture from the estate's feudal superior, something which was irrelevant to udal property.

For parliament's benefit, the complainants explained partition of inheritances in the Northern Isles, the method originating when 'the lands of Orkney and Shetland were under the crown of the kings of Norway'. After a landowner died, a meeting of prospective heirs led to the drafting of a letter of division, 'called in Denmark and Norway ane shownd bill'. In Nordic tradition, the *sjaund* was a funeral feast on the seventh day after death. Inheritance by means of the shownd, parliament was assured, was 'first established in Norway, allowed in Denmark, and embraced and received by an inviolable custom in this kingdom'.

There was something incongruous about an appeal to Scottish lawmakers on the basis of laws derived from another kingdom and applying nowhere else in Scotland. Perhaps conscious of this, the petitioners stressed how successive kings had chosen not to change the property law of Orkney and Shetland.

Patrick's strategy was to acquire udal property piecemeal, consolidating his holdings in places where his father's lands already lay, particularly the old bishopric estates of Birsay and the West Mainland. He also emulated his father's methods. A new rental for Orkney, drawn up in 1595, recorded more than forty instances of escheat, mainly for theft. Some dated from Earl Robert's time, but many were more recent. There was a suspicious number of criminal forfeitures – eleven – in the single district of Marwick, a small coastal section of Birsay. Where the law was rigidly applied, people could not justifiably complain. But even the rental suggests this was not always so. A scribal note on Quoysharps, South Ronaldsay, records that 'the udal men allege ane part of this pennyland to be wrongously taken from them'.[17]

Patrick Stewart had a talent for making enemies. He was at loggerheads over lands in the South Isles with his brother John, Master of Orkney. Patrick imprisoned him, along with their illegitimate half-brother James Stewart of Graemsay, and pulled down James's house. A financial settlement was reached in 1595, but the feud was already at boiling point.

In June 1596, John was accused of plotting Earl Patrick's murder, and put on trial before the Edinburgh court of justiciary. With his half-siblings James and William, and his servant Thomas Paplay, John was supposed to have devised various schemes for doing away with his brother – killing him at a banquet in Kirkwall, or creeping at night into his chamber in the palace at Birsay to murder him in his bed. Paplay allegedly had orders to poison Patrick. Most shockingly, the indictment accused John of seeking to bring about his brother's destruction by 'consulting' with 'ane known notorious witch'.

There had probably been rash talk, and perhaps a plot, but the specific allegations against John were extracted with unspeakable tortures. Paplay was arrested first and, over the course of a fortnight, flogged with ropes that 'left neither flesh nor hide'. He endured eleven days in the 'caschielawis' – a metal frame clamped over the leg and heated with fire. In his agonies, Paplay gave up the name of Alison Balfour, a married woman from Stenness, who was probably a servant of the local laird, Patrick

Bellenden, that long-embedded thorn in the side of the Stewart earls. Balfour was tortured next, and interrogated relentlessly by Patrick's Orkney factotum, the minister of Orphir, Henry Colville. She, too, was subjected to the caschielawis, and her elderly husband, son and seven-year-old daughter were all tortured in front of her: the old man pressed by weights, the boy's foot crushed in a device known as 'the boots', and the little girl's fingers broken in the 'pilliewinkies', a kind of thumbscrew. Small wonder that she eventually confessed to witchcraft, and agreed to implicate Bellenden. Alison Balfour was burned at Kirkwall on 16 December 1594; Paplay had been hanged already.

Yet even as its threads tightened, the case started to unravel. Before her execution, Balfour dictated a statement before a notary. She swore, 'as she would answer at the Day of Judgement', that she was as innocent of witchcraft 'as ane bairn new born'. The last words of the condemned carried a presumption of truth. Balfour swore 'she knew nothing to the auld Laird of Stenness but honesty'; a piece of wax, which Colville trumpeted as an instrument of devilry, was given to her by Bellenden's wife as a plaster against the colic. John Stewart was acquitted, successfully arguing that the charges against him were implausible because they had been extracted under torture. It probably helped that Patrick's enemy Laurence Bruce was a member of the jury. The earl could manipulate courts in Orkney, but not so easily in Edinburgh.

A reckoning swiftly followed. On 12 July 1596, assassins caught up with Henry Colville in Shetland. Several people were tried for the murder, but only two, one a servant of John's, were convicted and executed. John himself was suspected of complicity, and Colville's brother sought to implicate Bellenden, but those charges were dropped. There are no compelling grounds for crediting a later tradition that the killers ripped out Colville's heart and drank his blood. But the Balfour–Colville episode shows the simmering potential for violence in Orcadian politics, and the ruthlessness of which Patrick's opponents as well as supporters were capable. More than likely, the Master of Orkney did express a wish to be rid of a turbulent priest.[18]

In June 1597, feuding in Orkney and Shetland moved the privy council to impose on the principal landowners there a 'General Band' – a legal device, favoured by the Scottish crown, whereby notables pledged to keep the peace, or forfeit specified sums. The amounts recorded for Orkney – reflections of an individual's estimated worth – reveal the extent to which social and economic power had swung away from the traditional landowning families, including cadet branches of the Sinclairs, and towards the beneficiaries of Bishop Bothwell's feus and the favour of the Stewart earls.

Patrick objected to the General Band in principle, as inappropriate for noblemen and gentlemen living under the obedience of His Majesty's laws in 'ane civil country' like Orkney. The system, Patrick protested, was devised 'for the chiefs of clans ... who have their livings and lands lying in the borders or highlands of this realm'. Orkney was not that sort of place.

Except, of course, that sometimes it was, and Patrick helped make it so. Barely three months after the Band was imposed, he invaded Westray with sixty armed men, in pursuance of debts owed by Michael Balfour and his son Andrew, heirs of Gilbert, the builder of Noltland Castle. The place scarcely lived up to its impregnable reputation, for the earl's men ransacked and stripped it, and over the following months Patrick held courts and collected rents and taxes from the Westray tenants. The Balfours' rights to 'wrack and waith' were usurped to allow Patrick to claim profits – estimated at over £3,000 Scots – from a pod of twenty-nine whales, beached fortuitously on the island's coast.

Patrick's actions had a faint colour of legality, but their execution was brutally heavy-handed. Andrew Balfour was threatened with hanging from his own roof beams, and Patrick issued a proclamation to declare that anyone hearing 'evil report' of him (Patrick), and failing to reveal it within twenty-four hours, would forfeit life, lands and goods. The oppressions, Andrew Balfour's petition to the privy council averred, were of a kind not 'heard of in any reformed country subject to a Christian prince'.[19]

In centuries to come, Earl Patrick, 'Black Patie', acquired an almost mythic status in Orcadian folk memory. Even more than his unlamented father, he became a symbol of coercion, cruelty and tyranny – a tyranny readily characterised as 'Scots' by nineteenth- and twentieth-century Orcadian writers, inclined to see the Norse era as a lost 'golden age'.[20]

It is a caricature, but not a travesty. Patrick had no interest in destroying the inherited culture, law or institutions of Orkney as a matter of patriotic principle. But there is much evidence to frame a portrait of a man who, even by the generous standards of the Renaissance Scots nobility, was exceptionally ruthless, unscrupulous, ambitious, spendthrift, grasping and vainglorious. In April 1599, the privy council received yet more complaints from Orkney and Shetland itemising 'many heinous points of oppression and cruelty'. Patrick was ordered to appear, but simply ignored the summons. In 1597, he had indignantly asserted the islands' civilised and peaceable character; now he insisted they required full-time personal supervision, remaining in the north, as the English ambassador put it, 'upon pretence to keep the countries of Orkney and Shetland'.

In the end, Patrick had no choice but to come south, settle with his brothers and other creditors, and attend upon the king. All, it seemed, was readily forgiven. In the spring of 1600, royal charters confirmed him in possession of the earldom and the bishopric estates, with rights of ecclesiastical patronage and full powers of admiralty – jurisdiction over maritime affairs, and claim to shipwrecks 'and whatever is cast up by the sea'. Underpinning the wide-ranging grants was a recognition of the need to defend the islands 'from hostile foreign invasions'.[21]

Despite entrenched local opposition, and his own politically reckless behaviour, Patrick seemed at the turn of the new century to be standing strong. In 1601, he compiled the 'Uthel Buik', a listing of Orkney's udal landholders, with phraseology casting them as feudal tenants. At around the same time, he began a major work of construction, a project to outshine his father's edifice in Birsay. The 'New Work of the Yards' was to be an opulent, imposing residence in the heart of Kirkwall, a structure incorporating the existing episcopal palace as the secondary wing of an extensive palatial complex. On completion, it was one of the finest secular buildings in Scotland, its centrepiece a first-floor great hall, with two magnificent fireplaces, and fine French-style oriel or bay windows. Even today, in its state of ruin and decay, the palace feels both sensually overwhelming and oddly out of place in a small market town like Kirkwall.

The architectural elegance was a mask for menace and coercion. Like Patrick's contemporaneously built castle at Scalloway in Shetland, the palace was erected with forced labour. The earl's opulence was a dialect of domination. An anonymous contemporary *Historie of King James the Sext* recalled how, when Earl Patrick sat to dinner, three trumpeters blew fanfares between each course. His 'pomp was so great' he never ventured abroad, even to cross the street to the cathedral, without an escort of fifty musketeers.

The writer was not an admirer: Patrick's lust for power and wealth meant 'no man of rent or purse might enjoy his property in Orkney without his special favour, and the same dear bought'. But where later commentators were inclined to see the earl's 'tyranny' as an alien Scots imposition, the *Historie* took a different view. Patrick's harsh rule thrived under 'the shadow of Danish laws … more rigorous than the municipal or criminal laws of the rest of Scotland'.[22]

The anomaly of Norse jurisprudence in Orkney and Shetland was 130 years old in 1600, and four Scottish kings, several regents and a mercurial queen had lived with it contentedly enough. Yet, in the first years of the seventeenth century, momentous realignments were in motion, over-mighty subjects were coming under unprecedented scrutiny and the days of 'Danish laws' were drawing to a close.

Imaginary Bounds

In March 1603, Orkney's place in the world shifted. Kings of the Oldenburg dynasty continued to assert their sovereignty, but Danish leverage was weakened by a profound political reconfiguration in the British Isles. On the death of Elizabeth I, in default of other plausible Protestant candidates, James VI of Scotland became James I of England. A report from the Venetian ambassador, a couple of years after the accession, noted some tension between James and Christian IV over the status of Orkney, 'but Denmark will not press the point, being well aware that she would gain nothing and would lose the friendship of England'.

For centuries, Orkney had been a frontier zone, not just between Scotland and Scandinavia, but also with a frequently hostile England. In times of both war and nominal peace, English mariners harassed Orcadians, on land and sea, with distressing regularity. After 1603,

Scotland and England were, at least in theory, perpetual friends and allies. Unlike Elizabeth, James made no claim to be 'Emperor from the Orkney Isles to the Mountains Pyrenée', but his regnal title, set out in a proclamation of 1604, spoke a no less grandiose language of political geography. He intended 'to discontinue the divided names of England and Scotland', and adopt 'the name and style of King of Great Britain'.

The idea was slow to take hold among James's subjects, English or Scots, who retained their own parliaments and laws and rarely if ever thought of themselves as 'British'. James, however, saw himself as ruling an island realm that 'hath almost none but imaginary bounds of separation'. It had 'but one common limit or rather guard of the ocean sea, making the whole a little world'.

There was no recognition here that Great Britain was only one – if by far the largest – of the islands comprising the Stewart inheritance in the North Atlantic. James's feelings about the others were not immediately clear. But the king's eloquent fervour for unity implied that things might not carry on just as before in the kingdom's island fringes. In April 1605, a well-wisher in Edinburgh wrote to Earl Patrick to report complaints there against him, and to warn him that, since James acceded to the English throne, 'his Highness is of another kind of disposition than he has been in Scotland'.

The Edinburgh jurist Thomas Craig, an enthusiast for the king's vision of a unified Britain, argued in a treatise of 1605 that common legal systems were not essential to a successful union, albeit he sensed a harmony of principle and purpose sounding through Scots and English law. Craig considered it bizarre, however, that in Orkney 'Norwegian law has scarcely lost its force'. The Orcadians' enthusiasm for it was surprising, as under Norwegian law 'the least offence is punishable by total loss of property'.[23]

A more glaring disparity between England and Scotland concerned religion. The Church of England, though Protestant in theology, retained an old Catholic system of episcopal government. In the Kirk of Scotland, bishops had but a shadowy half-life, awkwardly assimilated to an egalitarian pattern of local and regional meetings of ministers in presbyteries, synods and General Assemblies. Many Presbyterians were eager to erase their presence entirely.

On both sides of the border, the union sparked aspirations and alarms. English 'Puritans' hoped that a king raised in the reformed rigour of the Kirk might prove willing to jettison old-fashioned ceremonies. Scots Presbyterians shared the ambition. Among writings inspired by James's accession was a Latin treatise, *De Unione Britanniae*, by the distinguished minister and jurist Robert Pont, a six-times moderator of the General Assembly. Pont was a rare Scottish enthusiast for the novel notion of Britishness, someone looking forward to a time when a Scot 'will not be known from an Englishman', with both united in their commitment to 'true religion'. Minor differences could be tolerated, but on issues of substance it was the English who needed to change. A time would come, Pont hoped, when 'he that inhabits the utmost borders of the Orcades', travelling thence into any part of the kingdom, 'is notwithstanding in his own country' – a revealing confirmation of the significance of peripheries at times of crisis and consolidation.[24]

Pont was an eternal optimist. Other Presbyterians worried that a monarch beguiled by notions of conformity might want to bring 'anglican' practices into his northern kingdom. They were right to be concerned. At the Hampton Court Conference of January 1604, where Puritan representatives met with church leaders to discuss contested issues, James confirmed his commitment to episcopal government, and famously articulated an intrinsic connection to his own royal authority: 'no bishop, no king'.

Orkney had been formally a 'no bishop' place since Adam Bothwell's death in 1593, and in practice considerably longer. Bothwell did not visit the islands in the last quarter of his life, his residual interests there largely financial. In the meantime, new church structures were devised for Orkney, and for Scotland as a whole. In 1581, the General Assembly divided the Kirk into presbyteries – decision-making regional meetings of ministers. A single presbytery was decreed for Orkney, and another for Shetland. The presbytery scheme was blocked by the government in 1584, and reinstated in 1586, when thirty-eight churches were listed as belonging to the Orkney Presbytery – an optimistic number, as no more than thirteen ministers were then serving in the islands.

It may have been several years before the Presbytery of Orkney first met. In the meantime, administration was in the hands of General

Assembly commissioners. Bishop Bothwell once held this position, and it was later exercised by his protégés Gilbert Foulzie and James Annand. In 1585, Thomas Swenton became commissioner and archdeacon of Orkney, on his appointment as minister of Kirkwall. Swenton, like Colville of Orphir, was an adherent of Earl Patrick and, like most of the clergy, a Scots incomer. But a sprinkling of Orcadians was still to be found.

Ninian Halcro, minister of South Ronaldsay, was in his life and lineage a bridge between the before and after of the Reformation. Like his combative brother Magnus, he was an illegitimate son of Provost Malcolm Halcro, an office Ninian himself held between 1590 and his death in 1608. James Cok, minister of Lady Kirk in Sanday, came from an old-established island family living at Runna Clett, at the north end of the parish of Burness. Sanday preserved a memorable tradition about the wealth of the Coks: their cows, returning from pasture in single file, formed a procession two miles long.[25]

The General Assembly, however, did not trust the clergy of Orkney to roam and graze unsupervised. In 1587, it put the islands on a list of troublesome Highland and northern regions where 'Jesuits and papists chiefly resort'. George Monro, chancellor of Ross, was charged to visit 'the bounds of Orkney', with instructions to depose unqualified ministers, secure subscriptions to the 1560 Confession of Faith and do all necessary for 'good discipline of the Kirk'.

There is no evidence of Jesuits, the shock troops of Rome, coming to Orkney in these years. But the islands did not entirely drop from the radar of the Counter-Reformation. In 1605, the Jesuit Superior in Ireland was urged by Rome to send priests to Orkney, but admitted he lacked the resources to do it. Meanwhile, a report on the state of Scotland, prepared in 1602 for the order's Superior General by the Scots Jesuit Robert Abercrombie, offered a scathing assessment of Earl Patrick: a man of great gifts who squandered his talents; his life so debauched and dissolute that neither Protestants nor Catholics wanted him in their Church.[26]

The earl of Orkney had little interest in Calvinist reform for its own sake, and neither did some of the ministers he promoted. Thomas Swenton began life as a reader (p. 131) in Perth, under a minister known to be sympathetic to bishops. Henry Colville owned books by the

German Lutheran Johannes Brenz, a critic of Calvinist theology. At the end of the nineteenth century, these were in the possession of the antiquarian James Craven, who examined Colville's marginal annotations. A manicule (pointing finger) was inserted next to a passage arguing for the lawfulness of pilgrimage, if undertaken to encourage faith in Christ. It was a view no self-respecting Calvinist could ever have endorsed.

Gilbert Body, freshly minted in 1590 as minister of Holm, is an especially intriguing figure. A man of the same name enrolled in 1583 as a student in the English Catholic college at Douai in the Spanish Netherlands, and left abruptly in 1585. The identification cannot be definitively verified, but a possible Catholic past did not incline Body to shun the limelight as a Protestant minister. He represented Orkney at the General Assembly in 1598, when a vote was taken on whether ministers who were also bishops should have seats in parliament. Presbyterians feared – with reason – that this was part of a royal plan to restore episcopacy in a fuller form. James Melville, nephew of the prominent Presbyterian theologian Andrew Melville, recorded in his diary that Body, 'a drunken Orkney ass', cast the first vote in favour. The drunkenness might not have been a slander: Body is supposed to have drowned in a loch in 1606, returning from a parish wedding feast.[27]

Efforts to take the island clergy in hand were entrusted to Robert Pont. He served as commissioner between 1590 and 1601, convening a synod at Kirkwall in 1592, when five Shetland ministers were deposed for non-residence. In 1596, he came north again, to lay charges against no fewer than nine Orkney ministers, who were summoned the following year for trial before the Presbytery of Edinburgh.

The accused included Body, Swenton, Ninian Halcro and James Cok, as well as Adam Moodie of Walls, son of the ambitious minister-laird William Moodie. All we know of the charges is they were 'crimes … meriting deprivation'. The case was not pressed, and the ministers remained in post, but Pont took seriously his responsibilities for what an eighteenth-century biographer called 'those remote corners'. A redoubtable defender of Presbyterianism and the Kirk's independence, Pont was not afraid, in 1591, to tell the king himself 'there is a judgment above yours, and that is God's put in the hand of the ministry'.[28] It was the kind of message James did not like to hear and thought he was unlikely to hear from the mouth of a bishop.

A spate of new episcopal appointments followed James's accession to the English throne, and in February 1605, Orkney once again became the seat of a bishop. James Law, minister of Kirkliston in West Lothian, was a royal chaplain who shared the king's vision for the Church, and had little time for the schemes of moral reformation advocated by the heirs of Knox. Critics remembered he was once censured by the Presbytery of Linlithgow for playing football on the Sabbath. They probably didn't know he owned a copy of the Catechism of the Council of Trent, gifted to him by his brother David, a Catholic priest. Law did not immediately head north in 1605, but he began to acquaint himself with the extent of his legal and financial rights.

King James's enthusiasm for episcopacy provoked opposition. Law, and the recently appointed archbishop of Glasgow, John Spottiswoode, were strongly associated with the policy, and in 1605 the Synod of Lothian and Tweeddale accused them of seeking to overthrow the Kirk. That summer, twenty-seven ministers from fourteen presbyteries convened in Aberdeen, in defiance of James's refusal to permit a General Assembly. Six ringleaders were convicted of treason, and expelled from the realm. A further eight, refusing to renounce the assembly, were banished into Caithness, Sutherland, Kintyre, the isles of Arran, Bute and Lewis, and Shetland. James Irving, minister of Tough in Aberdeenshire, was sentenced to confinement in Orkney, but flatly refused to go, and was allowed, under restrictions, to continue ministering in his parish.

In Devon, the Puritan lawyer Walter Yonge heard about these sentences 'for the cause of Aberdeen', recording in his diary how the rebellious ministers were all 'banished into Orcades'. Yonge was further informed that 'the barbarian people of that country', having heard one of the exiles preach, were following him around 'in great troops'.[29] If such an incident happened, it must have been somewhere else. Yet it was Orkney that lodged itself in the Englishman's memory, an exotic locale of stoic suffering and redemptive possibility.

The Coming of Law

For Patrick Stewart, the possibilities of redemption were starting to narrow. Not least due to his lavish building projects, he was in massive financial difficulties, and creditors were circling. Robert Monteith of Egilsay claimed in 1605 that Patrick owed him £40,000 Scots. The earl, according to the *Historie of James the Sext*, 'was drownit in debt'. The largest sums were due to Sir John Arnot, a wealthy merchant who served as treasurer-depute – as well as lender – to the king.

In 1606, Patrick had to mortgage to Arnot the bulk of his Orkney estates, with little immediate prospect of redeeming them. Arnot also brokered an agreement with Bishop Law, who was now to receive 4,000 merks annually from the bishopric revenues, as well as the thirds of the bishopric (pp. 130–1) as stipends for the Orkney ministers. The bitterest pill for Patrick to swallow was loss of the 'New Work of the Yards'. In anticipation of Law's arrival, he had to promise to make the buildings wind- and watertight, and to repair all 'doors, windows, lofts, roofs, slate, glass, iron and timber work'.[30] Patrick's astonishing palace complex, absorbing the episcopal mansion his father wrested from Bishop Bothwell in 1568, was finally completed only to be handed to Bothwell's successor.

Patrick's money problems exacerbated the tensions in Orkney. The cells of Kirkwall Castle filled up with people from whom the earl was trying to extort money, usually as a result of bands or contracts he had made them sign. Powerful enemies retaliated in kind. Michael Balfour, Robert Monteith, Laurence Bruce and others secured letters of 'lawburrows' against Patrick – a provision in Scots law that required sums of money to be posted by persons from whom there was fear of violent harm. Failure to produce the surety meant 'horning' (p. 59). Patrick could prevent this in Kirkwall, but not in the south: Sir John Arnot wrote to say it grieved him 'to hear your Lordship's name bladdit out at the market cross'.

Robert Pont, in his treatise of 1604, pointed an accusing finger at various human obstacles to the project of full British union: papists, border reivers, wild Irishmen and Hebridean mercenaries. They also included 'fierce and insolent governors and petty princes possessing large territories in the places most remote and abandoned of justice'. No names were mentioned, but the profile certainly fits. King and council were increasingly concerned about Earl Patrick, and taking greater notice of the quarrels in the north.

A particularly dangerous antagonist was a neighbouring petty prince: George, earl of Caithness, a great-grandson of the peer slain at Summerdale, and head of a branch of the Sinclair family never fully reconciled to losing the Orkney earldom or seeing it pass into the hands of Stewart interlopers. The privy council feared his quarrel with Earl Patrick would lead to 'trouble and inquieting of the country', and in 1599 it compelled them to find surety to keep the peace. George had already caused special swords and daggers to be made, in anticipation of a duel.

The earls trespassed on each other's turf. Patrick made cause with George's other bitter enemy, John Gordon, earl of Sutherland, sending a warship to the Cromarty Firth in 1602 to bring Gordon north for entertainment at Birsay and Kirkwall. Meanwhile, word reached the English court that 'Caithness is making ensigns and buying an isle in Orkney', actions guaranteed to 'put them by the ears together'.

The isle in question was Eday. Years earlier, Bishop Bothwell had feued it to Edward Sinclair, of the family's Roslin branch. By the turn of the seventeenth century, Edward and his son William were enveloped in debt, which their cousin, the earl of Caithness, agreed to clear in return for recognition of his lordship. To Patrick, this was an intolerable provocation, and in 1604 he sent the captain of Kirkwall Castle to seize the island. When the case came before the council, Patrick claimed to have acted on behalf of the aged Edward Sinclair, a hapless prisoner of his callous son. William had summoned 'vagabonds, broken highland men of Caithness', to snatch his father's lands. Patrick knew how to play the patriot card and evoke memories of Summerdale. 'There yet remaineth experience', he wrote, 'of the countrymen of Caithness their barbarous enterprises.'

Earl George exacted a petty but piquant revenge. In 1608, several of Patrick's servants were forced ashore by storms in Caithness. Before putting them back to sea, the earl filled them with drink and, 'in a mocking jest, he caused shave the one side of their beards, and one side of their heads'. Tampering with beards, a symbol of manhood, was in reality no joke in the early seventeenth century. Patrick was furious at the insult to his honour.[31]

King James had once been fond of his cousin Patrick, the hunting companion of his youth. But there was a growing perception that the earl was more trouble than he was worth. Bishop Law assiduously encouraged it, writing to the king in November 1608 to underline 'the many great and continual complaints of your Majesty's poor distressed subjects in those isles'.

Complaints were becoming not only more frequent but more serious. In 1606, the lord advocate drew up charges against Patrick of lese-majesty – treasonously usurping the king's authority. The accusations went back a decade and more, and included the capture of Noltland Castle, now reimagined as a formal siege 'with iron, fire and war-engines'. There was the execution of Thomas Paplay, and two other men, for the alleged plot to slay Patrick and 'the crime of any thought against him' (Alison Balfour went unmentioned). The most serious allegation was of forging a royal commission of military lieutenancy, and using it to levy troops and secure Kirkwall Castle.

James was content to leave the charges hanging in 1607, once Patrick had agreed to the compact with Bishop Law. His quarrel with the earl of Caithness was privately patched up – according to Robert Gordon, brother of the earl of Sutherland, this was 'lest they should reveal too much of each other's doings'. But, egged on by the bishop, the earl's enemies, Bruce and Monteith, continued to collect evidence against Patrick. The council decided he still had serious questions to answer, and in July 1609, he was confined in Edinburgh Castle.[32] He would never again experience real freedom.

With the earl secure in the south, the bishop finally went north. James Law was now a bishop in the traditional manner. In June 1610, a tame General Assembly approved a parallel system of ecclesiastical jurisdiction, with new courts of high commission under the archbishops of Glasgow and St Andrews. Bishops were to be moderators of synods in their dioceses, and to approve all presentations to parishes. In October, Archbishop Spottiswoode and the bishops of Brechin and Galloway went to England to be consecrated by members of the English episcopate. On their return, they 'laid hands' on their episcopal brethren: the Scottish bishops were now equivalent to English ones, sanctified links in an ancient 'apostolic succession'.[33] It was the antithesis of how Presbyterians like Andrew and James Melville understood the shape and meaning of Christian ministry.

Law arrived in Orkney in the early summer of 1611. On the eve of his departure, he sought assurances about rumours that Patrick was to be released and allowed to return north. Law trusted King James would be 'more careful to see justice and peace established, even in those remotest parts of this kingdom, than to satisfy the earl'. Law, with his fellow commissioner Arnot, had powers as sheriff and justiciar to 'punish all oppressions', and restore in the islands good government long 'interrupted, neglected and overlooked, partly by the iniquity of the bypast times, and partly by the rebellion of Patrick Earl of Orkney'. At a ceremony in Kirkwall on 6 August, the ministers of Orkney surrendered their positions to the bishop, and received them again at his hands.[34]

Law's visitation consummated an apparent constitutional revolution, of significance not just to Orkney but to the wider self-image of Scotland. In March 1611, the privy council issued a decree. It was the will of the king, as of his predecessors and their parliaments, that his subjects 'should live and be governed under the laws and statutes of this realm only, and by no laws of foreign countries'. The council, therefore, abolished 'the said foreign laws, ordaining the same to be no further used within the said countries of Orkney and Shetland'.

For the first time, legally and jurisdictionally, the islands were fully incorporated into the realm. Yet, as so often in the relationship between Scotland and Orkney, things were not quite what they seemed. In the first place, it is surprising such an ostensibly momentous change was enacted by an order of the privy council, rather than by parliamentary statute. Nor was it in fact true that parliament had already insisted upon legal uniformity; as we have seen (p. 129), parliament considered abolishing the laws of Orkney and Shetland in 1567, and decided not to.

What exactly was being abolished is also quite hard to say, as the sixteenth-century Orkney and Shetland lawbooks have not survived. They were evidently derived from Scandinavian models, but diverged from contemporary Norwegian and Danish usage. Commentators often refer to 'the abolition of udal law' in the Northern Isles, but that is not straightforward either. Scots and udal property law had operated alongside each other for a century and more, and there was no rush on anyone's part to challenge udal stipulations, which continued (and continue) to have protected customary status within the Scots legal framework.

The 1611 announcement was in truth a political one. 'Persons bearing power of magistracy' in Orkney and Shetland had, in pursuit of private gain, presumed 'to judge the inhabitants of the said countries by foreign laws, making choice sometimes of foreign laws, and sometimes of the proper laws of this kingdom'.[35] In other words, Norse law was an instrument of the tyranny of Earl Patrick; abolishing it a way of protecting royal subjects from oppression and exploitation. The sincerity or cynicism of this justification is difficult to ascertain.

While the bishop prepared to stamp his authority on Orkney, the earl fretted and fidgeted in southern confinement. In June 1610, a formal indictment was drawn up, based on complaints from his local enemies. A trial was scheduled to take place at the start of August, but was almost immediately pushed back, and then postponed on a further six occasions though the winter and spring of 1610–11. For the moment at least, the king was happy to let the earl of Orkney stew.

In October 1611, Patrick drew up a defiant set of answers to the allegations against him. They shed revealing but unflattering light on his policies and character. It was said he imprisoned and banished people, 'as if he were a sovereign prince'. But this was but 'the old form and custom observed within Orkney and Shetland' by the earls. If he had forced people to work without wages, it was no more than 'other noblemen and gentlemen of the country are accustomed to do'.

Patrick offered a muddled response to the charge – one made against his father in 1575 – that he prevented people travelling south without a licence. This, he said, was necessary to keep the plague out of Orkney, and stop allies of his enemy the earl of Caithness entering and leaving. More generally, there was a risk of criminals escaping punishment, 'the malicious nature and inclination of the people being so great'.

The wickedness of the people over whom he ruled was a depressingly central theme of Patrick's defence. He was accused of confiscating people's property for 'not marking of their goods, or for wrong marking of the same'. 'Goods' here meant livestock, and referred to the distinctive ownership marks cut into the ears of sheep grazing beyond the township hill-dyke. Strictness around this was prudence, 'the people being so accustomed to theft'. Another harsh-sounding regulation, forbidding Orcadians 'to succour or relieve any ships distressed with storms or tempests', was similarly essential. For if any vessel was wrecked, 'the people so miserably and unmercifully invaded the distressed persons, their goods and gear, that no knowledge could be gotten thereafter'.

Patrick begged the lords of the council to bear in mind 'the evil inclination and unhappy nature of the isles people'. Without strict ordinances, he contended, they could not be made obedient, and so 'the old Danish laws, by the which they were governed, was so straitly set down, that for small offences they lost their lands, goods and gear' – an unwitting confirmation of the government case against him.

Even when viewed through the lens of the defence, Patrick's regime in Orkney, like that of his father, appears, in the historian Willie Thomson's apt phrase, 'essentially parasitic'. The earls extracted as much as possible in rents and skat, leaving families working what was often rich agricultural land with little beyond the necessary means of survival. At a time of upheaval and renewal in the Christian world, neither earl regarded the Kirk as much more than a money-making device, and a means of remuneration for compliant servants.

The earls demanded their share of harvests from the sea, as well as the land. There were tolls on fishermen, Scots or Dutch, visiting Orkney waters, and a ruthless determination to monopolise the sometimes staggering profits to be made from the wreckage of ships, brought to grief by storms and hidden skerries on the increasingly frequented 'north-about' route to the Mediterranean or Indies.[36] There is no doubt that Patrick Stewart was a ruthless and exploitative ruler. Whether his critics, up to and including the king, felt any greater real concern for the welfare of ordinary Orcadians can legitimately be questioned.

The Enterprise of Orkney

The Stewart earldom of Orkney ended not with a whimper but a bang – an armed rebellion raised by Patrick against the crown, followed by another in the sequence of incursions and invasions to which Orkney was remarkably susceptible in its first century and a half of Scottish rule.

The earl's instrument was that perpetual accoutrement of ambitious Orkney landowners, an illegitimate son. Patrick had no children by his marriage to Margaret Livingston, widow of Sir Lewis Bellenden, but there were two girls and a boy – Robert, born in the early 1590s – from his relationship with a woman called Marjorie Sinclair. She was an evidently resourceful and resilient woman, who ten years later can be found suing for arrears of rent owed to Patrick from lands in Sanday. It was perhaps from his mother that Robert inherited the 'tall stature and comely countenance' that impressed contemporaries.[37] His relationship with his overbearing father was deeply dysfunctional, and proved the death of them both.

As Bishop Law returned south in 1611, the earl secretly ordered Robert to go north as his deputy. When the privy council found out, Patrick was sent back to the castle, ending the freedoms he had enjoyed in Edinburgh. Patrick admitted giving Robert a commission of sheriffship and justiciary, but only to collect debts and arrears. Law, however, learned that Robert had been holding courts, in defiance of orders discharging Patrick's agents from administrative office. In the early months of 1612, as Patrick experienced still closer confinement in Dumbarton Castle, Orkney seemed on the brink of rebellion. Robert's followers had seized control of the castle and palace in Kirkwall, and broken into the town's 'girnel' or grain store.

On 5 June, Law told the king he was ready to go north 'to pacify the apparent troubles'. Lamenting a flow of 'continual directions to Orkney', the bishop begged James 'to close up the earl in his chamber, and that no man (some one or two servants excepted) have access to him'. Without explicit directions from his father, the hapless Robert proved more malleable. Law managed to persuade him to surrender the castle and palace, and to return peacefully to Scotland. The young man appeared before the council to give 'his solemn and great oath that he shall not resort nor repair to Orkney nor Shetland without the King's Majesty's licence'.[38]

In October 1612, the era of the Stewart earls came formally to an end. By parliamentary act, the lands of the earldom were annexed to the crown, 'therewith to remain perpetually and inseparably'. Sir John Arnot was compensated with a payment of £300,000 Scots. Not everyone regarded it as the joyful toppling of a tyranny, nor was Bishop Law universally welcomed in Orkney as a liberator. A proclamation condemned 'idle and busy bodies ... feeding the ears of the simple multitude with sinister hopes and foolish apprehensions of alterations'. Such people were 'desirous still to fish in drumly [troubled] waters'. The waters in question were not just metaphorical, for one issue highlighted by the proclamation was rights to wrack and waith, which persons unnamed were said to be claiming. There was to be no change of entitlement here: items cast on the shore pertained to the late earl, and 'all that was formerly possessed by him is now due to his Majesty'.

The fall of Earl Patrick removed an ambitious aristocrat who regarded the islands as his private fiefdom. It was also an opportunity for the crown to make money. In a return to earlier patterns, the earldom lands were entrusted to a tacksman, Sir James Stewart of Killeith. Actual collection of the rents was delegated to his brother-in-law, Sir John Finlayson. As sheriff-depute, Finlayson pursued dues and arrears with an implacable rigour, issuing no fewer than 2,400 written demands over the course of 1613. For many ordinary Orcadians, Earl Patrick's severe rule had given way to a still more oppressive regime. The earl of Caithness would later observe of Finlayson that, in Orkney, 'man, wife and bairn hates him to the death'.[39]

Patrick remained hopeful that the king might take pity on him, agree to a personal meeting and restore him to royal favour. With his trial seemingly suspended permanently, he resisted pressure to renounce his rights in Orkney and accept an offer of honourable semi-retirement as keeper of a royal castle. In January 1614, Patrick finally agreed to give up his claims, and was allowed to return to Edinburgh. But in mid-May he was back in detention at Dumbarton, as alarming news reached the court: Robert Stewart, his oath to the contrary notwithstanding, had returned to Orkney.

The earl's confinement in Dumbarton Castle, it would later emerge, was not nearly as narrow as the bishop of Orkney wanted it to be. Throughout 1613, a string of servants and clients paid court to him there, including his son, Robert. It was not a happy reunion. Patrick berated Robert for surrendering the castle in Kirkwall; he was a 'false feeble beast', 'the wreck of him and his estate', 'he should hang him with his own hand'. He hoped Robert might 'prove a pretty man, if the house came again in his hands'.

Whether from shame, pride or desperate need for paternal approval, Robert agreed to undertake 'the enterprise of Orkney' – an armed rebellion, to pile pressure on the crown, and provide a bolthole when Patrick escaped from Dumbarton, as he assured his son he would manage to do. Robert later said he agreed to go out of 'great need and necessity', to recover the favour of a parent who would 'scarcely ken him, or give him any maintenance'.

Patrick did not fully trust Robert with the task, telling a servant that his son 'had not a spirit nor courage to follow any enterprise'. The man he wanted was Patrick Halcro – probably a younger son of William Halcro of Aikers, a junior branch of the upwardly mobile South Ronaldsay dynasty. In the summer of 1613, Patrick summoned Halcro and asked his opinion on whether he 'could come by his houses again in Orkney'.

Halcro thought it would be hard to pull off without popular support, prompting an outburst from Patrick of characteristic contempt for ordinary Orcadians – 'a company of false people'. Still, 'I mind to send my son there ... will you assist and take part?' Halcro hesitated – Robert was 'over feeble a captain to follow in such a purpose'. Eventually, however, he agreed to go. Another in the small circle of conspirators was sceptical from the start, remarking that 'the earl is an unhappy man. He knows Robert can do no good in Orkney. His purpose is to bring him to the scaffold, and to bring the slander of his blood upon the king.'[40]

Perhaps. Yet there is some evidence of serious strategic planning. Patrick gave Robert several 'memorials', and sent further written orders to Orkney. He was to beware the blandishments of Bishop Law: 'take care that your Holy Father betray you not, as he did before'. Robert was to send a ship to Bergen, laden with bere and salt meat, for its crew to buy gunpowder. The appearance of James's brother-in-law Christian IV at the English court in July 1614, for a diplomatic purpose no one could quite fathom, was almost certainly unrelated to events in Orkney. But the visit underlined the sensitivity of political unrest in a part of James's kingdom which was still claimed by Denmark.

Patrick – surprisingly perhaps – seems not to have schemed for Danish involvement, but he had given thought to the wider dimensions of his enterprise. Confined with him in Edinburgh Castle was Sir James MacDonald of Islay, a chieftain with his own festering grievances against the crown. They spoke of returning to their territories, 'to do for themselves, the ane in the West and the other in the North'. MacDonald did in fact manage to escape in 1615, joining a rebellion in Islay which overlapped alarmingly with the troubles in the north.

The king, Patrick was convinced, 'had other things to think upon nor Orkney'. It was not long since the death, in November 1612, of his heir, Prince Henry. There was the prospect of a Spanish attack on the Rhineland, where James's daughter Elizabeth was newly married to the Protestant Elector Frederick. For his part, Robert did not understand all the political complexities, but knew they involved 'practices of papists'. He blithely assured followers of 'many alterations' that would maintain them in their strongholds in Orkney.[41]

The first stronghold to be taken was the palace in Birsay. Robert and Patrick Halcro seized it in May 1614, seeing off with light exchanges of musketry Finlayson's attempt to take it back. Robert Stewart, Patrick Halcro and fifteen other gentlemen now formed a 'band of association', swearing to uphold it with hands upon the gospels, and adorning their signatures with a defiant suffix – 'to the death'. Those unable to write 'did swear upon their swords'.

The conspirators declared themselves for 'the commonwealth of this country', blaming Orkney's woes on 'frequent entrance of extraniers [outsiders]'. There was an affirmation, usual among rebels, of loyalty to the king, but also a pledge of allegiance to Robert, 'son natural to our native lord and master, Patrick, earl of Orkney'. It is a measure of how deeply unpopular the royalist regime had become that Patrick, a Scots patrician disdainful of humble Orcadians, could plausibly be reinvented as a champion of the community.

For several weeks, the rebels raised recruits in the West Mainland and planned their next moves. In Dumbarton, Patrick fretted about his son's slowness to advance against Kirkwall – 'the Devil stick his fool's head!' An exploratory raid was driven off on 17 July, but Robert returned eight days later with a larger force. Patrick Halcro seized the cathedral, and the host marched into Kirkwall 'with sounding of trumpets … in sign of triumph and victory'. Finlayson offered a merely token defence of the castle; as the earl of Caithness would later contemptuously remark, he 'beastly gave it over for four shot of musket'. The sheriff-depute was imprisoned for a time, and expelled from the islands.

By the end of July, the rebels controlled the whole of Orkney, and Patrick's rent arrears were collected as provisions for an army. Not everyone was cheering, but Robert's cause garnered substantial popular support. As many as 700 were reported to have subscribed the band, 'to die and live' with their young leader.[42]

News of the fall of Kirkwall Castle reached Edinburgh on 29 July and could not be ignored. The tacksman, Stewart of Killeith, had the presumptive duty of re-establishing control over the islands, but he lacked enthusiasm for the task. Patrick's old enemy Robert Monteith accepted a commission to lead forces against the rebels, but he also rapidly thought better of it.

In the end, the role fell to George Sinclair, earl of Caithness, spying an opportunity to divert attention from his own feuds and misconduct, and return himself to royal favour. There was also a chance to settle the score from Summerdale – in the words of Robert Gordon, to 'revenge old quarrels upon the inhabitants of Orkney'. Rather than turn away from the task, Bishop Law reported, Caithness 'has willed rather to be buried in his grandsire's grave'. His Orkney enemies promised that he soon would be.

The expedition sailed from Leith on 20 August, with a reluctant Bishop Law in tow, and two pieces of artillery wheeled down from the armoury of Edinburgh Castle. The earl collected reinforcements in Caithness, and on Tuesday 23 August landed on the Shapinsay coast, in sight of Kirkwall to the south. A couple of hundred Orcadians rallied to the earl's standard, but Caithness 'did not find that ready willingness which I expected'. On Friday, his army crossed to the Mainland, landing at the Bay of Carness north-east of Kirkwall, to skirmish with Robert's forces on the edges of the town.

The invaders dragged their cannon along the coast to prepare for an assault on Kirkwall's stone-built trinity of power: the cathedral, the Palace of the Yards and the castle, its walls in places eleven feet thick. Robert himself was in the palace, determined not to yield his father's 'house' for a second time. On 29 August, it was subjected to fusillades from musketeers positioned around the walls of the garden, but the only casualties were a serving boy wounded in the thigh, and Robert's mother, Marjorie Sinclair, 'shot through the hand' as she stood at her son's side. As the artillery came up into position, Robert reluctantly withdrew into the castle.

Robert's men in the cathedral surrendered on 31 August, in return for a pardon. The earl of Caithness thought the concession worth making, to gain the vantage point of the steeple, instead of having 'to beat it and the kirk down, with much travail, cost and danger'. An artillery bombardment, which would have left St Magnus Cathedral today as much of a hollowed-out shell as the adjoining palaces, was only narrowly avoided.

Kirkwall Castle was fashioned to withstand a bombardment. Caithness's letters describe artillery projectiles 'broken like golf balls' upon its walls. It was 'one of the strongest houses in Britain', a place that could not have been built 'by [without] the consent of the Devil'. Sieges were generally a matter of who would crack first, and it was not inevitable that this would be Robert. The earl of Caithness worried about shortages of munitions and supplies, while the castle's garrison sat on 'great provision of victual, powder and bullet'.

Yet the defenders were unsure of their next steps. Caithness refused Robert's request for an amnesty, and in his dealings with the rebels was urged by the council 'to give remissions sparingly'. Inside the walls, Patrick Halcro suggested making an escape by sea, but Robert, haunted by the earl's stinging rebukes, would not countenance it – 'my father will then say I have feebly given over his house. I would rather hold out while the house be dung down above my head.'[43]

The siege continued for a gruelling five weeks, until Robert surrendered on 29 September 1614. Bishop Law gave the credit to (another) Robert Stewart, minister of Hoy and Graemsay, who was sent into the castle to preach to the rebels 'for instruction and conversion of their minds and hearts'. More likely, Patrick Halcro – who opened secret communications with the earl of Caithness – talked Robert into giving up, in return for minimal guarantees he would be allowed to make his case before the privy council. The fact that Halcro's life was spared suggests a deal of some kind. Others were not so fortunate; twelve defenders were hanged by Caithness at the castle gate.

The surrender was celebrated with 'sounding of trumpets, roaring of all our ordnance on land and sea'. Earl George and Bishop Law entered the castle together, to 'set up his Majesty's colours upon the fastiges [gables] of the house'. There was perhaps a concern for it not to seem too much like a foreign conquest; Law was keen to have it known how all 'caroused after the Orkney fashion'.

The crushing of the rebellion was a triumph for Bishop Law, and a vindication for the activist style of episcopacy James VI hoped to see rolled out across Scotland. Days after the castle fell, the king confirmed a new settlement for the bishopric, an exchange of lands restoring episcopal influence undermined by the excambion of 1568 (p. 139). Historically, bishopric and earldom lands in Orkney lay intertwined in runrig with the holdings of the udallers. Henceforth, however, some parishes were to be earldom parishes and others (Holm, Orphir, Sandwick, Stromness, Shapinsay, Hoy and Walls, parts of St Ola) bishopric ones, where Law would receive all lands previously belonging to either bishopric or earldom. Cases from bishopric parishes were the remit of Law's own sheriff court. It met for the first time at Kirkwall in November 1614, when Stewart of Killeith formally registered a protest for the new court not to 'prejudge [infringe upon] the general gift of the sheriffship' – there was still some life in the ancient rivalry between bishopric and earldom.[44]

For his part, the earl of Caithness decided he had no long-term ambitions in Orkney. With the ghost of Summerdale laid, he was eager to be off, 'for the weather is both evil and variant, and so is the people'. There was, however, an important piece of business: what to do about the castle of Kirkwall, 'whether it shall be repaired or demolished?' On the one hand, the castle was 'ane ornament for the town', as well as 'ane place of refuge and security for the whole country in time of foreign invasions'. On the other, it was inessential for the ordinary service of the king, 'and may more easily be taken by foreign foes ... than recovered again from them'.

The emphasis on foreign foes was a tactful way of not having to raise the prospect of another overmighty subject, or nativist rebellion. The privy council was less mealy-mouthed. In early 1615, it ordered the destruction of a citadel that served no purpose 'but to be ane starting-hole and place of retreat for traitors and rebels'.[45] The castle was rendered indefensible, the symbolic corpse of a once autonomous earldom, and, for local builders, a ready source of finished stone. The last remnants of its once impenetrable walls were cleared in the mid-nineteenth century. Now, all that remains is a single worn armorial stone, set in the gable of a former hotel and recently closed branch of the TSB Bank, while below it a discreet plaque explains to visitors why they are standing in 'Castle Street'.

For Robert, there was only one possible fate. He was hanged alongside five accomplices on 6 January 1615 at the market cross in Edinburgh. David Calderwood, minister and chronicler, thought he died 'pitied of the people'. It was an assessment shared by another cleric, a young Englishman called Richard James, travelling to Orkney in the immediate aftermath of the rising. Robert was 'a gentleman much lamented of the people in Scotland'.

Richard James found a society nervously on edge after the fall of the earl. He noted the prevalence of 'deadly feuds' among island gentlemen, ever disposed to 'cut and stab with their dirks and swords'. It struck James as remarkable just how many of them, 'in Orkney and Shetland and in Denmark and long Norway', bore the name Sinclair. They were, he learned, 'anciently earls of those islands, until late years'. The 'late years' of the Sinclair earldom were nearly a century and a half in the past. As we have seen (pp. 64–5), Sinclair memories were both long and selective.

Patrick disavowed any knowledge of his son's actions. But as individuals involved in the enterprise of Orkney were interrogated, the extent to which the old earl was pulling the strings became increasingly clear. This new treason, rather than lingering charges from his tyrannical rule, sealed Patrick's fate. He was executed in Edinburgh on 6 February 1615. As befitted his noble blood, Patrick was beheaded rather than hanged, a victim of 'the Maiden', a precocious form of guillotine. Calderwood claimed that when ministers came to offer the earl spiritual counsel, they found him 'so ignorant that he could scarce rehearse the Lord's Prayer'. They petitioned the council to delay the execution for a few days, 'till he were better informed'. The story is sometimes dismissed as a calumny, or as a ruse by Patrick in hope of last-minute pardon.[46] Maybe. But there was nothing new in suspicions that the earl of Orkney was a man with little time for God.

Patrick Stewart was the architect of his own demise. His ambitions were unconstrained by cautious self-interest, still less by concern for the fate of others. His son Robert ultimately understood this. One terrifying night during the siege, as the earl of Caithness's cannon battered the castle gates, Robert declared his true mind about his father: 'there was never anything looked well yet which he devised – God give I had never kent his turns!'[47]

Humanity and Civility

The royal standard, flapping in a brisk breeze over the ramparts of Kirkwall Castle, betokened the unfurling of a new era in the history of the islands. There would be no more preening earls exploiting a hybrid legal system, raising stone symbols of quasi-regal authority, trampling the rights of the Kirk, protecting pirates and pursuing schemes of private foreign policy. At long last, Orkney was to be fixed in its proper place on the map of Scotland. And yet, as we shall see, the islands were not about to lose all of their 'edginess' – their capacity to challenge assumptions about politics and law, and to pose questions about attitudes and identity.

A small incident, shortly after Robert Stewart's rebellion, suggests how the imperatives of incorporation might be resented and resisted. In late 1616, King James learned about 'a rude custom, continued in some remote parts and uncivil places'. This was the habit of plucking wool by hand from the backs of sheep – called in Orkney and Shetland 'rooing' – rather than cutting it with shears. This abuse, James informed the privy council, was 'already reformed in our kingdom of Ireland'. A 1617 proclamation banned the cruel practice, 'not to be tolerated nor allowed in no well governed country'. James Alexander, an enterprising citizen of Stirling, secured a licence to collect the fines, and headed for Orkney to make money from enforcing the ban.

A fierce local backlash ensued, producing a petition to the council, and a commission to the bishop and sheriff to investigate. It was explained to the lords of the council that while pulling the wool of sheep 'in the in-country may justly be thought a barbarous and uncivil form', in the Northern Isles it was otherwise. Shearing animals who spent most of the year in remote pastures, 'and feed chiefly upon sea-ware', would mean huge losses of livestock. But their wool could be plucked without injury in early summer, when the old fleece was naturally shedding. Incomers practising the southern custom had learned their error 'through dying of all or the most part of their sheep'. The council relented, and Orcadians and Shetlanders secured exemption from the national regulation.[48]

What of Orkney's spiritual flock? 'All today are of the reformed religion.' That was the sanguine assessment of an early seventeenth-century *Nova Orchadum Descriptio Chorographica* (A New Description of the Orkney Islands), written to increase knowledge of the islands among an educated European audience. In a broad sense, the appraisal was surely correct. There was no minority of dissident Catholics, emboldened by visits of missionary priests, and no sprinkling of exotic Protestant sects, separated from the official Church. Reformed ecclesiastical structures were in place. In particular, Orkney parishes accommodated kirk sessions – the most local form of church court, where lay elders met weekly with the minister to exercise 'discipline' over the thoughts and deeds of the congregation. In November 1615, the sheriff-deputes ordered parish baillies to cooperate with ministers and kirk sessions in enforcing measures against 'transgressors and sinners'. Repression of sin had been long hindered 'by practising of foreign and uncouth laws'.

The happily named Bishop Law was in 1615 nominated by James to the archbishopric of Glasgow. Whether or not Law was 'the saviour of Orkney' – as a nineteenth-century Episcopalian historian gushingly christened him – he was certainly an exemplar of efficient top-down Kirk governance, in close concert with civil authority, something for which

Orkney's rebellion supplied a timely test case. Law's promotion, suggested James's secretary of state Lord Binning, offered further opportunities to 'reduce the Church government to that happy estate which his Majesty has long wished'.[49] Law's successor in Orkney, George Graham, bishop of Dunblane, was a man cut from the same cloth, though, as we shall see (in Chapter 6), without his predecessor's style and finish.

The *Orchadum Descriptio* was penned by a minister who arrived in Orkney during Graham's episcopate. It was first published in a 1654 atlas of Scotland by the renowned Amsterdam mapmaker Joan Blaeu. The volume's map of Orkney was older, drawn in the early 1590s by the cartographer Timothy Pont, son of Commissioner Robert Pont, and Blaeu needed new accompanying text for it.

Strong circumstantial evidence points to Walter Stewart, minister of Rousay (1635–6) and South Ronaldsay (1636–52) as the *Descriptio*'s author, and it was probably written in the early 1640s.[50]

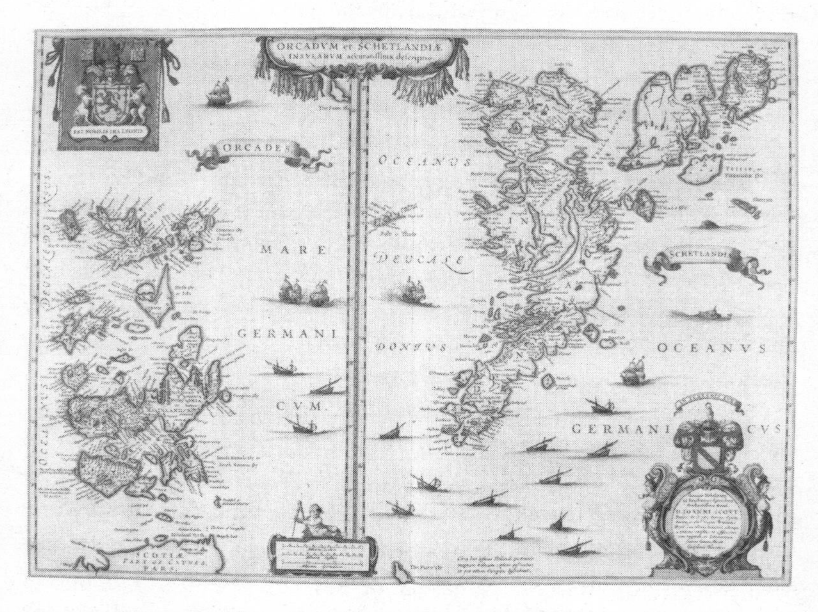

Figure 4.2: Maps of Orkney and Shetland from Joan Blaeu's 1654 Atlas of Scotland. Accompanying texts were probably written by the South Ronaldsay minister Walter Stewart.

Stewart was an incomer, the son of a Stirlingshire laird. He was hardly unusual in his southern origins. Only three of twenty-five ministers appointed to Orkney parishes in the first four decades of the seventeenth century were born in the islands. Like Jo. Ben., Stewart wrote with local knowledge, but from a place of cultural and educational distance: the people of Orkney remained 'they', not 'we'.

In the main, Stewart thought well of them. Orcadians were cheerful and industrious, usually intelligent, with sharp memories. They were kept fit and healthy by 'ignorance of luxuries, deriving from honourable poverty'. This was just as well, as there were no physicians in the islands.

The outsider's eye lands on details a native might not think to note. The women singing at their querns (hand mills), the sweetness of their voices softening the harsh grate of stone on stone. The profusion in the islands of the home-made baskets known as 'cassies', skilfully woven from straw, into which crops were loaded for export. The watchful inhabitants of Sanday, hoping for passing vessels to come to grief on their coast, 'so that they get a supply of material for fire from the wrecked ships'.

Regarding religion – and possibly mindful of his foreign readership – Stewart, the minister who oversaw the burning of an illegal statue in Burray (p. 120), refrained from charging Orcadians with ignorance or superstition. Dirty linen was best not washed in public. Stewart reported people coming to church in large numbers to participate in the Lord's Supper, observing without comment that they did so at Christmas and Easter – traditional holy seasons Calvinists were not supposed to make much of. Yet Stewart gave them credit for their attentiveness: 'even the country people listen carefully to sermons'.

One sentence in Stewart's qualified tribute to Orcadians particularly catches the modern reader's attention. 'They either express, or try to express, the humanity and civility which they have taken from Scots who live among them.' It was an unusually frank recognition of a society divided along ethnic lines. It was also, in its tone of gracious condescension, something like a declaration of victory. Thomas Craig, a lawyer with commercial interests in Orkney, had observed in 1605 that 'even in the Orkneys and Shetland, where in the course of this very century [i.e. the hundred years preceding] nothing but Norse was spoken, the ministers of God's Word now use English in church and are well enough understood'.[51] Craig was wrong to suggest that only Norn was spoken in sixteenth-century Orkney, but right to recognise it was in retreat.

A few decades on, Stewart noted 'the old Gothic or Norwegian language' to be still used in some, 'not in all', of the islands. It was 'most vigorous' in the Mainland parishes of Orphir and Harray. But it was now matched by 'the vernacular of the Scots, of whom a large number live among them, especially holding office in the Church and State'. Stewart scarcely needed to say Norn was never used in the formal deliberations of courts, in business contracts or land deeds. It was not heard from the pulpit, nor in the prayers, scripture and psalmody of the Kirk. The dice were loaded against the old language and in favour of the new.

Stewart regarded clergymen like himself as instruments of civility. There were now seventeen ministers in Orkney: eight in the Mainland, six in the North Isles and three in the South. He recognised the difficulties posed to their pastorate by the circumstances of an island environment. The people of Swona availed themselves of Stewart's preaching in his kirk in South Ronaldsay, 'when they conveniently can, in calm weather'. A Sanday minister had responsibility for North Ronaldsay, though between the two islands lay 'the most terrifying sea'. Despite these challenges, the ministers of Orkney nonetheless 'transact ecclesiastical business as in Scotland itself'.

This meant chairing kirk sessions, and convening in Kirkwall every month in the summer ('for in winter they do not venture to entrust themselves to the very savage sea') to attend presbytery meetings. Kirkwall was where the sheriff presided over sessions of the head court, and Stewart praised the fruitful cooperation of clergy and magistracy. Wrongdoers were punished on the basis of their confessions, or on the production of incontestable evidence, and after examination by ministers of the Church. Stewart then adds something odd. With the clergy's approval, such offenders are 'consecrated to Vulcan'.

It was not exactly a joke, though no doubt intended as a demonstration of shared erudition with a sophisticated international readership. Vulcan was the Roman god of fire, so it is hard to imagine what 'consecrated to Vulcan' might mean other than being put to death by burning. Heretics, in post-Reformation Scotland, no longer suffered this fate, though heinous moral crimes such as incest and bestiality were from time to time punished in this way. Murderers were sometimes sentenced to the painful death of being broken on a wheel – if their victim was of high status – and traitors might receive the relative clemency of decapitation if, like Earl Patrick, they were of sufficiently high status themselves. Most capital crimes – murder, manslaughter, assault, theft – led to the gallows, and a short drop at the end of a rope. Burning was not the penalty for ordinary criminals or ordinary crimes. Burning was for witches.

5
Devilry and Witchcraft

Cheust Folk

A short walk from the west door of St Magnus Cathedral leads to a small patch of grass, bounded on the east by fine Victorian houses, and to the south by modest twentieth-century ones, tokens of Kirkwall's expansion from its earlier string-like orientation along the waterfront. The approach from the west requires traversing a steep residential street known as the Clay Loan, the name identifying a site formerly supplying building-mortar, but originally just the Loan or Common Loan, a pathway through agricultural land. The point where the ground begins to level was 'the head of the loan', Orkney's place of public execution. A nearby cottage built in 1829 was christened Gallow Ha' (Hall), its location announced with literal-mindedness, and a touch of grim humour.[1]

In the centre of the little green lies a discreet plaque, laid in the spring of 2019 after a campaign by local heritage groups. The stone was placed 'in memory of those accused of witchcraft', and bears a poignant inscription in Orcadian dialect: 'they wur cheust folk' – they were just people. Today, the tranquility of the spot, disturbed only by a trickle of cars en route to Kirkwall Airport, belies the horror of what once took place there. 'Witches', having been convicted at trials held in the cathedral, were condemned to be escorted with bound hands to the Loan Head. There they were fastened to a stake and 'wyrried' – strangled – to death by the public executioner, 'thereafter to be burnt in ashes'.

It is not possible to say how many people suffered this fate, nor how many were accused of witchcraft in Orkney during the years (1563–1736) when it was a capital crime. My own trawl through the records found ninety-seven persons indicted or named as witches in the course of legal proceedings between 1594 and 1708, though in most cases we don't have the record of a trial, or know whether a sentence of death was carried out. Notwithstanding, and even over the space of a century and more, it is a notable tally for a small island community. Relative to population, around twice as many witchcraft accusations were levelled in Orkney as in Scotland overall.[2]

In a pattern common to most of Europe, the great majority of the accused – eighty-one – were women. Twelve were men, and the gender

Figure 5.1: 'They wur cheust folk' – memorial slab laid in 2019 at the Loan Head, the place outside Kirkwall where convicted witches were burned at the stake.

of four is uncertain. The statistics are certainly incomplete: there are, for example, no extant kirk session or presbytery records before the middle years of the seventeenth century, nor do we know the names of any of the 'incarcerated suspect witches' the ministers were invited to help prosecute in November 1658.[3] An accumulation of fears, suspicions and informal accusations remains forever immeasurable. Witchcraft was a fact of life in Orkney, but it was never unremarkable, familiar or tolerable.

Were those charged with witchcraft 'cheust folk'? They were usually simple, uneducated people, accused of crimes they could not possibly have committed yet could not plausibly deny. It is right to recognise the cruelty and injustice of proceedings against them.

Accusations of witchcraft were not, however, made at random. They emerged from intricate human interactions, where magic was indeed practised, and ratios of social and emotional power were not always fixed or predictable. It is fitting that the victims of witchcraft legislation be publicly commemorated. But others were victims, too. An honest engagement with the past demands an empathetic effort to understand the concerns of people who believed themselves or their loved ones to be the targets of an unspeakable malevolence and hatred.

Witch-hunting took place across Europe in the late sixteenth and seventeenth centuries. There were many common features, but also variations, quirks and peculiarities. The witch trials in Orkney, as we shall see, possessed unique features, in part related to the islands' distinctive past. The trial records open a window on Orcadian society at a moment of profound social and political change. They illuminate encounters – sometimes creative and often traumatic – between imported systems of explanation and deep-rooted intuitions about the hidden workings of the world. At the same time, they provide a wealth of incidental information about people, places, and the material objects and mental furniture of everyday life.

These tragedies were performed in local costume and dialect, but they link the islands to dramas playing elsewhere on the British and European stage. The trials took place not at a time of isolation or 'medieval' barbarity, but during an intellectually inquisitive century which saw Orkney drawn into first Scottish and then British systems of law, politics and religion. The witch trials were not a detour along decaying pathways of yesteryear; they were part of the journey on a road to modernity.

Naturally Inclined Thereto

We have encountered already the earliest datable execution: Alison Balfour (pp. 152–3), accused in 1594 of plotting against Earl Patrick. Yet she was by no means the first to suffer. The names of half a dozen Orkney witches, all but one of them women, appear in Patrick's rental of 1595. They are there since a conviction resulted in the escheat of land. The confiscations are undated, but some surely pertain to Patrick's father, Earl Robert. For a couple of cases, an execution is confirmed. In the township of Ronaldsvoe, South Ronaldsay, there was a parcel of ground 'pertaining to Jonet of Cara, quha was burnt'. In Walls, a woman with a Norse matronymic, Alison Margaretsdochter, forfeited a pathetic 'cowsworth' of land, a tenth or less of a pennyland, able to sustain just one beast. She, too, was 'burnt for witchcraft'.

Earl Patrick, by his own account, was the witches' worst enemy. He defended imposing heavy fines on Shetlanders for consulting with a sorcerer named Nerogar; they would otherwise 'all have become witches and warlocks, for the people are naturally inclined thereto'. Patrick was not alone in supposing the inhabitants of the Northern Isles to be predisposed to witchcraft. Scotland's most eminent authority on dark magical arts, James VI himself, excoriated witches in his 1596 treatise, *Daemonologie*: they were most common in 'wild parts of the world as Lapland and Finland, or in our North Isles of Orkney and Shetland ... where the Devil finds greatest ignorance and barbarity, there assails he grossliest'.

'The north' as a dangerously uncharted region of magic and superstition was an object of cultural prejudice in many lands.[4] Scotland itself had a reputation for witchcraft, which Shakespeare famously drew on in a play of 1606, designed to appeal to the new king of England. Orkney and Shetland were a north within the north – and a north for the north, onto which sophisticated Scots could project their own association with uncouthness and ignorance.

Witchcraft was also highly political. Sheriff-court ordinances of 1615, during Bishop Law's restoration after Robert Stewart's rebellion, declared 'heavy plagues' to have fallen on Orkney, due to 'lack of discipline and putting of the acts of the Kirk … to due execution'. Baillies were to publicise in their parishes a warning not to 'hide nor conceal any kind of theft, sorcery nor witchcraft'.[5] Parishioners evidently took note: fourteen individuals, eleven women and three men, were tried for witchcraft in 1615–16. The impression is of a backlog being cleared, of smouldering resentments receiving permission to ignite. Witchcraft persecution, in Orkney as everywhere, was a tango for two. It required a willingness on the part of authorities to prosecute, and a willingness on the part of ordinary people to report their suspicions and testify against their neighbours.

Prosecution gave a name to the unnamed. No one called themselves a witch. The path to the Loan Head began with mute suspicions which were often present in communities over long periods of time. Everything changed when these fears spilled over into public accusation. Some incident usually triggered an allegation that an individual was a witch – often, a quarrel followed by a misfortune. This encouraged others to come forward, though it is sometimes hard to know whether they always harboured mistrust, or whether the naming of a witch caused them to reinterpret past calamities in their lives.

A sense that some people knew how to work magic, and could use it to harm others, was old in Orkney, as in most places. The *Orkneyinga Saga* relates several stories involving sorcery. We know virtually nothing about witchcraft in later medieval Orkney, or how the Church might have dealt with it. In general, prosecutions were at a low level across western Europe before the middle of the fifteenth century, though cases were not unknown. At the far fringes of the Nidaros archdiocese, one of the last pieces of news to percolate from the colony in Greenland was the execution there in 1407 of Kolgrim, a sorcerer.[6]

The possibilities for turning local fears into legal process proliferated after 1563, when both the Scottish and English parliaments passed laws to make witchcraft a capital offence triable in secular courts. This followed closely on the political triumph of Protestantism, north and south of the border, though the relationship of witchcraft trials with the Reformation is far from straightforward, and persecution was equally intense in many Catholic territories.

In Scotland, however, parish kirk sessions created new opportunities for investigation and chastisement. These courts, parochial in every sense, encouraged denunciation of local wrongdoing and voicing of neighbourly grievance. Kirk sessions lacked the power to impose death sentences, but they could weigh evidence and draft the charge sheets known as 'dittays'. These laid out, in often fascinating and fantastical detail, what witches were alleged to have done – usually adding that the deeds were performed 'by your witchcraft and devilry'.

Parish ministers presided over the process, supplying theological wherewithal and liaison with the presbytery. As we know from Walter Stewart's *Orchadum Descriptio* (p. 178), the presbytery provided advice to the sheriff court. Presbytery minutes for June 1644 show Stewart himself producing 'Janet Smith, witch ... within the isle of South Ronaldsay, her dittays, which the brethren present examined and annotated', and delivered to the civil judge.[7]

Before the middle of the seventeenth century, nearly all Scottish witches were tried by 'commissions of justiciary', established by the privy council upon application to that body. Local authorities could not usually prosecute on their own initiative, and central government retained a remarkable degree of control. Uniquely in Scotland, witches from Orkney (and Shetland) were prosecuted as part of the ordinary business of the earldom and bishopric sheriff courts, rather than under special commissions.[8]

This was a concession to distance, rather than a recognition of special status, but it imparted an indigenous flavour to the process. Some of Europe's most vicious witch-hunts took place where local authorities could instigate proceedings without external oversight. Most of the time, that was not Orkney's experience. Rough checks and balances emerged from an interplay between the interests of secular authorities, the preoccupations of ministers, and the priorities of ordinary people. Kirkwall, site of the executions, supplied a remarkably small number of the suspects. But the body count was high enough.

First to appear in Orkney's legal records was the case of Marion Lening of Papa Westray, summoned before the sheriff court on 7 June

1615. She was married to a man called James Howieson; there is little to support the clichéd notion that witches were generally widows or aged spinsters. Her crime was 'witchcraft and consultation', echoing the wording of the 1563 witchcraft statute, which forbade seeking 'help, response or consultation' from persons using 'witchcrafts, sorcery or necromancy', or who 'give themselves forth to have any such craft'.

The act, as the historian Julian Goodare persuasively argues, was likely drafted by a zealous minister, possibly John Knox himself, and aimed to stamp out superstitions associated with unreformed Catholicism. It made no direct mention of the Devil, or pacts to serve him, and it criminalised 'charmers' and healers – purveyors of magical services in demand in local communities. With extraordinary ferocity, the act stipulated the clients of such practitioners should also face the death penalty. That seems never to have happened; Goodare suggests that, throughout Scotland, courts applied the legislation differently from the intention of its author. Charmers, whom parishioners and ministers saw as distinct from malevolent witches, were periodically disciplined by sessions and presbyteries, but did not face the capital penalties of criminal law.[9]

We hear nothing further about Marion Lening, or her 'consultation' in Papa Westray. In the magical economy of early modern Orkney, however, the distinctions between charmers and witches, healers and harmers, were never clear-cut. An astonishing number of people denounced as witches in Orkney purported to be able to offer cures, and accusers were often their former or current clients. There are thirty-nine cases where trial records survive to permit a detailed inspection of the accusations. In thirty-two (82 per cent), the dittays include charges not only of cursing but of healing – a remarkable statistic, which draws us into a world where medicine and magic, material and spirit, were inextricably entwined.

Witchcraft stories aimed to make sense of the inexplicable, but were told in different ways by diverse people: righteous judges, godly ministers, anxious or angry witnesses, the accused witches themselves. In the next part of this chapter, I unpick several of these stories in detail, tracing their patterns and following their threads to a variety of knotted conclusions and frayed loose ends. I then seek to deepen an understanding of how magic and witchcraft in Orkney 'worked', and what it can tell us about the concerns of island society. Finally, I turn to the prosecutors' perspective and the 'political' aspects of persecution; when and why it intensified, and when and why it declined.

Trowies, Devils and Rebels

On the day the sheriff court summoned Marion Lening, it also passed judgement on two women from the neighbouring island of Westray, Jonet Drever and Katherine Bigland. A guilty verdict was pronounced first on Jonet Drever, convicted on a single charge: 'the fostering of ane bairn in the hill of Westray to the fairy folk, called of her "our good neighbours", and in having carnal dealing with her, and having conversation with the fairy twenty-six years bygone, in respect of her own confession'.

This one remarkable sentence condenses an entire almanac of assumptions about the world. Belief in fairies – not diminutive messengers of winged blessing, but middle-sized and meddlesome magical creatures – was widespread in Orkney, as in Scotland and Scandinavia. If Jo. Ben. is to be credited, it was especially prevalent in Orkney's North Isles: Stronsay folk were said to believe 'people who die suddenly spend their lives thereafter with them'.

The fairies were 'good neighbours' not because they were intrinsically benevolent but because they were notoriously capricious, and it was wise to flatter them. Fairies were frequently associated with hills, and the seemingly imprecise description 'the hill of Westray' is significant – the topography of the island presents a contrast between the agricultural land of its southern spur and eastern seaboard, and a broken ridge of high moorland covering the north and west. More so even than elsewhere in Orkney, the hill-dyke of the Westray townships tracked a line between the familiar and the eerie. 'Conversation' with fairies involved crossing physical as well as metaphysical boundaries.

The meaning of 'fostering ane bairn' is not immediately obvious, though had the child been Jonet's it would surely have been described as 'her bairn'. One grim possibility is an ailing or disabled child deposited by Jonet on the hillside. It was common to suspect such infants were 'changelings', left by the fairies in spiteful recompense for stolen healthy children. A few years later, a suspected witch in Sanday told a sick child's mother the bairn could not be cured 'because he was taken away by the good wights in the cradle'. Maybe Jonet intended to persuade the fairies to take a child back. But there was no accusation of infanticide, or suggestion a child had died. Indeed, the sentence lends itself to an opposite interpretation – that Jonet fostered a child given her by the fairies, or believed she had. Her 'carnal dealing' with fairies, and socialising with them a quarter century before, suggests visionary experience of some kind, which she seemingly had not been shy to talk about.

Jonet's sentence was to be whipped from one end of Kirkwall to the other, 'and thereafter to be banished the country [i.e. Orkney] and never to return'.[10] It was scarcely lenient, but her life was spared. There was no explicit suggestion that Jonet's fairy folk were in reality demonic forces, or manifestations of the Devil, which some Calvinist ministers would surely have suspected. Jonet's confession of 'carnal dealing' invited such identification, for the Devil was known to entrap women with his sexual prowess. Competing 'languages' of the supernatural is a theme to which we will return.

By contrast, Katherine Bigland's fate was incineration at the Loan Head. The allegations were more varied, though some may have again involved fairies. Katherine stood in the entrance to the churchyard of Westray's Cross Kirk, 'with drawn knives in her hand', until a group including her mother came forth from the church, where they had spent 'the most part of ane night'. Was this a ritual of community protection? It was known fairies could not abide metal and naked blades (see Chapter 8). Westray's Holy Cross Kirk, perched above the rocky shore at the Ness of Tuquoy, on the island's south-west coast, is today a stately ruin, but in 1615 it was among the more imposing of the twelfth-century Orkney churches continuing in use after the Reformation. Whatever was going on in kirk and kirkyard that night, the participants evidently saw the place as a site of sacred power, not just a convenient location to go and hear sermons.

Katherine was also convicted over dealings with William Bigland of Swartmill, her 'master', and presumably her kinsman. Her crime was attempting to heal William, probably at his request. After he began suffering with a 'duyning' (wasting) sickness, Katherine went out at night to fetch water for washing his back. Waking from sleep, William cried out that he felt 'a thing like a ruche [unshorn] sheep above him'. Katherine told him not to fear, 'for it is the evil spirit that troubled you that is going away'. Next day, after sunset, she led him down to the shore, and washed him with salt water, a process repeated over the next five or six nights. As a result, 'he received health by her unlawful and devilish art of witchcraft' – an incongruous but ubiquitous formulation in the records of Orkney witchcraft.

It transpired, however, that Katherine took the sickness away from William and cast it on Robert Brown, his servant. Brown went almost mad, until Katherine, feeling his pulse and brow, and stroking his hair, assured him he would be well. This sent the sickness back to William; and, when challenged, Katherine reportedly said that 'if William Bigland lived, she would die, and therefore God forbid he live'. These words likely sealed the case, though no one seems to have died by the time the matter went to court.

There is a lot to unpack from this short, sad story – for one, the idea of sickness as bodily possession by an evil spirit. This was not just a belief of the 'ignorant' peasantry. Jo. Ben. wrote of a farmer's wife in Stronsay, who for a year and more was troubled by a spirit, becoming 'thin with grief'. The author understood it to be one of the sea-spirits called 'trows', and supplied a remarkable description: 'it is clad in seaweed, in its whole body it is like a foal, with curly hair, it has a member like that of a horse and large testicles' – an account with affinities to later Orkney folktales about a terrifying sea-creature, hybrid of man, horse and fish, known as the Nuckelavee.

Jo. Ben. helped the woman by urging her 'to pray, give alms and fast'. These were the biblically attested methods of Protestant exorcism, approved substitutions for the crucifixes, holy water and Latin prayers of the papists. Two spiritual healers, Jo. Ben. and Katherine Bigland, faced similar challenges, but one was an educated man and likely a Protestant minister, the other an illiterate woman of little social standing. Witchcraft suspicions attached themselves much more readily to the latter than the former.

The prominence of washing in the cure of William Bigland underlines a core principle of magical thinking: the transfer of invisible properties between substances through controlled physical contact. A 'liminal' setting, under the banks of a Westray beach, enhanced the efficacy. The principle was in tension with a deep-rooted island attitude. Orkney represents an acute case of what has been called a 'limited good' society – a place where wealth, health and happiness were only available on something akin to a zero-sum basis.[11] It was the disagreeable flipside of tightly knit, supportive communities, where people collaborated in wringing livelihoods from soil and sea. If someone was thriving, someone else must be in trouble. Sickness might not be cured unconditionally, but pass parasitically from one sufferer to another – a key reason why healing and cursing, the wise woman and the witch, were so closely and fatally connected.

On 19 March 1616, ten months after the burning of Katherine Bigland, the sheriff-depute pronounced sentence on Oliver Leask.[12] His story illuminates the 'professional' practice of a male charmer, and suggests how the practice of Orcadian magic may have been steeped in words and wisdom from the Scandinavian past.

Leask was originally from Groundwater in Orphir, but had become a vagabond, a wanderer. Complaints against him, ranging back a decade and more, came from Rendall, Rousay, Egilsay, Westray and Eday. He had also been to Shetland, where he was suspected of theft. Knowing of this, Andrew Keldall, a skipper from Holm, refused Leask's request to join a fishing expedition heading there in June 1615. Leask vowed 'he should repent it', and managed to make his own way north. Others on the voyage caught fish in abundance, but Keldall 'had never slain any'. Leask then promised to restore his luck, for a quart of ale and a shilling.

Leask's modus operandi involved travelling around, begging money, food and drink, and making veiled threats when his requests were denied. That he continued in this manner for so long suggests a reluctance to tangle with those thought to possess magical powers. Not everyone complied. During 'bere seed time', Leask came to John Craigie's farm at Swandale, on the east side of Rousay, 'boasteously' demanding grain. When Craigie refused, bitter words ensued, and Leask got a blow from 'the shaft of the clodmell' – a large mallet for breaking up clods of earth. Since that time, Craigie had found 'ane great number of his horses and beasts all dead'.

A more dignified denial was issued from Magnus Scott's wife in Rendall, at Hallowmas 1611. She had given Leask bread and fish, but baulked at milk and butter. 'If God be with me, what can ye do to me?' she asked. Leask retorted that whether God was with her or not, he would 'cast a bukie in her cheek'. It was likely a recognised curse; a bukie was a whelk shell, but also a facial disfigurement. Scott, too, suffered 'great loss' of goods.

Leask offered services in addition to making menaces. He learned that David Henry of Hackland in Rendall was long sick, and volunteered to heal him. In around 1605, Leask was in Westray, seeking alms from Janet Seater, wife of Patrick Harcus of Westbreck. Janet complained she 'wanted [lacked] the profit of her ky [cattle]'. It was a distinctively Orcadian expression. The 'profit' – of livestock, crops, meal, milk, butter or cheese – did not mean the value or expected monetary return, but something more like the benefit, health, quality and goodness; almost the living spirit of the thing. It could be lost, or stolen.

In return for two fistfuls of meal, Leask promised to restore the missing profit, and planted, or reinforced, a suspicion in Janet's mind. There was a neighbour, 'who had but one cow and had as muckle [much] milk as she had of her four ky'. Leask told Janet to send her servant to this woman's field, set a stoup [bucket] down and milk her cow. If the servant then lifted the stoup, pulled some grass from under it and dropped it in the milk, Janet would get her profit again. The neighbour was Katherine Bigland, later to lose much more than the inflated profit of her cow.

Leask revelled in his reputation as a practitioner of magic. When people accused him of causing the sickness of Thomas Anderson, he denied evildoing, saying he 'gaid with the good wights' – his powers flowed from association with spirits or fairies. He also operated under a pseudonym: 'Oliver Leask alias Walliman'. This was not a regular surname, but an audacious honorific, likely deriving from Old Norse *völva*, a female seer or prophetess. The claim to be 'Walliman' was an assertion of supernatural authority, rooted in ancient tradition. The records of Leask's case, like all trial records, are entirely in Scots, but he came from a parish where Norn was said to be 'most vigorous' in the early seventeenth century. It is likely he conducted much of his trade in that language.

The assize (jury) condemned Oliver Leask as a warlock, and declared him out of his wits. That probably saved his life – like Jonet Drever, he was to be scourged through Kirkwall and banished for life from Orkney and Shetland. Banishment was not uncommon in Scotland for less egregious cases of witchcraft. It seemed a particularly apt punishment in an island community with a strong sense of its defining 'bounds'. Seven witches convicted at the sheriff court in June 1616 were ordered to be banished, either from their home parish or the 'countries' of Orkney and Shetland.[13]

The week before Leask's trial in March 1616, a young woman called Elspeth Reoch appeared before the court. Her case demonstrates the potential for witchcraft to become ensnared with politics, and brings to the fore an element so far missing from our discussion: the agency of the Devil. Her detailed confession also tells a remarkable autobiographical tale – one which, stripped of the layers of interpretation her inquisitors draped over it, suggests how belief in supernatural powers could give shape and meaning to a troubled and unsettled life.

Unlike Drever, Bigland and Leask, Reoch was an immigrant to Orkney, a native of Caithness, where her father – or so she claimed – served as piper to the earl. Elspeth left home at the age of twelve, and went to an aunt and uncle at Lochaber in the West Highlands. There, as she waited to be ferried across a loch, two men approached her, one in black and the other in a green tartan plaid. The man in green said she was pretty, and promised to teach her 'to ken and see anything she would desire'. To acquire this 'second sight' Reoch had to wash her eyes on three successive Sundays with the 'sweat' of a roasted egg. She continued to wander through the Highlands, and within a couple of years, near Balvenie in Speyside, conceived a child. Elspeth gave birth in her sister's house, where the 'black man' from Lochaber returned to her.

He 'called himself ane fairy man, quha was sometime her kinsman called John Stewart, quha was slain by McKay at the down-going of the sun, and therefore neither dead nor living but would ever go betwixt the heaven and the earth'. Ghosts and fairies are usually thought of as distinct entities, but boundaries were not fixed in the popular imagination. 'Stewart' persuaded Reoch to lie repeatedly with him; who knows what real or imagined episodes of sexual experience – or abuse – underlay this startling confession. He also warned that 'gentlemen would trouble her', and took from her the power of speech. Nevertheless, in the words of the court, 'she still continued dumb, going about and deceiving the people, synding [ritual washing], telling and fore-showing them what they had done and what they should do'.

At some unspecified date, Reoch travelled to Orkney. She may have been trying to return to her Caithness home, found herself unwelcome, and just kept going. Before long, she was reportedly pregnant by a man called Patrick Traill. Her fairy-lover came north, too, and warned her to leave Orkney 'and go home to her own country because this country was priestgone' – which, he explained, meant 'there was o'er many ministers in it'. It would have come as news to members of the Orkney Presbytery, who always thought themselves chronically understaffed. But women like Elspeth Reoch were aware that the clergy disliked them. She perhaps recognised, too, that it was easier for ministers – those other men in black – to exert authority in sea-circled Orkney than in the diffuse, thinly populated parishes of the Highlands.

Elspeth Reoch should have listened to these imagined words. Her stories suggested to the court only 'the abominable and devilish crime of witchcraft'. There was no mystery who her supernatural lover was: 'the Devil, quhilk she calls the fairy man'. Elspeth's trial was a calamitous collision of rival 'cosmologies'. In her world, there was space for fairies and spirits of the dead, for gaps between heaven and hell opened fleetingly at the 'down-setting' of the sun, for visions and prophecies, for intercourse – in various senses – between denizens of the visible and invisible realms. For the sheriff-depute, Henry Stewart of Carlongie, his procurator-fiscal (legal officer), Robert Coltart, the assize chancellor or chairman, William Bannantyne of Gairsay, and the various ministers advising them, there was only God and the Devil.

We might wonder why Elspeth Reoch's relations with fairies were viewed through this unforgiving lens when Jonet Drever's were not: neither was accused of specific acts of magical harm. It perhaps had something to do with the charges against Elspeth, an outsider, not being filtered through a local kirk session. But there was also a political dimension. Elspeth's (human) lover, Patrick Traill, was a leader of Robert

Stewart's rebellion, one of those hanged by the earl of Caithness after the fall of Kirkwall Castle. Elspeth's foreseeing this featured in the charges against her: 'she saw Robert Stewart, son natural to the late Patrick, sometime earl of Orkney, with Patrick Traill ... and certain others with tows about their craigies' – ropes around their necks.[14]

Satan and sedition converged in a third trial of 1616. Jonet Irving was a woman from Harray, first seduced by the Devil – metaphorically and literally – more than thirty years earlier, as she kept solitary watch over some cattle. He taught her the secret of cursing: if Jonet wished ill to anyone, she should gaze at them 'with open eyes, and pray evil for them in his name'. As rebellion flared in the early summer of 1614, the Devil brought Jonet prophetic news: 'Robert Stewart had gotten Birsay and would get the castle and country.' She evidently sympathised with Robert Stewart's cause. The dittay claimed Jonet was 'angry with the old admission of the estate of the country' – that is, the restoration of the tacksman's regime. It inspired her to summon the Devil who, in April 1615, led her at twilight onto the ramparts of the soon-to-be slighted castle. There 'she cried three oyez' – the traditional call, meaning 'hear-ye', preceding public announcements – and petitioned Satan 'to put their new rulers away' in order that 'the kindly [native] country people might rule'.

For the authorities, this alarming report was at the same time oddly reassuring: they doubtless pressed Jonet to tell all she knew about Satan's sympathy for the traitor Robert Stewart. The Devil was father of sedition, instigator of rebellion among the angels; Orkney's spate of witchcraft trials in 1616 helped to confirm that God was on the side of the restored royalist regime.

During investigations into a coven of witches in East Lothian in 1590 – suspected of summoning storms to impede Anne of Denmark's progress to Scotland (p. 147) – one of the accused confessed to hearing the Devil himself complain that 'the king is the greatest enemy he hath in the world'. James was surely delighted by this flattering acknowledgement of his expertise in countering the dark arts. In Orkney, the Devil admitted to Jonet he lacked power to depose the islands' new rulers: they 'would bide so long as God would'.[15]

Jonet Irving and Elspeth Reoch were sentenced to be burned at the Loan Head. Prior to their deaths, they were probably confined in

'Marwick's Hole' – the name given (for unknown reasons) to a small bottle-shaped dungeon, accessible only from above, wedged between south choir wall and south transept chapel in St Magnus – unique among British cathedrals in housing so sobering a monument of coercive power, a mute witness to the alliance of Kirk and sheriffdom as guardians of moral and civil order.

There are no first-hand accounts of any executions at the Loan Head, so we must rely on imagination and comparisons with elsewhere. Almost certainly they were public occasions, town- and countryfolk attending in considerable numbers. The mood of the crowd is hard to gauge – sombre, sympathetic, vengeful or festive? Who paid is a vulgar but valid question. There were fees for the hangman, and costs of stake, ropes and fuel for the pyre – most likely peats, with oil or tar to accelerate combustion. Some Scottish towns made efforts to recoup these expenses from escheat of the witch's property, or tried to force relatives to pay. Failing this, costs might be shared between burgh and kirk session.[16]

Figure 5.2: The entrance to 'Marwick's Hole', a windowless cell placed between choir and transept in St Magnus Cathedral. Women (and a handful of men) accused of witchcraft were incarcerated here before their trials and while awaiting execution.

The killings were intended as instructive spectacles. Ministers were present, to lead prayers and urge malefactors to throw themselves on the mercy of God. Final words were never formally recorded; their disappearance from the historical record is especially regrettable. We will thus never know how many Orkney 'witches' internalised what they had been told about their own guilt, or how many took a last-minute opportunity to declare their innocence – as Alison Balfour did, more formally, before her passage to the Loan Head (p. 153). The fire made its own statement. It was an instrument of moral purification, and a prefiguring of the agonising torments of hell. The witch's body, reduced by the flames, was rendered ineligible and unfit for burial in kirk or kirkyard, forever cast out from the confines of community.

The witch's incinerated body certified her magic to be the fruit of Satanic conspiracy, a secret declaration of war against God, and against God's earthly representatives. But Orkney witchcraft was at the same time a matter of keeping the peace with moorland fairies, of failures at fishing, and milk stolen from a neighbour's cow. The enormity and banality of evil sit with each other in puzzlingly incongruous juxtaposition.

Walliman and Rigga

We can pursue the paradoxes into a couple of further cases, where the meanings of magic were contested between accusers and accused. On 11 November 1629, sheriff-depute John Dick held court in Birsay, in the old palace of the earl. Before him was a woman named for a nearby parish, Jonet Rendall, though the dittay adds 'alias Rigga', and 'Jonet Rigga' was what several witnesses called her. Rigga is a rocky point on the Rendall coast, formerly the name of a small farm, which may have been Jonet's birthplace. But it seems unlikely to be merely coincidence that *rigga* was also a Norn term of scornful address for a woman, implying lankiness or ungainliness.

Several Orkney cases hint at familiarity with Norn, and point towards an alternative world of belief and custom, to which Scots-speaking ministers and judicial officers had little possibility of access. We have already met Anie Tailzeour of Sanday (p. 1), with her Norn nickname 'Rwna Rowa', or Red Runa. Writing in the 1920s, the Orcadian linguist Hugh Marwick remembered an old Rousay man telling him of a witch who summoned a storm by her vigorous churning of milk. As it spilled from the mouth of the kirn, she declared, 'Tara gott! That's done!' The Norse phrase – '*þat er gǫrt*' – was probably employed as an incantation down the preceding centuries. In 1633, another Sanday witch, Marion Richart, confessed to performing a healing ritual where she took a bucket of water and 'aundit in bitt' – which, the clerk of the court added, 'to expone it into right language, is as meikle as she did blow her breath therein'.[17] Norn was never, for wielders of authority and makers of record, 'right language'. But it hovers like a sea mist over the practice of Orcadian magic.

The cards were stacked against Jonet Rendall. She was indicted on twenty separate counts, and fourteen witnesses (all men) from Rendall and Birsay bore testimony against her. The accused in Orkney cases, even when convicted, were not invariably found guilty on all the individual charges. Jonet was. She was sentenced to be strangled and burned for contravening the 1563 Witchcraft Act 'and for company keeping with the Devil, your master'.

Her acquaintance with the Devil began two decades earlier, 'above the hill of Rendall' – somewhere along the windy ridge joining the summits of Hackland, Enyas and Gorseness Hills, and separating a narrow strip of farmed coastline from a cultivated valley to the west. Satan arrived at a vulnerable moment; Jonet had 'sought charity and could not have it'. He made her an offer, and though the dittay uses the theological language of a 'pact', the Devil does not seem to have demanded anything in return. He promised to 'learn you to win alms by healing of folk, and whosoever should give you alms should be the better either by land or sea. And those that gave you not alms should not be healed' – indeed, 'whatever you craved to befall them should befall'. Despite this promise of vengeance on the mean-fisted, Satan, it appears, wanted to make Jonet Rendall into a magical healer.

It was, however, the prosecutor who identified the mysterious figure 'clad in white clothes, with ane white head and ane grey beard' as the Devil. Jonet had her own name for him: Walliman. It seems very likely that her life-changing encounter was with a human not a supernatural being: the showman-warlock Oliver Leask, whose professional technique was precisely as described in the offer to Jonet, and who frequented this part of Orkney in the first decade of the seventeenth century. Jonet said she met Walliman 'in Nicol Jock's house in Hackland' – the Rendall township where, as we have seen, Leask offered to cure the chronically sick David Henry. Henry must have recovered, for, at Candlemas 1624, he gave Jonet a silver coin to make sure his horses would be well in the coming year.

If Jonet Rendall learned her secrets of curing and cursing from Oliver Leask, their last meeting must have been prior to Leask's banishment in the spring of 1616. In the succeeding years, he evidently assumed in her mind a more than merely human status. Jonet's confession spoke of 'praying to Walliman', and, in admitting nearly all the charges, she emphasised his power rather than hers: 'Walliman took away the profit of the ky'; 'Walliman slew the mares'.

Jonet's career as a charmer-extortioner, like Leask's, was an extended one: the earliest charges dated back over twenty years, and other incidents took place four, five, six or eight years prior to the trial. Something tipped uneasy tolerance into vengeful prosecution, though one of the most recent episodes, around Whitsunday 1629, was hardly the most serious. At the farm of Midland, on the boundary between Rendall and Evie, Jonet was unhappy with her payment for putting a protective spell on John Turk's batch of new ale, and so 'Walliman took away the profit of it'.

There had been some earlier brushes with the law. In 1621, Andro Matches of Sundiehous in Hackland denounced Jonet to the sheriff after losing the profit of his milk. Nine months later, he took a complaint to the Evie Kirk Session. Matches was clearly unusual in his willingness to cause trouble for Jonet; a few days later, she remonstrated with him for 'always dealing with … and complaining' of her, and said he would repent it.

Sure enough, three days later, Andro went out of his wits. His wife's reaction may seem curious to us, but is one commonly met with both in Orkney and elsewhere: she sent for the woman she believed to have cursed her husband. As soon as Jonet arrived, Andro began to feel better, and Jonet was rewarded with a plate of meat. Before tasting it, she spat three times over her left shoulder, and Andro's wife, 'fearing ye had been doing more evil', sprang forward to strike her. 'Let me alone,' Jonet snapped, 'for your goodman will be well.'

The image of Andro's anxious wife, unsure if she was witnessing a ritual of healing or harming, captures the uneasy relationship between those who knew how to work magic and those who were eager to benefit from its effects. People in the farms of the West Mainland put up with Jonet's presence because she offered useful services and because they were afraid of her. At Howakow, a small dwelling on the lower slope of Hackland Hill, Jonet's arrival at Candlemas 1629 prompted young Patrick Gray, 'fearing your evil', to rush to the barn to fetch an offering of corn. Five years earlier, pathetic scenes attended Jonet when she sought alms at Gilbert Sandie's house in Isbister township. Sandie said he had neither silver nor meal to spare, 'but bade his wife give you three or four stalks of kale'. Jonet departed in high dudgeon, with Gilbert's wife in pursuit, begging her to take the kale. Two of Gilbert's horses later died.

Jonet's discontent with the offering at Howakow exacted a still higher price. Patrick Gray fell sick, and slowly wasted. The family summoned Jonet to cure him, but he had died by the time she came. As soon as Jonet entered the house, Patrick's corpse 'bled much blood as ane sure token that ye was the author of his death'. This, only weeks or days before the trial, was the likely tipping point.

'Cruentation' – spontaneous bleeding of a corpse in the presence of its murderer – enjoyed powerful customary status as judicial proof of guilt. A distinctively island version of it is found in a Hoy case heard before the sheriff court of the bishopric in January 1623. Theft charges against John Sinclair led to the dredging up of an old accusation; that he magically sank a boat to restore the health of his sister. When four drowned bodies were recovered, and Sinclair forced to lay hands on them, 'they gushed out with blood and water at mouth and nose'. Bishop Law, before whom the case was originally heard, was perhaps not entirely convinced, but Sinclair was fined for witchcraft and undertook penance in the cathedral and in Hoy Kirk.[18]

Jonet Rendall's confession to nearly all the charges against her is at first glance puzzling. Her callous comment on Patrick Gray was that he offered her 'but shillings' – corn not yet ground into meal – and so Walliman slew him, 'as he promised he was true to her'. Coercion seems unlikely, as Jonet denied the summary charge of 'practising of witchcraft and sorcery'. In a male-dominated society, where social power attached to tenure of land, and women's status to motherhood and marriage, Jonet Rendall was a person of little account. Her abilities, and her relationship with Walliman, made her a somebody. It is perhaps little wonder that she revelled in her power, even as her boasting sealed her pitiable fate.

None of this had much to do with the designs of the Devil, as Protestant theologians understood them. References to the controlling hand of Satan appear everywhere in court formularies and charge sheets – 'by the direction of the Devil', 'by the power and working of the Devil, your master': they are the approved spelling and punctuation of Orkney witchcraft cases, but they rarely supply their underlying vocabulary and grammar.

Orcadians were taught to believe in the Devil, and associations were occasionally made. One strand in the 1624 prosecution of a Birsay woman, Marabel Couper, was a banal argument with a neighbour over some yeast. Marabel swore, as her 'soul might never see the kingdom of heaven', that she had none to spare, but Elspeth Thomson found some in her cellar. She reproved Marabel for having 'given yourself so to the Devil', and Marabel retorted she 'had nothing to do with the Devil'. A dramatic 'gotcha' is then inserted by the dittay. When Marabel lay in childbed, 'your company came and took you away … they fetch you and ye are with them every moon together'.

Mentions of covens or 'sabbaths' are exceptionally rare in the Orkney sources. There were witnesses to most of the charges against Marabel Couper, but not – unsurprisingly – to this one. Nonetheless, she confessed it. Or rather, as a marginal note makes clear, she 'confessed the going with the fairy'.

This was one of several cases where suspects ingenuously admitted to powers acquired from the fairies; the phrase 'going with' implying a trance or a dream-like state. Thomas Corse of Westray was another 'vagabond warlock', who at his trial in 1643 confessed, like Oliver Leask, to going with 'the good wights'. Corse's career as a charmer began after the fairies came to him in his sleep when he was 'travelling in the hill'. Elspeth Cursiter, burned in 1633, similarly claimed knowledge from 'going with the fairy'. She directed Thomas Seatter in Birsay to the whereabouts of a missing trout, as she was 'with the good wights when they took it out of the pot'. Conversely, when a Kirbister couple found some ale missing, and suspected 'the good wights had drunken the same', Elspeth was able to assure them otherwise. In all these cases, the court translated suspects' words and understanding of their own experience into 'right' theological language: 'the working, craft and knowledge of the Devil'.

Explicitly 'demonological' items were nonetheless rare in the Orkney witchcraft dittays, compiled by clergy from the raw material of neighbourhood accusation. When they did appear, they were charges on which assizes sometimes acquitted. In 1643, Jonet Reid from Sandwick was condemned to death on multiple counts, but found not guilty on the most eye-catching item in the dittay – that she was overheard by the fire saying, 'I am drying this corn to the Devil', and that the corn and a hot stone were then seen flying through the house.[19]

In August 1635, another Mainlander, Helen Isbister, was acquitted of serving the Devil, being cleared on nearly all the charges against her. Her trial shines revealing light on the social dynamics of witchcraft accusations, and on what assizes composed of local farmers considered pertinent and plausible.[20]

Helen's story lent itself to the demonological lens, as it involved encounters with a 'black man'. He appeared to her four or five years before the trial, when she was a servant at Hunclett in Holm. The visitation was heralded by howling of the farm dog, who afterwards refused to eat and died. Helen was 'troubled' by the visitant for the next two or three nights – proof, the dittay insisted, that 'the Devil conversed and lay with you'.

Helen had her own take on what, or who, the apparition was: Mr James Wilson, lately deceased minister of St Andrews, as recognisable as when 'he stood in the pulpit'. Another servant, Christine Coupland, revealed what Wilson said to Helen: 'he would never get rest in holy muld [earth] till that sin of hers with Alexander Irving were punished'. The sin, which appears to have been Christine's, was presumably fornication. We can only guess at the reasons for its naming by a minister's ghost – a mechanism for acknowledging guilt, or a ploy to punish a jilting lover?

Protestant theology had no truck with wandering spirits of the dead. Souls were in heaven or in hell, awaiting Last Judgement. Appearances of 'ghosts' were delusions of the Devil, to lure people into his snares. The assize at Helen Isbister's trial, however, was prepared to confirm only the death of the dog, and the reality of her vision of a 'black man'; implicitly, it accepted her interpretation of what she had seen.

Helen was acquitted on several serious counts: of slaying the horse of James Linklater in Bimbster (Harray), after he reproved her for walking through his crops; of killing a servant by giving him a glass of milk. She allegedly used magic to make James Grimbister fall in love with her niece, and caused the young woman to get over her physical aversion to James by means of a piece of bannock placed overnight between his skin and shirt. John Sinclair suffered after his sister sat with Helen 'upon your consultations almost the whole day' by the side of the burn at Hestwall in Sandwick.

Other charges related to Helen's cure of Adam Leonard, from the township of Ireland in Stenness. Coming over the hill into the neighbouring township of Bigswell, Leonard beheld a company of people, in their midst the dead wife of William Louttit, and he heard her saying 'it were no fault that that man that rides so high were made to lie'. Leonard was ill for several weeks, while he 'consulted' with Helen Isbister. She declared he would never be well till he acquired William Louttit's chair, and sat in it three times. Inevitably, Leonard's gain was Louttit's loss – over the course of the following winter, Louttit's cattle, oxen and horses died. Moreover, when Helen came to Adam Leonard's house, 'hoping to have received favour and benefit for curing of him', she was disappointed to receive from his wife only 'a sup of her worst milk', and went her way 'very angry through his ky in the hill', one of which died.

The assize, its members drawn from across the Mainland, acquitted Helen of all these charges. The presence of two men named Isbister perhaps weighed in her favour, though the assize also contained a Louttit, who might have been related to one of the alleged victims. Probably the assizemen considered the charges insufficiently proven, a corrective to assumptions that witchcraft accusations must always have triggered panicked credulity. Some were noted to rest on 'great presumption' – that is, suspicion and inference, rather than solid evidence.

The assize voted to convict on only one item: that when Helen was a servant at Sebay in the East Mainland she cast a spell upon the mice, 'which went by her enchantment into a stack of corn, and when it was castin in [taken down] they were found all dead in the heart of the stack'. Helen denied the other accusations, but admitted 'she spake some words' to the mice. Despite the demonological determinations of the dittay, it was not, the assize decided, the sort of thing to be laid at Satan's door, and she was released by the judge without punishment.

Charmers and Witches

Helen Isbister was a 'charmer', a woman of low status but considerable social utility. There were many of them in seventeenth-century Orkney, offering a range of services, curing disease foremost among them. Walter Stewart might boast of islanders kept healthy by 'honourable poverty' and 'ignorance of luxuries' (p. 179), but the reality was otherwise. Orcadians, like other pre-modern people, were prey to a spectrum of often inexplicable ailments. In a subsistence economy, where all adults – and children, too – had indispensable roles, disease brought not just personal misery, but risk of family impoverishment.

Pre-modern people, it is commonly claimed, were well aware that illnesses might have natural or supernatural causes. In Orkney, the distinction was blurred, and perhaps largely meaningless. Everyone knew that hidden forces were at work in the world – forces which, with appropriate words and techniques, could be manipulated to advantage or vexation. It was a mindset rejected as incorrigibly superstitious by the university-educated clergy, who maintained that everything – even the malign activity of Satan – took place only by God's express permission.

Not all superstition was seen as witchcraft, whatever the 1563 act might say. But habits once perhaps winked at were treated more seriously in the middle decades of the seventeenth century. In 1640, Magnus Grieve from Rousay passed along a chain of authorities – kirk session, presbytery, sheriff court – before being required to swear never to repeat 'his superstitious and devilish practice': 'going backward in a harrow to see what wife he should have, and how many children'. Harrows – framed devices for breaking up the clods of earth in ploughed fields – were often hand-drawn in Orkney.[21]

A more homely divination device, known throughout Europe, was the sieve and shears. The method involved suspending a sieve from a pair of shears or scissors (usually made in a single piece, with rounded handle), with its blades inserted into the rim, and its handle held lightly by the fingers of two persons. The sieve was watched for movement in response to questions, which might relate, for example, to the identity of a suspected thief. The Orkney Presbytery fretted about sieves and shears. Three St Andrews parishioners were sentenced to public penance for using them in 1641, and, like Grieve, were ordered to swear to desist from doing the like and to 'enact' themselves in the book of the sheriff. Two years later, ministers throughout Orkney were asked to search for Janet Sutherland, 'fugitive from South Ronaldsay for turning the sieve and shears'.

In 1643, the Orkney Presbytery passed an act against the practice, for 'censure and punishment of charmers, consulters, abusers and persons guilty of such foul crimes … these being common in this land'. It was for 'terror of others' that Agnes Bernardson was, in 1644, made to stand in sackcloth before the pulpit in the church of Walls; Inga Flett was similarly exhibited in Shapinsay's kirk, 'for abusing of the people'.

The conventional terminology of 'abusing' – that is, deceiving – made the clients of the charmers sound like guileless victims. That was perhaps a convenient fiction: despite the prescriptions of the witchcraft statute, it was simply impracticable to punish everyone who consulted with charmers, which might in any case lead in unexpected directions. Sometime before 1634, Alexander Gara sought the services of the charmer, and suspected witch, Christian Gow. Gara was a servant of the aged minister of Westray, Alan Hutton, and wanted Gow to cure his master's horse. Even more remarkably, the minister's wife, Joan Gibb, gave Gow a bannock as a reward.

But amnesty was not guaranteed. David Watson, Westray's new-broom minister, ensured Gara and Gow (though not, it seems, his predecessor's widow) were hauled before the kirk session. In June 1643, Elspeth Linklater and Andro Velzean from Harray were summoned before the presbytery, 'the one as ane abuser, the other as a consulter'.

There was no predictable trigger for a charmer to become suspected as a witch, though it was an ever-present possibility. The earliest surviving session minutes outside Kirkwall, those for Shapinsay, reveal that in January 1649 a stranger, Margaret Greeg, was lately come into the island 'to abuse the people' and 'suspect of witchcraft'. She was perhaps an adept of the sieve and shears, for 'she can tell them who it is who takes anything that is stolen'. A few years later, the session referred Margaret Wick to the magistrates for 'practices tending to delude the people that may savour of witchcraft'. When Katherine Manson was summoned before the South Ronaldsay Kirk Session in April 1660, the charge was 'curing sick persons and beasts by charming or witchcraft'.[22]

The sometimes slow process of turning a charmer into a witch can be observed in the case of James Knarstoun, from Tuskerbister in Orphir, convicted in February 1633 on all but four of sixteen charges. They included causing death by transferring sickness, and stealing the profit of milk, geese and grass. Yet it is clear that he was principally a healer, with a considerable medical reputation.

Alexander Duncan's wife visited 'to seek her health of him', and Magnus Aith, father of an ailing boy, was 'counselled to come to the said James'. When Katherine Mowat became ill, 'she was informed her disease was called the baneshaw [sciatica] and that there were divers in the country that could mend it, especially the said James'. Katherine was married to Hew Sinclair of Damsay, a laird; resorting to charmers was not the sole preserve of the rural poor. Knarstoun's business crossed the lines between countryside and town: hearing that John Fogo, a Kirkwall cooper, was seriously ill, 'ye came to him and persuaded to cure him of that disease'.

James Knarstoun was a repeat offender who had been up before the sheriff in Earl Patrick's time for 'using of the like sorcery and charms'. In around 1627, he was called before the presbytery, and admonished 'to abstain fra the like practices'. He gave certification in the sheriff court – an undertaking he was charged with breaching in 1633. In the meantime, he was in trouble with the Orphir Kirk Session, having quarrelled with an elder, Jasper Flett, who he declared 'should never sit in judgement upon his cause'. Despite Knarstoun's combative self-confidence, a steady accumulation of resentments, some attendant on inevitable failings as a healer, in the end weighted the scales against him. On 28 February 1633, the assize remitted 'sentence to the judges and doom to the dempster [public executioner]'; there is little doubt that he was burned.[23]

There was perhaps a connection with the Sandwick witch Jonet Reid, burned in 1643. Reid similarly claimed expertise with the 'baneshaw', and used an identical technique, washing the patient's legs, and touching the joints with small stones steeped in the water. She also used the same procedure as Knarstoun for curing the 'hart cake', cardialgia or heartburn: pouring melted lead into a bucket of cold water through the circular handle of a pair of shears. It is difficult to say how far there was direct instruction in methods of healing, practitioner to practitioner, or how far such cures formed part of a common stock of island knowledge. Remarkably, Jonet was executed, though none of her patients actually died and she faced no accusations of transferring their illness to others.

The ability to heal was thus not necessarily proof of benign character or intent, and could even indicate the opposite. It made for a febrile atmosphere around magical cures. Marabel Couper, condemned and executed in 1624, was first tried in 1616 but managed to avoid serious punishment on that occasion. The records of the earlier case describe a quarrel with John Mowat's wife, Agnes Ingsetter, who subsequently sickened. The couple begged Marabel to cure her, but she failed to come 'until they promised to delate [denounce] her for a witch if she refused'. Marabel duly turned up and laid hands on Agnes, who 'convalesced and received her senses again, by her devilry and witchcraft'.

Threatening witches with the law was a risky but sometimes necessary course of action. It was a strategy Marabel's Birsay neighbours perhaps discussed among themselves. Andro Cooper's wife unwisely tethered her cow in Marabel's yard while she went to speak with an acquaintance, and later found it 'lying stricken on head and feet'. The beast was set down before Marabel's door, 'assuring her if the cow overcome not again, she should be burned as a witch'.[24] As an end in itself, the punishment of witches was probably not a priority for the country people of Orkney, who were usually more concerned with ensuring that magical forces worked to their advantage and not their detriment. But when other possibilities of redress were exhausted, they understood well enough that the laws of kirk and kingdom placed the practitioners of magic in an exposed and vulnerable position.

Not all charmers were principally healers, or feckless wanderers. Bessie Skebister, a respectably married mother with her own servant, lived in 1633 in Walls, at the southern tip of Hoy. She was what would later be known as a 'spaewife', a diviner or soothsayer. According to her dittay, 'all the honest men of the isle declared that it was a usual thing when they thought boats were in danger to come or send to you the said Bessie to inquire of you how they were and if they would come home ... a common proverb is used: if Bessie says it is well, all is well'.[25]

Her case draws us into the anxieties, and occasional agonies, of a small community on the perimeter of the Pentland Firth. It is a snapshot in a much longer story, for the township of Brims, where Bessie lived, houses the Longhope Lifeboat Museum, commemorating the eight local men who drowned answering a distress call in March 1969, a tragedy touching virtually every family in the parish.

In March 1632, storms delayed the return of Alexander Thomson's vessel, and when one of its oars was found on the Walls shore 'all the people thought that boat was lost'. But Bessie advised them to 'take ane good heart, for they are all well and will be home before they sleep'. So it transpired. A few months later, as some Walls boats were fishing off Sule Skerry, a man was alarmed when he came across Bessie in a state of distress – 'all is not well if ye be weeping'. She told him she wept 'for the trouble they were in, but not for their deaths', and the fishermen duly returned safely.

As a confirmation of hope, Bessie was a valued community resource. To calm Christine Mowat's fears for her sailor husband, Bessie cast a sixpence into a bucket of water: if the side marked with the cross landed uppermost, 'then they are well'. The cross fell face down, but Bessie instructed Christine to 'tak' heid again', dropped the coin 'in ane other manner', and all was well indeed.

Around harvest time in 1632, however, folk in Walls became concerned about a ship headed for Norway, and Bessie's messages about it were equivocal. The local men evidently did not return, for the dittay speaks of 'that time that the ship perished'. Bessie's fate was sealed by Isobel Sinclair, a rival healer and seer, with an itinerant practice across the Mainland and South Isles. At her own trial, Isobel confessed to giving a man in Harray special herbs to make his corn flourish, but said 'she knew no witchcraft by it'.

Isobel was another who communed with the 'good wights'. It was alleged that for seven years past, 'at the reathes [quarter days] of the year, she hath been controlled with the Fairy, and that by them she hath the second sight'. On two occasions, she advised Agnes Moodie, Bessie's near neighbour in Walls, to dissuade her husband James Sandieson from going to sea, saying she would not see him again if he did.

She also claimed to have seen the Devil lying with Bessie Skebister. Around the time of the shipwreck, and while she was 'with the Fairy', Isobel had a vision of Bessie lying down in her house, a great strip of cloth in her mouth. Shortly before her own trial, Isobel directly accused Bessie of being responsible for the deaths of 'that whole kippage' (ship's crew). The claim was repeated, 'her being at the stake' – a dispiriting attestation that women condemned for witchcraft sometimes used their final words to implicate others in the crime.

It made queasy sense that the people with foreknowledge of events had power to bring them about. Suspicions about the Norway ship probably swung opinion in Walls against Bessie Skebister, who had already fallen out with Agnes Moodie and James Sandieson over their tenancy of the farm neighbouring her own property. James fell ill, and claimed he was tormented in the night by Bessie 'carrying him to the sea and to the fire, to Norway, Shetland and to the south … with ane bridle in his mouth' – possibly a description of the experience of sleep paralysis. Agnes demanded that Bessie 'come and look upon him'. But Bessie was stopped from doing so by her own husband, who took the view 'if ye gaid and if James Sandieson should either die or grow better it would be called your deed'. He was cynically insightful: once suspicions took light, both good and bad deeds were fuel for the fire.

James did, it seems, recover and join the crew of the ill-fated ship; Agnes was a widow by early 1633. Isobel Sinclair had a premonition of his fate, and a vision of 'Bessie Skebister worrying him abune the craigie' (about the neck). The dittay's fifteen charges against Bessie did not include the sinking of the ship, and the assize convicted on just six counts, giving no credit to the nocturnal harrying of James Sandieson. But it was enough to justify a sentence of death on a woman 'reputed and holden ane abuser of the people and ane dreamer of dreams'. Bessie herself admitted 'the casting of the sixpence in the water', and denied all the rest.

Only the toss of a sixpence separated the healer from the curser, the wise woman from the witch. The people accused of witchcraft in Orkney were often socially marginal. But they were rarely complete outsiders, and more often than not were integrated into their communities in a variety of co-dependent ways. Of one thing the trial evidence leaves no doubt: witches and their 'victims' had a common understanding of the world's problems, and they shared a strong belief in the transformative power of magic.

Methods in the Magic

Practices of witchcraft and charming were tightly entwined with Orkney's physical environment of soil, stone and sea, all pulsing with latent magical powers. Katherine Caray, a widow burned in 1617, cured Oliver Hewison in Evie with earth taken from a field where a man had been killed. She also made use of a curative technique she claimed to have learned on a visit to Caithness. There, at sundown in the hills, she met 'ane great number of fairy men' and three women, one of whom took her into a little house where a man lay sick. The fairy woman 'took three stones, ane for the spirit of the hill, ane for the kirkyard, and ane for the sea and laid them in the fire under the ash'. The hill stone and kirkyard stone made 'great murmuring', but the stillness of the third stone allowed her to infer 'the sick person to be troubled by the sea'.

The diagnostic method was likely already known in Orkney, as it featured in the repertoire of the veteran charmer James Knarstoun. He similarly picked up 'ane stone for the ebb, another for the hill and the third for the kirkyard'; although, rather than putting them in the fire, he boiled them in water before placing them overnight above the lintel of the sick person's house – it was a staple of magical thinking that lintels, thresholds, hearths and windows were in need of protection, as, like the orifices of a human body, they were points of potential entry for malign forces. The following day, Knarstoun placed the stones in water again, 'using some words known unto himself', and by observing sounds and movement was able to tell 'what spirit it is that the person diseased has, and so to call them home again'.

In Rousay, in early 1640, as Robert Robson lay deadly sick, Katherine Craigie offered to tell his wife, Jonet, 'whether it was ane hill-spirit, a kirk-spirit or a water-spirit that so troubles him'. According to the dittay, Katherine's own curses caused Robert's illness, Jonet desperately promising she would 'never reveal anything upon you, if ye helped her goodman'. Katherine's method was in essence the same as Knarstoun's, though she placed the stones under the threshold in preparation for the water test which led to her declaring, 'Jonet, it is a kirk-spirit which troubles Robbie'. The remedy was to wash him with water wherein the stones had lain.

Katherine was acquitted of witchcraft in June 1640, but condemned at a second trial in July 1643. In between, she diagnosed others with the three-stones technique. They included Thomas Irving, troubled with 'the sea trow'. He followed Katherine at dead of night, in silence and 'sore afraid', to the shore at the Bay of Saviskaill, where she cast three handfuls of water over his head.

Katherine Craigie was another healer reclassified as a witch after suspicion of harming those who crossed her. Her client base was wide, and included Isobel Craigie – perhaps a relative – who was the widow of a local laird, George Traill of Westness. Katherine offered her a special grass, which, when used correctly, would cause her to win the love of Harry Bellenden – great-grandson of the famous Patrick Bellenden of Stenness. Unusually, Katherine confessed to the sources of her knowledge: a charm for stemming blood was taught her by her late husband, and she 'learned the charm by [involving] the using of the stones' from a woman called Elizabeth Linay.[26]

Stones and water are everywhere in the trial accounts of healing – they exemplify magical 'rules' of touch, transference and sympathetic action. Ailments passed from suffering bodies into the inert substances, rendering them potentially animate – and perilous. William Swenton was reported to have died as a result of James Knarstoun dropping before his door the soggy herbs with which he washed the limbs of Katherine Mowat. The theme is present in the first trial of which we have a full record, Katherine Bigland's in 1615, and also in the very last, that of Katherine Taylor before the Stromness Kirk Session in 1708. She threw into a 'slap' – a gap or opening in a dyke – the water used to wash a man called William Stensgar (about whom we will hear more later; see pp. 330–1). George Langskaill, passing through the slap a few hours later, 'was overcome by bodily indisposition'.[27] Ritual washing both symbolised and effected a flushing-out of the sickness-causing spirit. In a religious context, the action might be called sacramental, though any comparisons with baptism would have struck ministers as ludicrous and blasphemous.

The habit of garnering stones – from the ebb, kirkyard and hill – mapped onto the body itself the markers of Orkney's peopled environment, with its physical and figurative boundaries of shore, hill-dyke and kirkyard-wall. Sickness ensued when the boundaries were in some way transgressed. James Knarstoun's emphasis on 'calling home' diseased persons from kirkyard, sea or hill implies some form of supernatural kidnap and substitution – akin to the fear of fairies leaving changelings in place of abducted human infants. This conception would long remain a part of Orcadian culture. In the later nineteenth century, people in Rousay suffering from physical or mental ailments called for help to 'tak' them oot o' the hill'.

Was a water-spirit the same as a sea-spirit? Lochs provided another set of boundaries in the island landscape, and the ancient belief in spirits inhabiting water seeped into the veneration accorded to holy wells. One of Knarstoun's cures involved sending a man with a sick sister 'to the well called Mary Well in Kirbister betwixt midnight and cockcrow', to fetch water for washing her.

Hill-spirits were fairies, or trows, or the hogboons inhabiting mysterious mounds not yet understood as Neolithic burial sites. 'Kirk-spirits', the third and most recent addition to the triad, were likely ghosts of the dead. Jonet Reid advised a man 'sore troubled in his sleep with apparitions of his first wife' to visit her grave in the kirkyard and 'charge her to lie still and trouble him no more'.[28]

The mental world of the charmers and their clients was not, by any recognised standards, a respectably Christian one. But Christian concepts formed an important element of it, and there is little reason to doubt that the people accused of witchcraft thought of themselves as Christian believers. Trial records preserve snatches of charms: Katherine Caray was taught by 'ane little black dog' to heal injuries with a 'wresting thread' ('wrest' means a twist or sprain). This was to be applied to limbs with the invocation, 'in the name of the Father the Son and the Holy Ghost, bone to bone, sinew to sinew, and flesh to flesh, and blood to blood'. Christian Gow cured the minister of Westray's horse by telling it: 'Three things hath thee forespoken [bewitched], Heart, tongue, and eye, almost; Three things shall thee mend again, Father, Son, and Holy Ghost.' Similar invocations of the Trinity, in English or Latin, or of Our Saviour and St Peter, appear in other charms and incantations.[29]

Sites as well as words of Christianity radiated power. Kirks and chapels were implicated in the ways of magic, as we saw with Katherine Bigland (p. 190). Helen a Walls, one of the seven banished witches of 1616, argued with William Holland in South Ronaldsay and then 'ran to the Lady chapel hard by, and went thrice about it upon her bare knees, praying cursings and maledictions'. This was in all likelihood the chapel built by the Halcro family as a token of their Catholic devotion (p. 100). John Sinclair, suspected in 1623 of boat-wrecking to restore his sister's health, was also accused of carrying her backwards on a horse to the kirk, and going around it seven times. In Rousay in 1640, Katherine Craigie's healing of William Flaws involved persuading his wife to walk with her before sunrise around the Loch of Wasbister, and circumambulate the Corse [Cross] Kirk, a chapel and burial ground along its northern shore. John Budge appeared before the South Ronaldsay Kirk Session in 1660, and confessed that 'in time of his heavy sickness he went to the old chapel in Grimness called St Colm's Chapel', most likely by instruction of the healer, and suspected witch, Katherine Manson.[30]

Magic and religion were parts of a common system, the prescriptions of the charmers resonating with ingrained habits of pilgrimage to old kirks and chapels, habits the Protestant Reformation found exceptionally hard to shake. In 1615, the sheriff court ordered baillies to assist kirk sessions in 'suppressing all idolatry, specially of walks and pilgrimages', and four years later the ruling was echoed by the Kirkwall Session, threatening punishments for 'such as goes on pilgrimage' to the Bairnes of Brughe, or to a place that is barely legible in the manuscript but must be either Birsay or Damsay.

Jo. Ben.'s account of pilgrims at the Bairnes of Brughe in the late sixteenth century (p. 88) has them walking around the chapel, and 'throwing stones and water behind their backs'. The overlap with healing methods advocated by charmers can hardly be coincidence. The prime motivation for pilgrimage was the prospect of recovery from disease, and leaving stones was a means of shedding and depositing sickness at a place of hoped-for cure. The South Ronaldsay Kirk Session, in response to a recent inhibition from the bishop on 'consulting', resolved in 1666 to crack down on 'charming and charmers, haunting of superstitious places, as chapels or wells, or other the like unlawful means of their healths'.[31] The campaign against charmers fuelled the drive for Protestant Reformation in Orkney. The 'popish' instinct to assign sanctity to landscape features such as wells, or to buildings like old chapels, was not yet expunged from the hearts of the people.

Yet none of this was really popery in the sense some ministers may have feared, a principled preference for the proscribed teachings of the medieval or Counter-Reformation Church. Rather, it betokened a prudent and wary regard for long-hallowed places, words and objects. This was an inclusive instinct, which could find room for new media of magical power. In the aftermath of the revolution of the 1690s, Alexander Pitcairn, minister of South Ronaldsay, was accused of failing to visit the sick. He said this was because 'many of the vulgar desire the prayers of the Church upon a superstitious rather than a religious account, thinking that if they have any lingering disease they will thereafter either end or mend' – in other words, they did not grasp the clear distinction (clear at least to Pitcairn) between a prayer and a spell.

In 1664, South Ronaldsay witnessed an eye-catching illustration of adaption in the magical mindset, when the Peterkirk was obliged to acquire a new sackcloth – the rough fabric garment worn by miscreants for performing public penance – after the existing one was stolen (see Fig. 8.2). The culprit was Jonet Budge, 'making use of it to her own body and her ... distracted daughter'.[32] With superlative irony, a disciplinary object used by the Protestant authorities to punish peddlers of superstition was transformed into an instrument of magical healing for body and mind – an item infused, in Jonet's comprehension, with the sacral power of the Kirk.

The most remarkable instance of such appropriation relates to the island of Stronsay, and may or may not have taken place. The folklorist Walter Traill Dennison recorded a tradition about the Stronsay witch Scota Bess, who lived around 1630. She conjured up mists, and lured ships onto rocks, from her vantage point in 'the Maiden's Chair' – a natural seat cut into the cliffs of Mill Bay, on the island's east side. Conceivably, 'Scota' is a version of *skodda*, a Norse term for a sea-fog – further evidence of Orkney's witches occupying the shadow ground between old and new linguistic cultures. Bess fell victim to a shocking act of vigilante violence, lured by a group of men to a barn at the farmstead of Huip, and there beaten to death with their flails. Her murderers took the precaution of first dipping the flails in water with which the kirk's communion cup had been washed.[33] There is no contemporary verification of the episode, but local memory of it testifies to the deep-rooted belief in spiritual properties passing through the medium of water.

Other operative principles had little to do with the trappings of Christianity, Catholic or Protestant. Tying knots in a thread could mend sprains, or untying them unleash pent-up winds. The magical power of unravelling extended to literal letting-down of hair – an act which transgressed expectations of female public deportment. Helen a Walls pulled off her 'curche' or cap and 'shook her hair about her' during her argument with William Holland. Jonet Sinclair of Westray was observed acting suspiciously before her arrest in 1643, 'with her hair about her lugs [ears], carrying a stoup [bucket or pitcher] of water'. One of the allegations against Bessie Skebister was that, in the presence of Margaret Moodie, 'ye sat down and taking off your curche shook your hair loose'.[34]

Another accusation made against witches was of doing things 'withershins'; that is, contrary to the course of the sun, or anti-clockwise. William Scottie, a Rousay 'vagabond warlock' put on trial in 1643, was seen walking 'witherwards' around the house of William Okilsetter, whose wife fell deadly sick. In the manner we have observed already,

Scottie was called upon to cure her, and the sickness passed fatally into a 'young mare foal, that was standing upon the house floor' – a reminder of the close proximity in which humans and animals so often lived.

In another 1643 trial, the Birsay witch Marion Cumlaquoy 'turned herself three several times round witherwise about the fire' after coming to Robert Captain's house with an unexpected gift of milk for his mother. Afterwards, Robert's oats would not grind, and his bere turned blue and rotten, though 'both were fresh and good when he put them in the yard'. In Westray, about 1626, Jonet Forsyth was refused corn at James Rendall's house. She 'went to the barnyard, and faddomed [encircled with her arms] ane of the best stacks in the yard about, contrary to the sun's course'. Magnus Irvine, elder in Shapinsay, reported Margaret Wick to the session in 1659 after she was seen coming from his dwelling, bowing several times and 'glomering' [groping] with her hands. She departed, 'having gone against the sun about the town[ship]'. In South Ronaldsay, in early 1662, the session took seriously reports that a man had gone 'twice weather-ways and once sun-ways' around James Windwick's boat.[35]

Boats were items of value and pride: one of the benefits of witchcraft trials – for historians at least – is the light they shed on the life and culture of ordinary Orcadians, drawing our attention to objects on which people placed worth, or matters about which they felt keenly. Beyond the ever-present fear of sickness and death, we hear much about specific dates in the calendar, and associated activities whose success, never assured, was crucial for well-being and prosperity: seed time and harvest, the brewing of ale, grinding of meal and churning of butter.

There are recurrent references to ploughs, prized items in the inventories of Orkney farmers, but intricate and often surprisingly fragile objects. Early in 1616, Anie Tailzeour, alias Rwna Rowa, quarrelled with a North Ronaldsay father and son, William and Thomas Burwick, at work in the fields at 'fauchland time', when fallow ground was put to the plough. Begging for a handful of bere, she was rebuffed with taunts: 'Away, witch!' When William and Thomas yoked their oxen, they found 'the plough would not enter in the ground'. Worse, it came to pieces in their hands: 'culter and sock [the attached iron cutting blades] gaid off the plough by your witchcraft'.

Helen Isbister, allegedly, 'so distempered the plough and goods' (animals) of John Sinclair 'that he could get no labour of them'. One morning in Birsay, sometime before 1616, Geillis Sclater passed Robert Gray as he was yoking his plough, but she 'would neither speak to him nor bid him God speed; fra that time forth his horse did him no good'. Such neglect of neighbourliness probably seemed especially egregious in the season of ploughing, a communal requirement of Orkney's runrig field system. An element of guilt perhaps underlay an accusation against another of the witches banished in 1616, Agnes Tulloch. Robert Mowat refused her the loan of an ox, 'in respect it was the first day of his tilth' (ploughing). He said she might have it the next day, but Agnes never came back for it, and since then 'the whole strength of his tilth has decayed'.[36]

The ubiquity of livestock in trial records mirrors the priorities of a mixed pastoral–arable society, and the vulnerability of beasts that for much of the year were left to fend for themselves beyond the hill-dyke. Relatively few cases say much about fishing, though Anie Tailzeour, staying in the house of Thomas Muir, was said to have woken him from sleep one storm-blasted November night and 'bad him rise up and gang about the shore, and he should find fish'. On one of the sand-banks of Sanday's western coast, Thomas came across a beached whale – a piece of remarkable good fortune, 'foretold by your witchcraft and devilry'.

Sheep appear rarely in the dittays, probably because of their comparatively lower value, and their habitual grazing away from places of settlement. Of greater worth, monetarily and emotionally, were horses, mentioned in over 40 per cent of the trials for which we have a detailed record. This was sometimes because a witch was said to have cured a horse, but more usually because she was suspected of killing one; dittays sometimes name the 'best horse', to underscore the enormity of the crime and the pain of the loss.

Cows were more common still, featuring in over 60 per cent of the trials. As sources of milk, related dairy products and eventually meat, they were productive in ways horses were not, and many cases involved loss of the animal's 'profit'. Anie Tailzeour was quizzed by the minister of Cross and Burness, Thomas Cock, and the baillie of Sanday, Thomas

Sinclair, about 'how ye took the profit of the kyne and gave it to others'. She described taking three hairs from the cow's tail, three of her own pubic hairs and 'three of her paps'. Anie walked three times around the cow, withershins, before striking it on its left side, casting the hairs into a churn and saying, 'come, butter, come'. The churn's owner would thus get the whole profit of the herd to which the cow belonged.

Another Sanday witch, Marion Richart, had a method for getting milk's profit back, telling a woman in 1633 to go and count nine waves breaking against the shore. When the tenth came in, she was to gather three handfuls of water into her bucket, 'and when thou comes home, put it within thy churn, and thou wilt get thy profit again'.

The Orphir charmer James Knarstoun dispensed extravagant promises about profit and yield. He 'caused all his neighbours' wives to follow him to the field ... each of them to milk their own ky'. Knarstoun placed the milk in a large pot, and proceeded to 'yearn' it – that is, curdle it, the first step in the cheese-making process. But something went awry: when their individual shares were returned, the women complained James had removed the profit.

Knarstoun had cattle problems of his own. One of his cows being so sick it could not stand, 'when he came in, he struck the cow with his foot and said, "Rise up in the Devil's name, for I have been far enough seeking you – I have fetched you from the hills of Hoy!"'[37] Whether this was a serious invocation of the Devil, or a simple cry of frustration, is open to question. But Knarstoun evidently applied the same technique of spirit-healing – fetching home from the hill – to cattle as he did to people. He 'came in', rather than went out, to the beast – as an English visitor noted a few years earlier, families 'eat and sleep on the one side of the house, and their cattle on the other'.

Cow cures were various, and of sundry efficacy. In 1623, Katherine Grant's prescriptions anticipated those of Marion Richart in drawing on the restorative power of the sea and virtue of the number nine. A sick cow was to be led backwards in the water, until nine waves washed over it, with three handfuls of each cast over the beast's back, which was then brushed with a handful of burnt malt straw. More prosaically, Katherine Manson in 1660 advocated placing 'something like ground salt into ane clout' and putting it on the cow's back'. The cow died, perhaps because her client forgot an additional instruction to pull up some rushes and feed them to the animal.

Elspeth Cursiter's charm involved a cog of water from the burn running before William Anderson's door, into which three straws were dropped, one for Anderson's wife, one for William Coitt's wife, and one for William Bichan's wife. The straws danced in the water, as if it was boiling, and Elspeth placed them in the beast's mouth, with 'your arm to the elbow in the craigie [throat] of the cow'. The animal immediately rose up, 'as well as ever she was'. But, as so often, a price was exacted. At the instant William Anderson's cow was healed, his ox, grazing on the hill, collapsed and died.[38]

None of this had much to do with the Devil, or any teachings of orthodox Christianity. These were in all likelihood age-old beliefs and methods, which become visible to us at a particular historical moment. For the authorities, in an era of profound political and religious realignment, Orcadian folk beliefs about healing and cursing could not be allowed simply to coexist with the doctrines of the Kirk; they had to be made to conform. Maladies of person and beast, malice of neighbours, lost profit of bere, butter and milk; all were weapons for Satan's armoury in his primordial rebellion against God. Traditional remedies were not the cure, but part of the problem.

The Arc of Persecution

In April 1644, the Orkney Presbytery issued a decree: 'the Elf-belt to be destroyed, in respect it hath been a monument of superstition, as was declared by Mr George Graham and Mr James Aitken, and the silver of it being melted, payment to be given out of the box unto the owner for the same'.

Nothing else is known of this intriguing item, though it seems to have been a familiar one in seventeenth-century Orkney. George Graham was minister of Sandwick and Stromness, and James Aitken of neighbouring Birsay, so the Elf-belt's owner likely lived in the far West Mainland. An 'elf-belt' must have been a protective device, perhaps to guard against being seized by the trows; it was in the same category as the 'trow's glove'

Katherine Craigie laid on the neck of a Rousay woman in in 1640, after asserting her disfiguring boil was not the result of Katherine's curse, but of 'the trow that had gripped her'. That the elf-belt was inlaid with silver makes it unlikely, however, to have been the instrument of a poor, itinerant charmer; its owner was a person of sufficient note to be compensated, not prosecuted.[39]

A thought – no more – calls to mind the ornamented girdles adorning saints' statues in the later middle ages, and periodically lent to women to assist conception or for protection in childbirth. Could the Elf-belt have been such an object, removed from its setting at the Reformation, and endowed with new meanings through the succeeding decades? We cannot say, though its destruction marks a moment when the Orkney clergy were demonstrating a renewed eagerness to repress the old religious culture. In January 1644, George Johnston, minister at St Magnus, 'did rebuke some persons of the vulgar sort from the pulpit for their superstitious observation of festival and holy days'. Only months earlier, as we have seen (p. 120), the presbytery ordered the destruction of a statue of St Peter that had somehow survived in Burray. A few years later, in 1648, the Kirkwall Session was threatening public penance in sackcloth for anyone 'found to make use of that most fabulous letter brought hither from Norway, or superstitiously keeps the same up'. No further details of this mysterious item have been uncovered, though 'fabulous', meaning 'deceitful', was not a compliment. One possibility is a copy or extract from the notorious Norwegian genre of 'black books' – manuscript compilations of charms, counter-curses and medical knowledge that circulated widely in that country from the mid-seventeenth century.[40]

These spurts of local godliness in Orkney came at a time of seminal national developments to be explored more fully in the following chapter – the signing of a Solemn League and Covenant, by which Scotland agreed to help English Parliamentarians, in revolt against Charles I, implement a long-hoped-for pan-British reformed religious settlement. In such times, the rooting out of sin was of profound political importance, and almost no sin was greater than witchcraft.

More than a third of the known Orkney cases (thirty-one) can be assigned to just two years, 1643–4. This was, indeed, a witch-hunt – a moment when ministers actively sought information about suspects in their parishes, forwarding their findings to the sheriff-depute. Witches began appearing in batches, a sign they were naming each other, rather than manifesting as a result of independent accusations. In April 1643, the presbytery considered 'points of witchcraft alleged against ... Thomas Corse, Christian Marwick, Jonet Sinclair and Helen Hunter in the isle of Westray'. Two months later, it weighed evidence relating to women 'suspect of witchcraft' in Sheriff-Depute Harry Aitken's custody: Margaret Thomson, Helen Tailzour, Barbara Yorston, Janet Pekol. All hailed from little North Ronaldsay.[41]

The actual practice of magic in Orkney parishes was in all likelihood fairly constant across the seventeenth century. But the distribution of witchcraft cases reflects official anxieties and willingness to prosecute. As we have seen, there was an earlier clutch of prosecutions in 1615–16, when the new regime was making its mark, and possibly – though we cannot be certain about this – because the abolition of Norse law simplified the framing of capital charges.[42] There were few prosecutions through most of the 1620s, with a slight rise at the end of the decade, and a noticeable spike in the early part of the next one: 1633–5 witnessed nine trials.

The early 1630s were a traumatic moment for Orkney. A chronicler described the years 1631–3 as a time when famine was so severe in the Northern Isles that 'horse flesh was good cheer there'. The experience seared itself on popular memory: a full half-century later, the hungry year 1631 was said to be a point 'from which the common people usually reckon'. In April 1634, a petition to the privy council from the bishops of Orkney and Caithness reported 'such tempestuous and bitter weather' that the harvest was destroyed before it could be taken in. People were reduced to eating seaweed, and dogs, and 'some have desperately run in the sea and drowned themselves'. It was believed in the following century that 3,000–4,000 perished.

The privy council arranged for charitable collections throughout the kingdom; inadequate, according to a petition signed in February 1635 by thirty Orkney lairds and ministers. It made a powerful appeal as to why the state of Orkney, struck by 'a most fearful famine', should concern the king himself, considering 'the ancient note and worth of these islands, his Majesty's own property'. Orcadians 'are great supporters of the kingdom in many things', their homeland lying 'in the sight of all strangers trading and frequenting thither from all the northern parts of Europe to all parts of the world'. The northern periphery was a show window of the kingdom.

Famine may have contributed to an atmosphere of competition and suspicion, but no direct reference to dearth conditions appears in the trial dittays of the early 1630s. Perhaps when there was so much hardship and death, it was more difficult to attribute it to individual acts of malice. 'The heavy hand of God' was the preferred causal description in complaints emanating from Orkney at this time.[43]

Witchcraft trials, in Orkney as elsewhere, were in any case never straightforwardly barometers of the social or political weather. They had their own timetables and logic, with often unpredictable triggers. There was no uniform 'national' pattern of witchcraft prosecution, though the largest spikes in known accusations – so-called witchcraft 'panics' – came in 1629–30, 1649–50 and 1661–2, all periods when (with the partial exception of 1629) very few trials took place in Kirkwall. Orkney's own peak, as we have seen, was 1643–4 – not one of the largest panic years in the rest of Scotland, but undoubtedly a moment of heightened anxiety.

The General Assembly was showing renewed concern with witchcraft, passing edicts in 1640 and 1642 linking it to charming. The association was intensified in a third ordinance of August 1643: 'because charming is a sort and degree of witchcraft, and too ordinary in the land, it would be enjoined to all ministers to take particular notice of them, to search them out, and such as consult with them'. At the General Assembly in Edinburgh, the Orkney Presbytery was represented by its moderator, Walter Stewart, and Patrick Graham, minister of Holm.[44] Orkney's permeable line between charmers and witches perhaps helped turn a ministerial crackdown on the former into a comprehensive roundup of the latter.

Personal factors were also in play. Henry Smyth, minister of Shapinsay, was convinced that the sudden death of his mother-in-law in March 1632 was caused by 'hard speeches' from a woman called Marjorie Paplay. Belatedly, in November 1642, he urged Sheriff-Depute Henry Aitken to action. Aitken was called before the presbytery to confirm that when

Elspeth Cursiter was tried by him in 1633 she declared 'the said Marjorie was the greatest witch of them all'.[45]

Marjorie Paplay, however, was no wandering charmer or North Isles farmwife. She belonged to one of Orkney's oldest and most illustrious families, her father a prominent servant to Earl Robert. Prior to her widowhood, she was wife to two prosperous burgesses of Kirkwall. Her eldest son by the first of those marriages, the merchant James Baikie, was perhaps the wealthiest man in Orkney, owner of a fine landed estate in Tankerness, and a recently acquired grand Kirkwall townhouse. Smyth probably launched his action because he feared Baikie was about to sue him for slander.

Through 1642–5, Marjorie Paplay's fate divided respectable Orcadian opinion down the middle. Baikie, with a network of well-connected kinfolk, and support from the Kirkwall Session and a couple of clerical allies, used all possible legal and procedural expedients to stymie the investigation. Smyth's allies included most of his ministerial brethren, as well as the commissary of Orkney, John Aitken, son of the sheriff-depute and brother to the minister of Birsay.

The case widened in the autumn of 1643 with the arrest on suspicion of witchcraft of Barbara Boundie, an itinerant recently returned from Shetland. At the time of her imprisonment, she was 'giving herself out for a discoverer of witches'. John Aitken avidly questioned her on what she knew about Marjorie Paplay, and seemingly persuaded her to implicate others – two of Marjorie's sisters, her daughter Elspeth, and Ursula Foulzie, wife of Edward Sinclair of Essenquoy, a former sheriff-depute, and daughter of Gilbert Foulzie, first Protestant minister of the cathedral. These alleged witches were not friendless old women, but wives and mothers of Kirkwall's urban elite.

Barbara, confused and terrified by interrogations with grimly intimidating men, changed her story. Questioned by the presbytery on 9 November, she retracted these allegations, as well as a claim she witnessed the Devil 'having carnal copulation with Marjorie Paplay' upon the Ball-Ley – a patch of common land in Papdale, on the eastern edge of Kirkwall.

The ministers, or Smyth's allies among them, had put words into Barbara's mouth to incriminate others and make her confessions fit the expected pattern of forming a pact with the Devil. What she actually said was that 'the fairy appeared unto her beside the ball-ley' to tell her about Marjorie Paplay. During one interrogation, Patrick Wemyss, minister of Hoy and Graemsay, tried to link Barbara to suspected demonic activity in his own parish. He demanded to know 'if she was one of the fourscore and nineteen that danced on the links of Moaness in Hoy'. Barbara admitted she was.

We learn nothing more about these ninety-nine dancers, cavorting along a sandy beach on the north-eastern coast of the island, near the jetty where today a small ferry deposits foot passengers from Stromness. The implication is of a large 'sabbath' – a suggestion unique in the Orkney witchcraft material. But these dancers Wemyss heard rumours about were more probably fairies or other nature-spirits; Barbara was said to have 'been with the fairy' during her travels in Shetland. The presbytery was in any case sceptical: Barbara, by her own confession, was first deceived by the Devil six years ago, and it was 'eleven years, or thereby, since the dancers in Moaness were first spoken of'.

Our knowledge of Barbara's case ends ominously: a note in the presbytery minutes in November 1643 to apply to the privy council for a warrant 'to put Barbara Boundie to tortures' – the sole explicit mention of torture in the Orkney trial records. Another note from April 1644 insisted she remain in prison until dittays arrived from Shetland. Her fate thereafter is unclear. There is a telling contrast between her treatment and that of Marjorie Paplay. Despite lingering suspicions, Marjorie does not seem ever to have been formally indicted, and neither were any of the other high-ranking women named as witches by Barbara Boundie.

There would be further trials of witches in Orkney, and perhaps more than we know of, given the fragmentary state of the sheriff court books for the 1660s. But 1644 was a high-water mark, and perhaps a kind of watershed. For most Orcadians, witchcraft accusations based on naming by other witches, and on flamboyant demonic allegations, were intrinsically less plausible than charges rooted in experiences of malice and misfortune. And when the dynamics of 'chain-reaction' witch-hunts started to single out wealthy and important people, the authorities began to back away.

Slander and Imprecations

Witchcraft prosecutions did not completely come to an end in Orkney in the second half of the seventeenth century. Increasingly, however, when references to magic and sorcery appear in court records, it was accusers who were subject to discipline – for defamation. In 1672, for example, Elspeth Smith complained to the Hoy Session about Marion Mangie, 'for slandering her as guilty of witchcraft'. Marion put it about that Elspeth, in buying a pint of ale from her, 'did thereby take awa' the fruit and fushion [wholesomeness] of ane dozen of pints or thereby that remained in the vessel'.

Slander charges had always been a possibility. In November 1635, Elspeth Thomson took action in the Kirkwall Session against Patrick Murray after he was heard saying 'Elspeth was a witch ... her hail family were a batch of thieves and witches'. In March 1665, Alexander Flett of Evie accused Francis Auchinleck of slandering his wife, and the presbytery issued a summons to Auchinleck, calling his bluff with a demand 'that he may either intend process against the said Alexander's wife for witchcraft, or else he shall be reputed as a slanderer'.

Marjorie Paplay likely died in the late 1640s, but echoes of her case reverberated through the following decade. In May 1658, her son James Baikie complained to the Kirkwall Session against Margaret Robeson on behalf of his sister Elspeth – one of the women named by Barbara Boundie in 1643. Margaret had repeated a story circulating in the town that a suspected witch, Elspeth Hirdie, 'cast stones at the said Elspeth [Baikie]'s gate and said if she were burned for a witch the said Elspeth should be burned also'. Margaret was sentenced by the session to stand for two hours in the 'jogs' – an iron collar attached by a chain to a post or pillar. But success in actions against slander was never assured. A few weeks later, when Elspeth Hirdie was in custody, Alexander More and his wife Jean Gunn complained in the session against Helen Oback, for saying Jean 'consulted' with Elspeth Hirdie to seek the death of Jerome Aitken. Helen was able to produce four witnesses who heard Elspeth Hirdie say Jean had sought her services, and the session referred the matter to the justices of the peace.[46]

Witch-accusers being punished as slanderers points to a gradual decoupling of folk beliefs from the diminishing demonological preoccupations of the ministers and courts. In Shapinsay, in 1663, Magnus Cumming approached the session with accusations against John Stiaquoy, a man 'given to cursing and praying ill for folk's beasts'. Magnus had lost a horse, and a cow belonging to Marion Drever had 'turned mad and ran upon the sea'. But the matter ended with Cumming doing penance for slander. It echoed a case from two years earlier, when Margaret Gunn was ordered to make public satisfaction for slandering Margaret Gray 'in blaming her as having cast sickness on her cow'. The slander was confirmed by the testimony of two other women. This angered John Michal, who jumped up in the session to say he would 'not give a bubble for any woman's witness-bearing'. But the minister, George Smyth (son and successor to the witch-hunting Henry Smyth), was having none of it – women's testimony was perfectly admissible, and Michal was punished for his disrespect.

Fairies and trows would long remain part of the islanders' mental world (see Chapter 8), but after the Civil Wars they no longer triggered instant associations with the Devil. In 1657, Francis Brown was ruled a slanderer by the Kirkwall Session for saying James Twatt's wife was a witch and 'a trow in the hill witching his neighbours'. In 1668, the presbytery heard from the minister of Orphir about Hugh Moar, who confessed before the session to being 'troubled these three years bygone by evil spirits, namely, fairy folk'. Moar claimed to have seen in their company Barbara Hutchison, 'who from that time and hitherto has four dead children with the said fairies', and who had garnered 'the profit of all the Bu of Orphir'.

It was a rich brew of accusations, but neither Moar nor Hutchison was accused of witchcraft. The presbytery's priority, indeed, was for Barbara to 'be restored to her good name'. Moar was confined to Marwick's Hole (see Fig. 5.2), while the laird of Graemsay, with the session's advice, contemplated suitable punishment for him.[47]

Around the same time, at the Ladykirk in South Ronaldsay, another potential witchcraft prosecution was swiftly, if gruesomely, nipped in the bud. On 28 October 1666, three widows and a bereaved mother approached the minister and elders before Sunday sermon. Their menfolk had drowned fishing in Burwick Bay, a sheltered harbourage overlooked by the little kirk. After this tragically unexpected occurrence, the men's ghosts appeared to people in the island, claiming to have 'perished by wicked persons, instruments of Satan'. The distraught women wanted the minister, Edward Richardson, to arrange for the opening of the graves, so 'wickedness might be brought to light', and the witches punished.

Richardson, 'for quieting of the sad hearts of the widows', agreed. Parishioners were summoned on the following day to an inspection of two of the bodies, supervised by the baillie. Persons present when the bodies were found, and who helped prepare them for burial, were made to handle the disinterred corpses – presumably to test whether the lifeless bodies would bleed, emit water or manifest other accusatory signs in the presence of suspected murderers. They did not. Richardson declared the matter closed, and exhorted the parishioners 'to lay aside all sinistrous

Figure 5.3: The late eighteenth-century church of St Mary, built on the site of an earlier kirk near the shore at Burwick, close to the southern tip of South Ronaldsay. In 1666, the minister arranged a public exhumation in the churchyard, to prove that several drowned fishermen were not victims of witchcraft.

[malicious] or evil thoughts towards any persons'. If witchcraft allegations continued to circulate, their authors could expect chastisement. A couple of decades earlier, Orkney ministers, or some of them, fanned the flames of persecution; now Richardson was eager to see them doused.[48]

Those suspected of being witches resented it for various reasons, not least that witchcraft remained a capital crime, and rumours might develop momentum. In June 1660, the presbytery received an impassioned petition from some people in Stronsay: 'they are scandalised by several in the isle who reproached them as witches upon the affection of a woman who was challenged as a witch'. In other words, they had been named by a suspected witch, in the chain-reaction manner of the prosecutions of 1643–4. Sixteen years on, the instinct of the presbytery was not to follow up leads, but to seek ways 'for removing that and such-like scandals'.

An impression of an aftermath, an era of witch-hunting in the process of passing, is reinforced by a clutch of cases from later seventeenth-century Shapinsay. They involved actions for slander after someone was labelled, or libelled, as the offspring of a witch. Magnus Hutcheon performed public penance in 1672 for accusing various people of being a 'witch's thief's bird' – 'bird' was a common term for human as well as animal progeny. Magnus Irving was similarly punished in 1680 for saying David Rusland was a 'witch's bird or oy' (grandchild), and in 1687 William Williamson called Magnus Cumming 'a witch's bird and a warlock'. A further bill of complaint was received by the session in 1699, alleging David Anderson had smeared Cumming's wife as 'a witch's and thief's bird'. The slander was doubly invidious because of the suspicion that occult powers passed down along bloodlines.

The Shapinsay Session initiated no proceedings for the crime of witchcraft itself in these years, although in 1698 Elspeth Morray was reported for 'consulting' with a charmer called 'Highland Christian'. Elspeth had suffered some slight or injury, and while Christian was probably a wanderer from mainland Scotland, her technique for identifying who wronged Elspeth was characteristically Orcadian – three stones, placed above the lintel overnight, then dropped into a stoup of salt water. The stones had been marked with the suspects' names, and the one bearing the name of the guilty party was expected to make the water gurgle.[49]

Orkney's authorities, ecclesiastical and civil, did not entirely abandon the pursuit of witches after the Restoration. In 1669, the sheriffdom tax officials known as 'commissioners of supply' minuted their intention to secure a privy council warrant allowing the justiciar and JPs to proceed with 'putting of witches and incestuous persons to a trial' – it seems the earlier autonomy in these matters had been curtailed. In 1693, the Kirkwall Session received an accusation against Jane Seater, 'banished from Westray for suspected witchcraft'. She was given a deadline to produce 'a testimony from Westray of her innocent conversation [conduct]', or else remove herself from the town.

Just over the Pentland Firth, the Caithness Presbytery complained in 1698 that 'sorcery and witchcraft abounded much in the parish of Wick', since 'sorcerers banished out of Orkney lurked there'. The grievance suggests the continuation of a hostile environment in Orkney, though a patchy survival of kirk session records for the late seventeenth century makes it hard to say how many neighbourhood complaints were still being registered with the courts.

The Orkney Presbytery was certainly less concerned with the issue than once it had been. In November 1680, it simply referred back to the Rendall Kirk Session the case of Margaret Sclater, accused of witchcraft by John Flett and John Mowatt. She had apparently promised Flett, in return for gifts of cloth and money, to 'put his wife down [presumably, to kill her] and cause him get ane young wife again'. Mowatt complained that Margaret 'told whoever she pleased that she did ride him' – the kind of supernatural nocturnal subjugation (or nightmare) of which Bessie Skebister had stood accused in 1633.[50]

A rare reference to witchcraft in the 1687 Kirkwall Session minutes confirms the shifts in official attitude. Helen Paplay was summoned to appear 'for imprecating William Grimbister, who since her imprecation is sickly'. Helen confessed before the session and Bishop Murdoch Mackenzie that after Grimbister called her a witch she retorted, no doubt ironically, 'so might he thrive and bruk [enjoy] his health'. Rather than investigate this as an instance of malevolent magic, the bishop and session resolved to visit William to persuade him 'to forgive and pardon her and she to pardon him'. A week later, the elders reported Helen had sought, and received, William's forgiveness. The case was one of unchar-

1. The eleventh-century Cathedral of St Magnus in Kirkwall, viewed from the east. The church remains the most imposing building in the islands.

2. The cathedral north aisle. The modern statue of St Olaf was a 1937 gift from Nidaros (Trondheim) Cathedral to mark eight hundred years since the construction of St Magnus – a symbol of enduring ties of identity between Orkney and Norway.

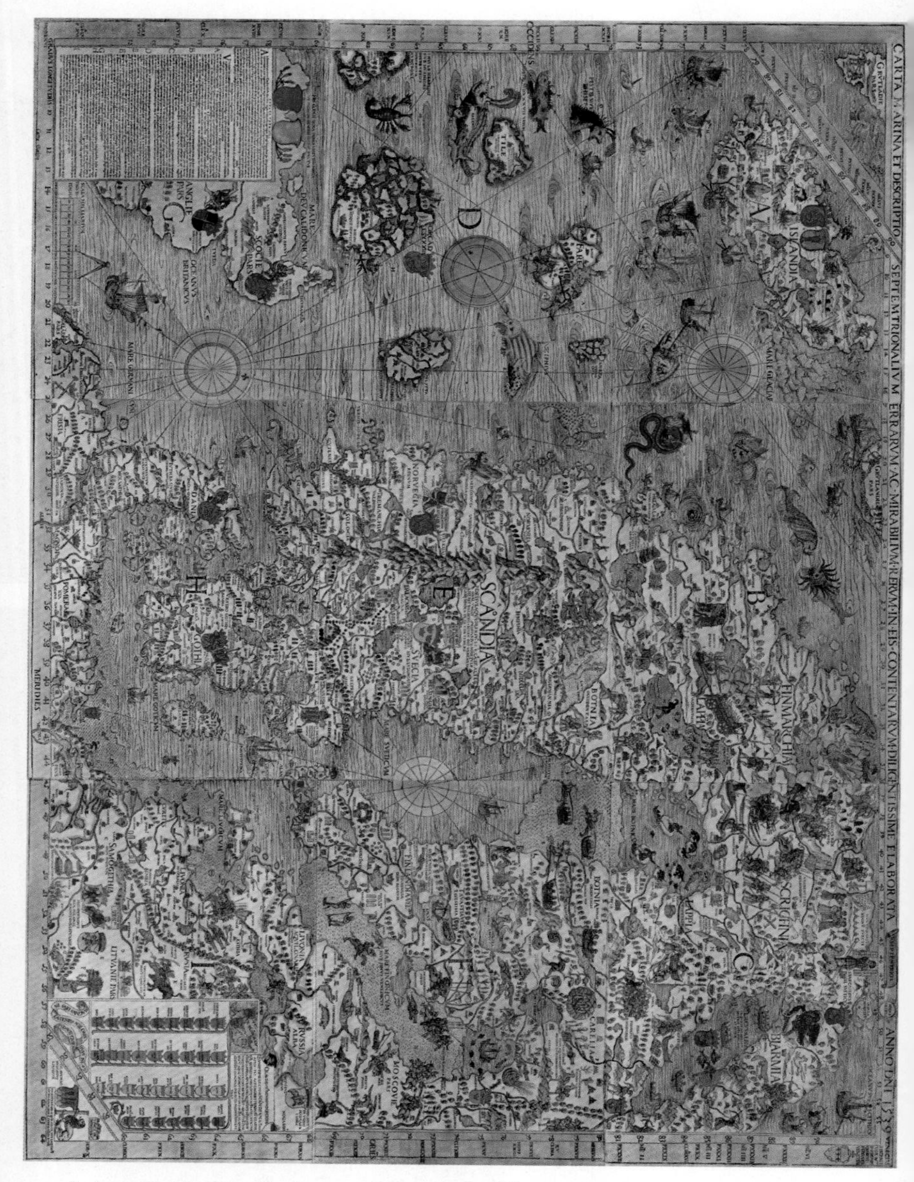

3. A composite from the two surviving copies of the *Carta Marina et Descriptio Septentrionalium Terrarum* (Maritime Map and Description of the Northern Land), created in 1539 by the Swedish-Catholic churchman, Olaus Magnus. This pro-Norwegian (and anti-Danish) depiction of the Baltic and North Atlantic world draws attention to islands, Orkney prominent among them.

4. Christian I of Denmark and his wife, Dorothea of Brandenburg, by an unknown artist of the late fifteenth century. Christian pawned Orkney to James III of Scotland in 1468, in his capacity as king of Norway. He had no consent from his Norwegian subjects for this action, and for centuries his successors would continue to assert their sovereignty over the islands.

5. James V, king of Scots, by the Dutch painter Corneille de Lyon, *c.*1536. James was the first reigning Scottish monarch to visit Orkney, in 1540, and the last to do so before the twentieth century.

6. The site of the 1529 Battle of Summerdale, on the boundary between the parishes of Orphir and Stenness. This view is from the east, looking past the farmstead of Oback, where the earl of Caithness was reputed to have been killed.

7. A stirring battle scene, painted by John Archibald Webb, from *The Fight at Summerdale* (1913) by the Orcadian novelist John Gunn. Like other Orkney writers of the nineteenth and early twentieth centuries, Gunn saw the Summerdale campaign as a struggle for island liberties against encroaching Scots 'tyranny'.

8. The Brough of Deerness, a rocky promontory at the eastern end of the Orkney Mainland housing a chapel known as 'the Bairnes of Brughe'. Pilgrims continued to visit here, scrambling up the cliffs to the flat summit, for centuries after the Reformation.

9. The Marykirk, or 'Our Lady of Grace': a now almost entirely vanished pilgrimage chapel on a sliver of land poking into Harray Loch in the West Mainland. Members of the defeated army of the earl of Caithness sought sanctuary here after the battle of Summerdale, and were dragged out to their deaths by the supporters of James Sinclair of Brecks.

10. Statue of St Magnus in relief, dating from the fourteenth century. This, and a companion statue of St Olaf, were spared from destruction at the Reformation and discovered in an upstairs room of the cathedral in the mid-nineteenth century.

11. Fourteenth-century statue of St Olaf, from St Magnus Cathedral. Olaf was invariably identified by his battle-axe, while Magnus was usually depicted holding the sword with which he was slain.

12. The round tower Bishop Reid added to the episcopal palace in Kirkwall, known locally as the 'Moosie Toor'. The figure in the alcove (a replica of the original medieval statue now in the Orkney Museum) has been identified as St Rognvald. The image was likely moved to this location in the tower to protect it at the Reformation.

13. The Earl's Palace, a masterpiece of late-Renaissance architecture in the heart of Kirkwall. It was built, with forced labour, by Earl Patrick Stewart as a symbol of his authority.

14. The beach at the Bay of Saviskaill, Rousay. In 1643, the accused witch Katherine Craigie was alleged to have brought Thomas Irving here, and cured him of 'the sea trow' by casting water over his head.

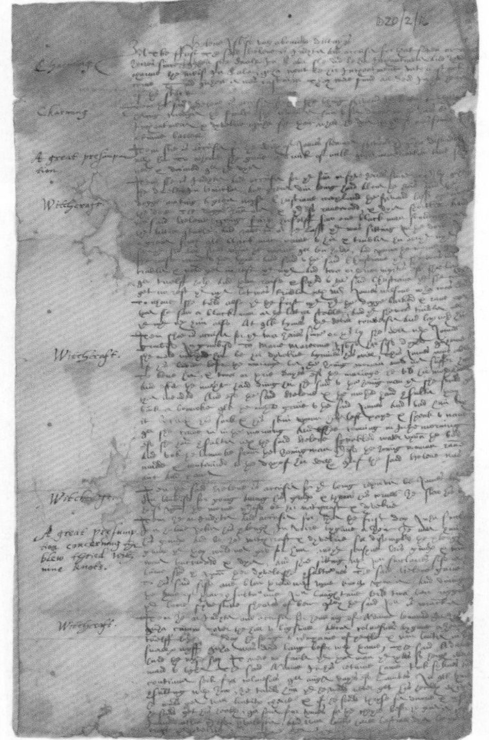

15. The 'dittay' from the trial of Helen Isbister in 1635, accusing her of various points of witchcraft. Helen was a charmer, who admitted using incantations to banish mice. Unlike most suspects in witchcraft trials, however, she was acquitted on virtually all charges and freed without punishment.

itable behaviour, rather than of Satanic spell-casting – though whether William fully agreed with that is open to question. In grudgingly accepting Helen's apology, he 'wished that God would forgive them that sins likeways'.[51]

The term used to describe Helen's words – 'imprecation' – is itself a semantic yardstick of change. Traditionally, it implied the invocation of evil, the very opposite of 'precation', a blessing or prayer. In that sense, imprecation was the defining activity of witches. The 1633 dittay against Katherine Grieve, a suspected witch from the parish of Evie, alleged, for example, 'that by your cursings and imprecations ye wrong and hurt man and beast'.

Not all imprecation was demonic witchcraft, even at the height of the trials. In May 1643, the Kirkwall Session ordered Marjory Carmichael and her daughter Isobel to make public repentance in the cathedral, clad in linen undergarments, after Margaret Baikie, sister to the merchant James Baikie, complained of their 'using many evil prayers, most fearful curses and imprecations' against her whole family. The presbytery, in April 1640, issued a similar sentence against Helen Groat in South Ronaldsay, 'for her abuse of God's name upon her knees in ane idolatrous manner, intending imprecation'. Another South Ronaldsay woman, Jonet Barclay, was disciplined by the Ladykirk Session in 1669 after coming to the house of Alexander Stewart and on her knees cursing all within it, especially one of Stewart's servants, whom she hoped 'might come ashore like whale blubber' whenever he set to sea.[52]

This was, at worst, a kind of semi-witchcraft, betokening a willingness and intention for evil to befall, without actual evidence it had taken effect, or use of special rituals to bring it about. Imprecation cases appear frequently in the session minutes from later seventeenth-century Orkney. It seems that people were often eager to relate to the minister and elders the evil words neighbours spoke against them, sometimes in memorable ways. 'Your tongue to the stone and your tail to the fire!' was the malediction cast at John Heddle by Walter Scott in the course of a Shapinsay quarrel in 1664. In Sanday, Richard Fotheringhame, himself an elder – and quite probably one of my ancestors – reported in 1704 the 'horrible imprecations' pronounced on him by Magnus and Joan Skea, 'willing that the Devil might tear the soul out of the said Richard's back'.

Outbursts like this were increasingly regarded as a species of reprehensible but hardly heretical moral offence, a matter of taking the Lord's name in vain, and of antisocial behaviour towards neighbours. Through the 1680s, the Shapinsay Session dealt with a demoralising succession of such cases. It disciplined William Williamson and Magnus Cumming in 1681 'for using imprecations one against another', and Magnus Steinson and Nicol Aim in the following year for choosing 'to live in malice and praying one against another'. In 1687, Katherine Drever was sentenced to stand in sackcloth in the kirk, for her 'graceless expressions and rash judgement', wishing a place in hell for 'the Mistress of Gairstay'. The defamed party belonged to the influential Irving family, an ancestor of the American author Washington Irving, whose story 'Rip van Winkle' took Orcadian motifs about the lure of living with hill-fairies, oblivious to the passage of time, and translated them to the Catskill Mountains of New York. John Guthrie, 'rebuked for his folly' by the session in 1682, was a man known for 'a habit of cursing and banning [swearing]'.[53]

'Cursing' hovers here on the threshold between older meanings of calling down supernatural power and more modern ones of giving gratuitous offence. Magnus Garioch, cited before the Holm Session in February 1700 for 'imprecations' against John Foubister, was hardly a scheming warlock. Foubister had displaced him as cottar (a precarious landless tenant) on the farm of Greenwall, and Garioch said only that he hoped Foubister 'might sit ill and rise worse out of the house'. Nicol Gorn and his wife, Elspeth Budge, were summoned by the same court in November that year for 'cursing and using imprecations against James Meall'. They wished 'there was a quick calf in his belly'. It was, one presumes, not meant literally, and neither (probably) was the malediction pronounced on Magnus Seatter by Sarah Moir in Orphir in 1710: 'God grant he might have as many bairns as there were hairs in his head, and that he might carry them to the kirk in a riddle' (a coarse-meshed sieve).[54]

The courts increasingly viewed imprecation as a matter of public order and personal morality, but complainants may not always have felt the same way. If ministers and elders were no longer much interested in prosecuting witches for trafficking with Satan, then some protection and redress might be secured by reporting them for cursing. George

Anderson's complaint to the Shapinsay Session in June 1678 was surely motivated by genuine fear for his life and his livestock – 'that John Work had used imprecations on him and his sheep likewise, in saying curse of God be on him and his sheep too!' In 1684, Margaret Murray 'did curse and imprecate horribly' against Margaret Hutcheon. Murray claimed Hutcheon provoked her 'by saying she had hid a woman's profit under a fail-peat [cut turf]'. Such allegations might once have contributed to a dittay, though Hutcheon protested the words were spoken 'only in sport'. A third Shapinsay case, from June 1686, saw Margaret Leith accused of 'cursing and praying for [*sic*] Thomas Heddle and his family'. Heddle's wife had called her 'witch and thief'.[55]

Not all cursing coming into the purview of the Shapinsay elders took place on the island itself. In August 1698, John Irving and John Swannay appeared before the session to deny they 'did use imprecations on the sea, in entreating the Devil to sweep the whole flow up'. But witnesses later confirmed they did use such imprecations, and they were put to penance.

Orkney fishermen were perhaps readier to curse in their boats than in their houses because of a sometimes misplaced assumption their words would not be reported back. But there is a sense, too, of the sea as a place of different moral order, where the writ of the Kirk did not fully run (see Chapter 8). The Devil's sway was greater at sea than on land; it was daring and dangerous to invoke his name there. David Rendall was questioned before the Shapinsay Session in 1705 for using 'horrid cursing and swearing in a boat at sea'. He was reported saying to his crewmate John Hepburn, 'Devil blow the liars in the air!' Another witness, on account of the strong winds, could testify only that 'he heard the said David name the Devil'. Hepburn's reply is difficult to interpret: 'the thunder thou heard in the air may fear thee to curse so'. Did he – like a good Protestant – think the thunder was God's providential rebuke of Rendall's impiety, or that the storm breaking around them was the Devil's doing, or the work of his earthly servants?

Storm-raising and ship-sinking do not figure very often as items in the Orkney witchcraft dittays. Such crimes were difficult to prove to the satisfaction of assizes, and victims of weather-magic might be passing sailors, rather than the local farmers and their wives who trooped to the session with stories of sick children and 'profit' stolen from milk and barley. Yet the ability to summon winds is a stock feature of later traditions about Orkney witches, and it would be surprising if seventeenth-century suspects were not thought to possess it. Adverse weather was hardly an unusual occurrence in Orkney, yet it caused jittery talk in Kirkwall in the late summer of 1676. Margaret Cromartie vehemently denied being 'one of the persons that raised the Devil at the last storm', and presented the session with a bill of complaint about 'the foul slander and imputation raised upon her'.[56]

Water and Stone

Both before and after the era of the witch trials, Orcadians believed in powerful correspondences between the visible and invisible, in hidden causes for unusual events, and in supernatural malice as a plausible explanation for personal misfortune. Witchcraft was not an everyday occurrence, but it was an everyday possibility. Magic was ingrained in the vistas of hill, farmland, shore and ocean, and in the multiple little mysteries of societal survival: ploughing, sowing and harvesting, ale-brewing and butter-churning, breeding and nurturing of animals, procreation, pregnancy and birth. Stories of such processes being corrupted by witchcraft seem to us strange and outlandish, if not downright repulsive. Yet they are a map of what mattered to ordinary folk.

Witches were almost never strangers, but people known to their victims, often over years. Their malice was calculated and considered, vindictive and vengeful. Fear of witches as 'the enemy within' was widespread in sixteenth- and seventeenth-century Scotland; perhaps, indeed, a core characteristic of witch-beliefs as a global and trans-historical phenomenon.[57] Yet Orkney's defined boundaries and co-dependent conditions of living lent a particular intensity to this dread of polluted neighbourhood.

The witchcraft trials stuttered to an end in Orkney around the beginning of the eighteenth century, as they did in Scotland more widely and in other parts of western Europe. They were tragedies for the individuals involved, and often traumatic episodes for the communities affected. Far from being a sign of the 'backwardness' of Orcadian society, the trials were a symptom of the islands' growing integration into the kingdom of Scotland and the wider Anglo-Scots polity. All the trials we know of were conducted within legal and ecclesiastical frameworks introduced to Orkney after the second half of the sixteenth century. An incoming clerical elite, educated at the universities of St Andrews, Edinburgh and Aberdeen, inflamed traditional peasant anxieties by insisting that fractious women (and a few men) with knowledge of the workings of magic were perfidious puppets of the Devil, just as they themselves were righteous instruments of God.

Once the clergy began rowing back from this certitude, 'victims' of witchcraft were thrown on their own resources. In the spring of 1670, Annis Lowtit, wife of a Stromness sea skipper, complained to Bishop Honyman that Katherine Cromertie had knocked her to the ground, asserting 'I had bewitched her', and struck her on the head with a stone so that 'the blood came down over all my body in great abundance'. Annis demanded that Katherine be punished, 'and that I may be restored to my honest good name'. The assault took place as Annis was on her way 'to make a visit to ane diseased person', strongly implying she was a healer or charmer, and Katherine an aggrieved ex-client. Annis portrayed the assault as attempted murder, but it was probably a desperate act of ritual counter-magic. Drawing a witch's blood from 'above the breath' was believed to break the power of her spell.[58]

Walter Traill Dennison, as a young man in Sanday in the late 1840s, was told by an old woman, herself the granddaughter of a notorious witch, about a ritual for acquiring the powers of witchcraft. Almost fifty years later, its details were published, after Traill Dennison sent them to the folklorist William Mackenzie.

The method involved going to the shore at midnight, on the way turning oneself withershins three times. Supplicants then lay on their backs, between marks of high and low water, a stone in each hand, another by each foot, one at the head, and two more across the chest and heart. With eyes closed, and arms and legs outstretched, the following words were to be slowly recited:

O, Mester King o' a' that's ill,
Come fill me wi' the warlock skill,
An' I sall serve wi' all me will.
Trow tak' me gin I sinno [if I shall not]!
Trow tak' me gin I winno [if I will not]!
Trow tak' me whin I cinno [if I cannot]!
Come tak' me noo, an' tak' me a',
Tak' lights [lungs] an' liver, pluck [viscera] an' ga' [gall],
Tak' me, tak' me, noo, I say,
Fae de how [cap] o' de head tae de tip of de tae [toes];
Tak' a' dat's oot an' in o' me,
Tak' hide an' hair an' a' tae thee,
Tak' hert an' harns [brains], flesh, bleud, an' büns [bones],
Tak' a' atween de seeven stüns [stones]
I' de name o' de muckle black Wallawa!

The spell thus spoken, the new witch or warlock would fling their seven stones into the sea, each accompanied by a specific curse. Traill Dennison's informant claimed to have forgotten the words for these, but, according to Mackenzie, 'he rather suspected she considered the imprecations too shocking to repeat'.[59]

This elaborate incantation is, to put it mildly, most unlikely to be an authentic oath of witches' initiation, passed from mother to daughter down the course of centuries. It is possible too, as the later Orkney folklorist Ernest Marwick once suggested, that Traill Dennison was tempted to 'improve' the text – the opening couplet in particular has a rather Robert Burns ring to it.

But, however it came to be composed, as a compendium of inherited knowledge about magic, the Sanday spell is a fascinating and invaluable text. It reminds us how power was unlocked at 'liminal' times and places, in acts of turning against the course of the sun, and in the magical capacities of water and stone – these, the enduring sacraments of popular magic in Orkney, just as surely as bread and wine were abiding offerings for the communion rites of the Kirk.

The idea of a formal pact with the Devil was learned, and incorporated into folk culture, from the preaching of the ministers. But the names cited in the process are of an older vintage. 'Trow' was the general Orkney term for a malevolent nature-spirit, whether of hill or sea; it had nothing much to do with the Devil of Christian theology.

And the 'muckle black Wallawa'? There may be an association with 'wallaway' or 'wellaway' – an exclamation of lamentation ('woe, alas'), common in Old English as well as Scots. But there is surely a connection with Walliman – Jonet Rendall's mysterious spirit-master and *nom de théâtre* of Oliver Leask – and to the ancient Norse figure of the *völva* or prophetess. This, too, may have morphed over time into what the clergy insisted it must be, a familiar name for Satan. In a glossary of Northern Isles dialect published in 1866, 'Wallawae' was noted as a Shetland term for the Devil.[60] Yet the long spoor of witchcraft belief we have tracked into the era of the Enlightenment and beyond reveals how three and more centuries of Scots law and lordship in Orkney, and twenty decades of Protestant piety and preaching, had failed to erase words and concepts brought to the islands by the pagan Norse.

Orkney's trials, in the word's emotional as well as legal connotations, harness the islands to wider patterns of historical development, but also articulate their distinctions and difference. The persecutions followed a broadly similar chronology to the rest of Scotland and much of Europe, and involve many of the same themes: a clash between elite and popular conceptions of magic; an anxiously competitive world of 'limited good'; the misogynistic assumption that witches were more likely to be women than men.

But Orkney's pattern was not a 'typical' one, and none of the various schema proposed by scholars to explain witchcraft accusations – the 'charity-refused' model, or the response to a subsistence crisis precipitated by changing climate – quite captures the essence of its experience. Orkney was drafted into service for the most notorious of twentieth-century theories about early modern witchcraft: the anthropologist Margaret Murray's claim that witches were organised practitioners of a surviving pre-Christian religion. She knew about Walliman from a 1903 printing of Jonet Rendall's trial, finding in the word proof that 'the god of the witch was not the same as that of the Christian'.[61] But she knew nothing of the name's genesis in furtive encounters among marginalised charmers, at once supplying the needs and exploiting the fears of apprehensive but resilient Christian communities.

The close association between healers and witches was a distinguishing feature of the Orkney witchcraft trials, as was the ubiquity of stones, seawater, beaches and other ecological elements of life in an interconnected island environment. Witchcraft was always an aberration, but its investigation makes visible the tenacious grip that magical belief exercised over society as a whole. In Orkney, its deep layers were compounded not just by the imposition of Protestant reform, but by the islands' slow and incomplete absorption of Scots language and lore.

For the ordinary people of Orkney, Scots law on witchcraft, bolstered by the theological certitudes of the Kirk, offered powerful possibilities of redress against supernatural malice and injury. But it also enflamed latent fears, encouraged suspicion and denunciations, and convulsed small communities that must previously have managed to deal with their anxieties about magic through informal mechanisms of appeasement and reconciliation. Agonising and terrifying deaths were dealt to dozens of hapless individuals, who may indeed have been difficult and disruptive people, but were in the end 'cheust folk'.

Orkney's witch-hunts were a domestic trauma, but also a painful experience of participation in nation-building, part of a wider pattern of conflict and crisis that was starting to engulf the larger archipelago in the decades after 1600. Through all the momentous years of Britain's seventeenth-century troubles, Orkney would find itself battered by strong winds from the south – and, at times, the very centre of the storm.

6
Revolution

Orkney's Flodden

The battle was lost almost before it started, and the islanders turned, fled and died in their hundreds. They had marched for a fortnight, since embarking at Holm and crossing the Pentland Firth to John o' Groats – an Orcadian invasion of Caithness, mirroring the northward incursions of 1529 and 1614. The first of those invasions led to the triumph of the Battle of Summerdale, Orkney's Bannockburn. The fleeting Battle of Carbisdale, fought in the late afternoon of Saturday 27 April 1650, was a tragedy and a disaster: Orkney's Flodden.

There were over a thousand Orcadians in the little army. Some came cheerfully, believers in the cause, but others had little say in the matter. A report from an English vessel watering in Orkney claimed that 'all from sixty to fourteen' had been forced into the ranks. Few had any fighting experience; even after weeks of training, the pikes and muskets felt awkward in their hands. With them marched a small corps of Danish and German mercenaries, 500 or so. The column was woefully short of cavalry; a mere forty horsemen screened the line of advance.

What the army did have was a lucky, audacious and charismatic commander: James Graham, marquis of Montrose, the king's Lord Lieutenant in Scotland. That king, Charles II, was a sovereign without a throne, in Netherlandish exile. His father, Charles I, had fought and lost a war against English Parliamentarians and Scots Presbyterians, signatories to a National Covenant opposing royal policies. The fighting engulfed Ireland, too – what historians used to call the 'English Civil War' is now recognised to have been a War of the Three Kingdoms.

Montrose, like many prominent Scots, once backed the Covenant. But he came to distrust the Presbyterian leadership, and took up the king's cause. In 1644–5, at the head of a ragtag army of Highland clansmen and Catholic Irish, Montrose conducted a brilliant military campaign in Scotland and inflicted several humiliating defeats on the Covenanters, but he was forced into exile in 1646, the year Charles I became a prisoner of his Parliamentarian enemies. In 1649, the king was tried for treason and executed, and England became a republic. Scotland's parliament was willing to recognise Charles II, but refused to let him enter the realm without guarantees for the supremacy of Presbyterianism. Charles was willing to talk, but decided to strengthen his negotiating hand with one last throw of the dice. In March 1650, Montrose sailed for Orkney, and the eyes of three kingdoms turned to the islands along their northern rim.[1]

The campaign began promisingly. An advance guard moved swiftly and secured the Ord of Caithness, a steep pass marking the Caithness–Sutherland boundary. But hoped-for support was not forthcoming from the Caithness and Sutherland clans, who had made their peace with the Covenanting regime. The nearest government commander, Archibald Strachan, was a tough cavalryman, zealously Presbyterian and fervently anti-Royalist, a veteran of Cromwell's New Model Army. Rather than profane the Sabbath, he pressed north on Saturday 27 April. Strachan had only half Montrose's number of men, many of them undependable Munro and Ross clan levies. But he had the advantage in cavalry, by a factor of five to one.

The armies met at the southern border of Sutherland, on level ground near a farmstead called Carbisdale. On Montrose's east flank was the Kyle of Sutherland, a broad river estuary flowing into the Dornoch Firth. There was forest to the west, and to the north the wooded slopes of Creag a' Choineachan. The hill's name means 'mossy crag', but through its similarity to *caoineadh*, weeping, the place came to be known as the Rock of Lamentation.

Strachan kept his cavalry concealed in vegetation along the shore of the Kyle, and advanced with a single troop. As Montrose was ordering his men into position, Strachan charged with his entire force. The Royalist cavalrymen fled into their own infantry, and though the Danish musketeers managed a volley, the outcome was decided. The Orcadian foot soldiers scattered. Two hundred drowned trying to swim to safety across the Kyle, along with one pursuing dragoon, the sole recorded casualty among the government force. Others sought refuge in the woods of Creag a' Choineachan. Here, through several grim hours, the Munros and Rosses hunted them mercilessly. Over 400 were killed, another 450 captured.

A hundred or so Royalists managed to escape. They included Montrose himself, who wandered for three days across the bleak moorland of Strath Oykel, in the company of an Orcadian officer, Major Edward Sinclair, before being discovered and sent to Edinburgh to suffer the inevitable fate of a traitor. Five standards were taken on the field, one, in funereal black, displaying the severed head of Charles I. Montrose's personal banner of white damask depicted a lion preparing to leap from rocks across an intervening river. The motto was 'Nil Medium' – no middle course, no compromise. It was an ironically apt epitaph for the farmers and fishermen who perished miserably in the waters of the Kyle.

A few straggled home. A petition of 1661, drawn up by the Kirkwall magistrates after Charles II's Restoration, recalled how ships from the town 'did transport such of the said army as retreated to them to the said burgh', many 'wounded and hurt'. A month after the battle, parliament considered the fate of 281 common soldiers in Edinburgh's Canongate prison. Forty were released, since they were conscripted and had wives and children in Orkney. Six Orcadians, skilled fishermen, were 'given' to David Leslie, the general who had defeated Montrose a few years earlier and who was about to assume command of the Covenanter armies; another six to Montrose's bitter political enemy, Archibald Campbell, marquis of Argyll. They were luckier than the half-dozen 'lusty young fellows' packed off to lead mines owned by the judge Sir James Hope. The other prisoners were handed over to Lord Angus and Sir Robert Moray, raising regiments for mercenary service in France.

Carbisdale was an Orcadian catastrophe. More men from the islands died that day, and in the immediate aftermath, than in the First World War. A 1662 memorandum from Orkney's landowners lamented that 'there was almost no gentleman's house in this country but lost either a son or a brother', not to mention 'the loss of near 2,000 commons'. We know a few names of the gentlemen, but hardly any of the commoners, though they included three conscripted teenage sons of James Morison, minister of Evie. Morison was unusual among Orkney ministers in his antagonism to Montrose, and his ability to petition parliament for redress. But his experience was typical of the bereaved parents, 'a sad aggravation of my grief and losses'.

The fate of men who marched with Montrose remained for years uncertain. Marion Wode's husband, Hew Rendall, 'went to the wars', and had not returned in October 1654 when Harry Marwick asked the Orkney Presbytery for permission to wed Marion. The ministers judged Rendall a man 'of whose death there is no certainty', and commanded the couple to wait 'until such time the business be cleared'. In June 1658, James Sinclair wanted to marry Jean Linay. But the Kirkwall Session 'knew not whither her husband was dead or alive'. Jean said it was eight and a half years since David Galliard went south, 'and never received word of him'. She produced two witnesses who had seen Galliard 'chased on the water by horsemen'. One was willing to swear that Jean's husband perished in the Kyle, but the other 'knew not certainly'. In the end, Jean was married – though not till 1662, and to a different suitor.[2]

Today, Carbisdale is largely forgotten in Orkney, and little remembered elsewhere. Yet it had profound consequences for the people of the islands. It was a moment of political possibility, when Scotland's 'periphery' imposed itself on the centre, and threatened to rewrite the national script. The middle years of the seventeenth century, for Britain and for Ireland, were an era of revolution. The story has been told many times, but never with Orkney, rather than London, Edinburgh or Dublin, as the hub and axle of a wheel of change.

Toward and Tractable People

The countdown to Carbisdale began in 1625, with the death of James VI and the accession of his surviving adult son to the three Stewart thrones. Charles I was born in Scotland – the last British monarch enjoying this distinction – but unlike his father was a thoroughly English ruler. A letter to privy councillors in Edinburgh early in the reign set the tone; it referred, revealingly, to 'your nation'.

James's passing presented royal obituarists with the usual problem: mourning a catastrophic loss, while rejoicing at the arrival of a flawless successor. One elegy was written by the Middlesex clergyman Richard James, who had visited Orkney in the wake of Robert Stewart's rebellion. It flattered James's self-image as the ruler of 'Great Britain', and watchful warden of its wild and lawless edges. The late king shone as an inspiration to

> The Redshank [bare-legged warrior] which frequents that
> northern shore,
> Where Neptune's waves against cold Orkney roar.

In Charles, God had provided a model sovereign for 'this sea-divided land,/Which runs from Kantium [Kent] to th' Orcadian strand'. Orkney, once more, set the scope of royal dominion.[3]

For the new king, Orkney was more a source of revenue and patronage than a fount of poetic inspiration. Stewart of Killeith, whose administration helped provoke the 1614 rebellion (p. 169), lost the tack in 1622 under a cloud of corruption. He was succeeded by John Buchanan, who shortly afterwards became Sir John Buchanan of Scotscraig, the tack being given to him in the form of a joint grant with his wife, Margaret Hartsyde.

It was a remarkable turnaround. Margaret, the daughter of a Kirkwall merchant, was a lady-in-waiting to Anne of Denmark, and went with her to the English court in 1603. In 1608, she and her husband were put on trial for theft of the queen's jewels – a murky episode which some thought had to do with court politics. John was acquitted, but Margaret was not, and both were banished to Orkney.

James VI, as we have seen (p. 161), favoured such banishments on both practical and symbolic grounds; they represented clemency, of a kind. Helen Erskine, a daughter of the laird of Dun, was exiled to Orkney in 1615, after being found complicit in the murder of a relative. Her sentence, tellingly, was to 'be banished out of this kingdom during her lifetime'.

But, in Orkney, Helen made a respectable marriage to a great-grandson of the cathedral provost Malcolm Halcro. The Buchanans thrived in Orcadian banishment, acquiring property and establishing good relations with other gentry families.[4]

The Orkney tack was a labyrinth of financial complexity. Actual collection of the rents was often sublet, and sums paid by the tacksmen frequently went directly to creditors of the crown. Towards the end of his reign, James wanted to see increased revenues from his Orkney estates, and considered the option of re-feuing them. It excited the interest of his cousin, John Stewart, former Master of Orkney, now Lord Kinclaven. John sniffed some possibilities for a revival of his brother Patrick's earldom. Neither James nor King Charles was willing to contemplate that, but in 1628 Charles compensated John by making him earl of Carrick. The title hints at the irrepressible ambitions of the Orkney Stewarts. It had royal associations, belonging once to the family of Bruce, whose former lands in south-west Scotland were in Kinclaven's hands.

John's pursuit of the title was politically sensitive, but with a subtlety not often on display in his exercise of kingship, Charles allowed his cousin to have it, on condition he rechristened his Eday estate 'Carrick', to differentiate his new earldom from the older royal one. In 1632, Charles granted the whole island to the earl, and a little settlement on Eday's north-east coast, looking over to the islet known as the Calf of Eday, was grandly designated 'the town of Carrick and port of Calf Sound', with burgh status, market and trading privileges. It was scarcely a serious rival to Kirkwall, but in the mid-1630s Eday became a major producer and exporter of salt, extracted from seawater in giant salt-pans heated by the island's seemingly inexhaustible supplies of moorland peat.[5]

Buchanan's tack, renewed in 1627, was almost immediately cancelled in favour of the Scottish treasurer-depute, Archibald, Lord Napier, a grandson of Bishop Bothwell's sister. His payment for it was mostly to go to George Seton, earl of Winton, who had settled on the king's behalf debts incurred by Robert Maxwell, earl of Nithsdale, commander of a Scots army raised to support Denmark in the Thirty Years War. Napier sublet the tack to William Dick, an affluent Edinburgh merchant descended from Alexander Dick, Malcolm Halcro's successor as provost.

It was a familiar pattern: outsiders with Orkney connections extracted rents, without the islands getting much in return. The deal with Dick encouraged corruption accusations against Napier, and in 1629 Charles granted the tack to Dick and his son John, with sums deducted for the earl of Annandale. Shortly afterwards, the remaining royal income from the tack was diverted to William Douglas, earl of Morton, a major royal creditor. The privy council remonstrated with Charles over his generosity: 'the best part of your property, such as Orkney, Islay, Kintyre, Dunfermline, Dunbar and Fife', was burdened with fees and pensions to the extent that the Exchequer could barely defray 'necessary affairs'.[6]

Those affairs included a desultory war with Spain, declared in 1625 after Charles failed to secure a Spanish marriage alliance. Between 1627 and 1629, England contrived to be at war with France as well. These Catholic powers were preoccupied with the Thirty Years War in Europe, and there was little fighting with British forces. Yet international conflict drew attention back to the vulnerable northern frontier, with John Dick, sheriff of Orkney, reminding the privy council in 1628 how 'these countries and isles lie open to the invasion of foreign enemies'.

Fears that Orkney and Shetland might prove the weak link in the kingdom's defences were not unjustified. As we have seen (p. 149), William Sempill, a Scots Catholic in Spanish service, drew up plans for an attack on Orkney in 1591. Thirty years later, Sempill was still advocating a northern strategy against the king of Spain's enemies. A 1619 memorandum proposed sending an Armada to harass Dutch and English fishing. The vast Dutch herring fleet convened mainly in Shetland, but the Netherlanders fished off Orkney, too; in the year Sempill wrote, twenty or thirty Dutch vessels were reported to be harbouring in Stronsay.

In 1628, Philip IV dusted off Sempill's proposal, and contemplated the seizure of Orkney and Shetland as the base for an invasion of England. Preparations were made to dispatch a fleet to Shetland, where the Spanish hoped to capitalise on local hostility to interloping Dutch fishermen.[7] The plans were not, in the end, implemented – any more than the German invasion of Orkney from occupied Norway, feared to be imminent in 1940. But had Catholic Spain decided to bring war to Protestant Britain in the 1620s, the storm would have broken along the kingdom's northern edge.

Perceptions of the islands as, simultaneously, frustratingly remote and alarmingly accessible prompted proposals for their reform – though we do not know the details of 'ane overture for the improvement of Orkney and Zetland', presented to the king in 1633. Another analyst, William Monson, a veteran English naval officer, worried about Dutch domination of the northern fisheries. With respect to Orkney and Shetland, he judged it a 'pity so good and civil people should inhabit no better a country'. The 'way to bring them to civility, wealth, and strength to defend themselves' was for the king to insist on receiving his rents in fish. To outsiders, then and later, it was a source of bewilderment that Orcadians were not first and foremost fishermen.

The Lanarkshire writer William Lithgow was also an advocate for 'improvement'. In 1633, on the occasion of Charles's Edinburgh coronation, Lithgow published moralising verses in which he bemoaned a lack of Scottish fortifications, arguing for new entrenchments at Sumburgh Head in Shetland, and the strengthening of Earl Patrick's forbidding tower house at Scalloway. With these improvements,

> And that Orcadian Kirkwall, eke rampired,
> With Calfa Sound, that harbour much admired:
> Then would these Isles Septentrion [northern] safer be,
> When made defensive 'gainst the hostile sea …

Lithgow was no great shakes as a poet, but for a lowlander he knew the Northern Isles well. He visited twice in his youth, and years later in 1629 took another trip to Orkney, recounted in an autobiographical travel narrative. The islands were one of the 'separated parts of the earth, which of themselves of old made up a little kingdom'. Where Monson beheld bleak and benighted landscapes, Lithgow perceived a storehouse of God's providence, teeming with wildfowl and livestock. 'I have seldom seen in all my travels more toward and tractable people (I mean their gentlemen), and better housekeepers, than be these Orcadians and Zetlanders.'[8]

Lithgow's suggestion that Orcadians – lairds in particular – were peaceful and 'tractable' might have raised eyebrows among those responsible for their governance. William and John Dick, Charles I's tacksmen, sought advice in 1631 about 'what punishment should be inflicted for mutilation' committed within the islands. A year earlier, the privy council, concerned at reports of unruly persons flouting the law in Orkney because of a lack of places of incarceration, ordered offenders to be detained in the Palace of the Yards, at least until the King's coffers, 'which are now straited and pinched with many urgent occasions, be better provided … for building of ane gaol'.

Orkney's gentry feuds confounded the capacity of local justice to contain them. William Sinclair of Sebay complained to the privy council that, in 1630, men 'armed with swords, staves, pistols' destroyed crops on his Deerness estate of Braebuster. The alleged leaders were Stewarts, assorted grandsons of Earl Robert in the illegitimate line: Harry Stewart of Graemsay, Patrick Stewart of Gyre, John, son of Sir James Stewart of Tullos. The feud remained lively in 1636, when Sinclair claimed the Stewarts and their allies occupied his house, broke his plough and harrow, and shot at him with a musket. The assailants included the local minister, Patrick Waterston, about whom more will be heard shortly.

There were evidently two sides to the story, as the council judged the charges to be unproven, and ordered Sinclair to compensate witnesses obliged to travel to Edinburgh. Parliament had already received complaints from Orkney about vexatious letters of 'lawburrows' – financial bonds to deter physical assault – taken out in the Edinburgh court of session. People unable to afford the journey south to post the requisite sums risked seizure of their property. In 1633, the council ordered lawburrows from the Northern Isles to be registered only in Orkney's and Shetland's court books.

Orkney's local leaders enjoyed an open relationship with political and judicial authority in Edinburgh, cosying up when it served their interests to do so. In November 1636, David McClelland, the chamberlain-depute charged with local management of the earldom estates, wrote 'in name and behalf of the whole body of this country' to ask the council to ban passing vessels from recruiting servants as crewmen. This common practice threatened to depopulate the islands, and make remaining servants, aware of their options, intolerably insolent. The 'problem', of course, was one of Orkney's adjacency to major sea-lanes, rather than of its impenetrable 'remoteness'.[9]

At other times, engagement with the centre seemed like a burdensome chore. In February 1628, at the traditional head court in Kirkwall, Sheriff-Depute Robert Monteith required the suitors of court – successors to the old roithmen (p. 34) – to select commissioners for the parliament in Edinburgh. After deliberation, they declared themselves 'not such persons as should appear in Parliament ... nor men of that quality to choose or make choice ... in respect of their estate and inability, being but mean gentlemen and farmers'. Kirkwall's status as a royal burgh also entitled it to send a representative, but James V's charter exempted the inhabitants from the obligation, 'in respect that they lie at so great a distance'. As late as the fourth decade of the seventeenth century, no Kirkwallian commissioner had ever gone to parliament.

Yet the burgh authorities were quick enough to invoke national authority when they needed local leverage. On 2 August 1637, the privy council weighed a supplication from the baillie and councillors of Kirkwall. It concerned a 'venel' or narrow lane, called Scholars' Campus Close, which ran from the common street down to the seafront, between tenements belonging to Robert Monteith, who had walled up the end of the venel, and placed a gate at the seaward head, 'intending to appropriate the same to his own private use'. Quite apart from this disregard of immemorial custom, his actions shut off the Grammar School students' only convenient access to water 'in case of any danger or accident of fire'.

To the assembled privy councillors, it surely seemed a matter of spectacular unimportance. Edinburgh was in a ferment. Just over a week earlier, a new form of service, based on the English Book of Common Prayer, was used for the first time in St Giles' Cathedral. It represented the culmination of Charles I's determination, encouraged by Archbishop Laud of Canterbury, to 'conform' Church of Scotland practices to those of the Church of England – which Charles and Laud were reshaping to emphasise ritual, ceremony and the 'beauty of holiness'. To many in St Giles' Kirk, the Prayer Book betrayed the principles of the Reformation. A woman called Jenny Geddes may or may not have thrown her stool at the minister. But there was certainly a riot, and before long it turned into a revolution.

The dispute over a minor right of way, in a little port town more than 200 miles to the north, was swiftly dealt with by the councillors, who referred the matter to local commissioners: the sheriff of Orkney and his deputes; the bishop, George Graham; and an Orkney minister, Mr Walter Stewart, the name of whose parish was left blank in the minutes, for filling in later when someone could be bothered to find out what it was.[10]

The minds of the councillors turned back to more pressing matters. Yet small asteroids can collide with larger planets and threaten to knock them from their courses. Within two years, Bishop George Graham would be notorious throughout Britain as a defeated symbol of the old order, and Walter Stewart of South Ronaldsay would stand on a national stage as his scourge and nemesis.

The Excrement of Bishops

Orkney's third Protestant bishop was a man not made for revolutionary times. Like his predecessors, George Graham was an incomer, the younger son of a Perthshire laird whose grandfather, William Graham, first earl of Montrose, ranked among the noble dead of Flodden. George was thus a distant cousin of James Graham, marquis of Montrose, whose own destiny would fatally entwine with the politics of Orkney.

George Graham was ordained minister in 1589, and served the Perthshire parishes of Clunie, Auchtergaven and Scone. In Scone, he moonlighted as tutor to the sons of a local laird, Alexander Smyth of Braco, and on their father's premature death, the boys, Patrick and Andrew, became Graham's wards. A kinsman, Andrew Graham, was nominal bishop of the Perthshire diocese of Dunblane, as James VI sought to re-energise episcopacy in the Kirk. Andrew's resignation in 1603 cleared the way for George to succeed him.

There was little to mark George Graham out as a zealot for muscular episcopacy. James's desire for bishops to sit in parliament was accepted by the General Assembly in 1600 with a series of reluctant 'cautions' – they were to draw diocesan revenues only once schools and kirks were provided for, and were not to deplete them. Bishops ought to remain ministers of specified parishes, subject to the General Assembly, and should cast no parliamentary votes without warrant of the Kirk. In the years following, embittered Presbyterians saw all these promises broken. Yet, in 1604, George Graham declared, 'I would he were hanged above all thieves, that presseth not to the uttermost to see the cautions kept, to hold out corruption, the pride and tyranny of bishops.' Some colleagues were sceptical. The minister of Falkirk, Adam Bellenden (a son of Sir John Bellenden of Auchnoul; p. 107), urged the minister of Scone to cast aside 'that unlawful place and calling which thou hast taken'. Bellenden was reported to have said that Graham, 'the excrement of bishops, had licked up the excrement of bishoprics'.

Precisely because Dunblane was an impoverished diocese, Graham hoped for better things. When the king moved Archbishop Spottiswoode of Glasgow to St Andrews in 1615, Graham petitioned for the vacancy. It went to Bishop Law of Orkney, and Graham had to be content with going north as Law's successor. His erstwhile critic, Adam Bellenden, became bishop of Dunblane – not ashamed, in the words of the Presbyterian chronicler David Calderwood, 'to lick up his excrements, and to accept that mean bishopric to patch up his broken lairdship'.[11]

The episcopal fabric in Orkney furnished finer cloth for the quilting of a family inheritance. Graham arrived in his diocese accompanied by his wards Andrew and Patrick Smyth, his wife, Marion Crichton, and the first of what would eventually be seven daughters and five sons. Graham's overriding principle was to see them all securely settled.

His strategy was shamelessly nepotistic, if technically legal: to continue Bothwell's policy of feuing the bishopric estates. Patrick Smyth of Braco, 'a faithful servant', benefited most. In 1617, Graham invested Smyth with estates in Holm and Paplay, including the little island of Lamb Holm, where centuries later homesick Italian POWs would decorate their Nissen-hut chapel (p. 67). Smyth became Graham's chamberlain and sheriff-depute, and, from 1618, his son-in-law, after marrying his eldest daughter, Catherine. He would be a force to reckon with in Orkney for decades to come.

Staring across at Lamb Holm from the Mainland side, Smyth built a new mansion, later to pass into the hands of the bishop's third son,

Patrick Graham, and be rechristened Graemeshall. Another fine property was a house the bishop rebuilt at Skaill on the West Mainland coast, close to the undiscovered remains of the Neolithic settlement of Skara Brae. The lands were feued to Smyth, with a sub-feu to the bishop's wife and reversion to their eldest son, David.

Graham oversaw construction of a third fine mansion at Breckness, a few miles down the coast, with fine views over the purple hills of Hoy. The Breckness estate had been acquired in the sixteenth century by the priest-cum-minister and laird William Moodie (pp. 133–4). In 1623, his grandson Francis transferred the lands to Patrick Smyth as a nest egg for the bishop's fourth son John. In Moodie family tradition, Francis authorised the sale in return for the bishop waiving punishment for his prodigious adulteries; the reality is that he was desperately in debt.

Nonetheless, stories proliferated about the bishop's dubious methods. One tale, said in the nineteenth century to have been 'handed down from generation to generation by the old inhabitants of the township', concerned Breckness House itself. Now a tattered ruin, it was once an imposing multistorey construction, whose main staircase rose over a little closet. Dwelling therein was the 'Brownie of Breckness', a small magical creature with prophetic 'second sight', who advised Graham on business matters and reported to him the sins and secrets of people in the district.[12]

In all likelihood, Bishop Graham did not employ an officious fairy as an agent of church discipline. But his rule recognised little separation between the spiritual and the worldly. The bishop's son Patrick, endowed with lands in Stronsay, Shapinsay and St Ola, followed his father into the Kirk. In 1634, the bishop appointed 'our well-beloved son' minister of Holm. A namesake, Mr George Graham, was installed as minister of Sandwick and Stromness around 1631. He is usually described as a cousin of the bishop, though it is possible he was a 'cousin' in the pre-Reformation manner, an illegitimate son. Henry Smyth, minister of Hoy and Graemsay, then Shapinsay, the accuser of Marjorie Paplay (pp. 223–5), may have been a kinsman of Smyth of Braco.

Clashes with other Orkney ministers arose from Graham's determination to maximise his profits from leases, teinds and 'glebes' – the parcels of land assigned to supplement ministers' stipends. One soured relationship was with the minister of Kirkwall, Patrick Inglis, a friend who came north with the bishop in 1617. Graham confiscated Inglis's glebe at Glaitness, nestled along the southern shore of the Oyce, the sheltered natural harbour bounding Kirkwall to the west. A century later, the incident was vividly remembered in Orkney. As the bishop stood in his window in the Palace of the Yards, his gaze fell upon Glaitness: 'Mr Patrick, I must have that room of Glatness from you, and I will give you the room of Corse for it, because it lieth in mine eye.' The farm of Corse, a little to the south, did not strike Inglis as fair exchange: 'Pick out that greedy eye, my Lord, that would take Gladness from me and give me Cross!' Inglis's own recollection was less droll. While he was busy preaching in the cathedral, Bishop Graham went with a dozen men to the glebe and seized the crops.[13]

Another critic was the provost of Orkney and South Ronaldsay minister, Daniel Callander. The details are murky, but Graham accused Callander of preaching against him. Callander also fell foul of an episcopal ally – the laird of Burray, William Stewart of Mains, whose daughter Anna would marry Graham's son Patrick. In November 1635, at a synod in Kirkwall, Callander confessed to slandering Stewart, and was afterwards deposed by the Scottish court of high commission – a body seen by critics as an insidious instrument of episcopal power against principled Presbyterians.

Callander's replacement was a man we have now met on several occasions: Walter Stewart. His relations with Graham were sticky from the start. Stewart would later claim that Graham failed to support him, against parish opposition, in enforcing presbytery decrees. He reminded the bishop that, 'as he was set over the ministry to censure them for their faults, so he was obliged to maintain them in doing the duties of their calling'. Graham in turn accused him of sedition.

It did not help that Stewart was an appointee of the crown; Graham was jealous of his patronage rights. The Kirkwall Synod of 1635 weighed the claims of Patrick Waterston to the vicarage of Birsay, on the basis of a presentation secured from the king. Graham declared the nomination invalid: 'the bishops of Orkney had and has the presentation, advocation and donation of all the vicarages in Orkney'. The synod blocked Waterston's appointment, and Inglis of Kirkwall went to Birsay instead.

Waterston was unusual among the ministers in being a native Orcadian; his father, another Patrick Waterston, was minister of Orphir. Patrick junior was also unusual – against stiff competition – in the extent of his litigiousness and general belligerence. Since 1634, Waterston had served as minister of St Andrews and Deerness, but hardly satisfactorily. The synod admonished him 'to agree better with his parishioners'. Asked if he exercised discipline, he said he did not, and knew not how. The synod ordered him to make a copy of the decrees 'whereby our kirk has been hitherto governed'; Waterston refused, saying contemptuously 'they were but a nose of wax'.

Nor did Waterston accept the Birsay rebuff gracefully. He appealed to the archbishop of St Andrews, the lords of session and the court of high commission, and made two trips south in pursuit of his claims. Graham warned him that a third 'would be over many in one year'. In his parish of St Andrews, Waterston allegedly took part in the assault on William Sinclair. The exasperated bishop considered the minister a 'swindger' – a rascal – and started proceedings to depose him. This was averted in February 1637 by an abject letter of apology. Waterston pledged 'to live for ever afterwards in all orderly manner as becomes the servant of Christ'. It was a hollow promise; Waterston's transfer to Rousay and Egilsay, sometime before the spring of 1639, was probably an attempt to minimise his capacity for trouble.[14]

Bishop Graham had his admirers. William Lithgow, after his visit to Kirkwall in 1629, praised 'the stately and magnifick Church of St Magnus'. But the town, he declared, was 'more beautified with the godly life of a most venerable and religious bishop'. Orcadians, 'a scattered people without a head', turned to him for leadership and justice, as their official governor (William Dick) was 'an alien to them, and a resider in Edinburgh'.

To give Graham his due, he was often resident in Orkney, dispensing secular justice through the sheriff court of the bishopric parishes. The non-survival of presbytery minutes prior to 1639 prevents any detailed reconstruction of Graham's ecclesiastical administration, but the records of a 1627 visitation suggest a generally conscientious approach to both the endemic and sporadic problems of Orkney's parishes.

The bishop restored Kirkwall's Mercat Cross in 1621, and gifted two silver communion cups to the cathedral. In 1630, 'out of his own liberal-ity', Graham gave 500 merks to the session to spend at its discretion. Another 500 had been collected for relief of the Huguenots of La Rochelle, and retained after the fall of the city to Catholic forces in 1628. The combined sums were invested to provide a stipend for the master of the Grammar School.

In the cathedral, Graham constructed an elaborate wooden gallery or 'loft' on the south side of the chancel. Its lower section housed an episcopal throne, and the upper part had seating for the bishop's extensive family. Carved wooden panels displayed the arms of the bishop, of Patrick Smyth of Braco, and of his other sons-in-law, Adam Bellenden of Stenness (a grandson of the formidable Patrick) and William Henryson of Holland in North Ronaldsay.

Cathedral spaces were tokens of privilege and power; living and dead contended over them. In 1629, the session denounced 'great incivility and rudeness of the basest sort … who being once set down in the chiefest seats in the kirk would not rise up to give place unto their elders and superiors'. One such superior, in his own estimation, was the merchant and session elder James Baikie, later to go to remarkable lengths to shield his mother from accusations of witchcraft (p. 224). Baikie secured permission for a family pew, and in 1631 unveiled plans to enlarge it. The Baikies' 'seat' was in a part of the north choir known as 'The Stewarts' Aisle', on account of the burial there of Earl Robert and his brother Lord Adam. Robert's eldest surviving legitimate son, John, earl of Carrick, sent his half-brother Edward Stewart of Burgh to the session to demand all seats be removed from the aisle and exclusive use of it be reserved to the earl 'and others of their name'.

Graham's verdict was 'to give unto those complainers contentment', but Baikie dragged his feet. At a session meeting in September 1631, the bishop delivered an imperious dressing-down. Baikie should reflect that 'it would come to his Majesty's ears how such persons did sit there, and trample upon his Highness's grand-uncle's belly'.[15] He needed, in a literal sense, to know his place. It was a slight the ambitious merchant would not forget.

The spats over space in St Magnus played out as people across Scotland pondered what churches were actually for. James VI had started to make Church of Scotland practices more like those of the Church of England

– not, as Presbyterians had hoped in 1603, the other way around. In 1618, a General Assembly at Perth was pressured into accepting five articles, authorising private communion of the sick, private baptism, confirmation by bishops, and observance of holy days. Most controversial was the first article, commanding people to receive communion kneeling – a practice leaving the taste of popish transubstantiation in the mouths of Protestant Scots. The 'Five Articles of Perth' were patchily enforced but heavily resented; especially in Edinburgh and southern Scotland, seeds of opposition began to sprout.

Charles I pushed on with the anglicising policies. The king was irritated by a 1628 petition from two Edinburgh ministers asking him to sanction seated communion. Instead, he ordered Archbishop Spottiswoode to enforce kneeling. Charles's visit to Scotland in 1633, the year Laud was installed as archbishop of Canterbury, increased anxieties about the king's intentions. Bishops at his Edinburgh coronation sported English-style surplices; soon, all ministers were ordered to wear them.

In 1636, as the Prayer Book edged towards publication, new canons (church laws) were issued for Scotland. They placed restrictions on preaching, and assumed the Kirk to be governed by bishops, with no mention of presbyteries or General Assemblies. Fervent 'Laudians' filled episcopal vacancies. Thomas Sydserf, bishop of Galloway, was a zealous enforcer of the Articles of Perth, and allegedly kept a crucifix in his chamber. John Maxwell, bishop of Ross, argued provocatively for *jure divino* episcopacy – the idea that bishops were not just useful instruments of governance, but required by God's law. William Forbes, of a newly created diocese of Edinburgh, regarded transubstantiation as a mistake rather than a heresy. He favoured reconciliation with Roman Catholics – anathema to most right-thinking Protestants.[16]

George Graham never reached quite this temperature of Laudian zeal, yet he enjoyed the trappings of office, and was keen for royal policy to be seen being enacted in the furthest-flung Scottish diocese. When the new code of canons was published, Graham defended it in eye-opening style from the pulpit of St Magnus Cathedral. One canon encouraged private confession to a bishop or minister along with sacramental absolution – only in being optional did this seem to differ from the Catholic practice. Graham, however, thought people should fear confessing their sins no more than a man suffering from urinary stones should 'be ashamed to discover his private parts to another to shoot up his finger through his fundament to grip the stone'. The bishop supposedly made his point by 'acting it in the pulpit with a filthy demonstration'.

In another sermon, Graham likened himself and the sheriff to 'a galled horse' – one with skin sores – 'and the people of the country, taxing them in their particulars, to wasps or gleggs [horseflies], sucking their galled back'. Graham, 'winshing [prancing] and turning himself about in the pulpit after the manner of a galled horse', swore that however much such folk 'hum and bum, and bizz and buzz, we shall fart and fling [kick] and fling and fart so that they shall not get their wamb [stomach] full of our rotten flesh'.

Some of the resistance Graham alluded to came from his own ministers. He was reported to have argued with Henry Smyth of Shapinsay and David Watson of Westray about surplices and the crucifix. He spoke admiringly of Bishop Maxwell, declaring 'the smell of the Bishop of Ross's arse was more savoury than all the ministers' mouths of Orkney'. He sought to levy a charge on his ministers to subsidise Maxwell's attendance at court, and when the brethren refused, took payment by force.

The strangeness of the new Prayer Book disturbed even Orcadians not fundamentally opposed to bishops or ceremony. Robert Henryson of Holland, a commissioner for Orkney at the parliament of 1617, was father-in-law to Graham's youngest daughter. Yet Henryson felt 'great grief' at the bishop's bullish defence of the book. Two copies of it survived in Orkney at the end of the nineteenth century, probably acquired for the cathedral.[17] It is doubtful the service was ever used in parishes outside Kirkwall. Charles I's liturgical revolution was only starting to be rolled out in Orkney just as a political revolution was preparing to roll it away.

A Curler on the Ice

The manifesto of the revolution was a National Covenant, drawn up in February 1638. It was a verbose document, cobbled together from recycled denunciations of popery, and a listing of parliamentary acts underpinning the Scottish Reformation. But its symbolic significance was immeasurable, and the mass swearing across the country of oaths to uphold it, by people of all social ranks, including some women, was an unprecedented display of popular political participation. In putting their names to the Covenant, they were pledging to uphold 'God's true religion' against schemes of 'wicked hierarchy'.

Having lost control of Scotland, Charles I had little choice but to concede the opposition's key demand: a General Assembly met in November 1638 – the first since 1618. Orkney's ministry had not been represented at the Perth Assembly, but in 1638 the presbytery sent David

Figure 6.1: Crowds in Greyfriars' churchyard, Edinburgh, in February 1638, swearing to uphold the National Covenant – an imagining of the scene by the Victorian Scottish artist William Hole.

Watson of Westray and Walter Stewart of South Ronaldsay as delegates to Glasgow. There they presented the Assembly with two papers itemising charges against Bishop Graham. Stewart was said to be 'his main accuser'.

The first document alleged breach of the 'caveats' – the limitations on episcopacy which Graham once professed to support. He voted in parliament without regard to General Assembly or presbytery. He took diocesan revenues without kirks or schools being properly provided for. He deprived ministers of their due, while setting 'even nineteen-year tacks to his own bairns, as to Patrick Smyth, his godson in Holm'. He did not personally serve as a minister for any particular congregation, but acted as 'ane constant moderator' of the presbytery, against the wishes of the clergyman elected to the office. 'He hath used great tyranny over his ministers.'

All this was grist to the mill of the Assembly's resolve – not just to clip episcopacy's wings, but to eradicate it from a purified Presbyterian Kirk. The second paper, signed by eleven Orkney ministers, cited more transgressions. Unsurprisingly, the bishop's son Patrick, and his 'cousin' George Graham, did not set their hands to it, and neither did a smattering of others. The clergy of Shetland were left out of the loop.

The indictment listed Laudian innovations, along with abuses of spiritual authority, and a style of speaking which 'savoured many times of lasciviousness'. The bishop permitted marriages without reading of banns, and allowed ordination candidates, including his son, to preach 'without trial'. Graham shamelessly justified his nepotism 'by the example of Christ, who preferred his own kinsmen to the ministry'. Sexual offences went unpunished. Moreover, 'he did many times – charmers and such like being brought before him in the presbytery – extenuate their faults so far as he could, saying that all their practices were but trifles and that he could do as much himself'. That relaxed attitude to the performance of magic, as we have seen (pp. 221–2), was to change in Orkney in the coming years.

Other grievances were scarcely theological, though Graham's depletion of the ministers' stipends was deemed to be 'oversight of sacrilege'. There were complaints about feus, manses and glebes, including Inglis's grudge over Glaitness. Patrick Waterston the elder reported pathetically how Graham took from him 'a maser [cup] of silver, worth an hundred pound or thereby, beside [in addition to] a stand [set] of table clothes of damask'.

The ministers were perturbed by Graham's enthusiasm for the Perth Articles, but even more so by his aggressive managerialism and his acquisitive eye on their manses, glebes and tablecloths. A national revolution provided the occasion for a local coup d'état – a consummate example of the 'periphery' harnessing the 'centre' to reorder its own affairs. There was also the matter of what Orkney could do for Scotland. In the campaign to first tarnish, and then abolish, episcopacy, the nation's edge became for a time the revolution's cutting edge. Without asking for them, the bishop of Orkney was about to experience his fifteen minutes of fame.

Graham's case was heard at the Assembly on 11 December, with a reading-out of the accusations catalogued by Stewart and Watson. A further charge was added, perhaps recalled from earlier in Graham's career. 'He was a curler on the ice on the Sabbath day' – a devotee of the winter sport of curling, in which Scotland still excels, and a flouter of God's command to preserve Sundays for pious contemplation.

All of the Scottish bishops were dismissed at the end of 1638. But while Archbishop Spottiswoode and most of his brethren were excommunicated, Graham – along with the compliant bishops of Dunkeld, Murray, Argyll, and the Isles – was 'merely deposed from the ecclesiastical function'. Graham wrote to the Assembly, promising 'obedience to all the acts thereof', and declaring he 'never loved the novelties obtruded upon the Church by the bishops'. It was enough for a stay of (spiritual) execution. The threat of excommunication would be lifted, however, only after 'serious signs of repentance'.[18]

Excommunication, unlike deposition, involved confiscation of goods and property, and this the ex-bishop of Orkney was determined to avoid. In October 1638, he had begun moves to insulate himself, transferring rents into the hands of the Smyth brothers. At the start of 1639, he was ready to say whatever was required of him.

On 11 February, Stewart and a select group of ministers – James Heind, presbytery moderator, Robert Pierson of Firth, and the bishop's son Patrick Graham of Holm – convened at Graham's house at Breckness.

They were witnesses to a remarkable declaration. Graham was 'sorry and grieved at my heart that I should ever, for any worldly respects, have embraced the order of episcopacy, the same having no warrant from the Word of God'. Recognising the 'evil consequences', in Scotland and throughout Christendom, Graham wished to 'disclaim and abjure all episcopal power and jurisdiction and the whole corruptions thereof'. He swore to obey all the acts of the Glasgow Assembly, and do his utmost 'in advancing the work of Reformation'. On 10 April, Graham's statement was formally registered at a synod in Kirkwall. A proud prelate had reinvented himself as a humble Covenanter – and the estates and houses at Skaill and Breckness stayed in the secure bosom of his family.

After the surrender of the bishop, Orkney's episcopal regime rapidly collapsed. In March 1639, the presbytery ordered ministers to ensure their congregations swore the oath to uphold the Covenant, which after November 1638 included an abjuration of episcopacy. In most instances, only elders were invited to swear, though James Haigie, minister of St Andrews and Deerness, and Pierson of Firth scrupulously put the oath to all their parishioners. There were a handful of holdouts. Walter Stewart reported the elders of Burray had not sworn, perhaps influenced by Graham's confederate, William Stewart of Mains. Among Pierson's parishioners, all swore except 'the old laird of Stenness' – Adam Bellenden, another ally and in-law of the bishop. The earl of Carrick at first refused, but by early June fell reluctantly into line.

In the meantime, Charles I went to war with his own kingdom of Scotland, or its Covenanting government. The first 'Bishops' War' ended in June 1639, a temporary cessation of hostilities. A General Assembly was to convene in August in Edinburgh, to consolidate the work begun at Glasgow. The summons arrived in Kirkwall at the beginning of the month, leaving barely a fortnight to select commissioners and dispatch them south. The presbytery again picked Walter Stewart as 'the most able man'. A 'ruling elder', 'chosen by plurality of voices', went too: Patrick Smyth of Braco, factotum and son-in-law to the recently deposed bishop. This appointment, along with the thinly attended Breckness meeting, alerts us to a distinct whiff of collusion around Orkney's transfer of power. Braco went to safeguard the ex-bishop's interests, and to guarantee he would not rock the boat.[19]

On 17 August 1639, the Breckness Declaration was read in the General Assembly, where it caused a sensation. The moderator, David Dickson of Irvine, thanked God and wished others 'might take the like course'. Orders were given to enter it into the official record, 'ad perpetuam rei memoriam' – as a perpetual memory of the thing. Graham's recantation was printed in a 1641 denunciation of 'lordly prelacy' by the English Puritan activist William Prynne, and in pamphlet form in both Edinburgh and London. The chronicler James Gordon, minister of Rothiemay in Banffshire, termed it 'a piece whereof few patterns are to be found'. Other bishops in history had retired from their office, but Graham, uniquely, 'abjured it as ane antichristian function'. This, Gordon concluded, 'rendered him very detestable to the episcopal part'.[20]

In 1640–1, episcopacy's future was a burning issue in both England and Scotland. Hostilities between the kingdoms resumed in the Second Bishops' War of August–September 1640, and royal forces were humiliated. Charles had no option but to summon an English parliament, the first since 1629, whose members had Laud and other bishops in their sights. Laud himself knew of Graham's abjuration, condemning his 'ignorance' in a letter of November 1639 to Joseph Hall, bishop of Exeter, who was preparing a defence of episcopacy in the face of mounting Presbyterian attacks. The inspiration for his book, Hall wrote, was a report he received from the recent General Assembly 'that one M[aster] G. Graham, Bishop of Orkney, had openly before the whole body of the Assembly renounced his episcopal function ... it had been much better to have been unborn, than to live to give so heinous a scandal'.[21]

Graham himself begged to differ. His last years, before his death in late 1643, passed in comfortable retirement between Skaill and the family estate at Gorthie in Perthshire. William Dick acquired a tack of the bishopric estates. Reputedly the wealthiest man in Scotland, Dick lent large sums to the Covenanting government to maintain its armies in the field. In 1641, the tack passed to Edinburgh's burgh council. Graham was happy to answer in loquacious detail questions from the magistrates about rents, feus, stipends and the character of udal landholding – though there were hints of his bitterness at being turfed out of the palace in Kirkwall: 'I hear it is both ruinated with the weather and not well used.' Another lost property was the fine townhouse, a former cathedral

Figure 6.2: Tankerness House, Kirkwall. Originally, a pair of manses for the Catholic clergy of the cathedral, it became home to the first Protestant minister of St Magnus, Gilbert Foulzie, and was acquired in 1642 by James Baikie, who bestowed its current name. The building now houses the Orkney Museum.

manse, opposite St Magnus. In 1642, the ex-bishop made it over to Smyth of Braco, who sold it to James Baikie. The merchant was a coming man, with a laird's estate in the East Mainland. Renamed Tankerness House, his Kirkwall dwelling was a certification in stone of his victory over the bishop who had humiliated him a decade before.[22]

Revolutionary Orkney

By 1641, Charles I's authority had been undermined in all three of his kingdoms. In the summer, following his defeat in the Second Bishops' War, Charles travelled to Edinburgh to ratify a treaty which confirmed the abolition of episcopacy, and left the Covenanters in control of the government. In October, a Catholic rebellion in Ireland intensified Protestant fears about popery, and about what Charles might want to do with any army raised to crush the rising. The breakdown of trust led to the outbreak of civil war in England in 1642. That in turn generated an alliance between Scotland and the rebel English parliament. A 'Solemn League and Covenant', drawn up in 1643, committed the Scots to military aid for the Parliamentarian war effort, in return for Presbyterianism becoming established across all three kingdoms.

Royalists, Parliamentarians and Covenanters sought allies where they could, and Orkney once more became a negotiable asset. Christian IV of Denmark resented a lack of English support against the Holy Roman Emperor in the Thirty Years War. Yet Charles was his nephew, and rebellion against him could scarcely be condoned. In 1640, Christian sent emissaries with offers to mediate between the king and his Scottish subjects. In treaties of 1621 and 1639, Christian had promised not to raise during his lifetime the question of Orkney and Shetland. But circumstances had changed. He now empowered his ambassadors to discuss military and financial assistance in return for Danish repossession of the islands.

Charles was willing to consider it. He offered to pawn Orkney and Shetland to Denmark for 50,000 gold guilders. Christian thought this too much. His counter-proposal was more radical. He would seize Orkney, fortify it, and with a combination of Danish and mercenary troops use the islands as a base to gain territory in mainland Scotland. In this way, the Covenanters would be forced to come to terms with their king.

By the time the offer was made, Charles had lost the Second Bishops' War. But schemes involving the Northern Isles continued to be considered. A Royalist envoy sent to Denmark in 1643 to acquire arms and munitions was authorised to offer Orkney and Shetland as security against payment for the weapons. Christian's conflict with Sweden, however, prevented him making any significant financial or military interventions in Britain's Civil Wars.

In 1643, as security for loans to the crown, Charles mortgaged the earldom of Orkney and lordship of Shetland, with heritable sheriffship and rights of church patronage, to William Douglas, earl of Morton, who, as we have seen (p. 245) was already enjoying income from the tack.[23] The grant was redeemable on a payment of £30,000 sterling, which there was no immediate prospect of the king finding. Morton fancied his chances of keeping possession as earl in all but name.

Within Orkney, political tensions became entangled with local rivalries. In the spring of 1639, John Craigie of Sands, ruling elder in St Andrews and Deerness, complained 'in name and behalf of the whole Covenanters' about another Deerness laird, John Cromartie of Skae, reported to have 'called all Covenanters traitors'. Craigie was at odds with the Royalist earl of Carrick, who, 'taking violent advantage of the troubles of the times', seized possession of the fields and links [coastal ground] of Sands in Deerness, along with the ware collected by Craigie and his servants 'for gooding of his said lands'. Seaweed, as we have seen (p. 62), was a contested resource on the coast of the East Mainland. Even in his old age, Carrick was an opponent to take seriously. In correspondence with the Edinburgh magistrates, ex-Bishop Graham referred to him slyly as one of 'the sons of Zeruiah' – warrior nephews of King David, who found them 'too hard for me' (2 Samuel 3:39).

In November 1639, the presbytery addressed a situation Bishop Graham had long tactfully overlooked. The minister of Deerness, James Haigie, was ordered to admonish Carrick to separate from Helen Lindsay. The earl had an English wife, Lady Elizabeth Howard. In a 1643 petition to the House of Lords, she described his 'having deserted her twenty years ago', to remain ever since 'in the remotest parts of Scotland'.

Carrick's response was unsatisfactory, and a month later Haigie was deputed to return with two of his elders and reiterate the presbytery's order. Carrick was then cited, twice, before the Deerness Kirk Session, like any common fornicator. In February 1640, he complained of procedural irregularities, and Haigie was ordered to restart the process. But this confrontation between established political power and the revived authority of the Kirk ended in qualified victory for the latter. Carrick did appear (perhaps privately) in front of the session, and the presbytery declared itself satisfied with his promise to send Lindsay away. In revolutionary times, no one was too great to evade scrutiny for their sins.[24]

Yet even as the Orkney Presbytery seized the reins from a discredited episcopal regime, it found its authority hobbled by faction and infighting. In the summer of 1641, matters had reached a point where a petition was dispatched to the General Assembly from Walter Stewart, Patrick and George Graham and five other ministers lamenting that 'the discipline of our Church is altogether shaken loose' and it was hardly possible to hold meetings of the presbytery.

The chief source of trouble was the truculent minister of Rousay, Patrick Waterston. Days after Bishop Graham's declaration was registered in April 1639, a furious Walter Stewart was demanding action over a 'foul imputation'. Waterston had been telling people that Stewart 'was partial in his proceedings, specially anent [concerning] the receiving of the late pretended bishop's repentance' – indeed, that no repentance took place. Waterston apologised, and the ministers were for the moment reconciled. But the insinuations of a stitch-up perhaps contained enough substance to touch raw nerves.

There were further squabbles over the expenses of commissioners going to the 1639 General Assembly. In April 1640, Harry Stewart of Graemsay declared, 'in name of himself and the rest of the Covenanters', that he would contribute nothing towards the costs of Patrick Smyth of Braco, 'in respect he had embraced ane office clean contrary to the Covenant'. A complaint, subsequently found to have been drafted by Waterston, Watson of Westray and Thomas Abercrombie of Cross and Burness in Sanday, was sent to parliament, prompting a counter-petition by Smyth's supporters, who denied any reason to doubt 'his affection to the common cause'.

The ostensible issue was Smyth's appointment, on 8 November 1639, as a justice of the peace, alongside James Baikie of Tankerness, and a dozen other notables. The Orkney moderator was tasked with writing to the Edinburgh Presbytery to ask whether it was contrary to the Covenant for an elder to serve as a JP. The reply was that bans on secular office-holding applied only to 'preaching elders' (that is, ministers), not to lay 'ruling elders'.[25] It is in fact doubtful that Harry Stewart's attack was motivated by such scrupulosity. Smyth of Braco was a power-broker of the old regime, stabilising his position under the new; long-time rivals looked for opportunities to unseat him.

An ally of Smyth, and a sworn enemy to Harry and a brood of other Stewarts, was William Sinclair of Sebay. In April 1640, he and Smyth complained to the presbytery against Patrick Waterston. The Rousay minister had slandered Sinclair, saying he had stolen some gold and hid it in his 'sark lap' (the fold of his shirt). Waterston had also spoken derisively of Patrick Smyth: 'is that curdie [small boy] your commissioner, that sowter's [cobbler's] son?' He challenged people to single combat, including Harry Bellenden, brother-in-law to Bishop Graham's daughter Anna. On one occasion, Waterston lay in wait to ambush and murder James Haigie, his successor as minister of Deerness. He called the presbytery moderator, James Morison, and its clerk, George Moodie, knaves, and termed the General Assembly in Glasgow 'ane knaves' assembly'. He was, in brief, 'an incendiary and tumultuous person'.

In June 1640, Waterston ignored a formal summons from the presbytery, sending a message through Patrick Halcro of Wyre (husband to the exiled murderess Helen Erskine) that 'certain enemies convened in Kirkwall, who had resolved to take his life'. In July, Waterston agreed to submit to the presbytery's judgement, but only if ministers he regarded as biased were disbarred: Patrick Graham, James Morison, Walter Stewart, James Heind, Robert Pierson, James Haigie. The presbytery agreed to recuse Patrick Graham from complaints concerning Patrick Smyth (his brother-in-law), but otherwise held firm. Waterston appealed to the General Assembly.

His petition was subscribed by a surprisingly large group of clerical and lay allies. In addition to his father, three ministers – Thomas Abercrombie of Sanday, Henry Smyth of Shapinsay and Alexander Somerville of Stronsay – set their names to it, as did Harry Stewart of Graemsay, Robert Monteith of Egilsay, Patrick Halcro of Wyre and five other lairds. The petitioners lamented 'the great division which has been these two years almost in our presbytery', and requested it entertain no charges against them until matters were resolved by higher authority. The General Assembly considered this reasonable, and referred the dispute to a commission, to convene at Thurso in Caithness, under the presidency of the earl of Sutherland.

The commission never met, and over the following two years appeals and counter-appeals flowed from Orkney to the General Assembly. Waterston complained of persecution by 'enemies in favour of the late pretended prelate', and in 1641 persuaded the Rousay elders to send a letter lauding him as 'second to none in the country in bringing on the blessed work of reformation ever since the first moment the Covenant was heard of'. Another petition, sent in July 1642, was signed by nineteen ruling elders from Westray, Papa Westray, Sanday, Stronsay, Rousay, St Andrews, Holm, South Ronaldsay, Rendall, Firth and Harray, as well as John Craigie, ruling elder of Deerness, Harry Stewart, ruling elder of Graemsay, and the sheriff-depute, Thomas Buchanan, as ruling elder in Kirkwall. The missive extolled Waterston's 'fervency in beating down all corruptions in religion', while conceding that, to some, 'he might seem, with Elijah, to be an incendiary and troubler of the land'.[26] The Old Testament prophet Elijah (1 Kings: 17–19) perturbed the Israelite political authorities of his day with his violent denunciations of false gods, and was remembered as a harbinger of the Messiah and end times.

Like Ancient Israel, Orcadian society was riven with feuds and factionalism. That was not unusual for the shires and regions of seventeenth-century Britain, but the finite boundaries of the island world imparted a particularly claustrophobic character to its quarrels over land, office and resources. The Revolution of 1638, perhaps even more than the Reformation of 1560, injected ideological fervency into these rivalries, as self-appointed champions of the Covenant uncovered machinations of clerics and laymen with links to the episcopal regime. To what extent religious ideology was the real substance of the quarrels, and to what extent a convenient veneer, is not always easy to discern.

The parish of Birsay is a case in point. In 1641, it required a new minister. The presentation belonged to the crown, for confirmation by the presbytery and the parishioners themselves – this 'call' to a minister from his parish was dear to the mentality of Presbyterianism. The nominee was James Aitken, graduate of Edinburgh and Oxford, and, at the time of appointment, chaplain to the marquis of Hamilton, a leading royal adviser. Aitken, born and raised in Kirkwall, was also the son of Henry Aitken, commissary to Bishop Law and latterly sheriff-depute.

In June 1642, Alexander Johnston and Robert Linklater, commissioners from Birsay, informed the Orkney Presbytery that all parishioners regarded Aitken as 'blameless for conversation [conduct] and able to edify them'. However, William Moncrieff and John Sinclair claimed that they were the rightful parish commissioners, and that Aitken was someone 'we will never grant to receive with our hearts'. They alleged Aitken 'had spoken against the Reformation', producing Patrick Waterston as their key witness. He recalled telling Aitken how episcopacy was antichristian, and receiving the tart reply that 'many good divines thought the contrary'. Aitken, in turn, accused Waterston of being the 'chief instrument in drawing away the hearts of the people from receiving him'.

The presbytery admitted Aitken, but his troubles were not over. In 1643, Aitken reported Sinclair for calling him 'base fellow' and saying his wife 'lied like a whore'. Some parishioners refused to participate in the Lord's Supper, after Sinclair spread rumours that Aitken was adding water to the communion wine. This dilution was an ancient Eucharistic custom (perhaps symbolising Christ's dual nature), abandoned in Scotland at the Reformation, other than in some conservative pockets of Aberdeenshire. The practice was admired by adventurous Laudians, and denounced as popish by strict Presbyterians. Aitken dismissed it as 'calumny', but it was a strange story to invent. More than ceremonial niceties were at stake in the sniping. Sinclair was married to the widow of Francis Lidell, former minister of Birsay, and was suing Aitken to recover sums laid out by Lidell on purchase of a glebe.[27]

Factional quarrels also shaped – and were shaped by – a phenomenon discussed already: the spate of witch-hunting gripping Orkney in these years.[28] In particular, the shadow of politics fell across a question polarising Orkney opinion in the mid-1640s: whether Marjorie Paplay, mother of James Baikie of Tankerness, was a notorious witch (p. 223). Henry Smyth of Shapinsay, who believed Paplay responsible for his mother-in-law's death, was a one-time ally of Patrick Waterston. But other ministers zealous for the prosecution of Paplay – James Aitken, George Graham, James Haigie and Walter Stewart – all had troubled history with the Rousay minister. At the height of the controversies, in March 1643, Waterston was forced to seek forgiveness for calling them 'ill-affected, and of the episcopal faction'.

Waterston himself was the presbytery's most vocal defender of Paplay, 'an honest woman'. James Baikie, his former parishioner in St Andrews and Deerness, was an old associate. At the time of the furore, Baikie was at odds with Waterston's successor, James Haigie, over attempts to extend his glebe. Like that other declared champion of the Covenant, John Craigie of Sands, Baikie had clashed with Waterston's Deerness antagonist, William Sinclair of Sebay.[29]

Networks of affinity and enmity spanned the Orkney Mainland. Baikie was related by marriage to William Moncrieff, Aitken's opponent in Birsay. William's elder brother Thomas was married to Baikie's sister Elspeth, one of the women named as a witch by Barbara Boundie (p. 224). Waterston relied on Baikie for practical advice, passing him a draft of the petition sent to the General Assembly in 1642, 'that it may be corrected if it shall seem to require'. Baikie's quarrels with Haigie about the glebe, and his guerrilla warfare over the allegations against his mother, antagonised the presbytery, encouraging it to try (unsuccessfully) to fine him, 'according to the law concerning the vilipendence [insulting] of ministers'.

Baikie hit back. At the 1643 General Assembly in Edinburgh, the presbytery was represented by two leaders of the 'episcopal faction', Walter Stewart and Patrick Graham. On their return, they reported being hauled before a committee of the Assembly to answer accusations that 'our Presbytery of Orkney would not admit ruling elders, nor give them liberty to voice when they came'. Efforts 'to see by whom that calumny had come abroad' pointed to Baikie, who all but admitted making the complaint.

The status of ruling elders was a sensitive matter. Before 1638, elders did not usually attend Scottish presbytery meetings. After the adoption of the Covenant, they were expected to play more significant roles, and 'ruling elder' came to mean a delegate to the presbytery from a kirk session. Ruling elders in Orkney did attend presbytery meetings, but never from a full roster of parishes, and there may have been some truth to the charge that 'ruling elders had no voices when their ministers were absent'. In the early 1640s, Orkney had become a dripping tap of allegation and insinuation. Kirk authorities were to understand that, despite having delivered the scalp of Bishop Graham, the Orkney ministers – or most of them – were not true believers in the Covenant.

The convulsions over Paplay's guilt or innocence effected one notable conversion. James Morison, son of an Aberdeenshire minister, was presented to Evie and Rendall by Bishop Graham in 1620. He was presbytery moderator in 1639–41, again in 1643, and regarded by Waterston as one of the 'episcopal faction'. But Morison grew disillusioned with his colleagues and with the ecclesiastical patronage of the earl of Morton. In

May 1645 – after his brother ministers undertook some sleuth-like comparison of handwriting – Morison was revealed as the author of a paper about Marjorie Paplay delivered to the presbytery by James's brother, John Baikie. This, and other documents penned by Morison, seemed 'full of foul imputations on the presbytery'. He was, his brethren concluded, a person of 'manifest ambition and presumption', 'vehement and unmanly passion'. The ministers suspended him.

Yet Morison refused to stop ministering in Evie, and when the presbytery formally deposed him in 1646 he appealed to the General Assembly. The Assembly – by now surely in despair at the dysfunction of the Orkney Presbytery – judged the deposition unsound, and restored Morison to the exercise of ministry. But Morison was also at fault, for preaching under suspension, and should acknowledge his offence. Final resolution was relegated to a future commission, with the Assembly candidly admitting to insufficient knowledge of the facts. A decade and more later, 'the case of Mr James Morison' was still being cited in printed debates in Scotland about the relative authority of local presbyteries and General Assemblies.

Physical distance from Edinburgh or Glasgow left much practical decision-making in the hands of local clergy. In 1646, in an attempt to rectify 'great abuses and disorders', the General Assembly ordered the Presbyteries of Orkney and Shetland to join an annually convening Provincial Synod of Caithness and Sutherland. No synod records from this period survive, but there is little to imply that Orkney ministers ever attended in significant numbers. In 1648, the General Assembly excused Shetland from the arrangement, belatedly recognising 'the great distance of that isle by sea'.

Throughout the 1640s, the General Assembly decreed 'fasts' – national days of preaching, prayer and penitence – to mark occasions of thanksgiving or threat. The notifications of these sometimes arrived in Orkney after the day itself was past, or ministers became aware of an impending fast but decided not to advertise it 'until such time as we get certain information concerning the causes'.

Some were missed completely – like the fast to allay 'dangers imminent to the church and kingdom', mandated in September 1643 as Scotland adopted the Solemn League and Covenant. It was a moment of national crisis, but the Orkney Presbytery did not meet that month at all. When Walter Stewart, back from another stint as Assembly commissioner, demanded an explanation, he received some testy justifications from his brother ministers why meetings could not be held 'in these our remote and far distant parts': 'tempestuousness of the weather'; 'impossibility of transporting by ferries, in respect the people were busied about their harvests'; 'few brethren were at home in the country'. The Solemn League and Covenant was eventually sworn as required, but not until 17 December in Kirkwall and the turn of the year elsewhere.

Difficulties in hearing from the south, and of travel around within Orkney, could sometimes be turned to pragmatic advantage. On 1 April 1646, Sheriff-Depute Thomas Buchanan approached the presbytery with a 'Humble Remonstrance' from the earl of Seaforth – a notorious vacillator between Covenanters and Royalists, shortly to throw in with Montrose for the last gasps of his Highland campaign. Seaforth and the earl of Sutherland had together hatched a scheme to invite the king to Scotland and declare national unity on the basis of the 1638 Covenant, rather than the Solemn League. It was hard to know if this would succeed, and the presbytery deferred discussion to 'a more frequent meeting of the brethren, of whom many were absent, by occasion of the intemperance of the weather'. In fact, eight ministers, only a couple fewer than normal, were present at the meeting.

Three weeks later, after conferring with the gentry, the ministers decided to 'reject and refuse the foresaid demands, finding them in direct opposition unto our National Covenant'. In August, in a pulpit roundup of news from the south, Kirkwall's minister, George Johnston, publicised Seaforth's excommunication by the General Assembly.

Through a decade of crisis and war, Orkney's political and religious leadership professed loyalty to the Covenant, and compliance with the General Assembly – though dissident voices whispered, and sometimes shouted, that this was a sham. A lack of true enthusiasm for the cause of the Covenant is suggested by the apparent ease with which troops were raised in Orkney in 1648 for the army of the 'Engagement' – an ill-fated attempt to free Charles I from English captivity on the part of a group of nobles led by the duke of Hamilton, former patron of James Aitken of Birsay.[30] The scheme was supported by parliament, but opposed by the

Kirk, since Charles refused to compel anyone to swear the Solemn League and Covenant, or promise to impose Presbyterianism permanently in England.

As he announced Seaforth's excommunication in the cathedral, George Johnston could have had no notion that before long he himself, with all but a couple of his clerical brethren, would be similarly shamed and excommunicated. Having helped to set Scotland's revolution in motion, Orkney was about to become the epicentre of efforts to reverse it.

Montrose

On 5 September 1649, an old, battered vessel docked at Kirkwall, having weathered storms in the North Sea and dodged the attentions of Parliamentarian warships. It was the forerunner of a Royalist fleet, mustering in ports around the Baltic, where agents of the exiled marquis of Montrose were recruiting men and purchasing supplies and weapons. Frederick III, who succeeded his father Christian IV as king of Denmark in 1648, was appalled by the judicial murder of his cousin, Charles I, and allowed the Royalists access to Danish and Norwegian ports; discreet support from Queen Christina of Sweden made Gothenburg on the country's western coast into Montrose's centre of operations.

Sixth months earlier, at Breda in the Netherlands, Charles II, new king of Scotland and England, had broken off negotiations with the Scots commissioners urging him to embrace the Solemn League and Covenant. A sharp strike, Charles believed, would force the Covenanters to moderate their demands. In July 1649, Montrose issued an uncompromising declaration. He intended to 'enter the kingdom of Scotland, through which I will march into the kingdom of England'. Rebels laying down their arms would be spared, otherwise 'I will with all violence and fury pursue and kill them, as vagabonds, rogues and regicides'.

The blow would be struck from Orkney, long seen by strategists of various stripes as the optimal base for an invasion of Scotland. Montrose expected to be welcomed in a locality never regarded as a hotbed of Covenanting, and had officers in his entourage with Orcadian connections. Chosen to lead the main invasion force was Lord Eythin, a thirty-year veteran of the Swedish army and a resolute Royalist commander in the earlier civil war in England. Eythin, ennobled by Charles I in 1642, was born James King, at Warbuster in Hoy, his father a sheriff-depute in the time of Earl Patrick and his mother a daughter of Lord Adam Stewart.

An advance guard of fourscore officers and a hundred Danish mercenaries was placed under the command of George Hay, earl of Kinnoull, a veteran of Montrose's Highland campaign. Kinnoull, as a grandson of Orkney's proprietor, the earl of Morton, was seen as the conduit to local support. Morton's own Royalist credentials were impeccable. Between 1635 and 1643, he captained the royal bodyguard, and when Charles I placed himself in the custody of the Scots army in 1646, Morton went to attend him. After the Scots handed Charles over to the English, Morton retreated to Kirkwall, taking a lease of the Palace of the Yards from the burgh of Edinburgh. In May 1648, during the Engagement, he was eager to 'further the levies in Orkney', and suspected of preventing the Kirk's declaration against the Engagers from being read there.[31] Morton, however, died in August 1648, and his tack and title passed to his son, Robert, already resident in Orkney.

On arriving in the islands, Kinnoull wrote to assure Montrose that 'Your Lordship is gaped after with that expectation that the Jews look after their Messiah'. But the new earl of Morton had already shown wariness of 'suspicion of being a malignant' (the Parliamentarian term for a supporter of the Royalist cause), and promised support only when 'we would show ourselves to be in a capacity to reduce the country'. Kinnoull responded to this half-hearted endorsement by dispatching soldiers to the palace in Birsay 'requiring a positive answer'. One of his officers put it bluntly: 'my Lord Morton was forced either to join with us for the king, or else to quit the islands'.

Morton himself, in a self-pitying letter to Patrick Smyth of Braco, called it 'God's blessing that we parted without blood', and complained of being ordered to convene 'the whole gentry of Orkney' at Harray Kirk to hear a letter from the king. He felt insulted at Charles having conferred leadership on anyone other than himself. Kinnoull and his officers opposed having Morton in overall authority, but, for the sake of harmony, Kinnoull resigned to his uncle his commission to command an Orkney levy, 'which he was pleased to accept of before the gentlemen of this country'. Further squabbles were averted by a strange and inauspicious concurrence: Morton and Kinnoull died suddenly within days of each other in November 1649.

The authorities in Edinburgh had meanwhile become aware of Kinnoull's presence. General David Leslie, one of the Covenant's most successful commanders, and an old foe of both Montrose and Eythin, was dispatched north, but he lacked resources for a seaborne assault. Leslie wrote to Kinnoull, a past acquaintance, and advised him 'to make his retreat into some other country', but Kinnoull ordered the letter to be burned by the hangman. The General Assembly convened a 'committee for Orkney business', which wrote to the presbytery in Kirkwall denouncing this 'new war against religion and the Covenant'. Ministers were to shun the rebels, use their pulpits to dissuade people from enlisting, and threaten Montrosian leaders with excommunication if they didn't present themselves before the General Assembly.

These instructions were sent out in hope rather than expectation. The ministers seized their moment to settle old scores, forcing James Morison and Patrick Waterston from their parishes. Pleas to sustain them 'with that love and respect as becomes brethren' fell on deaf ears. In February 1650, the Assembly heard how Morison and Waterston were still 'kept from their charges by the insolencies of the rebels'.[32]

Through the winter, the Royalists levied and trained troops, and waited for the marquis to arrive. Not all in the camp was well. Colonel Thomas Ogilvie wrote at the start of March to warn Montrose that commissions were being 'put in some young hands who truly have not wit to govern themselves', though discipline had improved since the arrival in January of William Hay, Kinnoull's brother and heir, bringing with him 1,500 pikes, 2,720 swords, 100 sets of cavalry armour, 1,536 muskets and twenty barrels of musket balls.

Montrose travelled overland from Gothenburg to Bergen in February 1650, and sailed the following month to Kirkwall with a couple of hundred additional mercenaries. It was scarcely a Scandinavian annexation, and Montrose was frustrated by the prevarications of Frederick III. 'What Your Majesty shall please to do will be doubled by being done soon. In such affairs a refusal that sets us free to act is better than a promise that ruins us!' Nonetheless, there must have been a frisson around the arrival of Danish troops in a territory claimed by Copenhagen, and there to implement the very strategy proposed by Christian IV in 1640.

Few Orkney voices were raised in opposition. Patrick Smyth of Braco claimed that, when Kinnoull arrived, he fled to Caithness, returning only on discovering that 'all I had would be plundered, and my house and rent meddled with' – but this was an exculpatory letter to General Leslie, written after Carbisdale. In March 1650, Montrose was feted by the Kirkwall magistrates, who equipped a company for the army, and placed the town's fleet at its disposal. The gentry subscribed an obsequious bond of loyalty, as did the ministers. The declaration drawn up by the presbytery denounced 'that unnatural rebellion, maliciously hatched and wickedly prosecuted against his late sacred majesty'. The ministers invoked God's blessing on 'this present expedition of his Excellency, James, Marquis of Montrose', and swore never to cease exhorting their congregations against 'the rebellious faction'.

On the eve of his departure for Caithness, Montrose received word from Charles II that he had reopened talks with the Covenanters at Breda. The king did not expect this to 'give the least impediment to your proceedings', urging Montrose to continue 'with your usual courage and alacrity'. It was a cool, cynical betrayal. In expectation of a negotiated settlement, Charles ordered a halt to the reinforcements under Lord Eythin, whose troops remained stranded in a scattering of Baltic ports.

On 1 May, Charles signed the Treaty of Breda, agreeing to swear to the Solemn League and Covenant, and prepared for a peaceful return to Scotland. In the meantime, Montrose marched to his fate at Carbisdale. In the days before his execution, he agreed to spare the king embarrassment by concealing his royal commission. Four days after Montrose's hanging, a letter from Charles was read in parliament, 'showing that he was heartily sorry that James Graham had invaded this kingdom, and how he had discharged him from doing the same'.[33]

Aftermath

As survivors of Carbisdale struggled north, Orkney's counter-revolution stuttered to its inglorious end. A veteran soldier, Sir William Johnston, had been left behind as governor in Kirkwall. Through the first days of May, Johnston planned a protracted defence. He expected reinforcements from Shetland and begged the Orkney gentry to honour their pledges of allegiance.

The lairds felt they had done enough. On 11 May 1650, 'the greatest part of the gentry', meeting under the presidency of Bishop Graham's old supporter William Stewart of Mains, reminded Johnston they had 'advanced both men and monies for an army, levied for his Majesty's service and the defence of the islands of Orkney'. But, with the army defeated, they were 'unable to contribute any such assistance as may maintain his post'. Johnston replied bitterly he could see they planned 'to shift for yourselves', and announced his intention to do the same.

Taking some artillery pieces, and 'my Lord of Morton's whole jewels and plate', Johnston boarded ship for Norway. The evacuation was hurried and messy, and in the course of it Johnston was reported to have shot dead a junior officer with his pistol. Leading supporters, including Montrose's illegitimate half-brother Sir Harry Graham, and the minister James Aitken, departed with the governor, but lower-rankers were not so lucky. A Welsh officer, John Gwynne, complained in his memoirs that Johnston fled 'leaving some of us behind a sacrifice to Leslie'.

Orkney nearly exacted a sacrifice of its own. Johnston's ship got stuck on the Skea Skerries, a rocky cluster south of Westray, and only just floated clear. It encouraged Orkney's Covenanters to show their hand. A letter to the sheriff of Caithness (its signature unfortunately torn away) urged him to secure a warrant from Leslie to arrest Johnston and his party, 'sure [safely secured] upon the skerry', promising 'ye will not want aid in this country'. Two of Johnston's officers, George Drummond and Patrick Melvill, rowed across to Westray, and were promptly imprisoned in Noltland Castle. Half a dozen 'gentlemen of the isle' had seized the stronghold 'for the ends of the Solemn League and Covenant'. On 18

May, they asked Leslie for a hundred soldiers to help keep possession – Westray's loyalties were evidently divided.

Leslie dispatched a troop of horse under Captain Cullace, one of the cavalrymen of Carbisdale. The general had written already to the Orkney lairds, informing them of the agreement with the king, and asking them to apprehend remaining officers of 'James Graham'. Patrick Smyth of Braco was willing enough to cooperate, but he worried that Leslie's wish for them all to come south and appear before parliament would 'lay the whole country open to the invasion of pirates, foreign ships, and the people belonging to James Graham'.[34]

Cullace's men arrived too late to apprehend Johnston, but they arrested Drummond and Melvill, rounded up other followers of Montrose and ransacked Morton's house for incriminating papers. In their petition of 1661, the Kirkwall magistrates remembered the troopers showing the town's inhabitants 'much prejudice because of their loyalty to your Majesty'. The lairds likewise complained how Cullace 'violently quartered' his cavalrymen throughout the islands, 'destroying and eating, trampling and abusing the growing corn in the fields'. They soon returned south, though Drummond proved a troublesome prisoner; in Caithness, Cullace 'caused him to be shot at the post'.

In the last week of June 1650, Charles II landed in Scotland and appended his grudging signature to the Solemn League and Covenant. In Paris, it was believed that Charles had gone to Orkney, though his landfall was at Speymouth near Elgin on the north-east coast. His plan was to invade England, defeat Cromwell's New Model Army and regain his father's throne. It required an expanded army, and parliament sanctioned sweeping new levies. Orkney was to contribute 375 infantrymen (only seventy-five fewer than Edinburgh) and fifty-four cavalry. In August, parliament's powerful committee of estates sent a Major Campbell to Orkney with a demand for 400 foot, followed shortly by a Major Robert Stewart with orders to levy 300 horsemen. When Stewart discovered Orkney to be a place with 'no horse in it', the committee demanded a payment of 300 merks per cavalryman in lieu.

These were truly punitive burdens. Orkney contributed soldiers to the Covenanter army of 1640, the Engager army of 1648 and the ill-fated expedition of 1650. Barely four months after that last army was obliterated, the islands were expected to provide further regiments of foot and horse, under the command of the victors of Carbisdale. The lairds knew 'all our men were spent', yet since the army was in the king's service, they willingly cooperated – or so, at least, they claimed. Ordinary Orcadians had other ideas.

Between the late summer of 1650 and April of 1651, Orkney was in a revolutionary ferment. The government in Edinburgh, and the sheriff in Kirkwall, lost control of the islands to an insurrectionary force of commoners. It is an extraordinary episode about which we know frustratingly little – virtually our only source is a self-serving petition presented to the earl of Morton by the gentry in 1662.

The lairds termed it a 'whiggamorish insurrection'. The name derives from the Whiggamore Raid of 1648 – a march on Edinburgh by supporters of the marquis of Argyll, which overthrew the Engagement and restored the Covenanters. Yet there is little to suggest that the Orcadian rebels were motivated by religious fervour. Used at the Restoration, 'whiggamorish' may simply have meant to convey a seditiously egalitarian and peasant character to the disturbances – the petition called the insurgents 'clowns' and 'base fellows'.

Clowns or no, the rebels possessed a charismatic leader – an Orcadian Wat Tyler or William Tell. He was called Currey, and we know very little about him, not even his first name. Yet he was able to keep a force of 600 in arms for over six months, take control of Kirkwall and expel or imprison lairds attempting 'to debate public business'. The rebels made short work of Major Campbell and his fellow officers, the committee of estates hearing how the commanders 'that went to Orkney to take up the levies are imprisoned, and their party forced home to Caithness'. The like fate awaited a Major Melvill, who arrived in January 1651 with fresh demands for troops of dragoons.

Orkney's experiment in communal government came to an end in April 1651. Sir James Douglas, uncle to the young earl of Morton, arrived with a royal commission to raise a new regiment. He was resisted, but the lairds rallied to him and managed to arrest Currey and other leaders and 'quiet the mutiny'. They later observed with regard to the rebels that Sir James, 'if he had pleased, might have caused hang them all', which implies Currey and his confederates escaped capital conviction. Currey disappears from the historical record as anonymously as he entered it, and Orkney – normally fertile soil for local mythmaking – preserves no memory of him. There was, in the end, no escaping the demands of other people's wars: 600 men were raised, equipped and sent south – the same number, and no doubt many of the same individuals, who comprised Currey's insurgent force.[35]

Meanwhile, Charles II's restoration was starting to unravel. Cromwell crossed the border in July 1650, and in September, at Dunbar, east of Edinburgh, he soundly defeated Leslie's army. The English occupied Edinburgh, and their warships prowled around the Scottish coast – several islands in Orkney were raided. Against Leslie's better judgement, Charles insisted on taking the fight to England. At the Battle of Worcester on 3 September 1651, a year to the day after Dunbar, Cromwell again comprehensively bested the Scots-Royalist army. Charles II, after a famous interlude in an oak tree, managed to escape to France.

Remaining Scottish resistance was extinguished by Lieutenant-General George Monck, who set up garrisons in places of strategic importance. The Kirkwall magistrates later boasted that their burgh 'was the last part of the kingdom that yielded', yet this had more to do with the chronology of English advance than with dogged Orcadian resistance. As news of Worcester filtered north, Patrick Smyth wrote anxiously to his son in Edinburgh for advice about which way to jump, 'for I desire neither to be first nor last in taking course'.[36]

The Inglishes

English troops crossed over the Pentland Firth at the end of January 1652, a time when, newsletters reported, 'there was the greatest storm there hath been known for many years in these Northern parts'. First to arrive were two companies of Colonel Thomas Cooper's regiment, later joined by additional forces under Robert Overton, one of Cromwell's senior commanders.

Overton, back in Edinburgh in March, updated Cromwell about the situation in Orkney. His men had possession of the cathedral, and the Palace of the Yards, 'where we can upon occasion very conveniently and entirely lodge a regiment of men'. On the shore, just east of St Magnus, he built a fortification whose position, its guns trained out to sea, meant '100 sail of ships may safely ride from the annoyance of any enemy'. Within weeks, an order went to the Tower of London Armoury for twenty pieces of artillery to be shipped to Orkney. A second fort was built on the facing side of Kirkwall Bay, and additional companies were sent north from Colonel Matthew Alured's and Colonel Ralph Cobbett's regiments. Orkney's garrison was more than symbolic, its presence an indication of the islands' strategic and political importance.

Overton described 'fair comportment from the generality of the people', though some soldiers had been attacked. Their assailants were 'beaten into good submission', and the incident supplied 'occasion to disarm the whole island, which, as it is reported by some of the best affected inhabitants, are able to raise an army of 5 or 6,000 fighting men' – the same (probably inflated) figure cited by Jo. Ben. some sixty years earlier (p. 25). The enforced disarmament was recalled bitterly by the Kirkwall magistracy: 'the usurpers ... did take all the arms and ammunition belonging to the burgh, which was near twa thousand stands [a soldier's complete set of equipment], with twa great iron guns [and] the whole colours and drums'.[37]

Figure 6.3: The fort built by Cromwellian soldiers overlooking the harbour in Kirkwall, photographed in the late nineteenth century when it housed a battery of the Orkney Artillery Volunteers.

Encounters with 'Inglishes' were nothing new in Orkney, but never before had several hundred Englishmen been billeted in the islands as an occupying army. A fortuitously surviving paybook for Captain John Gillott's company provides lists of names that were surely exotic to Orcadian ears: Richard Hornbuckle, William Musgrave, Abraham Shockley, Thomas Testwood, Humphrey Witcherley. The soldiers themselves felt posted to the ends of the world. In March 1653, a satirical communiqué in the Republican newspaper *Mercurius Politicus* described Orkney as 'not much inferior to the wilderness wherein the Israelites continued for forty years, only it hath more plenty of water'. Its inhabitants were 'as happy as any other nation under the sun, because they never knew better, but to all strangers it is Purgatory, if not a Hell itself'.[38]

The author of the report might have been the soldier who spent his free time in the late summer of 1652 writing two long satirical poems. One, a pastoral idyll, seemed to praise the wonders of 'our rare Pomonia (which the natives style the Mainland)'. But the final stanza confessed to composition in an 'ironic strain':

> Thus ends Pomonia's praise, which might well vie
> With all the world, was Scotland not so nigh.

The other piece was a viciously inventive libel. The poet had landed in a place 'not unlike a turd in a full chamber pot'. With a mixture of facetiousness and pride, he concluded that

> Had we not conquered Orkney, Cromwell's story
> Had cleared [manifested] no more of honour in't and glory
> Than Caesar's, but with this conquest fell
> Under his sword, the forlorn hope of Hell.

It is a notion we have come across already – that the islands were beyond civilisation, yet somehow essential to a proper understanding of it. It was in this vein that a rather greater poet, John Milton, wrote in 1654 in praise of General Robert Overton: 'even the remotest Orkneys confess your humanity and submit to your power'.

'The Character of Orkney' was dated 9 September 1652, 'from my cave called the Otter's Hole, in the third month of my banishment from Christendom'. The name seems intended to conjure the damp, dirt and disorder, and foul commingling of animal and human, that for the fastidious poet defined Orcadian domestic life.

> Have you ever been
> Down in a Tanner's yard, and have you seen
> His lime-pits when the filthy muck and hair
> Of twenty hides is washed and scraped off there?
> 'Tis Orkney milk, in colour, thickness, smell,
> Every ingredient, and it eats as well.

The name 'J. Emerson', written in the same hand, is attached to another poem in the manuscript volume, and it seems almost certain the author was James Emerson, an officer in Alured's regiment, who later served as commandant of Duart Castle in Mull.

Emerson's diatribes about diet and hygiene interwove themselves with misogynistic and sexualised abuse of Orcadian women. His landlady was a particular target – Emerson buried an anagram of her name (which I have failed to decode) in a descriptive passage of sustained nastiness:

> ... paps of a brindled cow,
> Behemoth's legs, bear's feet, an ape's arse, and
> A stinking shoulder of mutton for each hand ...

We can easily recognise the toxic masculinity of the dislocated soldier, here perhaps rooted in personal bitterness – Emerson was divorced from his wife, Jane, who in 1651 was whipped and expelled from the Leith garrison for serial fornication. Emerson was a jaundiced but nonetheless acute observer. He had heard about a charm for churning butter, whose components will sound familiar: nine Ave Marias and 'three heated stones' were used

> To fetch their nasty butter up, which, when
> They've done, the witches conjure down again.

Allusions to witchcraft permeate the poem. Perhaps, a few years after the spate of trials in Kirkwall, there was still much talk of it locally. But Emerson seems to have arrived with preconceived notions of Orkney as a site of dark magic – beyond Christendom, and just this side of 'Oberon's and Mab's black land'. His poetic invocations read like exorcisms, or curses:

> ... Orkney, be thy poisonous name
> Abhorred, may none but witches use the same.
> To every hag that speaks it, let it be
> A mark upon her of discovery.
> He that the nauseous name of Orkney hears
> And spews not, sorrow faw [clean out] his ill-bred ears.[39]

If such attitudes were typical of the occupiers, it is little wonder that local resentments flared. Kirkwall's magistrates remembered that inhabitants speaking loyally of the king in 1652 'were scourged by the hand of the hangman'. Their own authority was annulled, and justice administered by JPs and sheriff, sitting jointly with a military governor – Colonel Cooper in the first months of the occupation, succeeded by his regimental officers Roger Sawrey, Henry Watson and Henry Powell.

The combative old Covenanter John Craigie of Sands was hauled before the court in March 1653 for 'seditious and treasonable words, to the prejudice of the Commonwealth of England'. Not all vocal dissidents were arch-Royalists, and not all were men. Similar charges were laid against the 'goodwife of Aikers'. Jean Halcro, daughter of Hew Halcro of that Ilk, and wife to her kinsman Harry Halcro of Aikers, was as indigenous an heiress as they came – a great-granddaughter of Earl Robert, and great-great-granddaughter of the priest-patriarch Malcolm Halcro. Other pillars of lairdly society were more circumspect: Patrick Smyth assured his son in September 1653 that he had destroyed all 'papers or letters ... that either I or any friend had fra Montrose, Sir William Johnston or any of their accomplices'.[40]

After the Restoration, litanies of iniquity were laid at the door of the occupiers. They were said to have broken down the wall of St Magnus kirkyard, to use the stones for their forts and a barracks. They dismantled the cathedral pulpit and smashed seats for firewood. They destroyed the tomb of Bishop Thomas Tulloch, a traditional site for making agreements and settling debts, and conveyed away its marble and copper.

Local tradition maintains that iron rings in the four supporting piers of the cathedral crossing were put in for the soldiers' horses. Improbable stories about the stabling of horses by Cromwellian troops are attached to several cathedrals and other churches in England, but there may be some substance to the belief in Kirkwall. As the garrison was drawn from infantry regiments, these were presumably for officers' mounts; the placement suggests an intention for ad hoc tethering rather than fixed accommodation. Most likely, the English respected existing arrangements in the cathedral, its choir reserved for worship and its nave available for a variety of secular and semi-secular purposes. The New Model was a 'godly' army, but there was no wave of fanatical iconoclasm – the carved wood of the choir stalls, and of Bishop Graham's 'loft', was left intact.

An Orcadian writer in the eighteenth century believed Kirkwall 'had formerly a castle of very great strength, which was taken and destroyed by Oliver Cromwell'. In Orkney, as throughout Britain, local memory credited Oliver with destructive deeds for which he had a secure alibi – in this case, having never come to Kirkwall and being only fifteen when the castle was demolished. A nineteenth-century visitor smiled at how people still believed all historical acts of vandalism were committed by 'the Englishes', Cromwell's soldiers 'nearly as useful as scape-goats in Kirkwall as the cat is in lodging-houses'.[41]

Some more positive traditions about the occupiers were set down at the start of the nineteenth century by the Shapinsay minister George Barry. Though 'guilty of several irregularities and oppressions', Cromwell's soldiers introduced 'the planting of cabbages, which were before that time in great measure unknown'. Other legacies of English know-how were the use of marl (a deposit of clay and calcium carbonate) as a fertiliser, and the manufacture of domestic locks.

Over the course of an eight-year occupation, ways were found of living together, sometimes literally. Kirkwall's register of baptisms and marriages starts in 1657, and records fifteen weddings between Orcadian women and garrison soldiers up to 1660, along with a smattering of illegitimate births from irregular liaisons. Some soldiers stayed on in Kirkwall after the Restoration, setting themselves up in respectable trades.[42] Orkney's capacity to colonise its colonisers worked with the English as well as the Scots.

The Desires of Orkney and Shetland

Local accommodations were part of a wider story of national realignment. The Cromwellian conquest, Scotland's most thorough-going experience of subjugation, produced the first real political union among the nations of the British Isles, if not quite the happy and confident 'Great Britain' envisaged by James VI.

In April 1654, an Ordinance of Union declared Scotland 'incorporated into … one Commonwealth with England'. Charles II's claims as king of Scotland were nullified, there was to be religious toleration (Catholics and Episcopalians excepted) and the nation would, with Ireland, send representatives to a Commonwealth parliament at Westminster. The ordinance acknowledged a special status, or at least separateness, of the Northern Isles: incorporation was of 'all the people of Scotland, and of the Isles of Orkney and Zetland, and of all the dominions and territories belonging unto Scotland'.

The process began with a 'Tender of Union' from the English council of state. It was presented to representatives of the Scottish shires and burghs in February 1652, at a meeting with English commissioners at Dalkeith, and voted on in August at Edinburgh, where deputies were selected for the consolidated parliament. The union was subscribed late by Orkney and Shetland's parliamentary delegates, Arthur Buchanan and Hugh Craigie, who 'by storm at sea came not to Edinburgh till a day after the election'.

Scottish agreement to the forced political marriage was grudging and incomplete. In the end, fewer than half the shires and burghs formally endorsed the Tender. At Dalkeith, deputies were permitted to present the English commissioners with grievances and proposals – the 'Desires' – of their locality. Mostly, these focused on protection for the Kirk at a time of toleration for Protestant Dissenters, on withdrawal of English troops and release of prisoners, and on reversing property sequestrations.

Only one submission evinced any enthusiasm for the union. The 'Desires of Orkney and Shetland' fretted that 'seditious and divisive spirits' might manage to disrupt 'the Union settled in our shire in subordination to the Commonwealth of England'. They asked for the expulsion of 'persons lately crept into our land whose fortunes lie in the south, labouring to disturb the peace of the land and to interrupt the begun union'.

It seems unlikely that Orkney's lairds felt emotionally drawn to union with England, though a commitment to ideals of national independence may have been weaker than elsewhere in Scotland. A more audible strain is the harmonious ideal of the 'community of Orkney', and the old tune of setting one external authority against another, in order to protect established interests from predatory incomers not yet acclimatised to island ways.

Like other shires, Orkney and Shetland asked that 'our prisoners be released and the burdens of our cess eased' – Cromwell's land tax to cover the costs of occupation fell heavily across Scotland. Orcadian grievances, however, were aimed more at the old Scottish regime than the new English one – 'the sad and deplorable condition of our land, which hath been overburdened with heavy taxations and impositions from the Estates of Scotland for the promoting their several engagements in war'. Particular resentment attached to a levy exacted in 1651 by Sir James Douglas, 'which he did apply to his own private use'.

Other priorities were equally domestic: the speedy appointment of a sheriff, 'in regard we lie remote from the public seat of justice'; the recovery from private hands of monies to maintain a schoolmaster in Kirkwall; and the provision of licences to export grain to Norway, where 'our shire in all times bygone has had commerce and trading'. Politics were imperilling the traditional Norwegian trade. In 1656, there was short shrift for two Orcadian skippers complaining to the governor of Bergen about the seizure of their ships in the vicinity by a privateer from the Spanish Netherlands. The fault, the governor told them, was 'upon yourselves, who have made Cromwell your king'.

Orkney's first 'Desire' was to know 'what order shall be established for the replanting of the kirks of our shire, they all being vacant for the present except two'.[43] In the weeks after Carbisdale, the ejected ministers James Morison and Patrick Waterston exacted a high price from their 'brethren'. They sent petitions to parliament and the General Assembly, demanding restitution and reprisal. Both bodies agreed that Walter Stewart and other ministers should be summoned south 'to hear and see themselves fined and censured'. The presbytery's position was shaken further in the last week of May 1650, when one of Montrose's fugitive ships, its crew having mutinied in Bergen, sailed into Leith. The vessel was laden with incriminating documents, including the ministers' oath to Montrose; another copy of this 'vile blaspheming piece' had already been sent to Edinburgh after Carbisdale by the chaplain of Archibald Strachan's regiment.

Morison and Waterston excepted, all the ministers of Orkney, fifteen in total, were deposed and excommunicated – an imposition of spiritual sanctions more swingeing than in any place of comparable size in Britain. For 'speaking in defence of the truth and discovery of the enemies thereof', Morison and Waterston received compensation from the stipends of the vacant benefices – 6,000 and 4,000 merks respectively. Enforcement of the edict was, at their request, granted to Thomas Buchanan of Sound, David MacLellan of Woodwick, James Moodie of Melsetter, and James Baikie of Tankerness, 'well-affected gentlemen'.[44]

Several further demands of Morison and Waterston were approved by the General Assembly's Orkney Committee: 'apostate ministers' to be shunned, and if possible brought to trial; 'none of these who have proven active in this rebellion shall present their children to baptism'; 'the sacrament of the Lord's Supper be not administered to the people until they be purged'; kirks to be replanted anew with 'good able preachers'.

'The great defection of the people of Orkney', as the General Assembly styled it, had grave consequences. In October 1650, a Perthshire laird with Orkney affiliations, George Drummond of Blair, husband to Bishop Graham's daughter Marjory, and father of the officer shot by Captain Cullace, wrote plaintively to Robert Douglas, minister of St Giles' in Edinburgh. Celebrations of communion had ceased in Orkney, 'a great grief to all godly persons', and no weddings had been celebrated since the troubles began, 'how clear and innocent soever the parties desiring the marriage have been'. Drummond had matches lined up for his daughters, one of them with Patrick, son of Robert Monteith of Egilsay. Patrick, he assured Douglas, was free of all imputation – 'except only in what he did with the rest of the gentry in Orkney'.

'Replanting' of the Orkney Kirk was easier to announce than implement. In 1651, General Assembly commissioners identified two suitable 'expectants' – candidates for the ministry – John Gibson and David Kennedy, and sent them to the parishes of Holm and Birsay. A sprinkling of additional ministers was recruited over the course of the 1650s, and an Orkney Presbytery was in 1654 reconstituted in skeletal form. But most of the deposed clergymen carried on living in their manses and preaching in their kirks – the Cromwellian authorities had little interest in policing the Church of Scotland's internecine disputes.

In time, ministers' situations were regularised, and they were restored to their parishes – 'reponed' in the jargon of the Kirk. George Graham in Sandwick, John Balvaird in Rousay and George Johnston in Orphir were readmitted in 1658. The following year, William Watson was restored in Hoy, and James Douglas of Kirkwall transferred to Sanday. But in places there was real local disruption. The new presbytery worried about 'the desolate condition of the kirks of South Ronaldsay and Burray this long time bygone for want of a minister'. Walter Stewart died in January 1652, and not until 1657 was a new preacher found. In 1659, the young earl of Morton bemoaned 'the deplorable condition of many desolate congregations here in Orkney'.

Patrick Graham, son of the ex-bishop and brother-in-law of Patrick Smyth, was quickest off the mark for rehabilitation; he was raised in a family whose most sacred principle was to thrive and survive. In January 1651, the commissioners referred to the Presbytery of Sutherland Graham's petition to have his sentence remitted, and asked the Sutherland ministers to 'bring him to a sense of the offences for which he is excommunicated'. No doubt Graham supplied the required assurances. Shortly afterwards, the commission granted designated visitors the authority to meet with ministers and, 'after evidence of their repentance', relax their sentences. The Orkney visitation never got underway, but the initiative reflected the diminishing influence of a hardline 'Kirk party', and dominance of the

1651 Assembly by 'Resolutioners' like Robert Douglas of Edinburgh – so-called because they backed resolutions to allow former Royalists and Engagers to join the army against the English.

James Morison, the hero of May 1650, had his knuckles rapped in March 1651. Anna Stewart, wife to Patrick Graham, complained about Morison's refusal to baptise their child. The commissioners found his explanation 'exceeding strange', and criticised his reluctance to permit marriages, admonishing him to 'walk more circumspectly', lest he 'bring the ministry of the gospel under contempt'.[45]

Morison's fellow stalwart, Patrick Waterston, was behaving yet more strangely. He gave up attending meetings of the presbytery, which in 1658 began proceedings 'in relation to his withdrawing from them', and later threatened him with penalties as 'a condemner of the judicatories of the Kirk of Scotland'. A minister seemingly addicted to dissent, Waterston was in near-continuous conflict with the authorities – whoever they happened to be – for over twenty years.

This latest bout of truculence may have been theologically motivated, with Waterston abandoning orthodox Presbyterianism for Congregationalism or 'Independency', the model of Protestantism that regarded individual congregations as self-governing churches. Independency was rife in the New Model Army, where Oliver Cromwell was its most famous adherent. Waterston was perhaps influenced by the 'Englishes', or at least took advantage of their presence to defy presbytery mandates and run his parish as he saw fit. In July 1659, he put his name to a petition to the Commonwealth parliament, calling for religious toleration of the kind ferociously opposed by respectable Presbyterians. Other signatories included Baptists, Quakers and radical groupings from across Scotland.

Quakerism was the most dynamic of the new religious sects. 'Several of our Friends were moved to go to the Isles of Orkney' wrote the movement's founder, George Fox, in a manuscript account of its remarkable missionary achievements. It was almost inevitable that Quakers would want to demonstrate their spiritual mastery of Britain by making converts in the Northern Isles. During Britain's political and religious upheavals, Orkney quite often seemed not so much a backwater as a battleground.

The Quaker preacher John Bowron stayed in Orkney in 1656, en route for Barbados and Surinam, and a Quaker seaman, William Plumley, visited the following year. A sympathetic pamphlet published in London in 1659 reported the case of three Quakers 'banished from the Isles of Orkney by George Monk's command': Richard Ismead, George Wilson and George Fox the Younger – the latter was a namesake of the founder, so-called because 'younger in the truth'.

'Two Quakers, Inglish men', turned up at the Peterkirk in South Ronaldsay on 4 September 1659 and, in the words of the session register, 'troubled the congregation and abused the people before the minister came'. There was also a public confrontation with James Morison and other Orkney ministers. George Fox (senior) documented the episode under the sarcastic heading 'these be the principles of the priests of Orkney'. The account was provided by 'G. W.' (George Wilson), and witnessed by 'Luke Liddall, soldier', presumably a convert from the Kirkwall garrison. Morison's view of the Quaker doctrine of inner illumination was that 'all who said they had Christ within them, were devils and witches, and so bid the people take notice'. Comparing Quakers to witches was a strategy of their critics throughout Britain, though whether these strange Englishmen looked or sounded like witches to bemused Orcadians may well be doubted.

Meanwhile, another threat to Presbyterian supremacy was making its first discreet advances. In around 1653, Vincent de Paul, French founder of the Lazarist Order, asked permission from Rome's Congregation for the Propagation of the Faith to send missionaries to the Highlands and Hebrides. One of them, Thomas Lumsden, focused his efforts principally on Caithness, until obliged to retreat by a Cromwellian edict asking magistrates to hunt for missionary priests. Prior to this, in October 1657, Lumsden proudly reported that 'I even went to the Orkney Islands'.[46] In Orkney, the Kirk's monopoly of God was beginning to fray at the edges.

Where the Herring Fisheries Are

No sooner had Orkney's representatives pledged assent to union with England than the islands were pitched once more onto the front-line of political and military conflict. In July 1652, arguments over international trade boiled over into unlikely war between Cromwellian England and Protestant Holland. Its major battles took place in the Channel, and in the southern part of the North Sea. Yet a stream of intelligence raised anxieties about the Northern Isles. Mindful of 'the fishing which the Dutch have every year about Orkney and Shetland', the council of state recommended reinforcement of the Kirkwall garrison, and captains patrolling the upper reaches of the North Sea asked for more ships.

Robert Lilburne, the army commander in Scotland, felt badly stretched. In September 1653, he responded cautiously to Cromwell's instructions to bolster Orkney's defences. Faced with rebellion in the Highlands, led by the earl of Glencairn, Lilburne had few men to spare. He reckoned a garrison of around 800 would be required to defend Orkney, 'if the Dutch and Middleton do agree, and that they come over anything considerable, as it may be expected they will'. John Middleton was Charles II's military commander, a former Covenanter and Engager, who crossed to Scotland to take control of Glencairn's forces, but the rising collapsed amid internal recriminations in 1654.

Royalists, many of them exiled in the Netherlands, eagerly courted Dutch support. In September 1653, Middleton sent two memoranda to the States-General of the United Provinces. Stressing 'the readiness and ability of the Scots to fight for the king', he asked for arms and ammunition – the highlanders, he implausibly claimed, had only bows and arrows. In return, the Dutch would be permitted to build forts along Scotland's west coast for protection of their commerce, and the king would undertake 'to assign to the estates the revenues of the Orkneys, amounting to 80,000 livres, as well as to put those islands into their hands ... until their expenses are defrayed'. The proposal had been mooted in Royalist circles for some time. In April 1653, the Venetian ambassador in Paris knew that the king of England had 'offered to the Dutch the Orkney Islands, where the herring fisheries are'.

Charles I had been prepared to surrender Orkney to the Danes, Charles II was willing to give it to the Dutch – neither seemingly regarded the archipelago as an inalienable part of their patrimony. It was perhaps easier for monarchs to imagine risking, losing or gaining islands than it was to do this with other types of territory – something which was not necessarily incompatible with their enduring importance as measuring rods of royal power and achievement.

The Dutch were not, in the event, interested in such a reckless scheme. It would have angered Frederick III, whose support Charles also avidly sought. Denmark, looking to make its own accommodation with Holland, and under pressure from Sweden, was eager to be involved in the Anglo-Dutch negotiations that produced an end to the war in April 1654. During the discussions, Cromwell's government received information from Korfitz Ulfeldt, a renegade Danish councillor (and brother-in-law to Frederick III) now in Swedish exile. The treaties between James VI and Christian IV, he revealed, contained a clause allowing Denmark to recover Orkney for a payment of £13,000 sterling. Ulfeldt suggested that the peace between England and Holland should include an explicit repudiation of Denmark's claims. The proposal was not a sufficiently high priority for the Commonwealth regime and went no further.[47] Danish demands were dormant, but not yet dead.

Orkney tells its own story of Britain's Civil Wars. The main battles took place elsewhere, but the islands were no calm and peaceful refuge, and suffered the repeated bruising of factionalism, rebellion, invasion and occupation. Orcadians lived in both big and little worlds: they were aware of being participants in arguments affecting the nation, but were often concerned primarily with the issues of their own community, and highly alert to how what was happening in other places could be levered to their own advantage, or to the detriment of rivals. Their experience in these years invites us to reflect on how the local, the national and the international are inextricably entwined, and exert powerfully reciprocal pulls.

Orcadian society weathered the hurricane, though not without considerable storm damage. Countless young men went off to war, and many did not return. Fines, levies and sequestrations caused hardship and bitterness among the survivors. There was, however, ultimately no complete overthrow of the existing social order. James Baikie of Tankerness endured the disdain of a bishop and the enmity of a presbytery to cement his wealth and status. In 1657, he received from the burgh of Edinburgh – no doubt with quiet satisfaction – a tack of the bishopric estates, and oversight of the ministers' stipends.

Baikie was among the lairds tasked in 1653 with creating a new record of the landed wealth of Orkney. Such a listing was one of the 'Desires' presented the previous year to the Commonwealth commissioners, 'seeing that since our last valuation some men's estates within our land are bettered, others totally ruinated'. Other assessors included the minister Patrick Graham, sporting the lairdly title 'of Rothiesholm', as a result of former church lands in Stronsay feued to him by his episcopal father; Patrick Monteith of Egilsay, now satisfactorily married to the heiress Margaret Drummond; Robert Stewart of Burgh, a namesake and grandson of the sixteenth-century earl; and the ubiquitous Patrick Smyth of Braco. All were wealthy men in receipt of copious rental income – in meils of malt, marks of oil and barrels of butter.[48]

Smyth of Braco was the perfect specimen of a representative type – an outsider who made his home and fortune in the islands, and whose willingness to twist and turn allowed him to prosper almost equally under Episcopal, Covenanting, Royalist and Commonwealth regimes. He would doubtless have welcomed the return of the king, but did not live to see it. His son entered a note in the family bible: 'The 28 of April 1655, it pleased the Lord to remove my father, Patrick Smyth, being Saturday, coming from Stronsay in the night-time.'

The Smyths left an extensive archive of family papers, and it is through the eyes of people of his class we must usually look for perspectives on seventeenth-century Orkney. The voices of the powerless are mostly mute, or audible only in muffled and distorted echoes. One such echo is a tradition, written down in the nineteenth century, about the drowning of Patrick Smyth. He went to Stronsay to collect his rents, but one of the tenants, a poor widow, was unable to pay. The grasping Smyth was determined to have his due, and seized every animal belonging to the woman and herded them onto his boat. When a gale blew up in the Stronsay Firth, the agitation of the cattle made the vessel unmanageable, and it capsized and sank.[49]

Such stories are the slow-tempo vindications of the exploited and oppressed. They speak to an elemental longing for justice and fairness, in a world where injustice seemed so often to triumph. There was something distinctively Orcadian, too, in the tale's blend of a restorative comeuppance, a fateful role for domestic animals and a final remorseless judgement of the sea. Through an age of commotion and cacophony, deeper and slower rhythms still reverberated in the island world.

7

The Earldom and the Kingdom

Restored to the King of Denmark

The king's ambassadors were insistent. The treaty must confirm that 'the Orkney Islands belonged as an inalienable dependency to the kingdom of Norway', and should be 'restored to the king of Denmark' without delay. The summer of 1667 seemed an auspicious time to make the demand, as Denmark had just landed on the winning side in a short, sharp European conflict. The Second Anglo-Dutch War began promisingly for Charles II, but the tide soon turned. Louis XIV entered the war on the side of the Netherlands; so too did Frederick III of Denmark, tempted by subsidies and cancellation of debts. Louis thought an attack on Orkney by the Danish fleet might follow, though his preference was for it to join battle in the Channel. The help was scarcely needed: in June 1667, a Dutch fleet sailed up the Medway and burned English ships laid up in their anchorages.

The warring parties, with Sweden as honest broker, met at the castle of Breda, where, in July 1667, England signed three bilateral peace treaties. As late as the middle of that month, it was rumoured that the Dutch 'stood upon high terms', and 'would have the Orkney Islands'. A separate rumour reached Charles XI of Sweden: a Danish admiral was preparing to move against Orkney; six Dutch frigates and a thousand men stood ready.

No mention of Orkney or Shetland appeared in the final text of the Anglo-Danish treaty. The English ambassadors, like Scots ones before them, resorted to obfuscation. They had no instructions about the islands, and could find no reference in earlier treaties. Crucially, neither the French nor Swedish delegations supported Frederick, and his envoys reluctantly gave way. They wanted it on record, however, that Denmark's claims stood 'until a better opportunity presents itself, whether soon, or after a long time'.[1]

The time never came. The Breda summit was not the last occasion when the Danish-Norwegian claim to Orkney would be mentioned, but it would never again be seriously pursued through diplomatic channels. At his coronation in 1648, Frederick III issued a charter promising to restore Orkney and Shetland to the kingdom of Norway, just as his predecessors did. Frederick, however, was the last monarch to make such a pledge – not so much because the claim was abandoned as because the Danish crown no longer needed to strike bargains with Norwegian nobles. Under the 'King's Law' of 1665, the Danish sovereign was endowed with unrestricted authority, free from obligations or oaths.[2] It was Europe's purest expression of 'absolutism', a system usually associated with the Sun King, Louis XIV.

For Charles II and his brother and successor James II – in whose kingdoms monarchical power had been extinguished in the mid-seventeenth century – 'absolute monarchy' exerted a seductive appeal. Yet the kings' efforts to rekindle and then fuel their authority produced only further rounds of rebellion and revolution. Out of the royal failure to establish absolutism, 'modern' Britain started to take shape. The earldom of Orkney, transplanted from the wistful memories of a Scandinavian kingdom to the uncertain promises of a British one, is an articulate, if accented, witness to what that process involved.

This chapter charts the tumultuous political and religious history of Orkney, and its interplay with national and international events, from the Restoration of the monarchy to the aftermath of the Jacobite rebellions. It examines how people in the islands, individually and collectively, addressed legacies of the past and framed hopes for the future; how external pressures for change were received, resisted or renegotiated; and how the experience of becoming 'British' did little to dilute distinctively Orcadian interests and identities.

Restorations

Charles II's return was supported by the army commander in Scotland, and opposed by the army commander in Orkney. The death of Lord Protector Cromwell in September 1658 was followed by the ineffectual brief rule of his son Richard. General George Monck, secretly in contact with Charles, took his army south in December 1659 to prepare for a royal restoration. Captain Henry Watson, governor of Orkney, was the sole superior officer in Scotland to dissent. Monck's actions, he wrote, would 'let in the boars of the forest into the Lord's vineyard'. The mutiny was swiftly crushed. Monck's deputy sent troops to Orkney, Watson and his officers were imprisoned, and his company disbanded.[3]

Watson complained of being 'far from news', yet word of the transformation in British politics arrived in Orkney with astonishing rapidity. On 27 May 1660, two days after Charles II landed at Dover to be greeted by Monck, the event was celebrated at South Ronaldsay's Peterkirk: 'glad tidings being certified concerning his Majesty's safe and happy return ... the minister touched at length upon the matter to stir up the people unto the duty of thanksgiving'.

Edward Richardson had arrived in South Ronaldsay in August 1657 from Forteviot in Perthshire – a parish where he had been in trouble for drinking to the 'unlawful engagement' (p. 271). In 1660, former Engagers – conviction Royalists – had a spring in their step. Stalwart Covenanters, like James Morison of Evie, were pleased to see the back of Cromwell's regime, but were otherwise filled with foreboding.

'Restoration' brought with it destruction. One casualty was the union between England and Scotland, which Orcadian lairds had accepted readily enough but which Charles II saw as a potential restraint upon his power. Orkney's leading landowners met in Kirkwall to choose a commissioner for the parliament scheduled to convene in Edinburgh in January 1661. They decided to grant him £10 sterling for clothing, 'the better enabling him to render himself in a condition in some measure suitable to other members of parliament of his rank'.[4] No one wanted the gilded worthies of Edinburgh sneering at the shabbiness of the envoy from Orkney.

Restoration also came with a reckoning. The 'regicides' were ruthlessly brought to justice. But every locality in Britain had its own experiences of suffering, compromise and betrayal. In Orkney, rejoicing soon gave way to recrimination, as lairds and magistrates fell over themselves to assert their loyalty, and denounce the backsliding of rivals.

The loudest voice was that of William Douglas, ninth earl of Morton. The 1643 royal grant of the islands to his grandfather (p. 263) was in 1657 annexed to the Commonwealth, and at the Restoration Morton moved swiftly to assert his rights. From 1661, he was acting as sheriff, and warning the presbytery against appointing ministers without his permission.

Morton was resurgent but also insolvent. It explains the oddness of a parliamentary act of 1662, mortgaging the crown's rights in Orkney and Shetland not to Morton but to his uncle, George Villiers, Viscount Grandison. Administration of the earldom on Morton's behalf was vested in a trust headed by John Middleton, the former Royalist exile last seen offering Orkney to the Dutch (p. 291). Now earl of Middleton, and royal chief minister in Scotland, he had recently become Morton's father-in-law. The trust employed Morton's distant kinsman, Alexander Douglas of Spynie, as chamberlain, with a brief to 'follow forth all actions of reductions of vassals, infestments of the said earldom, lordship and udal lands'.

This tactic – converting independent udal lands into dependent feudal tenures – was the policy of the 'tyrant' Stewart earls. Morton asserted udal right to be 'merely a possession and no kind of right or title by charter'; many udallers, believing written documentation would offer security, bowed to pressure and purchased Spynie's feu charters. The percentage of Orkney held by owner-occupiers declined precipitously. In the decade before 1630, 29 per cent of 'sasines' – deeds recording land transfers – were made by udallers. In the 1660s, it was 16 per cent, and by the end of the century a mere 6.[5]

Another obstacle in the way of the restored earl of Morton was the burgh of Kirkwall, and its provost, Patrick Craigie. During the occupation, the town came to an accommodation with the Cromwellian authorities. In 1654, it resumed administration from the justices of the peace, though Morton countered with his own claims under the 1643 grant. In 1658, Craigie, then town baillie, went to Edinburgh to explain to the governing council of state 'the true position of the burgh'.

In spite of their record of pragmatic collaboration, the magistrates were quick off the mark in 1660, presenting to the king moving accounts of Kirkwall's support for Montrose and its stalwart resistance to the 'usurpers'. The strategy worked, at first. In May 1661, Charles II confirmed the charter of privileges originally granted by James III. Morton struck back, seeking to persuade the privy councillors that Kirkwall's alleged status as a royal burgh rested on an illegitimate grant from the Republican government; the town was a mere 'burgh of barony' under his authority.

Seating in St Magnus Cathedral was a flashpoint for rival claims, as it had been in the past (p. 254). Morton alleged that Craigie and his confederates broke down his pew and 'erected a seat for themselves in place thereof'. He also contrived to portray the proclamation of the annual fair as a provocatively rebellious act – the 'pretended provost did in ane hostile, seditious and tumultuary manner pass through the town with two persons beating drums'. Doubt was successfully sown. In May 1662, parliament suspended Kirkwall's jurisdiction, and the council summoned Craigie to answer for recognising 'treacherous and bloody rebels'.[6]

Kirkwall, until 1638, was also the seat of a bishop. In 1660, it was a surprisingly open question whether restoration of the monarchy meant the restoration of Scottish episcopacy. Initial assurances were taken as a royal commitment to preserve Presbyterianism. The king acted cautiously, aware of the deep hostility to bishops felt by many Scots. But Middleton was eager to 'settle the Church upon its old foundation', and proposed a 'recissory' act, annulling at a stroke all legislation subsequent to the parliament of 1633, the year Charles I was crowned in Scotland. Royal authority was thus rinsed from the taint of the Covenant, and bishops restored by default.

A man much cheered by this was Thomas Sydserf, former bishop of Galloway, the sole survivor of the pre-war episcopate. At the end of 1660, Sydserf sent a memorandum to Charles, making the case for bishops as safeguards against the Presbyterian heresy that 'kings may be deposed from their thrones'. Sydserf was hardly a disinterested observer; he encouraged the king to 'take special notice' of those serving the crown faithfully during the Interregnum. Throughout the summer of 1661,

Sydserf was in London, lobbying to be appointed archbishop of St Andrews. Samuel Pepys noted in his diary the bishop of Galloway's readiness 'to admit into orders anybody that will' – a casual approach to the episcopal power of ordination that annoyed the English bishops. Charles valued loyalty and service, but only to a point. His choice for St Andrews was James Sharp, a leading light of the 'Resolutioner' party (pp. 288–9), only recently converted to the virtues of Episcopalianism.[7]

Orkney was Sydserf's consolation prize. Eighty years old on his appointment, he never visited the islands, and Edward Richardson, as dean, was summoned to Edinburgh for instructions. Richardson presided over a synod in St Magnus in November 1662, and conveyed Sydserf's displeasure at learning of civil courts meeting in the cathedral. Reverence for sacred things, persons and places was the order of the day.

It belied frantic jockeying for position among the clergy. Some scrambled to declare allegiance to the new regime, but two – Alexander Lennox of Kirkwall and Arthur Murray of Lady Kirk, Sanday – followed their consciences and departed. There was no sweeping purge of pro-Covenant clergymen, as happened in the south-west of Scotland, but letters trickled in from deprived Episcopalian ministers, or their widows, attesting to sufferings in the king's cause, and asking for compensation.

Some sought not just recompense but revenge. James Douglas (p. 288), restored to his position at the cathedral, sent a petition to parliament to complain of 'cruel persecuting' at the hands of Patrick Waterston and James Morison – guilty, as Douglas could testify, of 'treasonable speeches'.

Waterston was already gone from Orkney. Perhaps emboldened by the withdrawal of the Cromwellian garrison, the presbytery deposed Waterston in March 1660 for seeking 'to erect himself into ane congregational church'. In the summer, Sheriff-Depute Patrick Blair arrested him for treasonous talk, and he was sent to face justice in Edinburgh. He then seemingly absconded: Douglas described Waterston in 1661 as 'fugitive to Holland', where he was reported to have died the following year.

It was not the end of the story. Nearly twenty years later, Waterston's daughter Isobel and her husband, Edward Brown, were in debt to the baillie of Kirkwall, David Drummond. The three petitioned the privy council against the Kirkwall merchant Harry Erburie, who in 1660 had been 'caution' for Patrick Waterston – he posted bail guaranteeing Waterston's judicial appearance. Erburie was believed to have got 'considerable sums' from Waterston, which the petitioners hoped to recoup, having loyally informed the king of his entitlement to a decades-old forfeiture.

The petitioners also wanted the council to know that Erburie was an Englishman, 'who entered this kingdom ane foot soldier sworn to the colours of the late usurper' – one of the members of the Cromwellian garrison who stayed, and thrived, in Kirkwall after the Restoration.[8] Erburie's willingness to stand caution for Waterston deepens a suspicion that contact with New Model Army soldiers hastened the minister's conversion to Congregationalism. Decades after his death, Patrick Waterston was still causing trouble.

James Morison's story had a no less memorable denouement. Morison, the towering figure of the Orkney Presbytery in the Interregnum, was willing to conform, but his colleagues weren't inclined to make it easy for him. In 1662, he was summoned to answer articles 'relating to his going forth of the presbytery in the time of Montrose'. In the following year, the synod examined the 1650 act rewarding Morison and Waterston at the expense of loyalist brethren (p. 287), and Sydserf decided to suspend him.

It brought out the ageing minister's fighting instincts. He produced a seven-point manifesto to prove the case legally unsound, not least as contrary to the 1662 Act of Indemnity and Oblivion, which offered amnesty for offences committed during 'the late troubles'. Suspension, Morison protested, would be 'ane starving of souls committed to me'.

Morison won his case, and despite increasing ill health resumed his duties in the presbytery. In August 1664, we find him asking his brethren to make diligent search for a parishioner, Anna Cathness, who 'had fallen in fornication and was now fugitive'.

It was an ironic foreshadowing of his own fate. On 21 June 1666, the presbytery received astonishing news. Two days earlier, Morison had gone to the sheriff court 'and charged himself with grievous crimes'. A demand for details elicited an extraordinary confession: over the course of his long ministry, Morison had committed 'diverse adulteries'. They included 'incest' with a married woman called Margaret Trott – Morison had had sex with her mother, Helen Odd. Other sexual partners were now deceased, and Morison did not want to name them, 'thinking it a burden to his conscience to leave a reproach upon the posterity'.

We will never know what induced a sudden confession to a secret double life, but it was not entirely a bolt from the blue. Back in 1643, Patrick Waterston – who else? – had hinted at a relationship between Morison and a woman called Kate Davie, but he later backtracked, saying he had mistaken the woman's name, and did not mean to imply 'reproachful uncleanness'.

Morison was deposed by the presbytery, and sentenced to shameful public penance: 'in sackcloth, bareheaded and barefooted, at the church doors both of Evie and Rendall'. It was seemingly a humiliation too far. When the presbytery reconvened to fix timings for his repentance, Morison was found to be 'gone over ferries'.

He was, in fact, in Edinburgh, with Elizabeth Ogilvie, his surprisingly devoted third wife. In April 1668, she petitioned the privy council on behalf of an 'indigent' husband and large family of 'poor children'. It is not certain how impoverished Morison really was: in that year, he secured admission as a burgess of Edinburgh, and published a work of piety, *The Everlasting Gospel of the Everlasting Covenant Discussed.* Addressing his readers, Morison assumed knowledge of his fallen circumstances. But he was determined to make his vice a necessitous virtue, trumpeting God's favour to penitent sinners: Nebuchadnezzar, Solomon, King David, St Peter.

Morison was not so quick to forgive his own debtors. He obtained a judgement in the court of session against the notary John Spence, and against other Orcadians he claimed owed him money. In 1669, the redoubtable Elizabeth Ogilvie went north to enforce it, and was met by Spence and town officials, armed with swords and halberds, 'of the special direction and command of Patrick Craigie, Provost'. According to her complaint to the privy council, they dragged her through the streets and threw her in prison, only releasing her when she confessed her husband was actually in debt to Spence. Ogilvie and Morison appeared personally before the council in January 1670, but Spence, Craigie and other witnesses (including Harry Erburie) failed to show up – a familiar story for the central courts in handling Orkney cases.[9]

Provost Craigie and his allies had reasons to be cheerful, for they had just triumphed in their long battle with the earl of Morton. In January 1669, Craigie appeared at the Mercat Cross brandishing a copy of Charles II's charter. A few weeks later, the court of session annulled Grandison and Morton's grant of the Orkney earldom. That summer, a representative from Kirkwall, James Moncrieff, went for the first time to the Convention of Royal Burghs – a body meeting in parallel to the Scottish parliament. In the following year, Moncrieff became Kirkwall's first ever commissioner to parliament, which proceeded to ratify the town's burgh status.

The restored privileges of Kirkwall betokened a spectacular eclipse of the earl of Morton. In December 1669, parliament confirmed the court ruling against him and annexed the earldom of Orkney to the crown. It was, the act noted, not the first such annexation. Alas, royal generosity and the importunity of suitors had induced the king and his father to grant the earldom away. That was now found to be contrary to law, and to the welfare of Orcadians. Their interests were better served by 'immediate dependence upon his majesty'.

The Annexation Act was a royal love letter to Orkney and Shetland, 'a jewel of their crown', part of 'his Majesty's ancient kingdom', over many ages 'the occasion of much trouble and expense of blood and money for maintaining thereof against the invasion of foreigners, and recovering the same out of their hands'. It was an inventively heroic retelling of the prosaically fiscal, and relatively recent, way the islands had come into Scotland's possession.

Presented as statesman-like reform, the annexation was actually the expression of avarice and pique. One thing Charles II knew about the Northern Isles was that they were a source of profit in the form of shipwrecks. In December 1664, a merchant ship of the Dutch East India Company, the *Kennemerland*, foundered on the Out Skerries in Shetland. The earl of Morton's agents recovered both coin and cargo from the wreck, on the basis of his authority as 'vice-admiral' of Orkney and Shetland. But the outbreak of the Anglo-Dutch War persuaded the king to claim the windfall as crown property. A complex case, pursued through the court of exchequer, ended in 1668 with a judicial conclusion that Morton had attempted to defraud the crown, and owed Charles twenty-four bags of gold ducats, as well as two brass cannon reportedly spotted in the Palace of Birsay.[10]

The Annexation Act acknowledged the exotic otherness of the Northern Isles – 'of a great and large extent of bounds, and so remote and at such a distance from the ordinary seat of justice'. But it also saw in them a test case for restored royal and ecclesiastical governance. A direct connection to the king was established by making Orkney and Shetland a 'stewartry' – the jurisdiction of a royal stewart, rather than a sheriff. This provision had been included in the act of 1612, but not apparently implemented. In announcing the stewartry, the 1669 act was careful not to 'prejudice the bishop of Orkney of his patrimony and privileges'. Orkney had a proven record of loyalty to the House of Stewart, and of effective episcopal administration in alignment with royal objectives. Far from Edinburgh, and further still from the royal court in London, the islands might yet bear witness to the reach of royal absolutism.

Absolute Orkney

From 1665, Orkney was once again home to a bishop: Andrew Honyman, archdeacon of St Andrews, came north to replace the absentee Sydserf, who had died in September 1663. The restored Scottish bishops did not preside over anything very conspicuously 'Episcopalian'. There was no attempt at reintroducing the ill-fated Prayer Book of 1637, and Sunday worship continued much as before, though with encouragement for congregational recital of the Creed and Lord's Prayer. The moral discipline of session and presbytery remained, bishops assuming the role of moderator.

Yet there were changes in tone and pitch, amplified in Kirkwall by the survival of an imposing medieval cathedral, and a fine episcopal palace, from which Honyman managed to evict the earl of Morton's adherents. Sydserf and Honyman employed the seal used by bishops before the Reformation, depicting St Magnus standing in the doorway of his cathedral. The cathedral was now better staffed, the Restoration regularising an arrangement whereby a principal minister exercised the 'first charge' (a word here meaning pastoral office), assisted by a deputy holding the less remunerative 'second charge'. Both at the cathedral and elsewhere, Honyman encouraged regular celebrations of Holy Communion, seldom performed in Orkney over the preceding decades, largely as a result of concerns about congregational 'ignorance'. In presbytery minutes, the more stately word 'Church' replaced workaday 'Kirk'.

For Honyman, the fabric of the cathedral was a recurrent concern. Sydserf had bequeathed money for repairs in his will, and Honyman launched an action to recover it from foot-dragging executors. Calamity struck in January 1671, when the steeple was hit by lightning and consumed by fire. Quick action by the townspeople saved the bells (p. 94), which dropped onto piles of earth heaped into the transept. They were rehung in 1679, though the largest of them, cracked in the fall, was sent for recasting – the contract with the Amsterdam bell-founders stipulated that Bishop Maxwell's arms, and the portrait of St Magnus, should be retained.[11]

Respect and reverence for the past – even the popish past – was exemplified in the new minister of the cathedral's first charge. James Wallace was an Aberdeen graduate, who worked as a schoolmaster in Fortrose just north of Inverness before being presented by Honyman to Lady parish, Sanday, in 1669. His removal to St Magnus in 1672 was contentious – not on account of his character or opinions, but because of disagreement over who had authority to make the appointment. Provost Craigie and the magistrates – fresh from their triumph over the earl of Morton – disputed the bishop's right, and were mollified only by being allowed to draft their own parallel grant of presentation.

The twice-chosen Wallace proved a popular preacher and pastor, a man of wide learning and intellectual curiosity. In 1684, he wrote 'An Account of the Ancient and Present State of Orkney', a fuller and better informed survey of the islands' social, political and natural history than any attempted hitherto, a version of which would be published posthumously in 1693. Also in 1684, Wallace laid the foundations for what can claim to be the oldest public library in Scotland – lineal ancestor of the institution in which this book was principally researched.

The kernel of the collection was a set of 160 volumes belonging to William Baikie, third son of the venerable James Baikie of Tankerness. William bequeathed the books, of a largely religious nature, to his 'dear friend' Wallace, and probably intended them for study by the Orkney

Figure 7.1: A drawing of St Magnus Cathedral, from the manuscript of James Wallace's 1684 'Account of the Ancient and Present State of Orkney'.

ministers. But Wallace, adding further volumes of his own, inscribed them as belonging to 'the Bibliotheck of Kirkwall' – a public amenity for the burgh, and a reminder to posterity never to regard Orkney as a regressive backwater.

By the time the library was established, Orkney was on its third post-Restoration bishop, after Honyman's death in February 1676.

Murdoch Mackenzie was already advanced in years, but proved an energetic, level-headed and tough-minded administrator, a legacy of his service as a Swedish regimental chaplain during the Thirty Years War.

The positive impression Mackenzie made on Orcadians is preserved in the tradition, documented by the late nineteenth-century antiquarian and Episcopalian minister James Craven, that when he landed at Scapa in August 1677 he was presented with an ancient and enormous drinking vessel, 'the cup of St Magnus'. New bishops were invited to drink from this vessel, brimming with strong Orkney ale, as an augury of good fortune. Mackenzie gratified hopes by draining the cup in one go, and asking for a refill.

Sadly, it seems unlikely to have happened. The cup's existence is first recorded in George Buchanan's 1582 *History of Scotland*, alongside cynical comments on how Orcadians justified their habitual drunkenness by pretending to have a relic of St Magnus. Wallace knew the story of the cup from Buchanan, but gave no indication of its continued presence, at Scapa or elsewhere. It is uncertain whether it ever in fact existed, though in the 1640s Walter Stewart claimed it was taken to Glasgow by Bishop Law. Untrue stories are often more revealing than real ones. The tale points to the imaginative hold that St Magnus, centuries after his death, exercised on ordinary Orcadians, as well as to possibilities among them of esteem for episcopacy.

Mackenzie was no prissy Laudian; indeed, in his youth he had the reputation of a zealous Covenanter. But he shared with his predecessors a sense of the Church's sacred character, and the respect due to him as God's local deputy. In 1684, Mackenzie complained to the privy council against the authorities of Kirkwall for using the cathedral as a guardhouse during town markets. The councillors agreed it was 'ane high abuse', and ordered the magistrates to desist.[12] Conscientious and resident bishops, in a territory privileged as a royal stewartry, offered possibilities for 'peripheral' Orkney to shine forth as a beacon of absolutist rule.

From the outset, however, there were obstacles. One was ongoing lawlessness among the Orkney lairds. James Moodie of Melsetter in Walls, and his eldest son, William, were primordial offenders. In 1664, William had been involved in a violent altercation with Morton's agent

Alexander Douglas. Yet again, the cathedral was a flashpoint. Douglas was met there by Moodie and a score of followers, 'drawing their swords and presenting their pistols to give fire within church doors in the very time of divine service'.

Through the 1660s and 70s, the Moodies feuded with a neighbouring laird, David Sinclair of Ryssay. The authorities heard about seizures of corn and livestock, unlawful imprisonment, armed incursions into Sinclair's lands 'with ane piper playing before them in warlike posture', and theft of that critical item in Orkney farming communities, the mill-stone for grinding corn. Court rulings counted for little, Sinclair lamented in 1668, for 'the laws cannot be put in execution against these persons living in remote islands'.

Another tenacious troublemaker was Robert Sinclair of Sebay. Between 1669 and 1671, he toured the Mainland and North Isles with an armed gang, demanding money and livestock from defenceless farmers. The thin veneer of legality for his actions had a distinctively Orcadian colour. Thirty or so years earlier, a ship, the ironically named *Good Fortune*, came to grief on the Deerness coast, close to the house of Robert and his father, William Sinclair of Sebay – with whose own addiction to feuding we are already acquainted (p. 247).

It was notorious, said Sinclair's opponents, that he and his father had looted the wreck, and used its cargo of timber to repair houses at Braebuster and Sebay. One of the ship's owners, the laird of Philforth in Aberdeenshire, obtained an order from the court of session for Robert Sinclair and others to pay compensation. But Sinclair came to an arrangement with Philforth to make the order over to him, and used this authorisation to extort payment from people living miles from the site of the wreck. It was another instance where the complexity of the case, and difficulties of summoning witnesses to Edinburgh, persuaded the privy council to delegate to a commission headed by the bishop. But local justice encountered its own headwinds: by October 1672, three separate commissions had not succeeded in resolving the matter.

Descriptions of Orkney as a jewel in his crown did not prevent Charles II reverting to the policy of subletting the earldom to a succession of tacksmen, now graced with the title of 'stewart-principal', but principally interested in profit. The tacksman for much of the 1670s was Captain

Andrew Dick, a grandson of the Edinburgh merchant William Dick (p. 244), whose heirs were left saddled with heavy debts. In 1685, Captain Dick was found by the council to have 'grossly malversed in his trust'. Complaints spoke of tenants ruined by extraction of 'pretended arrears', and of forced and unremunerated labour on the salvage of wrecks.

In another money-making scheme, Dick signed a contract in 1677 to supply 150 soldiers for the regiment of Colonel John Kilpatrick, of the long-established Scots Brigade in the army of the Netherlands. Unable to drum up sufficient recruits, Dick resorted to impressment, bringing 'a general consternation upon the whole country'. It reportedly involved tyrannies worthy of the Stewart earls: aged fathers seized and imprisoned until sons offered themselves for service, inhabitants 'forced to flee to caves and rocks'.[13]

Refuge in caves was associated elsewhere in Scotland with dissident Presbyterians unwilling to accept the return of episcopacy and abolition of the national Covenants. The Covenanters were a religious movement with profoundly political aims. They wanted royal influence and lay patronage banished from the Kirk, and their open-air 'conventicles', presided over by ejected ministers, were regarded by the authorities as dangerously subversive.

There were no conventicles in Orkney, and sparse evidence of sympathy for Covenanters. The islands lay far from the Covenanting heartlands of Scotland's south-west, and Orcadians were not involved in the agitation leading to the abortive Pentland Rising of 1666, or a later upsurge of militancy crushed at Bothwell Bridge near Glasgow in June 1679.

Orkney managed nonetheless to echo and amplify a national struggle between aspirant royal absolutism and uncompromising Protestant faith. In July 1668, the people of Scotland were shocked, or thrilled, by an assassination attempt on Archbishop Sharp. On the High Street in Edinburgh, a radical Covenanter fired a pistol into the archbishop's coach, hitting in the arm his friend and travelling companion, Bishop Honyman of Orkney. Honyman had recently completed the first part of a work attacking the Covenanters, and his wound, though serious, did not prevent the appearance of a second part the following year. In later pro-Covenanter writings, Sharp and Honyman were often paired as persecutors and episcopal timeservers.

It seems likely the incident influenced a privy council decision of 23 July 1668. Alexander Smith, an ejected minister convicted for conventicling, had just returned to Edinburgh after four years' enforced exile in Shetland. He was now sent away again – to North Ronaldsay, and ordered 'to confine and keep himself within the limits of the island'.

Banishment to Orkney, as we have seen (pp. 144, 161), was a time-honoured political punishment. In 1661, the veteran Presbyterian minister Robert Baillie pleaded with the secretary of state for Scotland, the earl of Lauderdale, to prevent the execution of radical Covenanting ministers: 'send them to Orkney, or any other place where they may preach and live'. It was a fate not quite worse than death. In September 1665, Archbishop Alexander Burnet of Glasgow, a scourge of Covenanters, reported with satisfaction to Archbishop Sheldon of Canterbury that, after a private meeting with the king, he had authorisation 'to confine disorderly ministers to some of our northern islands'.

On arrival in North Ronaldsay, Alexander Smith drafted a letter to Sheriff-Depute Blair, announcing an intention to preach to the islanders the following Sabbath – they had, he said, 'received me with much joy'. With greater zeal than tact, Smith requested that 'the rotten-hearted old man get not liberties to vex these poor people that are not pleased with his dead way'. Whether by this he meant Bishop Honyman, or William Cochrane, the Sanday minister with nominal responsibility for North Ronaldsay, is not quite certain. The officer transporting Smith to the island challenged him about his intention to preach, but Smith reassured him it would not be at the kirk, as that was 'the king's house and he his Majesty's prisoner'. Rather, he would keep 'family exercise [worship]', whether any came or not. Smith lived in North Ronaldsay for the next four years.[14] The presbytery does not seem to have investigated any conventicles. But it would be surprising if such a fervent reformer, on an island notoriously neglected by the ministry, did not exercise some kind of pastorate.

Alexander Smith's sojourn in North Ronaldsay left no documented legacy, but the Covenanter rebellion of 1679 – which followed a second, successful attempt on the life of Archbishop Sharp – was destined to be forever associated with Orkney. A hard core of 257 prisoners taken at Bothwell Bridge were condemned to penal servitude in the Caribbean,

16. James Graham, Marquis of Montrose (1612–1650), attributed to Willem van Honthorst. Montrose's invasion of Scotland from Orkney in 1650 drew to the islands the attention of all Europe, but ultimately spelled disaster for countless Orcadian families.

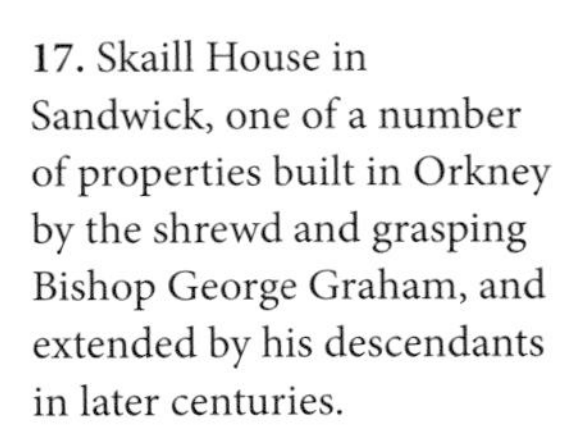

17. Skaill House in Sandwick, one of a number of properties built in Orkney by the shrewd and grasping Bishop George Graham, and extended by his descendants in later centuries.

18. The Covenanters' Memorial, Deerness, erected in 1888 near where the government vessel the *Crown* sank in December 1679, drowning over 200 political prisoners en route for the Americas. Seventeenth-century Orcadians were much less sympathetic to the Covenanting cause than their Victorian descendants.

19. The gravestone in St Magnus Cathedral of Elizabeth Cuthbert, died 1685. She was the wife of James Wallace, minister of St Magnus and chronicler of Orkney, who was buried beside her on his own death in 1688. The 'Glorious Revolution' which followed brought political and religious upheaval to the islands.

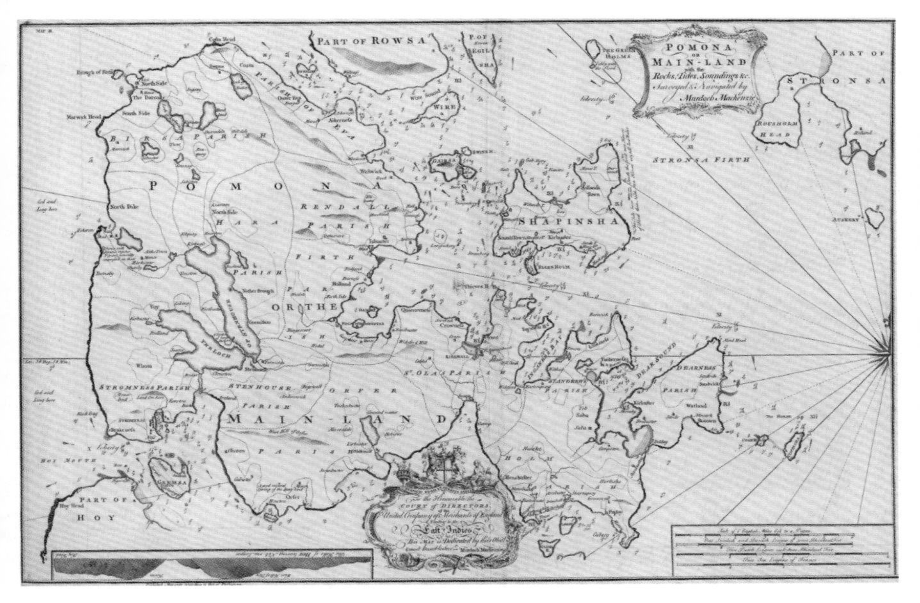

20. Map of the Orkney Mainland, from the cartographer Murdoch Mackenzie's *Orcades* of 1750. Mackenzie's charts of the land and sea of Orkney were of unparalleled accuracy and awed his contemporaries. His introductory essay to the volume glossed over Orkney's role in the recent Jacobite Rebellion.

21. James Douglas, fourteenth earl of Morton, and his family, painted in 1740 by the Anglo-Scottish portrait artist Jeremiah Davison. Douglas was the last earl of Morton to seek to control Orkney on the basis of a crown grant of the earldom estates. Despite his eventual victory in the gruelling 'pundlar process', in 1766 he sold his interest in the islands to Sir Lawrence Dundas.

22. Seals basking on the Orkney shore. The presence throughout the islands of these cautious but curious mammals gave rise to tales of selkies, supernatural creatures able to shed their seal-skins and walk in human form.

23. Chambered cairn at the Head of Work north-east of Kirkwall. Neolithic burial mounds of this kind littered the Orkney landscape. In the centuries before their archaeological excavation, they were sites of sometimes perilous enchantment, the haunts of fairies and 'hogboons'.

24. The ordination of elders in a Scottish kirk, an 1891 painting by John Henry Lorimer. Apart from the beards, the scene would not have looked substantially different a century or more earlier. Elders were not necessarily elderly, but they were men of substance in the parish, given divine blessing to police the activities of their neighbours.

25. The twelfth-century St Boniface Kirk in Papa Westray, Orkney's only intact example of a medieval parish church. In the early eighteenth century it was the site of an epic battle for control between the local laird and the minister with spiritual oversight of the island.

26. The Devil's Clawmarks: a curiously grooved stone on the exterior staircase of the ruined kirk of Lady Parish in Sanday. Folklore maintains the marks were made by Satan, in frustration at failing to grab a wicked minister escaping into the building – an example of how, long after the Reformation, churches remained places of sanctity in the popular imagination, and of how physical features of the environment preserved snatches of historical memory.

27. William Daniell's 1821 depiction of the lighthouse at Start Point, Sanday, the first in Scotland to be fitted with a rotating lamp. Lighthouses made navigation of the seas around Orkney safer, but were resented by some islanders as an infringement on their profits from wrecks.

28. A 1784 cartoon satirising Charles James Fox's election as MP for the Northern Burghs (including Kirkwall), while also claiming to have been elected a member for the constituency of Westminster. Orkney's unexpected prominence in the national rivalry between Fox and William Pitt the Younger unleashed a stream of Tory invective about the remoteness and barbarity of the islands.

29. Kirkwall, with the cathedral and ruins of the Earl's Palace, viewed from the Ayre across the sheltered harbour of the Oyce. This aquatint print was based on a drawing made by William Daniell during his trip to Orkney in 1818. It depicts the islands' capital as Walter Scott had encountered it on his visit four years earlier.

30. A portrait of Sir Walter Scott by Sir Thomas Lawrence undertaken between 1820 and 1827, when Scott was at the height of his fame. Scott's visit to Orkney in 1814 provided material for his 1821 novel *The Pirate*, which sought to embed the experience of the islands into his understanding of British history.

and in November 1679 were loaded onto a vessel in Leith, the *Crown* of London.

On 10 December, the ship found itself in a storm off Orkney, caught in what Robert Wodrow, great chronicler of the Covenanters, called 'as a dangerous a sea as perhaps in the world', and was driven onto rocks on the coast of Deerness. The crew managed to get ashore using the ship's broken mast as an impromptu gangplank. But the captain, Thomas Teddico, 'a papist', ordered the hatches chained shut as the storm rose, ignoring the prisoners' pleas to be set ashore. A few dozen managed to free themselves and scramble to land, but the others were all drowned.

A thirty-foot memorial, in the incongruous likeness of an abandoned industrial chimney, now stands near the clifftop, a few hundred yards from the headland of Scarva Taing where the *Crown* sank (see Plate 18). Erected by public subscription in 1888, it expresses a poignant sense of loss and injustice, reflected too in Orcadian traditions about escapees sheltered by locals and settling in the islands. It seems in fact unlikely that the wretched survivors enjoyed much of a welcome. On 17 December, the Kirkwall burgh authorities issued a proclamation noting that prisoners from the wreck had gathered in the town 'to prejudge [injure] and wrong the inhabitants' and authorising citizens to seize them. The forty-six known survivors were recaptured for shipping to Barbados or Jamaica as slaves.[15]

Just over two years later, another shipwreck, far from Orkney, almost succeeded in altering the course of British history. In May 1682, HMS *Gloucester* foundered on a sandbank off the Norfolk coast and 200 crew and passengers were drowned. One who managed to survive, however, was James, duke of York – duke of Albany in Scotland – Charles II's younger brother, heir to his thrones, and royal commissioner to the parliament in Edinburgh, where the *Gloucester* was supposed to be taking him.

James had been a polarising presence in English and Scottish politics since it became known, almost a decade earlier, that he had converted to Roman Catholicism. Encouraged by lurid revelations from a fictitious 'Popish Plot', opposition figures in the English parliament tried, unsuccessfully, to exclude him from the line of succession. Supporters of the government began to call them 'Whigs', in derisive appropriation of a nickname for radical Covenanters, and the designation stuck. In Scotland itself, anxieties about the future king's faith fuelled further religious agitation. The 1680s brought a crackdown on dissent, and bloody suppression of the extremist 'Cameronian' faction among the Covenanters. In 1681, at James's instigation, the Scottish parliament passed an 'Act anent Religion and the Test', requiring clergymen and secular office-holders to swear an oath acknowledging royal supremacy in matters civil and ecclesiastical, and disowning any obligation to observe the National Covenant or the Solemn League and Covenant.

Even in Orkney, there was hesitancy. Bishop Mackenzie presented the oath to a presbytery meeting in early November, and 'the brethren, taking the matter to their serious considerations … think fit to delay the subscribing and taking of it till the last Wednesday of December'. But when Christmas 1681 came, all duly swore, the bishop making 'a pious, grave and pathetic [moving] speech concerning the lawfulness'.

Ripples of dissent caught Mackenzie's watchful episcopal eye. In 1683, he informed the privy council about James Arbuthnot, a recent arrival in Kirkwall, who was teaching students without having taken the oath. In this, Mackenzie complained, he was countenanced by the Kirkwall baillie, Harry Erburie – 'ane Inglishman, who carried charge under the late usurper, and is still ane enemy to the present established government'.

Orkney's lairds, meanwhile, fretted about requirements for an annual militia muster – not out of political objections, but because of the expense, and in frustration at the government's failure to appreciate archipelagic logistics. Charles Murray, Dick's successor as stewart, had to explain to the council in 1683 that 'it is impossible for the whole militia of Orkney and Shetland to rendezvous in one day and the same place'. Even assembling and transporting to Kirkwall the hundreds of men assigned as Orkney's quota was a massive undertaking. It took sixteen months of back-and-forth negotiation for the council to agree to the commissioning of officers to go to the isles and train men *in situ*, so that Orkney might be 'put in a readiness to oppose the invasion of all foreign enemies'.[16]

The prospect of invasion was neither hypothetical nor hyperbole. James II's accession in February 1685 lit a fuse of conspiracy and rebellion, in which Orkney was to become conspicuously involved. There were to be coordinated risings in Scotland and England, led by Archibald Campbell, earl of Argyll, and James Scott, duke of Monmouth. The latter was Charles II's eldest illegitimate son and long favoured by his father. Monmouth commanded the Royalist army at Bothwell Bridge, but later became ensnared in conspiracies with prominent Protestants who encouraged him to assert his own claim to the throne. By the early 1680s, both noblemen had taken refuge in the Netherlands, where the stadholder, William of Orange, though married to James's eldest (Protestant) daughter, Mary, was committed above all else to frustrating the ambitions of James's ally, Louis XIV, now at the height of his 'absolutist' power and seeking to dominate the politics of western Europe.

Argyll set off first with three ships, planning to sail north around Orkney, and raise an army among his Campbell clansmen at Kintyre in the Highlands. But gales and poor visibility left the flotilla fogbound in Scapa Flow. It was decided to engage local pilots, and seek to enlist Orcadian recruits. Argyll's chamberlain, William Spence, had an uncle in Kirkwall and on 6 May 1685 volunteered to go ashore, accompanied by the physician William Blackader, son of a prominent Covenanting preacher. Bishop Mackenzie and the Kirkwall magistrates promptly arrested them as 'servants to a rebel', and sent word to the privy council.

Argyll and his advisers, anchored at Cairston (Stromness), considered a full-scale assault on Kirkwall to rescue the officers, but decided instead to seize hostages. The bishop refused to negotiate an exchange, and Argyll, his plans now thoroughly compromised, sailed on to Kintyre with his prisoners, who included assorted Grahams and a South Isles laird, James Stewart of Graemsay.

A story was soon circulating that, before leaving the Netherlands, Argyll consulted a soothsayer, who promised him that 'James Stewart, of the blood royal' would fall into his hands. Argyll's captive took pleasure telling him he was deceived in thinking this signified the sovereign: 'for I am of that name, and descended of the Earl of Orkney, who was King James V's bastard son'. It was a good tale; too good, perhaps.[17]

In Kintyre, support proved patchy. Argyll was captured by government militia, and in June 1685 executed in Edinburgh by 'the Maiden' (p. 176). Three weeks later, the insurrection of the duke of Monmouth came to a bloody end at Sedgemoor in Somerset, and he too forfeited his head.

It was reported in the late seventeenth century, and repeated by numerous subsequent historians, that the nobleman executed after Sedgemoor was at one time duke of Orkney, Charles II only afterwards altering the title to Monmouth. The reality is that in 1662 the king planned to elevate James Scott to the peerage as Baron Fotheringay, but then changed this in the patent to Tindall. French editors of an influential historical dictionary, compiled by the priest-historian Louis Moreri, seem to have garbled 'Fotheringay' as 'Arkeng'. When an English version appeared in 1694, the translator took his best guess and entered the word as 'Orkney'.[18] It is, as they say, a non-story. Still, it reveals something about the islands' status and known links to the dynasty that educated people could believe 'duke of Orkney' a plausible distinction for a king's son.

Giving up the Ghost

In February 1688, Bishop Murdoch Mackenzie died in Kirkwall. The town notary and diarist Thomas Brown was mistaken in thinking him 'near ane hundred years of age', but he was certainly old as well as venerable. His death heralded the passing of an era, for the islands and the nation. A successor, Andrew Bruce, was nominated, but never visited during his short tenure.

Mackenzie's funeral sermon was preached by James Wallace. He too was spared from witnessing events he would doubtless have found distressing. Wallace passed away on 18 September 1688, and was buried in St Magnus next to his wife. His funeral sermon was preached by John Wilson, minister of the cathedral's second charge, who took for his text a plaintive passage from the Book of Job (14:10): 'yea, man giveth up the ghost, and where is he?'

Just over a week later, James II issued a proclamation warning of imminent 'sudden invasion from Holland', which threatened 'to embroil this kingdom in blood and ruin'. He exhorted all subjects to resist the invaders. In early November, the Kirkwall magistrates purchased gunpowder, and put out a call to arms against 'the common enemy'. Yet only weeks after William of Orange's landing at Torbay in Devon, James

Figure 7.2: Murdoch Mackenzie (1600-88), bishop of Orkney after 1677, from a portrait at one time in the possession of the Baikie family of Tankerness.

seemed to have given up the ghost of his fighting spirit. Where was he? In December 1688, the king of England, Scotland and Ireland fled ignominiously to France, allowing opponents to argue he had abdicated the throne.

The so-called 'Glorious Revolution' of 1688 was a political coup d'état made possible by armed foreign intervention, rather than a revolution in the conventional sense. Its 'gloriousness' lies firmly in the eye of the beholder. Claims of glorious bloodlessness hold true only for England. In Ireland and Scotland, brief but bloody civil wars flared between

'Williamites' and 'Jacobites', dividing society at every level, and leaving legacies of bitterness and suspicion.

Orkney was spared any actual blood-shedding, and James's overthrow, along with the offer of the crown to William and Mary by a Scottish Convention of Estates, at first brought little dramatic upheaval. The commissioners charged in March 1689 with levying money to raise troops for 'security of the Protestant religion' were a roll-call of Orkney's lairdly establishment: Stewart of Burray, Bellenden of Stenhouse, Baikie of Tankerness, Graham of Breckness, Craigie of Gairsay.

Bishop Bruce, meanwhile, got on with episcopal business, appointing Wilson as Wallace's replacement in Kirkwall, and in June 1689 confirming Alexander Pitcairn as dean and provost of the diocese. That, however, proved to be that. In the following month, parliament put a definitive end to the cohabitation of bishops and presbyteries in the Kirk of Scotland. In abolishing episcopacy, the act claimed to be opting for the form of church government 'most agreeable to the inclinations of the people'.

No one checked on the people's inclinations in Orkney. In Shapinsay, a privy council proclamation for a fast on Sunday 20 October 1689 was read in the kirk that very day, with apologetic explanations of how it 'came not to the minister until the 18th'. The congregation, like others across Orkney, was required to ask God to 'bless and preserve King William and Queen Mary', granting success to their armies 'that we may be delivered from the tyranny and slavery of the papists'. An additional supplication, 'that the Lord would bless the fruits of the ground', likely resonated more with Shapinsay farmers. The minister, James Heart, added a petition of his own: for God to 'heal our land, that is broken by sad divisions'.[19]

In Orkney, those divisions were about to get worse. On 19 October, a royal commission was granted to Colonel Robert Elphinstone, appointing him stewart and tacksman of the earldom and bishopric estates, replacing the previous holder, William Craigie of Gairsay.

The colonel was an exiled Orcadian with points to prove and scores to settle. He was the eldest son of John Elphinstone of Lopness in Sanday, and – unusually among Orkney lairds – a zealous Covenanter. In Charles II's reign, Robert Elphinstone went to Holland, where in 1675 he married

Clara van Overmeer of Utrecht. He took part in Argyll's rebellion, and was tried and condemned *in absentia* for treason. In 1688, he returned triumphantly in the army of William of Orange, and began to plan a foray to the north.

Clara van Overmeer headed the advance guard: her arrival in Kirkwall in July 1690, 'with her retinue ... being ten in number, herself, bairns and servants', was unsympathetically recorded by the Episcopalian diarist Thomas Brown. Elphinstone followed a few weeks later, took possession of the Palace of the Yards and summoned the Orkney gentry to receive 'further orders'. He was soon up to his neck in litigation over bishopric rental records with William Craigie and the former bishop's sons, Sir Alexander and William Mackenzie.

Elphinstone acquired a reputation for grasping and arbitrary behaviour, encapsulated in an episode recorded by Brown. In January 1691, an unusually large pod of 102 pilot whales was driven ashore in Stenness, a remarkable gift of divine providence for the community. Yet these were 'intromitted with [illegally commandeered] by Robert Elphinstone of Lopness', and taken to his storehouses in Kirkwall.

Other 'intromissions' were ecclesiastical. In October 1690, Elphinstone ordered John Wilson to stop preaching in the cathedral. Wilson refused to go quietly, and was supported by the new minister of the second charge, John Cobb – perhaps the last episcopally appointed clergyman in Scotland. There were farcical scenes in January 1691, when Elphinstone caused a fast to be proclaimed in St Magnus without consulting Wilson, and Wilson turned up later to make his own announcement. A few months later, Elphinstone abolished the cathedral second charge. Cobb acceded, then discovered Elphinstone lacked privy council authorisation, and applied successfully to be 'reponed' or restored to his position. The council did, however, agree in 1694 to the deposition of Wilson, a known enemy to the religious and political settlement.

By Elphinstone's account, the opposition he encountered reflected ideological objection to the Revolution, rather than resentment at his own high-handedness. In a letter of December 1693, he reported a rumour reaching Orkney earlier that year: William III had been killed fighting the French, and the old king had been recalled by the English nobility. It produced rejoicing among 'all the gentlemen of this country', with

Provost Hugh Craigie going 'up and down the town of Kirkwall drinking King James's good health and beating all such who would not do the same'. In response, Elphinstone sent word to adjacent parishes for the commons to assemble in defence of the government – 500 reportedly did so, crying, 'God save King William!' Further trouble was defused by news of William's safe return. But Elphinstone took the opportunity to remind the government of the strategic importance of Orkney, 'of moment to the enemy' on account of its profusion of harbours, and potential as a jumping-off point for incursion into staunchly Jacobite Caithness. He had learned that emissaries had several times been sent to explore the possibility. 'When Montrose invaded Scotland, this was the method he took.'[20]

Elphinstone's emphasis on the loyalism of Orcadian commoners may or may not be reliable. But he certainly enjoyed little support among the lairds. Like the earl of Morton before him, Elphinstone was heavily in debt. His Orkney estates were mortgaged to the Baikies of Tankerness, who rented them back to him but foreclosed when he fell into arrears. Elphinstone's attempts at forceful repossession led to further costly appearances before the privy council.

For Elphinstone, the best hope of staying financially afloat was to exploit the tack to its full potential. But in 1693 he was outbid in his attempts to renew it, and replaced as stewart by Sir Alexander Brand, a wealthy Edinburgh merchant, supplier of furniture to the aristocracy and armaments to the Scottish government. Elphinstone appears to have left Orkney the following year, and disappears into obscurity. Brand, however, soon had cause to regret his expensive investment.

The 1690s witnessed a succession of disastrous harvests. Famine, redolent of the terrible years of the 1630s (p. 222), returned to Orkney, and hundreds were reported to have perished. 'In the memory of man', Kirkwall's magistrates wrote in 1697, 'this country was never in so bad a condition.'

These were times of hardship throughout Scotland, 'King William's Ill Years'. In Orkney, they were remembered as 'Brand's Years', when an inability on the part of the earldom tenants to pay their rents, in bere, oil or butter, coincided with a ruthless determination on the part of the tacksman to exact them. Decades later, and long after the tack had passed into other hands, Brand was involved in litigation over arrears. Reduced crop yields in Orkney led to arable land going out of cultivation, and to tenants abandoning farms. A lack of surplus products for export devastated profits among Orkney's merchant-lairds and the mercantile community of Kirkwall. By 1696, a half-dozen ships previously owned by burgh merchants had all been lost or sold, and commerce with Norway and Leith was at a virtual standstill.[21]

War was a further obstacle to trade. William of Orange's triumph in 1688–9 brought his new kingdoms into the coalition assembled to confront Louis XIV's France. For Orkney, that meant resuming position as a ragged front-line for irregular naval conflict. French privateers energetically harassed shipping around Orkney's coasts, and in June 1694 two French ships seized three English ones near the eastern entrance to Scapa Flow. The Frenchmen proceeded to occupy for a week the little island of Lamb Holm, plundering the farm on the island, and crossing to the Mainland to ransack the fine girnel or grain store belonging to the Graemeshall estate, a building which still looks stoically out to sea in the little village of St Mary's. Brand travelled to Orkney to take stock, and on his return voyage suffered the indignity of being himself captured by the French and dumped in Shetland, while the privateers burned his ship and made off with his rents.

War also meant taxes: Scotland endured a tax on household hearths in 1690, and then a succession of poll taxes on individuals, in 1693, 1695 and twice in 1698. Whether it proved possible to raise all the poll taxes in Orkney is uncertain, but lists of taxpayers were drawn up. There was also liability for the cess or land tax. According to a petition sent to parliament by landowners in 1700, the tacksmen's failure to meet their share of this resulted in a constant quartering of troops on Orkney parishes. The petition painted a distressing picture of 'the extreme calamity of these five last years', and railed against 'the rigidity and oppression of the tacksmen', whose selfish misrule meant 'thousands of our people have been killed and starved'.[22]

For Britain as a whole, the Revolution saw the beginnings of limited constitutional monarchy and some first stirrings of religious toleration. But for the people of Orkney, beset by exploitative tacksmen and their trade decimated by war and recession, there was probably little that seemed particularly glorious.

No Constituted but Confused Church

The Church of Scotland, unlike the Church of England, was transformed in 1689. Not only was episcopacy removed, but an act of parliament of the following year abolished the rights of lay patrons to nominate ministers to parish livings, vesting the responsibility in elders and local heritors. For several years, however, the Church in Orkney was an Episcopal Kirk carrying on gamely in default of a bishop. Not before June 1697 was a presbytery formally reconstituted, when, by order of the General Assembly, John Cobb, now of Stronsay, James Graham of Holm and Thomas Baikie, new minister of St Magnus, came to a thinly attended meeting in Kirkwall.

In the meantime, with the exception of Wilson, existing ministers stayed in post, with Alexander Pitcairn, the former dean, remaining an influential force. Indeed, Orkney was a refuge for disaffected clergymen. Alexander Keith came to serve in Sandwick in 1693 at the invitation of William Craigie of Gairsay. Previously minister of Tillicoultry in Clackmannanshire, he was driven out when parishioners there took against him for 'preaching prelacy'. Alexander Mair, who succeeded to Kirkwall's 'second charge' in 1694, was illicitly ordained by the deposed bishop of Moray, William Hay. Patrick Guthrie, admitted to St Andrews and Deerness in 1695, was ordained by ex-bishop James Ramsay of Ross. Alexander Burnett, minister of Firth and Stenness from 1695, had in 1689 been deposed by the privy council from Abercrombie in Fife for refusing to read the declaration offering the throne to William and Mary.

It was not to last. In January 1699, the General Assembly assured King William that his suggestions for 'the planting of the North' had been taken in hand, with commissions having been dispatched 'to the remotest parts of the kingdom, and even to the Isles of Orkney and Zetland'. A committee for visiting Ross, Sutherland, Caithness and Orkney made its way through the northern Highlands, and opened for business in St Magnus Cathedral on 7 June 1698.[23]

Over the course of a fortnight, damning evidence was heard against a succession of Orkney ministers. James Leslie of Evie was 'not of sound principles', and a notorious drunkard. George Spence, 'about the time of the late happy revolution', abandoned Firth to 'intrude' into more profitable Birsay. He was, moreover, a shameless fornicator, who smuggled a pregnant mistress to Caithness. Both ministers 'demitted' – resigned – as did Thomas Fullerton of Westray, a maker of 'reproachful rhymes against the present ministry'.

The top scalp was that of Alexander Pitcairn of South Ronaldsay. Due to his relentless pursuit of teind payments, he had made enemies in his parish, and was tagged with a long litany of faults and failings. In addition to preaching little but 'rebellion and sedition', Pitcairn was accused of neglecting to visit the sick, to catechise, celebrate communion, baptise reverently, enforce the discipline of the Sabbath, or to treat parishioners with basic decency and respect. He instead called them 'villains, thieves etc., upon the slightest occasions'. The committee deposed him.

Others found ways out of escape. Patrick Guthrie apologised for 'intrusion' into St Andrews, and declared that, 'after prayer, reading, meditation and conference', he believed Presbyterianism 'to be of divine right and institution'. Just before the committee's arrival, Alexander Mair threw himself on the presbytery's mercy for his ordination from 'the pretended bishop of Moray'. The commissioners allowed him to transfer to Hoy and Graemsay, after fulsome assurances of 'grief for his compliance with prelacy'. A few years later, Mair was asked how he had managed to reconcile his conformity with his conscience. His reply had at least the virtue of honesty: 'What will not a man do for his bannock?'

James Heart of Shapinsay admitted he had not sworn the oath of allegiance to King William, but insisted the oversight was not due to disaffection: he had prayed for the king, and 'obeyed all fasts and thanksgivings'. Under presbytery supervision, Heart was allowed to remain in office, despite claims from parishioners that he 'winks and connives at sin'. One allegation was of 'witchcraft and sorcery used amongst that people, not taken notice of nor censured'.[24] Failure to persecute witches was no longer a disqualification for ministerial service in the Kirk of Scotland, but it remained a black mark in the eyes of some ordinary people in Shapinsay.

A second General Assembly committee arrived in 1700 to pick up where the first had left off. It aimed principally to order affairs in Shetland, but was authorised to continue 'planting churches' in Orkney, and spent a month in the islands assisting the presbytery and waiting for weather to allow it to progress north. One of its members was the minister of Borrowstounness in West Lothian, John Brand. On his return, Brand published an informative account of the geography and society of the Northern Isles. He hoped readers would understand why 'I speak of Orkney or Zetland as not in Scotland'; in this, he was simply expressing himself 'as the country do'.

Brand chose not to discuss the committee's deliberations, or say much about politics, though he recognised that Orcadians 'greatly cry out of the oppression they groan under, by reason of the frequent change of stewards'. He learned, too, that rents from bishopric lands were sent south, whereas before the Revolution they were spent locally in Orkney. In consequence, 'the Presbyterian government is made the less acceptable to many of them'.

The committee began its deliberations in Kirkwall in April 1700, and winkled out a few more unsatisfactory clergymen. The resignation of the aged minister of Walls and Flotta, Alexander Dalgarnock, was readily accepted. A harder case to resolve was that of Alexander Burnett in Firth and Stenness. Burnett was eager to continue, but, confronted with the question of whether Presbyterianism was the sole true government of the Church of Christ, would say only that 'he should live peaceably under the government and should do nothing to subvert the same'. Under increasingly rancorous questioning, Burnett blurted out that 'he would rather go to the grave or the Indies before he went one step further'. He did not need to; the committee deposed him as 'unfit to be a minister of the Gospel'.

The Kirk Revolution in Orkney came late but cut deep. When the (metaphorical) blood dried, it became clear that the extent of the amputation was greater than that produced by the Reformation of 1561, or, in practice, by the events of 1650. After the suspect James Heart of Shapinsay resigned in 1703, only two ministers remained who had served the Kirk before 1688: James Graham of Holm and Richard Mein of Cross and Burness in Sanday. Graham – himself the grandson of a bishop – was sufficiently rated by the Presbyterian authorities to be co-opted onto the Shetland–Orkney Committee. Mein was another matter. The 1698 commissioners judged him 'not of sound principles', and heard some eye-popping complaints about his conduct. In a parishioner's house at Stove, near the southern tip of Sanday, Mein reportedly acted in a play, standing on a chair with blackened face and hands aloft, crying, 'Cape and gloure! [Gape and glare!] Who would have or kiss me now?' In 1700, the committee readmitted the pliant and repentant Mein, rebuking him for 'former evil courses'.[25]

Not all the deposed ministers went quietly – a further contrast with 1561. For the first time in Orkney, an alternative religious observance was available, as 'Episcopalianism' stepped forward to challenge the monopoly of the Presbyterian Kirk. It involved an epic, and at times comic, battle for the soul of Kirkwall.

The protagonists were the cathedral minister, Thomas Baikie, and his deposed predecessor, John Wilson. Baikie was unusual among Orkney clergymen, both as an ardent Presbyterian and as a native Orcadian. He was the younger son of a nouveau Sanday laird, James Baikie of Burness, a nephew of the formidable James Baikie of Tankerness. Thomas Baikie's wife, Elizabeth Fea, was the daughter of another North Isles laird in the cadet branch, Patrick Fea of Whitehall in Stronsay, an irascible former privateer.

Baikie was ordained by the Aberdeen Presbytery in 1697, and soon after received a 'call' to the cathedral from the Kirkwall congregation. From the start, supporters of the old regime made his life difficult. Baikie complained of being 'maliciously slandered and reproached' in 'wicked and atheistical verses'. In Kirkwall, he was a local prophet without honour. 'My birth and education amongst them, together with my present youth-hood, makes my exhortations to be despised'. Baikie pleaded in vain for the committee to transfer him 'unto an easier charge in this country'.

For Wilson, post-eviction life was proving no easier, and in 1702, after nearly eight years 'deprived of the public exercise of the ministry to the which I was devoted from the womb', he reached out to the presbytery. Wilson asked the ministers to intercede with the General Assembly, so he might be permitted to preach, and work with them 'in promoting reformed truths'. What the presbytery wished to know was whether Wilson 'acknowledged the Presbyterian church government … to be the only government of this Church'. Wilson made a sufficiently non-committal reply for the presbytery to draft a cautious letter of recommendation, hoping he might 'yet be useful in the Church'.

Offered this reconciliatory inch, the former minister ran off along a mile of entitlement. On Sunday 3 January 1703, Wilson learned there was to be no service in St Magnus, since Baikie lay sick in bed. Seizing his moment, he asked the kirk officer David Seater to ring the bells to summon the congregation, and ascended the pulpit to preach. It was Baikie's wife, Elizabeth Fea – according to an ally of Wilson's – who prevented matters going further. Baikie 'came with his night-cap on, guarded by his wife, and they two violently pulled Mr Wilson out of the pulpit'.

After a subsequent presbytery investigation, Seater was sacked, while the town baillie, William Liddle, was accused of conniving at Wilson's usurpation. The Kirkwall magistrates agreed to prevent further attempts, though they would not make Wilson 'answer for former irregularities'. Wilson's own appearance before the presbytery on 7 January rapidly descended into acrimony. The ministers were unconvinced by his assertion of rights to the pulpit as 'second minister', pointing out he resigned that role on his promotion to the first charge. Nor were they moved by Wilson's claims of moral responsibility towards the inhabitants of Kirkwall, or how he 'frequently called them his people'.

What mattered, as Moderator Alexander Grant of South Ronaldsay put it to Wilson, was that his actions were 'directly contrary to the acts and appointment of the present constituted Church'. 'It was', Wilson blurted out, 'no constituted but confused Church.' As for the presbytery, it 'never had any being till after Christ 1542 years it was clecked [hatched] at Geneva … and that Calvin himself was only midwife thereto'. To ministers who regarded Presbyterianism as a blueprint from scripture, this was gratuitously offensive.

After this bridge-burner of an interview, Wilson ignored further summonses, and began circulating written justifications. A sentence of suspension was posted on the door of the cathedral, but Wilson pulled it down and replaced it with a 'false, lying and calumnious paper against the government and presbytery'. When, in January 1703, the presbytery's suspension was read in St Magnus, Wilson turned up to challenge it, and 'occasioned great tumult and confusion'.

A few weeks later, the stakes were raised further. On Sunday 21 February, Wilson preached in his house, a few hundred yards north of the cathedral, having 'called and convened the most part of the town'. His congregation included one of the baillies and the 'dean of guild', the official responsible for trade and building regulations.[26] It was a decisive parting of the ways. Episcopalians had now become what they once accused Covenanters of being – 'conventiclers', schismatics, separatists. The site of worship was a 'meeting-house', rather than a church, but it was the place of meeting for what increasingly looked like a separate religious denomination.

Union and Disunion

Episcopalians had wind in their sails at the start of the new century. In March 1702, William III died of pneumonia, contracted while recovering from a fall after his horse tripped over a mole-hill. William's demise was preceded in 1694 by that of his childless wife, Mary II. In September 1701, James VII and II had passed away in Paris. His son James, born on the eve of revolution, was now, in the eyes of supporters, the rightful sovereign. Foes called him a 'pretender', and later 'the Old Pretender'. The crown was given in 1702 to Anne, James II's other Protestant daughter by his first marriage. After a heart-rending succession of miscarriages, still-births and deaths in infancy, Anne had no living heir. To pre-empt another crisis over the succession, the English parliament in 1701 passed an Act of Settlement. It designated as Anne's future replacement on the English and Scottish thrones Sophia of Hanover – a daughter of James VI's daughter Elizabeth, and a person chiefly qualified by her Protestant faith. The Scottish parliament was not consulted. Though it was neither obvious at the time, nor inevitable in retrospect, the countdown to its abolition had begun.

For Orkney, the unravelling of events leading to the union of Scotland and England involved a tight entanglement of national and local concerns. One persistent thread was the determination of the earls of Morton to recover their grip on the islands. In 1673, William Douglas, ninth earl, had been discharged from obligations to the crown arising from the *Kennemerland* affair, granted a pension from the Orkney estates and authorised to collect unpaid rents. Charles II, however, wanted reassurances from William's successor, his uncle James. In 1684, payment of the pension was paused until the new earl and his son formally renounced their claims to Orkney.

James Douglas died in 1686, so it fell to his son, James, eleventh earl, to navigate the choppy waters of the Revolution. Ancestral loyalties to the Stewarts were laid aside, and in March 1689 Morton joined other Scots nobles in welcoming William and thanking him for 'preserving to us the Protestant religion'.

Morton cherished his Protestant religion, but cared more about the dues and rental income from Orkney and Shetland, 'without which that noble and ancient family, that he now represents, must necessarily go to ruin'. In a petition to parliament of June 1693, Morton asserted the illegality of the seizure of his family's 'proper patrimony' in 1669, and claimed that his own recognition of the annexation was extorted 'by a most visible and forcible concussion [extortion by threats]', taking advantage of his 'straits and difficulties'. Crown lawyers were unconvinced, and parliament rejected the petition. But a marker had been put down, and a renewed connection with the islands was strengthened in 1696 by the appointment of Morton's brother Robert Douglas, as stewart and tacksman. Morton doubtless flinched when, at the beginning of that year, Lord George Hamilton, general of the army, and husband to the king's former mistress, was created earl of Orkney, Viscount Kirkwall. Though the Hamilton connection with the islands was never more than nominal, it was a confirmation that the crown had no intention of resurrecting the old title for anyone else.

Morton continued to ingratiate himself with the new regime. He served on the privy council, and subscribed to the Association of 1696, a collective oath, following a failed Jacobite plot, to exact vengeance should William be assassinated. In 1700, Morton's petition to rescind the 1669 annexation was again considered by parliament, and in 1702 it recommended to the new queen that Morton be allowed to pursue his case, noting an 'early and zealous appearance for the late King William'. In the meantime, Morton was awarded £1,000 sterling per annum from the Orkney and Shetland estates, and appointed collector of the stewartry rents – arrangements ratified by Queen Anne in 1705. In the following year, Orkney and Shetland were removed from the jurisdiction of the Admiral of Scotland, and Morton was gifted rights to one of the isles' most lucrative commodities: wrecked ships. The *Kennemerland* was forgiven or forgotten.[27]

Morton's steady restoration culminated in February 1707, when new legislation repealed the annexation of 1669, and, for a modest annual payment of £6,000 Scots (£500 sterling), handed back to Morton jurisdiction over the earldom of Orkney on the same terms as the grant of 1643.

Orkney and Shetland's 'dissolution' from the crown was one of the last acts of the Scottish parliament, which a few weeks earlier had voted itself out of existence. In January 1707, an Act of Union, mirroring a similar measure in the English parliament, decreed that from 1 May the two kingdoms would forever 'be united into one kingdom, by the name of Great Britain'.

The union was a marriage of necessity, rather than any kind of love match. From the English side, it was driven primarily by a need to secure the Protestant succession, and prevent Scotland falling into the hands of the Stewart, or Stuart, pretenders. The ongoing war with Louis XIV's France (after a brief hiatus in 1697–1701) made that a real possibility, despite victories such as Blenheim (1704), where the titular earl of Orkney played a decisive military role. From Scotland's side, economic decline, exacerbated by the disastrous attempt to establish a colony at Darien in Panama, increased the appeal of access to English ports and markets. There were reassurances to Scotland about its separate courts and legal system, and – crucially – the independence of its Presbyterian Church. Bribes were liberally dispensed by royal commissioners to lubricate the act's passage.

Orkney, like Scotland as a whole, saw its political representation shrink as a result of the union: a mere forty-five representatives from Scotland would henceforth go to the parliament at Westminster, to sit alongside 469 from England and Wales, even though Scotland's population was roughly a fifth of its southern neighbour's. Orkney and Shetland would now send a single 'county' MP in place of two earlier commissioners. Kirkwall, after less than forty years of dispatching a delegate to parliament, was amalgamated into a composite, unwieldy constituency of 'Northern Burghs', along with Tain and Dingwall in Ross, Dornoch in Sutherland and Wick in Caithness.

For some enthusiasts of union, however, Orkney was symbolically constitutive of the new state. A thanksgiving sermon preached in Hackney on 1 May 1707 declared, 'We are all this very Day become one Body under One Government, from Dan to Beersheba, from the Orcades to the South Channel'. Dan and Beersheba were the northern and southern extents of biblical Israel, lending a sacred dimension to the preacher's patriotism. Poets struck similar notes, lauding an indissoluble union 'from Dubris [Dover] South/To Northern Orcades'; 'As far as to the very Orcades'; 'even to the Orcades her verge shall reach!'[28]

This was a new iteration of an old rhetorical device, paying attention to the islands on account of the very improbability of their presence. But in some practical ways, too, Orkney – in the person of James Douglas, earl of Morton – contributed to the constitutional creation of Britain. The 'dissolution' of February 1707 was clearly recompense for Morton's efforts on behalf of the court and the union. The earl was among the Scottish commissioners chosen to conduct negotiations with English counterparts, and in parliament he voted for all twenty-five articles of the resultant treaty. Many Scottish burghs opposed the union, but Kirkwall's commissioner – Morton's brother Robert Douglas – was firmly in favour. The commissioner representing the stewartry had reservations, but in the end voted for union under persuasion from Morton. Sir Alexander Douglas of Egilsay was a grandson of Alexander Douglas of Spynie, enforcer for the ninth earl. His mother, Marjorie Monteith, was the daughter and heir of Patrick Monteith of Egilsay and his wife Marion Smyth, who was herself a daughter of Patrick Smyth of Braco. The commissioner was a genetic case-study in the ability of nimble incomers to embed themselves in island society.

Morton's grip appeared firm, but Orkney in the first decade of the eighteenth century was no pliant and biddable fiefdom. In 1702, what turned out to be the very final election to the Scottish parliament was a bitterly contested one. William Craigie of Gairsay was returned initially, but then forced to withdraw, accusing Morton and his brother of improperly influencing the result. The candidate eventually selected alongside Alexander Douglas was the stewart, Sir Archibald Stewart of Burray. He missed the crucial votes in 1707, probably because of illness, so we cannot say how he would have voted. But he was certainly an Episcopalian, and possibly a Jacobite.

Ironically, it was to Sir Archibald as stewart that the presbytery complained in January 1703 about 'a lamentable and seditious design'; the ministers demanded he arrest everyone involved. It was 'Ane Address to her Majesty', asking her to permit parishes where most heritors and inhabitants were 'of the Episcopal persuasion, to call, place, and give benefices to ministers of their own principles'. The petition was gathering subscriptions throughout Orkney, and 'carried up and down through the town of Kirkwall'. Stewart coolly dismissed the ministers' concerns, saying he knew nothing about the matter, and anyway, 'it was lawful to address'. It seems doubtful that Sir Archibald was uninformed, as the organiser of the petition was his own stewart-depute, Henry Leggat. To Presbyterians in the south, it was notorious that 'busy agents' pretending to have Queen Anne's authority were active in Orkney, convening the people, assuring them episcopacy was to be re-established, and threatening those unwilling to sign with the displeasure of a new bishop.

Orkney's Episcopalian agitation was part of a wider effort in Scotland, inspired by the accession of a sympathetic Anglican queen. A national petition was presented to Anne in March 1703, via her secretary of state, the duke of Queensberry; Anne assured the petitioners of her favour and instructed the privy council to protect peaceable Episcopalians. Queensberry was also told to bring a bill to parliament, allowing preaching by Episcopal ministers willing to take the oath of allegiance. In the face of intense Presbyterian hostility, the act failed to pass; a submission from the standing Commission of the General Assembly, empowered to act on its behalf between meetings, declared that toleration for Episcopalians would 'establish iniquity by a law'.

Tolerated or no, Episcopalian ministers were on the loose in Orkney. Two clergymen ousted in 1698 – George Spence and James Leslie – remained in the islands, officiating at baptisms and weddings. In 1704, the presbytery became aware that Sir Archibald Stewart kept a chaplain in his house, a young convert from Presbyterianism called Thomas Rind. Two delegates from the recently established Synod of Caithness, Orkney and Shetland were 'appointed to speak to Burray'. But their attempts to travel to the island were defeated by 'boisterousness of the weather'. Stewart was not the only laird maintaining an Episcopal chaplain: William Craigie of Gairsay, John Stewart of Burgh, Robert Baikie of Tankerness and James Graham of Graemeshall were all reputed to do so.

In 1706, the synod was alarmed to discover that Sir Alexander Mackenzie, a former sheriff and a son of the late bishop, had taken it upon himself 'to convocate the people both in the bounds of Caithness and Orkney, and had read the Book of Common Prayer, explained Scripture and preached'. Laymen were not supposed to do such things, but desperate times demanded daring measures. Growing use of the Anglican Prayer Book among Orkney's Episcopalians signalled a parting of the ways with Presbyterians, and a growing fellowship with like-minded believers in the southern parts of 'Great Britain'.

Not all Episcopalians were Jacobites. Morton, indeed, aimed to persuade the chief minister Robert Harley in 1713 that the islands were loyal to Queen Anne precisely because, as he believed, 'there is [only] six Presbyterians in all the country of Orkney and Shetland'. It didn't look that way to Robert Wodrow, custodian of the Covenanting tradition. In 1711, he was hearing disturbing reports about Orkney, where the justices of the peace 'are all malignants and Jacobites'. In particular, unjust accusations had been made against a minister there, James Sands of Birsay and Harray.[29] Those allegations, of ministerial sheep-stealing, might at any time have raised eyebrows; in the fraught opening years of the eighteenth century they ignited a political and legal firestorm.

The Mutton Covenant

The sheep were two ewes in Birsay belonging to William Stensgar and his wife, Katherine Brown. James Sands contended they were impounded legally, with approval of the parish baillie, after Stensgar defaulted on vicarage teinds. He was, in any case, a disreputable person, justly punished 'for charming and consulting with charmers' (p. 213). On Stensgar's telling, Sands vindictively disciplined him after he remonstrated about sheep 'thievishly' taken by the minister's servant. The incident itself took place in 1703, but six years later Stensgar lodged a report with the justices of the peace.

He was likely put up to it. Shortly before Stensgar's complaint, Sands had written a letter attacking the teachings of James Lyon, Episcopalian minister in Kirkwall, and circulated copies of it throughout Orkney. Lyon arrived in 1708 to continue the ministry of John Wilson, who had left to serve a meeting-house in Edinburgh. His very presence, with lodgings adjoining Thomas Baikie's manse, was a provocation to the ministers of the presbytery – now known as the Presbytery of Kirkwall, after the establishment in 1707 of a separate Presbytery of the North Isles.

Denunciations of the interloper flowed from the pulpit of St Magnus. On one occasion, Lyon recalled, Baikie took exception to the Prayer Book formulary 'Lord have mercy upon us, Christ have mercy upon us'. This, Baikie alleged, 'derived from the worshippers of Baal', and he proceeded to alarm his congregation by excitedly calling out, 'O Baal, save us!' As a cathedral guest preacher, Sands mocked Orkney's Episcopalians for drinking healths to the Church of England: 'they have much need to do so, for truly 'tis a weak and sickly Church'. The minister of Cross and Burness, Murdoch Mackenzie – himself the grandson of a bishop – went too far, however. In August 1709, he told worshippers leaving Lyon's meeting-house that 'he wished the Service Book was burned, and all these that made use of it'. As Mackenzie might have paused to consider, users of the Book of Common Prayer included Her Majesty, Queen Anne. There were demands he be arrested for treason, though the stewart-depute decided not to prosecute, on the only relatively exonerating grounds that Mackenzie was drunk when he spoke.

In the meantime, the presbytery resolved to silence Lyon, planning its strategy at a meeting in January 1709, which elected Sands as moderator and swore members to secrecy as they embarked on a business 'of weight and moment': carrying out a directive from the General Assembly Commission to suppress 'schismatical meetings'. The presbytery lobbied Kirkwall's provost, David Traill of Sebay, but he prevaricated, warning there were people in the town 'who would not be so easily gained to let fall that affair of the meeting-house'.

That was an understatement. Kirkwall's upper echelons were riddled with Episcopalian sympathisers. The presbytery disciplined John

Dunbar, excise officer, for having 'reviled the ministers' and using his position to lure people to the meeting-house. He promised to amend, but was soon accused of saying 'it was a dangerous thing to hear a Presbyterian minister'. Donald Groat, procurator-fiscal – the stewartry's legal officer – got into an argument with Hugh Clouston, session elder. Attending the meeting-house, said Clouston, was against the government. Groat snapped back, 'Shame fall thee, and thy government!' – though he afterwards claimed, with lawyerly inventiveness, 'he meant only Hugh Clouston's government of himself'. Andrew Young of Castleyards – shortly to assume office as provost – considered the ministers 'a pack of knaves'.[30]

Having got nowhere with the magistrates, the ministers acted unilaterally, posting on the cathedral door an order banning Lyon from preaching. The General Assembly Commission forwarded to the ministers copies of letters from the lord advocate to the sheriff and provost, instructing them to close the meeting-house, and the presbytery decided to prosecute Lyon in the justiciary court in Edinburgh.

This was how things stood when, in November 1709, two justices, Captain James Moodie of Melsetter (son of the irascible William Moodie, pp. 307–8) and James Gordon of Cairston, took up Stensgar's complaint about the sheep and examined witnesses at the old palace in Birsay. Sands, they concluded, persuaded the baillie to supply a warrant for confiscation of the ewes only after the complaint was made to the JPs. The minister was arraigned for trial before the quarter sessions in Kirkwall.

At the same time, the campaign against Lyon was hotting up. In January 1710, the presbytery received copies of legal letters from the General Assembly Commission and lord advocate, advocating banishment for Lyon and fines for the Kirkwall magistrates, shown to be 'uncapable of public trust'. The town council buckled, and banned Lyon from ministering in the town. He left Orkney temporarily, though not before travelling to the East Mainland to baptise a child of Robert Baikie of Tankerness, one of the JPs driving the prosecution of James Sands.

With relations between the presbytery and justices at breaking point, a truce of sorts was negotiated. On 11 March 1710, 'Articles of Agreement' were signed at Kirkwall by eighteen JPs and eight ministers of the presbytery. The ministers agreed to expunge from their records 'aspersions upon the justices' and to 'carry themselves respectfully, Christianly and kindly' towards them. Mutual promises were made 'to forgive all offences'. The presbytery was not to usurp 'power of jurisdiction' properly belonging to the justices, nor the justices 'encroach upon the power of the presbytery'.

Another, more specific demand was agreed to: the justices wanted recital of the Lord's Prayer in public worship, when any JP or heritor requested it. This seemingly minor liturgical detail was a totem of controversy between Episcopalians, who favoured the practice, and Presbyterians, who generally opposed it. A recent national campaign on the issue ended with a vague recommendation from the General Assembly for congregations to use the 1645 Westminster Directory of Worship – the liturgical manual drawn up by Puritan English Parliamentarians, in conjunction with Scottish allies, during their period of political ascendancy. This included the Lord's Prayer, but there was little difficulty in ignoring the stipulation. The Kirkwall Presbytery, however, was willing to recommend the practice, hoping it would prove 'a mean to gain and bring in' dissenters from Presbyterianism.

The Articles of Agreement did not say so, but many believed that, in return for this concession, the JPs had promised a satisfactory outcome in the case against Sands. In July 1710, at a sessions in Birsay, Sands was acquitted by the justices. He reported with satisfaction to the presbytery how his chief accuser, Stensgar's wife, Katherine Brown, confessed to 'unjustifiably libelling him ... put upon it by ill-willers to the ministry'.

If there was a deal, it soon began to unravel. Later that year, an anonymous *Letter from a Gentleman in Orkney* was printed in Glasgow. It contained a hostile account of Sands's conduct, the text of the Kirkwall concordat and suggestions of a corrupt compact. The agreement, the author revealed, was 'commonly called in our country the *Mutton Covenant*'. A tradition recorded by James Craven ascribed the nickname to the fare on offer in a tavern where the agreement was concluded. But it was surely a knowing reference to Sands's alleged offence.

Details of 'the notorious Mutton-Covenant' were soon emerging in pamphlets and in national and provincial newspapers. The embarrassing disclosures, it was alleged, prompted the General Assembly Commission to order the Orkney ministers to 'vindicate themselves from the aspersion' by prohibiting the Lord's Prayer, a breach of promise which caused the justices to reopen the case against Sands. In May 1711, he was summoned 'de novo' before the quarter sessions, convicted and fined. The injustice of 'base and malicious slanders' rankled with Sands, and in the summer he headed for Edinburgh to seek legal redress. The General Assembly was already in receipt of a supportive petition from the Caithness Presbytery, lamenting how Orkney ministers 'groan under the unprecedented tyranny of these justices'. Tyrant-in-chief was Captain James Moodie of Melsetter in Hoy, perhaps – in a close competition – the most querulously distempered of all Orkney lairds. The ministers lamented his 'hellish malice'.[31]

In an age of easily available cheap publications, the squabbles of a small island community could become fuel for larger fires: Sands and Lyon aired their disagreements in print before a national audience. Matters of high principle were involved – the divine origins of episcopacy, and the respective merits of Presbyterian and Episcopalian worship. But there was room for low blows. Sands mocked Lyon as a clerical vagrant, 'a pastor without a flock'; Lyon showered Sands with ironic and punning praise for diligence in 'teinding your sheep'.

Lyon's own legal troubles diminished in June 1710, when he appeared in the Edinburgh justiciary court, and the judges dismissed the case on grounds of witnesses being improperly summoned. The following year, to the delight of his congregation and the consternation of the presbytery, Lyon returned to Kirkwall. It was a harbinger of national political change. The 1710 general election was a sweeping victory for the Whigs' adversaries, who had become known as Tories (from a term originally referring to Irish rebels). The Tories were opponents of the war with France, and for the most part fervent Anglicans with scant sympathy for the concerns of Scots Presbyterians. In Orkney, the earl of Morton's candidate, Alexander Douglas, was re-elected unopposed, and quickly transferred his allegiance from the outgoing Whig chief minister, the earl of Godolphin, to the incoming Tory one, Sir Robert Harley.

Toleration of Episcopalians, rebuffed by the Scottish parliament in 1703, was in 1712 granted by the British one, with freedom of worship assured to 'all those of the episcopal communion in that part of Great

Britain called Scotland' willing to swear allegiance to the queen. At a stroke, the ground collapsed from under the Kirkwall Presbytery's long efforts to exclude 'intruders'.

Undaunted, Thomas Baikie took to the pulpit and preached on a verse in Psalm 112 – 'He shall not be afraid of evil tidings.' This provocative commentary on the Toleration Act brought the affairs of Kirkwall back onto the national stage. The General Assembly Commissioners met to confer 'anent Mr Baikie', upon learning that Captain Moodie had cited Baikie, Sands and John Keith, Moodie's own minister in Walls and Flotta, to appear in the justiciary court. The 'pretended crimes' of Keith and Sands, the commissioners averred, were secular matters: alleged forgery and sheep-stealing. But Baikie's was 'a point of doctrine'. He was charged with blasphemy and irreligion, in the recent sermon, and in his mockery of the English liturgy, 'Baal, save us!'

The lord advocate believed Baikie's case might be thrown out as properly belonging 'before ane ecclesiastical judicatory'. But Baikie's own eagerness to have his day in court generated fierce argument among the commissioners – 'there were consequences of such processes to the whole Church'. Some thought that magistrates had a right to check for blasphemy and treason in sermons. Others invoked the martyr-spirit of the Covenanters, 'the practice of ministers, in former times, to suffer upon this'.

In the end, Baikie decided attack was the best form of defence. In July 1712, without consulting the rest of the presbytery, he, Sands and Keith, together with the minister of Dunnet in Caithness, launched an action for slander against Moodie and other Orkney JPs motivated by 'implacable enmity against our legal Church establishment'. Captain Moodie, they alleged, referred to ministers as 'giddy-headed Gospel-mongers'. Moodie loftily responded that he and other justices had 'been obliged for the good order and peace of the country to check and curb the growing indiscretion of some of the meaner sort of the clergy'.[32]

Episcopalianism was one strand in the opposition of lairds like Moodie to the absentee lordship of the earl of Morton. The hostilities, however, were never simply ideological. In the spring of 1713, another row erupted, about a wrecked Dutch merchant vessel, the *Rijnenberg*, with Morton aggressively asserting his rights against local salvagers. The general election of 1713 was – for the first time in Orkney – formally contested. Morton's candidate, his brother George Douglas, won with 64 per cent of the vote. But elections in Orkney, as elsewhere in Scotland, reflected wealthy men's interests not the will of the people. Douglas had the support of nine of the fourteen electors; three defections would have swung it.

The defeated candidate was Captain James Moodie – not the bombastic laird of Melsetter, but his uncle. The elder Captain Moodie was a distinguished Royal Navy officer, who returned to Orkney in 1713 when the Treaty of Utrecht ended the long-running war with France. He would soon take possession of the Melsetter estate, once the extent of his nephew's debts and mismanagement became clear. But the younger Captain Moodie was the moving force of the election campaign, and in its wake he drew up 'Articles of Association' under which he and several other Orkney gentlemen, including Robert Baikie of Tankerness and Andrew Young of Castleyards, pledged to oppose the earl of Morton's interests.

The earl died in December 1715, to be succeeded by his brother Robert. A more momentous death, that of Queen Anne, had taken place the preceding summer. Since Sophia of Hanover had narrowly predeceased her, the crown passed, under the 1701 Act of Settlement, to Sophia's son, Elector George. It was a moment of crisis, or opportunity. Subjects who had stayed loyal to James II's daughter might think twice about a German-speaking 'Hanoverian' successor, connected to the House of Stuart by the thinnest of dynastic threads.

George I landed in Britain on 18 September 1714. An anonymous London poet, invoking a litany of the nation's great rivers, gushingly remarked how

> Ev'n those that far as distant Orkney lye
> Rush out to view Great GEORGE, and leave their Channels dry.

Among actual Orcadians, there is scant evidence of such fast-flowing enthusiasm. Two days after the king's arrival, David Traill, heir to a Sanday estate, wrote from Kirkwall to his father, John Traill of Elsness. He advised him to come quickly to town to recertify as a justice of the peace: 'if ye omit, ye may assure yourself of being informed against as disaffected against the king'. This mattered because JPs were to have oversight of wrecked ships. John Traill should understand that Morton and his brother Robert – whose supporters 'will slip no opportunity to do you harm' – were working hard to get their own creatures 'fixed as justices'.[33]

Early eighteenth-century Orcadian politics, ecclesiastical and secular, were fractious and sometimes frenetic. The quarrels mirrored, and sporadically influenced, national religious and political divisions, in an era which has been dubbed 'the first age of party'. They were often, at heart, battles about fundamentally Orcadian issues, fought on home ground with locally manufactured weapons. But events outside the islands were shaping the allegiances and increasingly determining the outcomes.

Jacobites and Pirates

On 29 September 1715, outside St Magnus Cathedral in Kirkwall, an excited crowd gathered around a bonfire near the town's Mercat Cross, where edicts and proclamations were customarily announced to the people. The proclamation that day, read by a Mr Drummond, affirmed the right and title of James VIII as king of Britain, France and Ireland. In attendance were the Episcopal ministers James Lyon and George Spence, who offered prayers for King James, and drank his health. Orkney Jacobitism had finally dared to speak its name.

The Jacobite Rebellion of 1715 was the closest the exiled Stuarts ever came to recovering their lost thrones, and a moment of potential unravelling for the precarious union of 1707. A month before the proclamation in Kirkwall, John Erskine, earl of Mar – a Tory minister removed from office at the accession of George I – travelled secretly to Scotland. On 6 September, he proclaimed James VIII at Braemar, and placed himself at the head of a large Highland army. On 13 November, at Sheriffmuir, just north of Stirling, he confronted a smaller government force under the duke of Argyll, missing the opportunity to crush it. Among the Jacobite casualties of the battle was John Lyon, earl of Strathmore, perhaps a relative of the Kirkwall minister. James Lyon acknowledged Strathmore as his patron, and dedicated to the earl his book of 1710 defending the divine origins of episcopacy.

For a few weeks, the Jacobites dominated northern Scotland, but by the time 'James VIII' landed in Scotland in December, the tide had already turned, and on 5 February 1716 he left again for France. Ten days earlier, James's title was proclaimed at Wick in Caithness. Thirteen gentlemen of the county took part, nine of them predictably named Sinclair. An Orkney contingent was also present, including – inevitably – James Lyon.

On 11 February, Lyon was back at Kirkwall's Mercat Cross, 'publishing the proclamations in the pretender's name'. This was probably James VIII's declaration of 25 October 1715, which denounced possession of the throne by 'a foreign family', and promised to restore the kingdom – that is, Scotland – to its 'free and independent state'. On 12 February, Jacobite declarations were read at the kirks of Sandwick and Stromness, and probably at others in Orkney: 'ane express with proclamations' was sent over to Sanday.

Kirkwall, in early 1716, was in Jacobite hands. In the town were 120 soldiers from the earl of Mar's army, including the marquis of Tullibardine, Lord Duffus 'and several other lords' and earls' sons'. These, however, were defeated and retreating men, in search of ships to take them to Gothenburg – a reverse trajectory to that of Montrose in 1650. Among the officers was John, Master of Sinclair, eldest son of the Caithness nobleman Lord Sinclair. In his memoirs, John described his adventures getting to Kirkwall, and beholding there 'the melancholy prospect of the ruins of ane old castle, the seat of the old earls of Orkney'. It prompted reflection on how his ancestors came to lose 'so great and noble ane estate'. Sinclair memories, as we have seen (pp. 64, 175), were ruefully long.

Failed rebellions have few acknowledged parents. How many Orkney lairds or ordinary folk were active Jacobite supporters is hard to say for certain. Lord Duffus promised Mar in 1715 he could raise a large force in Orkney. He sent an envoy to the islands, but his own aide-de-camp was sceptical: Orkney was not a Highland clan society, organised for war. It is possible, too, that the dispiriting experience of 1650 cast a long and painful shadow.

Only three Orkney gentlemen were said to have been with Lyon at the proclamation in Wick: Sir James Stewart of Burray, son and heir of Sir Archibald; Robert Baikie of Tankerness; and one Borthwick, collector of the customs. In April 1716, Borthwick was on an arrest-list of persons 'instrumental at the proclamation at the Cross', along with Drummond, Lyon and the Kirkwall merchants Robert Donaldson, Magnus Meason and Thomas Bell. If others were involved, they managed to cover their tracks. Harie Graham of Graemeshall rejected as 'groundless aspersions' claims that he ordered proclamation of the pretender at kirks in the West Mainland. 'There are too many of our good friends', David Traill lamented, 'who now repent their actings.' He feared his own father had succumbed to 'bad advice'.[34]

For the ministers, it was a moment of vindication, and for vengeance. The presbytery deposed James Lyon, declared to have 'deserted the Protestant cause, and espoused the interest of a popish pretender', along with George Spence and James Leslie. Lyon evaded capture, perhaps because the lieutenant-deputes charged with securing him – William Liddle, Alexander Douglas of Egilsay and Harie Graham – were Episcopalians with hearts not fully in the task. As late as February 1717, Lyon was reported to be preaching at Graemeshall. But, metaphorically at least, Orkney grew too hot for him, and he left later that year, never to return. George Spence continued to minister in Kirkwall through to his death in 1720.

The Orkney Kirk received scant reward for its conspicuous loyalty. In March 1717, the ministers of the Kirkwall and North Isles presbyteries drafted a petition to King George, asking for revenues from the bishopric estate to be diverted to the cathedral fabric, parish schools and augmentation of ministers' stipends. One portion would be set aside to cover 'receiving intelligence, dispatching expresses', and whatever was 'necessary for the support of the government in that remote corner' – the ministers fancied themselves the eyes and ears of Hanoverian rule in an unreliable part of the realm. It was a slap in the face when the profitable tack was instead granted to James Moodie the Younger, suspected Jacobite and scourge of Orkney's 'giddy-headed Gospel-mongers'.

In the aftermath of the rising, life became uncomfortable for religious dissidents. In November 1723, the government ordered Morton to convene all non-Presbyterian pastors in Orkney and Shetland, and imprison them and close their meeting-houses if they did not conform to the letter of the law about oaths and prayers for the royal family. But the Toleration Act itself was not rescinded, and after a while a meeting-house in Kirkwall was reopened. From the presbytery's perspective, its new minister, William Harper, was no improvement on Lyon: 'unqualified, a declared Jacobite ... prays not for our Protestant King George'.

In April 1731, the presbytery learned that 'Mr Harper goes to other parishes in this country, particularly to that of Burray'. There, on the orders of Lady Burray, the kirk officer 'charges the people of the parish to come to hear Mr Harper and not the minister of the parish'.[35] The minister was Richard Mein, formerly of Cross and Burness; now, at nearly seventy-five, he was probably no great enthusiast for the choppy boat trip across from South Ronaldsay.

Lady Burray was the imperiously Episcopalian Anne Carmichael, daughter of an Aberdeenshire laird, and wife to Sir James Stewart, Jacobite cheerleader of 1715. She was managing the business of the estate while her husband was away from Orkney, hiding from the law and negotiating for a safe return. The story of his travails, painstakingly reconstructed by the historian Ray Fereday, demonstrates how island politics, well into the eighteenth century, remained a matter of personal feuds, visceral and violent. It shows, too, how local alliances traversed national lines of religion and politics, even as they sought to navigate those lines for material and social advantage.

Stewart was not, in fact, on the run as a result of the rebellion. Although Episcopalian and a Jacobite, he was a sworn adherent of the Hanoverian earls of Morton. Robert Douglas, twelfth earl, had been Stewart's guardian on the death of his father, and protected him in the aftermath of the rising. For his part, Stewart supported Morton in the parliamentary election of 1722, when the earl's brother George Douglas came back as a candidate to challenge the incumbent, James Moodie the Younger, who had managed to take the seat in March 1715.

Moodie was determined to hold it – not least because membership of parliament conferred immunity from prosecution. At an election meeting in April 1722, he protested about ineligible voters, supporters of

Douglas, being added to the electoral roll by the stewart-depute, Robert Honyman. To intimidate the electors, Moodie brought from Scotland a band of thirty armed highlanders, and billeted them in the Palace of the Yards. But the majority of lairds were now against him, including his uncle, who had consolidated a grip on the Melsetter estate. The defeated younger Moodie stormed out, and soon left Orkney to embrace the life of an unrepentant Jacobite exile, converting to Catholicism and serving in the army of Bourbon Spain.

Both Moodies were ill-disposed towards Robert Honyman, one of the sizeable number of Orkney lairds descended from a seventeenth-century bishop. They managed to keep out of his grasp a lady called Christiana Crawford, widow of William Bellenden of Stenness, on whose lands Honyman exerted a claim. In an unregistered, probably Episcopalian ceremony, Christiana married the septuagenarian James Moodie the Elder, and in February 1723 gave birth to a son, Benjamin – though which of the Moodies was actually the father was a matter of salacious speculation.

For a few weeks at the start of 1725, Honyman had other things to worry about than the enmity of Captain Moodie. On 10 February, his house at the Hall of Clestrain, on the west coast of Orphir, looking across to the little port of Stromness, was ransacked by pirates. The pirates' captain, John Gow, grew up in Stromness, and since the middle of January his ship had been anchored there, masquerading as a merchant vessel. Gow told people in Stromness his ship was called the *George*, but its real name was the *Revenge*.

It was the *Caroline* prior to November 1724, when second-mate Gow led a mutiny at Tenerife, murdered the captain, and embarked on a short campaign attacking ships off the coasts of Portugal and Spain. Gow came to Orkney thinking it a safely anonymous place to restock – and, like other pirates before him, he judged its coastal communities to be easy pickings for determined and ruthless men. The attack on the Hall of Clestrain followed exposure of the ship's true identity, after one of the crew deserted and alerted the Kirkwall authorities. Gow sailed next to Calf Sound on the north-east coast of Eday, where, a century earlier, the earl of Carrick had established his 'burgh' (p. 244). On the way, the earliest accounts agree, the crew kidnapped two or three young women, and repeatedly raped them – a corrective to any temptation to glamorise the actions of pirates in general or these ones in particular.

Gow's intention was to raid Carrick House, home to the laird James Fea. The plan went badly wrong. Gow expected Fea to be away, but his wife's indisposition found him in residence. Gow's ship sailed too close to the shore of the Calf of Eday and became stranded. After a few days, and the exchange of some remarkably courtly letters, he felt forced to surrender to Fea – a shrewd manipulator of the pirates' weaknesses, and by chance Gow's old schoolfellow. Gow was hanged at Newgate that summer, his deeds widely publicised in a history of pirates that may have been written by Daniel Defoe. In the meantime, leading gentlemen travelled to Eday to congratulate Fea on his triumph. They included James Stewart of Burray, but not James Moodie of Melsetter.[36]

The Stewart–Moodie feud had long been simmering – stoked by quarrels over lands and ministers' stipends, as well as an absence of empathy between the refined Jacobite baronet and the blunt Hanoverian sea-dog. It came to the boil in the year of Gow's capture, when Sir James's brother Alexander Stewart landed without permission to shoot wildfowl on the Moodie estates in Walls. In a fit of anger, the captain had him beaten. According to tradition, Stewart and his servant Oliver Irving were stripped and thrashed with tangles – long-stalked seaweed – from the shore.

Shortly afterwards, Captain Moodie came to Kirkwall to renew his appointment as a justice of the peace, ignoring warnings that the Stewarts were bent on revenge. On 26 October 1725, Moodie set out for St Magnus Cathedral, in the company of Sheriff-Depute Honyman, who had given guarantees about his safety. The Stewart brothers, armed with sword and pistol, were waiting in a house opposite the cathedral, intent on delivering a beating. There was a violent confrontation, in the course of which Oliver Irving discharged his pistol and hit Moodie in the shoulder. Witnesses heard Sir James call for him to fire again, 'for the old dog stands yet'. Honyman made no serious effort to detain the assailants, and six days later, Moodie died of his wound.

The murder caused a furore, and, coming so soon after news of the depredations of Gow, probably fuelled perceptions of Orkney as a desperate and lawless place. On 23 November, James Traill wrote from

Edinburgh to his brother-in-law, David Traill of Elsness: the affair 'makes a terrible noise here, and a reflection on our country' – the country of Orkney.

Honyman, at the instigation of Moodie's widow, Christiana Crawford, went on trial in Edinburgh for complicity in murder and negligence in his duties. The charges, on a majority verdict, were 'not proven'. The Stewarts and their servants fled, and no one was ever brought to book. In May 1731, Sir James Stewart received a royal pardon – a result of lobbying by Robert, earl of Morton, and by his brother George, who succeeded to the title on Robert's death in January 1730.

It was to be expected that Sir James would show gratitude to his benefactor. At the General Election of May 1734, Stewart, along with Gow's captor, James Fea of Clestrain, asserted a right to be included on the electors' roll and voted for Morton's candidate. This was the earl's son, Robert Douglas, who had already taken over the seat upon his father's elevation to the peerage – the Mortonian attitude to parliamentary representation was nothing if not candidly nepotistic.[37]

Sir James Stewart, however, would soon reveal himself to be committed to his own rather than his patron's interests. His challenge to Mortonian power would draw him into physical confrontation with the earl himself, and escalate into a social insurgency that lasted for well over twenty years and generated one of the most complex and rancorous cases in Scottish legal history. It drew Orcadians into sustained reflection on the islands' Scandinavian past, and showed that, while Denmark's claims might be moribund, Orkney was not yet ready to become just another Scottish county.

The Prince and the Pundlar

It was, naturally, a matter of money. Not long after returning to his mansion at the Bu of Burray, Sir James Stewart came to the conclusion he was paying too much in 'superior duties' – the mix of rent on earldom land, skat and other miscellaneous charges owed to the earl of Morton by the heritors of Orkney. Rents and fees were nearly all settled in kind, so it was literally a matter of how the payments were weighed. Orkney's traditional weighing-beams – the bismar, used for measuring butter, oil and cheese, and the pundlar, for heavier quantities of bere and malt – were supposed to be of a standard pattern. Concerns about them had been expressed before (p. 141), but variation and inaccuracy were now endemic. People suspected wily Kirkwall merchants of keeping two sets of bismars, one for buying and one for selling.

The trouble was triggered by procedural changes brought in by the earldom's factor or estate manager, John Hay of Balbithan. In around 1730, he began having butter weighed in the earl of Morton's storehouse, rather than accepting it by the barrel. He also insisted on his own servants weighing the superior duty of bere, instead of subcontracting this to a 'king's weigher' on each of the islands.

Stewart protested, and started putting it about that weights had over many years been deliberately increased to the earl of Morton's advantage. He hired as his own factor and legal adviser James Mackenzie – a great-grandson of the bishop, and an indefatigable antiquarian who began amassing huge quantities of evidence about weights, measures,

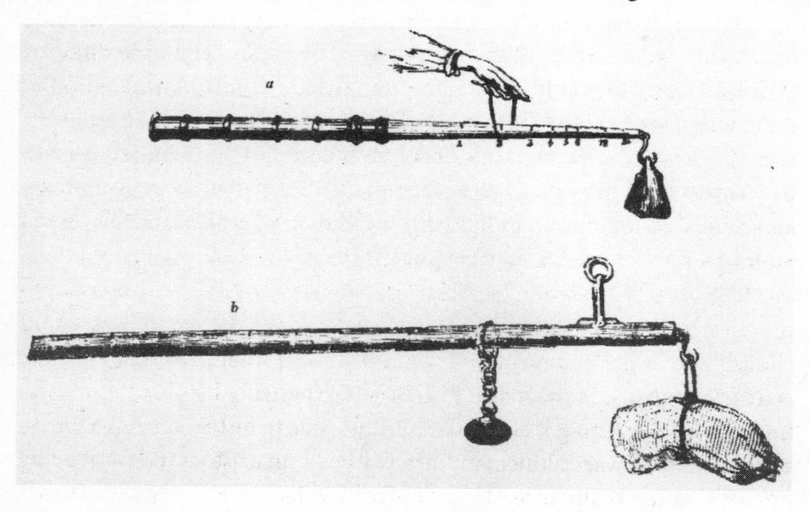

Figure 7.3: Orkney's controversial weighing beams – the bismar (above) and pundlar (below), from an original drawing by the minister and historian George Barry. The 'pundlar process' convulsed Orcadian society between 1735 and its final settlement in 1759.

laws and duties from the Orcadian past. In 1735, Stewart initiated an action in the court of session with the aim of reducing the superior duties by more than half, and thereafter withheld his own contribution. Within a few years, another seventeen Orkney lairds had followed suit and undertaken to oppose the earl of Morton, prominent among them the pirate-catcher, James Fea of Clestrain.

In the summer of 1739, the absentee earl of Morton came to Orkney to rally support. James Douglas had succeeded to the title only the year before, but he was familiar with Orkney affairs. Throughout the 1730s, he had advised his father on Balbithan's management of the estate, and on ecclesiastical patronage. As Lord Aberdour, Douglas had several times been chosen as Orkney's representative elder for meetings of the General Assembly. Morton's commitment to Presbyterianism whetted the edge of his conflict with the assertively Episcopalian Stewart, and he gathered evidence of 'hardships and severities' suffered at the hands of the baronet by tenants in South Ronaldsay. Stewart was incensed at this interference in his estate management, and on 1 August 1739, with three boatloads of armed men, confronted the earl as he was about to dine at Graemeshall.

It was very nearly a replay of the 1725 Kirkwall homicide. Stewart seized Morton by the throat, and servants on both sides were injured in an exchange of gunfire. The matter ended in the court of justiciary, where Stewart and his entourage were convicted of assault and obliged to pay heavy fines to the earl in compensation. Morton made a symbolic statement about crime and order by donating £200 sterling from Stewart's forfeit to the Kirkwall Council for a new town gaol.

Morton feared that the 'pundlar process' might have the potential to loosen his grip on Orkney. In a worst-case scenario, the crown could choose to reannex the earldom, or grant it to another proprietor. At the start of the 1740s, Stewart and his allies, 'perpetually hatching stories to hurt me', were putting it about that Orkney was to be taken from Morton and given to Stewart's kinsman Lord Garlies – an aristocrat described by the wit Horace Walpole as 'a man indecently Jacobite'.

The earl, however, had connections in the British state that the baronet of Burray lacked. His brother Robert, re-elected in 1741, remained Orkney's member of parliament, and Morton was on good terms with both the 'Court Whig' prime minister, Robert Walpole, and his 'Patriot' rivals, William Pulteney and Lord Carteret. No objections were put in the way of a private act of parliament, passed in 1742, which removed the royal right to redeem the islands and granted them 'heritably and irredeemably for ever' to Morton and his heirs. This, Morton believed, would allow him, with the agreement of the heritors, to 'settle the weights and measures and stop all grounds for that clamour'.

This proved unduly optimistic. A series of meetings held in Kirkwall in 1743–4, convened by the earl's stewart-depute Andrew Ross, summoned a plethora of witnesses and closely inspected weights, pundlars and bismars. But far from winning over the dissident lairds, the proceedings simply escalated the levels of rancour. The vigour of protest and petitioning by Thomas Traill of Westove in Sanday was such as to cause Ross to issue a warrant for his arrest. Throughout the process, Morton and his supporters, while indignantly denying weights had been tampered with, were ready to concede the imperfections of the old bismars and pundlars. They advocated replacing the whole system – the arcane Norse measurement in lipsunds, setteens, meils and lasts – with standardised weights and quantities used in the rest of Scotland. Not unreasonably, the protestors assumed that reform was intended to work to Morton's advantage, and pledged themselves to the old ways.[38]

While the lairds of Orkney bickered, French privateers prowled round their coasts. From 1743, Britain and France were once more pitted against each other, as antagonists in the War of the Austrian Succession. In May 1745, the war exacted a notable casualty – Robert Douglas, MP for Orkney and Shetland, the earl of Morton's brother, was killed at the Battle of Fontenoy. It was a personal tragedy and a political peril: the anti-Mortonians had been discussing ways to get more supporters onto the electoral roll and prise the parliamentary seat from the earl of Morton's grip; a by-election would be their opportunity.

The by-election was postponed, however, swept aside by greater events. In the summer of 1745, supporters of James VIII and III, the 'Old Pretender', made another attempt to seize back the throne. On 19 August, James's son, Charles Edward, the 'Young Pretender', raised his standard at Glenfinnan in the western Highlands, and began assembling an army from the disaffected Scottish clans. He was soon in control of Edinburgh, and after beating a Hanoverian army at Prestonpans a few miles east of the city, advanced into England.

Orkney's part in the Rebellion of 1745–6, the last major military campaign fought on British soil, was peripheral but not inconsequential. The Jacobite leadership had hopes for the islands, and was aware of how the prospects for support there were linked to local disputes. An agent sent to the north of Scotland in 1741 reported that the earl of Morton had the royal revenues of Orkney and Shetland, and 'could be persuaded to join' if he was guaranteed continued possession of them. But even without him, Sir James Stewart of Burray, 'with Grahams, Douglases, Gordons and others', could be counted on to supply a thousand men.

There was, in the event, no such detachment forthcoming, though a Jacobite newsletter announced satirically in October 1745 that the prince's army had been reinforced by a strong contingent of Orkneymen, bringing hundreds of ferocious dogs trained to fight by instruction from the bagpipes. As the campaign moved south, the islands were left further from the action, and cut off from the prince's forces by loyalist troops under the earl of Sutherland and Lord Loudon, who controlled the coastal region between the Ord of Caithness and Inverness.

The start of 1746, after the retreat of the prince's army into Scotland, witnessed another Jacobite victory at Falkirk Muir, but the war in Orkney remained one of words. On New Year's Day, the Kirkwall Presbytery announced a fast to petition God against the Jacobite cause, noting darkly that 'there are not only in the nation but even in this corner thereof a party inclining [to] the interest of a popish pretender'. At the Bu of Burray, meanwhile, Sir James and his wife sported white cockades, and entertained the officers of a Spanish brig sheltering in Scapa Flow en route to deliver arms to the Jacobite army.

Sir James Stewart's contribution to the Jacobite cause, in 1745 as in 1715, was more about performance than substance. Not so with the captor of Gow, James Fea of Clestrain. Early in 1746, Fea travelled south, securing an audience with Prince Charles, and offering to raise an Orkney regiment. Fea's motives were undoubtedly mixed. Like many Orkney lairds, he was instinctively a Jacobite, but he saw in the return of the House of Stuart a means to confound the earl of Morton, and to settle the pundlar process in the complainants' favour. This quid pro quo was put directly to the prince, who told Fea 'he had heard of the Orkney grievances', and undertook to redress them. Charles was more reluctant, however, to agree to Fea's appointment as commander of the Orkney regiment. That distinction surely pertained to Sir James Stewart, 'a man of the greatest fortune and consideration in the country'.[39] Social rank, rather than talent, was the key criterion for leadership in the Jacobite army.

The northward retreat of the prince's troops drew the potential of Orkney to both government and rebel attention. In late March, the duke of Perth, Prince Charles's second-in-command, led his men into Sutherland, as rumours swirled that 6,000 French soldiers were about to land in either Caithness or Orkney. On 31 March, Lord MacLeod, one of Perth's regimental commanders, arrived in Thurso and sent a company under Alexander Mackenzie of Ardloch to Orkney to raise money and recruit soldiers. At the same time, a Royal Navy frigate, HMS *Sheerness*, sailed into the harbour at Stromness, and seized pro-Jacobite ships.

There was no direct confrontation, but during a brief military occupation, the Jacobites made their presence felt and encouraged supporters to show their hand. The leading loyalists, Morton's sheriff-depute Andrew Ross, and the provost of Kirkwall, James Baikie, fled to Shetland, as rebels plundered the grain stores of the earldom estates.

If the Jacobites expected a flood of eager volunteers, they were to be disappointed. Few came forward, despite assurances that Jacobite victory would spell the end of Morton's feu duties. James Fea sought to persuade a reluctant Ardloch to press men into service, blurting out in frustration: 'the Orkneymen were slaves and, by God, they should be slaves'. Fea was also heard complaining about the ineffectual Sir James Stewart – 'no cause could thrive or prosper wherein he was embarked'. The Jacobite lairds of the North Isles did, however, advance the handsome sum of 100 guineas, and they drafted a letter to Lord MacLeod explaining that, while Orcadians were not natural soldiers, they made ideal sailors, and would be willing to serve in that capacity when called upon to do so.

The Jacobite troops left just under a week after arriving in Kirkwall, returning to Caithness via Stromness and Hoy, a route made safe after the departure of the *Sheerness* with its prizes. For the Orcadian Hanoverians, there was a sharp parting blow – in Walls, the highlanders looted Melsetter House, taking all of value they could carry and destroying what they could not. The young laird of Melsetter, Benjamin Moodie, was serving as an officer in the government army. He had no doubt that this was the malign doing of his father's murderer, James Stewart of Burray.

Moodie did not have long to wait for revenge. Ten days after Mackenzie of Ardloch pulled out of Orkney, Stuart dreams of restoration ended in carnage on Culloden Moor. The day before the battle, government militia defeated Jacobite forces in Sutherland. In retaking Dunrobin Castle, they discovered incriminating papers, including the offer of manpower for a Jacobite navy, signed by Archibald Stewart of Burgh, William Balfour of Trenaby (Westray), John Traill of Elsness (Sanday) and John Traill of Westness (Rousay). The duke of Cumberland, vengeful victor of Culloden, was determined they should all be punished.

An additional motive for sending ships and men to Orkney was Cumberland's concern that the islands might supply an escape route for the fugitive Charles Edward Stuart, just as they had for the Jacobite leaders in 1716. Rumours swirled through the spring and summer of 1746 that the prince had indeed gone to Orkney.

In fact, Prince Charles was in the Outer Hebrides, but the reports were not wholly groundless. A vessel was hired at Stornoway in Lewis to take the Young Pretender to Orkney and then on to France, but the scheme was abandoned when news of his presence became known in the vicinity. An alternative plan, to attempt the journey to Orkney in an open boat, fell foul of the Hebridean rowers' refusal to venture so far from their homes.

Orcadian Jacobites who accompanied the highlanders south were now also fugitives. William Balfour, Patrick Fea of Airy and others returned discreetly in the early part of May, but Patrick's kinsman James Fea remained in concealment in Caithness. He was right to be wary. A squadron of four warships was dispatched to Orkney, with a warrant to Captain Lloyd of HMS *Glasgow* directing him to destroy by fire the house at Sound in Shapinsay of the 'notorious rebel', James Fea of Clestrain – a task performed with grim relish on 10 May 1746.

This was only the beginning, and Sir James Stewart's hope that he might rest a little more easily in Burray after the departure of Lloyd's flotilla proved to be misplaced. A second warrant was issued to Christopher Middleton, an officer acquainted with Orkney and captain of a sloop with an appropriately predatory name, the *Shark*. His orders were to arrest Stewart and the signatories to the letter taken at Dunrobin, 'and to lay waste, take and destroy' any lands, cattle, houses or boats belonging to them. With Middleton travelled a contingent of marines, their commander hand-picked for his loyalist zeal and local knowledge – the newly promoted Captain Benjamin Moodie.

What ensued was as much the pursuance of a generation-old gentry feud as an orderly suppression of rebellion. Stewart left it too late to flee from Burray, and on 25 May was taken by Moodie and his marines in the course of attempting to do so. Moodie was determined to savour his cold dish of revenge, parading Stewart and other prisoners along the 'Broad Street' running in front of the cathedral, where twenty-one years earlier his father was fatally wounded, before incarcerating him in the tolbooth the baronet had been obliged to pay for after his 1739 fracas with Morton. Stewart was then marched across the Orkney Mainland to Stromness, for shipping to London and trial for high treason.

Stewart, though, had the last laugh, after a fashion. Two weeks after his confinement in the New Gaol, Southwark, he contracted typhus and died. There was no treason trial, and no forfeiture of his estates. These passed – along with leadership of the pundlar process – to his heir, the former Lord Garlies, now earl of Galloway. Moodie and other Mortonians suspected that Stewart took his own life just to thwart them.

The other wanted men remained at large. Balfour, Stewart of Burgh and the two Traills ignored an ultimatum to present themselves in Kirkwall, and so Moodie set off on a cruise of vengeful destruction through the North Isles of Orkney, looting and torching in turn the fugitives' houses of Westness in Rousay, Trenaby and Cleat in Westray, and Elsness in Sanday.

The lairds themselves evaded capture; the midsummer light made it impossible for boats to approach the isles unobserved. The fugitives banded together and spent the winter of 1746–7 hiding in a cave, forty feet down a precipitous cliff-face on the south-east coast of Westray, their daily supplies delivered by the loyal servant of a local tenant. Within a few years, the cave was being pointed out to visitors. Along with another cave in Westray that the lairds possibly used, it became known as 'Gentlemen's Ha' [Hall]' – a characteristically Orcadian blend of reverence and irony.[40] It encapsulates the tendency in Orkney for the landscape to serve as a repository of historical memory, the transient deeds of the islanders absorbed into the enduring elements of the earth.

The Brothers Mackenzie

Over the course of a discordant century, the history of Orkney played in lively counterpoint to the dominant refrains of Scottish and British politics, sharing their key- and time-signatures, but with themes and melodies attuned to local ears. National developments were followed closely in a cluster of islands that were rarely as 'remote' from events as contemporaries sometimes supposed. But Orkney was never merely a 'locality', where decisions taken elsewhere exerted a greater or lesser impact; rather, it was an active arena of political contest and competition, with its own agendas, actors and agents.

The defeat at Culloden, and the ensuing brutal suppression of Jacobite clans and Gaelic Highland culture, ended all realistic prospects of the Revolution Settlement of 1688–9 being overturned. But it did not extinguish hopes among Morton's Orkney opponents that they might yet prove victorious in reducing the superior duties.

After 1746, the pundlar process was energetically continued under the leadership of the earl of Galloway, with James Mackenzie continuing as chief adviser and advocating ever more ambitious lines of attack. The 'pursuers' now argued that skat was no part of superior duties, since it was in its origins a land tax, and long superseded by subsequent government impositions. Not only had scales been tampered with, but Orkney's weights and measures themselves, particularly the 'mark', had been wilfully miscalculated by Morton and his agents. The arguments were laid out in multiple submissions to the court of session, and in Mackenzie's *General Grievances and Oppression of the Isles of Orkney and Shetland* (1750), which told a poignant story of the islands' centuries-long sacrifice at the hands of tyrannical and self-interested court favourites.

Orkney, Mackenzie insisted, was never conquered by Scotland, but passed to it by treaty. It meant that Norway's laws and customs had remained in force, and that uniquely Norwegian weights and measures still rightfully pertained there. Making the case involved much meticulous discussion, not just of bismars and pundlars, but of roithmen, lawrightmen and the lawthing, of udal landholding, and of the tenacious tenure of the Norn language itself – several witnesses were called to testify to hearing it spoken in Orkney, either currently or within recent memory. To prove that 'skat' meant tax or tribute, Mackenzie quoted the Bible in Danish-Norwegian (Matthew 22:17) on rendering unto Caesar, 'Er det tilladt at give Keyseren skat, eller ey?' A written submission from the *borgermester* of Bergen was produced in court, and five Norwegians appeared in person, attesting to how a Norwegian mark was well known to be substantially lighter than a Scottish one.

Mackenzie disavowed any intention 'to contend for the rights of the King of Norway', but so radical an assertion of legal, linguistic and cultural distinctiveness was a bold strategy to pursue before a Scottish court, particularly in light of the Jacobite track records of several of the parties involved. The earl of Morton's defence team mocked Mackenzie's 'gothic learning', and questioned whether it was even legal to advocate regulation of weights 'by a different standard from that which is used in the country' – meaning Scotland, rather than Orkney.[41]

In June 1759, almost a quarter of a century after its commencement, the court of session case concluded with judgement in the earl of Morton's favour, and an order for crippling legal costs to be borne by the pursuers. 'Some of those heritors who were induced to concur in the process', Morton commented drily, 'begin to repent of their rashness.'

Morton, a busy president of the Royal Society, with many interests to pursue elsewhere, was growing tired of querulous Orkney lairds. In 1766, he accepted a cash offer of £63,000 for the earldom from Sir Lawrence Dundas, an Edinburgh merchant who had made a fortune as a military contractor. Soon afterwards, Dundas purchased the earl of Galloway's Burray estate, and acquired a lease of the bishopric lands. A substantial weight of social and political power in Orkney was, for the foreseeable future, to remain in the hands of an absentee southern 'proprietor'.

One lasting legacy of Morton's tenure was his patronage of James Mackenzie's elder brother. Murdoch Mackenzie was a former master of Kirkwall Grammar School, and a soon-to-be renowned cartographer and hydrographer – a measurer of the seas. The maps reproduced in Mackenzie's *Orcades* of 1750 were of astonishing and unprecedented accuracy. They employed innovatory techniques of triangulation to produce precise surveys of the land, reckonings designed to make for safer navigation of the surrounding waters. Less overtly political than the maps we considered at the start of this book (pp. 16, 21), Mackenzie's charts – which plot the winding course of old hill-dykes, and record the names of ancient townships – are an invaluable asset for social historians of Orkney.

In their parallel publications of 1750, the Mackenzie brothers put forward sharply contrasting visions of Orkney's past, and of its place in an emergent notion of the nation. James placed his emphasis on the historic rights, privileges and exemptions of an independent-minded provincial society; on the importance of formal recognition for the distinctive character of an ancient 'community of Orkney' (pp. 33–4), albeit one that had absorbed and nativised successive waves of incomers, like his episcopal great-grandfather. Only when external forms of oppression and exploitation were addressed, and redressed, could the islands fulfil their potential as a 'valuable limb of the British dominions, if so indeed they may rightfully be counted'.

Murdoch Mackenzie was, like his brother, a patriotic Orcadian, but one who felt less ambivalently about a supporting and subordinate role for the islands in the Hanoverian-British state. In his introductory essay to *Orcades*, Murdoch glossed over the gruelling pundlar process as a quarrel on 'slight causes'. With considerable inventiveness, he spun the failure of the Jacobite expedition to raise recruits in Orkney as a proof of loyalty 'not to be paralleled in any part of North Britain' – a description of Scotland favoured by keen unionists. Orcadian manners and customs, he believed, 'resemble those of the southern, rather than of the northern parts'. Mackenzie lamented how 'these islands are little known, and therefore concluded to be a very insignificant part of the kingdom'. In hopes of remedying this, he itemised how, with their natural harbours, and their advantageous position for trade with northern Europe and the

American colonies, the Orkney Islands might be 'rendered of great service and benefit to Britain'.[42]

Whether, over the course of the preceding ninety years, Britain had been of any great service and benefit to Orkney is another question. By the mid-eighteenth century, the islands were no longer a bargaining chip in negotiations with foreign powers, and had in various ways become more closely integrated into a national political community. But to powerful people in Edinburgh or London – inclined to regard the distant isles only as a useful financial resource, and an occasional military or strategic headache – the real concerns of islanders registered but faintly.

The ordinary people of Orkney, rather than their local leaders, have remained relatively quiet in this chapter. In the one which follows, we will seek to redress that imbalance, and to discover if a century of dramatic political changes in the islands was also one of profound and lasting transformation in society, culture and belief.

8

Sin and Selkies

The Oyce

Her baby was taken by a creature that came out of the sea. That was the story Margaret Linay told to the ministers of the Orkney Presbytery in June 1628. It was her second appearance before them: she had been accused in the previous year of committing adultery with John Sclater, udaller in Harray, a married man and her employer. The case was referred to the presbytery by the kirk session in Margaret's home parish of Evie, at the instigation of the minister, James Morison – as we have seen (p. 301), no stranger to adultery himself.

Margaret had confessed the sin, and admitted to being with child. But as her time of delivery approached, she fled to the parish of Firth. When she was discovered, and hauled back in front of the presbytery, the question was what had become of the baby. At first, contradicting her earlier confession, Margaret denied ever being pregnant. Then she revealed to them a terrible secret: 'that ane like ane man came out of the sea and took her bairn from her'. The ministers concluded there were reasons to suspect her of 'the maist barbarous, odious and unmerciful murder of her ane bairn', and referred the case to the civil magistrate. In January 1629, Margaret was put on trial for her life in the sheriff court at Kirkwall.

At the hearing, Margaret changed her story once more, saying the baby's abductor 'came out of the hill'. She added that 'he was clad in grey clothes and took her to the hill with him, and that there was many folk in the hill with her'. Margaret went back under the hill a further four times in the following days, even though 'she was sick as women use to be when they are after their childbirth'.

The assize remained unmoved, and found Margaret guilty of infanticide. The judgement of the court was that she should be taken with bound hands 'to the Oyce mouth'. The Oyce – from Old Norse *óss*, entrance to a river or lake – was the sheltered harbour inlet to the west of Kirkwall, known today, in its reduced extent, as the 'Peedie [little] Sea'. Margaret was to be marched to the end of the Ayre, the spit of land separating the Oyce from the wider expanse of Kirkwall Bay, 'and there drowned in the sea to the death'.[1]

It is an unbearably sad little story. Margaret Linay was in every likelihood poor, young and vulnerable, a servant girl seduced or pressured into sex by her master. No records survive to say whether Sclater was disciplined for his moral lapse, but the incident reeks with inequalities of power, and the double standards that could drive a shamed and frightened young woman to desperate deeds and astonishing explanations.

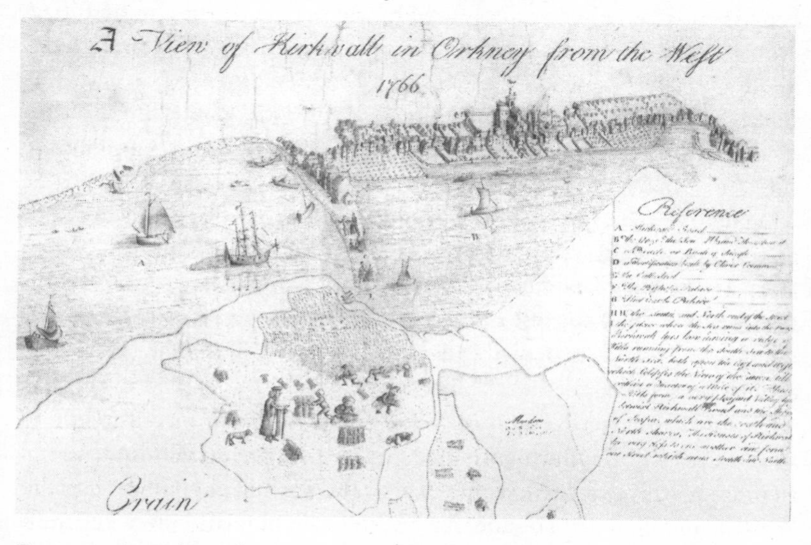

Figure 8.1: An eighteenth-century map of Kirkwall showing the harbour to the left and the Oyce (today's 'Peedie Sea') to the right. In the centre, figures walk along the Ayre, the route down which Margaret Linay was escorted to her death in 1629.

We will never know if Margaret Linay believed her own version of events, but her stories were authentic irrespective of whether they were true. Her fate is an instructive study in how the accidents of individual lives became entangled with structures of authority, and in how narratives became charged with perilous meaning, in a society exposed to pressures of profound change.

Margaret was not the first Orcadian to tell a sceptical courtroom about a child stolen by malevolent spirits. In May 1624, little James Clouston disappeared in Orphir, five weeks before his second birthday. He had been left in the care of his grandfather, another James Clouston, while his parents went to plough. James disliked his daughter-in-law and had 'often bidden her go hame to her father'. A few weeks after the child's vanishing, his bloodied clothes and a piece of his 'puddings' (entrails) were found beyond the hill-dyke. Clouston was convicted of murder, and hanged at the Loan Head (p. 182). But in court he repeated what he had told James's mother the day of the disappearance: 'he wist not [did not know] what became of him – the fairy or the serpent had ta'en him'. The identity of this 'serpent' is uncertain, but there were many Orkney legends of monstrous creatures prowling around the encircling seas.[2]

To call these stories about mysterious hill-dwellers and amphibious abductors 'folklore' is to distance and diminish them. They were part of a living and changing system of perception and explanation, ways of accounting for good and ill fortune – though we should be wary of imposing on them any neat 'functionalist' explanations of the kind once favoured by scholarly anthropologists, sometimes guilty of focusing on the putative social purpose of beliefs at the expense of their meanings for those who held them.

The beliefs and behaviour of Orcadians came under increasing pressure in the centuries during which Protestantism put down roots and the islands were drawn further into the legal and political embrace of the Scottish and then British state. As we have seen in relation to witchcraft, the mostly Scots ministers serving in Orkney had their own ideas about sacred and supernatural power. They were eager to effect thorough and lasting 'reformation', in doctrine, deeds and demeanour, of the people committed to their charge.

The primary instrument of this was the parish kirk session. The session records of the seventeenth and eighteenth centuries illuminate what Buckham Hugh Hossack, eminent Victorian historian of Kirkwall, derisively designated the 'rule of the Church' – a prurient and tyrannical policing of the most intimate aspects of everyday life. The story this chapter tells, however, is not simply one of what historians call 'acculturation' – an enforced assimilation of popular culture to the values of social and religious elites. Session and other records reveal complaint as well as compliance, and a sometimes remarkable resilience in patterns of thought and behaviour. They show, too, how the character of island communities formed and shaped, as well as reflected, the pronouncements and priorities of the Kirk.

Out of the Sea

The greatest constant was the presence of the sea. Every Orkney parish, bar Harray, had boundaries of meandering coast and horizons of blue-grey water. For Orcadians, the sea was a barrier, a highway, a storehouse, and a graveyard. The dimly remembered tragedy recounted at the start of this book (p. 23) was regularly re-enacted over the succeeding centuries. In December 1663, powerful north-east winds suddenly struck the waters off South Ronaldsay and – as the session recorded – 'an sad accident fell out by loss of ane boat and four men'. In the summer of 1683, there were collections throughout Orkney for seven widows in Westray, 'whose husbands did shortly perish by sea … having the number of twenty-three children amongst them'. A still greater disaster befell Graemsay in December 1739: a boat carrying fourteen men on a seal-fishing expedition to Sule Skerry was lost, and 'there are above 60 more orphans and widows in that island'.[3]

It was wise to treat the sea with respect. People thought it foolhardy to launch a vessel 'withershins' (p. 216). Further precautions included sprinkling boats with 'forespoken water' (a magical preparation to guard against 'forespeaking'; that is, bewitchment), or at Halloween marking them in tar with the sign of the cross. Such practices were reported with dismay by the General Assembly's commissioner to Orkney, John Brand, after his visit in 1700.

When it came to the sea, the wary egalitarianism of Orcadian peasant culture was strongly in evidence. In October 1660, William Rosie was 'delated [denounced] for superstition' to the South Ronaldsay Session, after preventing James Windwick from catching any more fish 'till his boat should take as many'. Windwick, whom we have seen already being protective of his fishing (p. 217), was a kirk elder, and perhaps affronted by the impiety as well as the impairment of his profits. But the instinct ran deep. James Emerson, the caustic Cromwellian chronicler of Orcadian society, described how fishermen rowed out in pairs, yet assiduously divided each day's catch:

Should they neglect to give the boat her share,
The next day's fishing would not be so fair.
And if, while they are undivided, first
A priest but sees them, all the fish are cursed.

It was believed throughout Europe that having a clergyman near boats or on the sea brought ill luck, but the notion is especially well attested for the Northern Isles. Later folklore collectors reported a reluctance of Orcadian sailors and fishermen even to speak of 'the minister', preferring such euphemisms as 'upstander', 'gentleman with the black coat', or 'white-throat'. An alternative was 'baeniman', just as the equally taboo 'kirk' could safely be rendered the 'baen-hoose' or 'baeni-hoose'. In Orkney (and still more in Shetland), Norn terms and phrases disappearing from everyday speech peppered the vocabulary of cautious and conservative fishermen; John Brand required explanations of 'Norish words' for coastal features, like 'voe' and 'oyce'.

Reasons for the ministerial taboo were never explicitly articulated, but perhaps reflected a longstanding perception (p. 233) that the sea did not really 'belong' to the Kirk or the Christian God; its ruling powers should not be needlessly provoked. An unsettling belief likewise widely associated with the Northern Isles was that it was unwise to assist anyone in danger of drowning. The sea, denied its expected sacrifice, would inevitably demand a replacement.[4]

Not all those lost at sea were drowned; some were seized by its magical residents. Sea-trows we have heard about already (p. 191) – malevolent spirits with an ability to enter into human bodies. The name was also applied to 'great rolling creatures, tumbling in the waters', which Brand described tearing the nets of fishing boats. The crews would sometimes say, 'it is the Devil in the shape of such creatures' – a sign, perhaps, that by the end of the seventeenth century the doctrinal prescriptions of the ministers were making a mark on the people.

Sea-trows were seemingly not the same as other oceanic beings with a potential resemblance to humans. The mermaids and mermen who from the time of classical antiquity had tantalised European imaginations took a distinctive form in the island world of the North Atlantic. Here, they were associated with the region's teeming population of carnivorous marine mammals – the inquisitive, dog-like creatures who track the progress of walkers along coastal paths, and swim ashore to bask contentedly on rocks and pebble beaches.

'Selkie', which was simply the common Scots word for a seal, was made by nineteenth-century folklorists into the proper name for a member of a supernatural species, able to shed its skin and walk in human shape. Whatever they were originally called in Orkney, belief in them was ancient and widespread.

It was in the selkies' nature to approach and cross boundaries: between sea and land, animal and human, spiritual and material – and between licit and illicit sex. Many stories were collated, and possibly embellished, in the nineteenth century by Walter Traill Dennison: the selkies lived on some distant outlying skerry; if a young woman was lost at sea, she was likely not drowned, but taken there by the selkie-folk; prudent island mothers would paint the sign of the cross on their daughters' chests, before travelling with them to the Lammas Fair in Kirkwall.

There were male and female selkies, both liable to assume human likeness during certain configurations of the tide. The males came 'to indulge unlawful love ... handsome in form and attractive in manners, they often made havoc among thoughtless girls, and sometimes intruded into the sanctity of married life'. The selkie-women's beauty caused men to fall in love with them, but they were not all-powerful sirens. A tale existing in several poignant versions tells of a selkie forced to marry an Orkney farmer, and bear him many children, after he discovers and conceals her divested sealskin. Only when the hidden pelt comes unexpectedly to light is she able to return to her own kind.[5]

There was also wary talk of 'fin-men'. Traill Dennison, and the later Orkney folklorist Ernest Marwick, believed these to be a wholly different species from the selkies, possessed of stronger magical powers and dwelling at the bottom of the sea, in a magnificent underwater haven known as 'Finfolkaheem'. It seems unlikely such clear distinctions were insisted on by seventeenth-century Orcadians – though we are often obliged to infer what they believed from the testimony of nineteenth-century descendants. Fin-folk, like the selkies, were shapeshifters, capable of initiating emotional and sexual relationships with humans, and of violently abducting them.

The name itself was Norse. Orkney's first Norwegian settlers used the term *Finnr* for the nomadic Sámi people of northern Scandinavia. The Sámi had the reputation of being magicians, able to transform themselves into wolves and bears, whales and walruses. A connection with seals was most likely made in the Northern Isles themselves, and may represent the merging of Norse beliefs with pre-existing magical traditions.[6]

The fin-men were more than memories from an ancient past. James Wallace heard that in 1682 one was seen off the southern tip of Eday, 'rowing up and down in his little boat'. Inhabitants of the isle rushed to see, but he quickly fled. Another was spotted from Westray in 1684. Brand, in 1701, wrote that 'there are frequently Finmen seen here upon the coasts' – one the preceding year in Stronsay, and another within the past few months in Westray, fishing with lines in a boat 'made of seal skins or some kind of leather'. The mysterious mariner again fled at the locals' approach.

These clergymen of the early Enlightenment did not believe they were documenting supernatural phenomena. Brand assumed them to be sightings of fishermen from Finland, and marvelled at their ability to travel so far in such tiny vessels. Wallace offered no theory on 'these men which they call Finmen'. But when his manuscript, edited by his son, was published posthumously in 1693, an explanation was attached: 'these Finmen seem to be some of these people that dwell about the Fretum Davis' – the Davis Strait between Greenland and Baffin Island. Fin-men, in other words, were stray Inuit, who had crossed the Atlantic in their kayaks.

The historical geographer Jonathan Westaway suggests the addition was an initiative of the Scottish Geographer Royal, Sir Robert Sibbald, who corresponded with Wallace in the course of collecting materials for an extensive survey of Scotland. The idea that adventurous Inuit were probing the waters around Orkney has gained traction in modern times, but there is little evidence for this pattern of reverse colonisation. A kayak reported by Wallace's son and by Brand to be on display in Physicians' Hall, Edinburgh, was said to have come from Orkney. The second edition of Wallace's history, published in 1700, added that 'there is another of their boats in the church of Burray'. But these objects (no longer extant) may well have been brought to the islands as part of a ship's cargo.

Scotland in the 1690s, about to establish its ill-fated colony at Darien in Panama, was starting to imagine itself an imperial power. It suited Sibbald, and fellow intellectuals linked to the Royal Society, to regard Orkney as, in Westaway's words, a 'site of trans-oceanic convergence and circulation'. In 1695, the naturalist Hans Sloane, secretary of the Royal Society, published a paper on a topic that intrigued both Wallace and Sibbald: 'strange beans, frequently cast on shore on the Orkney Isles'. These attractive heart-shaped 'Molucca Beans' – still a prized find along the Orkney shoreline – were seeds from vines growing in the West Indies and Central America, tangible symbols of the connective and instructional power of the Ocean.

Islanders themselves were more concerned about power than provenance. In the Hebrides, and perhaps the Northern Isles too, the seeds were valued as amulets, offering protection in childbirth. The problem with the fin-men was that they chased fish away from Orkney's coastal waters – a belief reported by Wallace and Brand and still talked about in the nineteenth century.[7] For most ordinary Orcadians, the supernatural was not an elevated sphere of metaphysical speculation, but an unruly borderland lying on the edge of everyday experience.

Superstition

The creature from the sea who stole Margaret Linay's baby – fin-man, selkie or sea-trow – was readily reimagined by her as a being from 'under the hill'. In old Orcadian culture, suggests the folklorist Alan Bruford, 'everything on land, including the fairies, has a counter-part in the sea'.

Orcadian fairies were often called 'trows' or 'trowies', words deriving from the menacing trolls of Norway. But Orkney's fairy-traditions likely reflected a mishmash of Scots and Scandinavian, and of still older, Celtic or Pictish, beliefs. The islands' numerous Neolithic burial mounds were, in all probability, already in the Iron Age seen as strange and unworldly locations. They became known as the haunts of hogboons or hogboys: spirits of the dead thought to be keeping wary guard in places they had ruled over while living – a domestication in the Orcadian landscape of the horrifying *haug-búi* (mound-dweller or barrow-wight) of the Norse sagas. Other solitary spirits were more domestic in character, and more likely Scots in origin – like the prying brownie in the closet of Bishop Graham (p. 251).

Margaret Linay was aware of 'many folk in the hill with her'. Trows and fairies were sociable creatures living in community, albeit beyond the lines of ordinary human habitation and the limits of normal human perception. Prior to the later seventeenth century, witchcraft trials, as we have seen in Chapter 5, provide the best evidence – a sometimes illuminating, sometimes distorting, lens on belief in trows and fairies.

As the trials ran out of theological and judicial steam, the attitudes of ministers towards stories of contact with fairies started to evolve. They now tended to see in them instances of quaint error meriting disdainful description, rather than of Satanic transgression demanding furious eradication. Wallace had relatively little to say about fairy-lore, though it pops up in a story he tells about a rock near Noup Head, on the north-west point of Westray. The inhabitants swore that 'if a man go upon it, having any iron upon him (an it were an iron nail in his shoe) the sea will instantly swell in such a tempestuous way that no boat can come near to take him off, and that the sea will not be settled till the piece of iron be flung into the sea'.

It was believed in Orkney, as in other places, that fairies were vehemently averse to iron – some scholars (controversially) theorise that the origins of fairy-belief lay in folk memories of aboriginal inhabitants, driven from places of settlement by metal-armed invaders. Wallace proposed 'to make an experiment' of the belief about Noup Head and offered a man a shilling to go onto the rock carrying a piece of iron, 'but he would not … on any terms'. The minister gave a similarly 'enlightened' response to reports of strange nocturnal lights in the parish of Evie, which caused coastal hillocks to seem ablaze, and illuminated the parish kirk of St Nicholas 'as if torches or candles were burning in it'. This, Wallace remarked, 'amazes the people greatly', but 'possibly it is nothing else but some thick glutinous meteor [luminous atmospheric phenomenon] that receives that light from the stars'.[8]

John Brand entertained ambivalent feelings towards Orkney's fairies. On the one hand, he knew that 'evil spirits, also called fairies, are frequently seen in several of the isles dancing and making merry'. Brand was given an account of the people's 'wild sentiments' about them, but – tragically for us – he decided that 'with such I shall not detain my reader'. Fairies, to the devoutly Calvinist Brand, were nevertheless important as a marker of Protestant reform. Wherever the Reformation rooted itself, 'the brownies, fairies, and other evil spirits that haunted and were familiar in our houses, were dismissed and fled'. Admittedly, Brand reflected, 'this restraint put upon the Devil was far later in these northern places than with us' – that is, with Lowland Scots like himself. Nonetheless, thanks to Gospel preaching, 'yet now also do these Northern Isles enjoy the fruits of this restraint'.

The fairies, scholars have observed, seem always to be 'in retreat'. From medieval times onwards, people associated them with nostalgia, childhood and the past, with the ethos of an old world fading and passing at the arrival of a new. Brand's announcement of their imminent demise in Orkney was part of a pattern, and it proved predictably premature.

In 1768, a young Forfarshire minister, George Low, arrived in Orkney as tutor to the children of Robert Graham of Redland. A gifted naturalist, Low took a tour through the islands in 1774, the year of his institution as minister of Birsay and Harray. Low, who became an industrious informant to learned contemporaries about matters Orcadian, wrote an account of his travels, a history of Orkney and an accompanying description of the islands and their inhabitants. Though not published in his lifetime, these works provide an invaluable picture of the beliefs of eighteenth-century rural Orcadians, and of the exasperations of an 'enlightened' Scots clergyman.

'So firmly is the belief of fairies implanted among the country people that it will be a very difficult matter to persuade them from it.' Low spoke with an Orkney farmer who, on Christmas Day 1773, saw 'a large company dancing and frolicking' atop the Knowe of Burrian – a mound in Harray housing the ruins of a Pictish broch. The man became annoyed when Lowe tried to persuade him that his vision was nothing 'but the effects of the cakes and strong ale he had for his Christmas breakfast'.

Whenever women were in labour, Low reported, steps were taken 'to hinder the child or mother from being carried away or changed by the fairies' – the waving of a burning rag around the mother and the placing into her hands of a knife and a bible. The knife affirmed ancient beliefs about the fairies' aversion to iron; the bible suggests a newer appreciation of the protective power of Christianity's holy book, though hardly in a manner of which a minister like Low could be expected to approve.

Languor rising from sickness was habitually ascribed to the sufferer being carried off by the fairies, 'or as they express it, "in the hill"'. An understanding of disease as a result of spiritual abduction by malevolent trows had seemingly changed little from the time of the witch-hunts.

Low was at once fascinated and dismayed by 'elf-shot' – the name given to disinterred flint arrowheads, which the fairies, intending to cause sickness or death, were thought to have shot at humans and livestock. Belief in the phenomenon occurred across England and both Lowland and Highland Scotland, as well as in Scandinavia, and it may have reached the islands along multiple routes of transition. Yet Low believed that there were as many credulous tales of elf-shot in Orkney 'as could be picked up in all Scotland'. Orcadian victims were usually cows, though Low met a gentleman, 'in other things little given to scruples', who heard an elf-shot rattle on his staircase and bent down to pick it up. On a visit to Stroma, Low was shown a collection of trowie arrowheads, but the inhabitants could not be persuaded to part with any: 'they imagine as long as they possess one of these, the fairies can have no power over their cattle or persons'.[9]

A well-weathered word summed up such habits and attitudes. 'Superstition' was something clergymen were apt to see everywhere, though its meanings changed over the centuries. The term originally implied an excessive or superfluous devotion, but in the era of Reformation it developed darker associations with demonic power and cosmic strife, seen by anxious Protestants as inextricably entwined with Catholic beliefs and practices. For later seventeenth-century ministers like Wallace, however, the meaning had evolved again. 'Superstition' still connoted ignorance, credulity and error – not necessarily harmless, but usually no longer regarded as an existential threat to religious and political order. Increasingly, it served as a marker of distinction – between the affluent, educated and cultured, and the impoverished, illiterate and vulgar.

Superstition, Wallace thought, was rife among Orkney's lower-class inhabitants, an opinion with which Brand and Low fully concurred. Brand considered Orcadians even more superstitious than Shetlanders. It is important to recognise, however, that the concept is not some neutral category of analysis. 'Superstition' can often mean little more than beliefs of which the person describing them disapproves, or which lack respectable institutional endorsement.

Wallace was breezily dismissive of the 'vulgar legends' and 'idle fables' of the Orcadian peasantry – for example, about the Dwarfie Stane in Hoy, or Cubbie Roo's Castle in Wyre (pp. 29–30). Yet he had implausible tales of his own to impart, albeit dignified as episodes of divine providence. There was the case of James Linay, sole survivor of a fishing boat capsized near Auskerry, who floated for four days on the upturned keel: 'a memorable example of God's providence in delivering people from almost desperate hazard'. Another story, from the time of Bishop Graham, involved a lad called William Garrioch, executed for minor theft at the instigation of a wicked uncle craving his inheritance. Later, as the uncle walked past the young man's grave in the cathedral kirkyard, the bishop's dog tore out his throat: 'a sad monument of God's wrath'.

Such tales could engage both lay and clerical imaginations, and suggest that the gulf between them was not always unbridgeable. Wallace repeated a strange story from Stronsay about a man called John Smith, who regularly fished with three other islanders. After he had gone back out to sea on several consecutive days, his wife determined that he should rest, and blocked the light from their window so that John over-slept. His companions went off in the morning without him, their boat inexplicably overturned, and all were drowned. John's wife was out when she heard the news, and rushed home to tell him what had happened. A terrible sight awaited her. John had tripped in rising out of bed; his head had landed in the chamber pot and he had drowned in urine. Wallace considered it an instructive illustration of how no one could escape whatever fate God had appointed for them.[10] For the fishing-folk of Stronsay, the death was likely laced with different meanings, proving the necessity of strict parity in profit, and the pitiless tribute exacted by the sea.

Divine providence, like vulgar superstition, lay confidently in the eye of the beholder. Brand cheerfully relayed a story about a boat which capsized off the coast of Eday, coming safely to shore on the west side of Sanday, after the sailors spent their hours adrift 'comforting and refreshing one another with places [passages] of Scripture'. From Richard Mein, minister of Cross and Burness, Brand learned of a strange incident in North Ronaldsay, which took place in about 1693. Helen Thomson had given birth to a baby with a hugely elongated neck, and nose, eyes and mouth to the back as well as front. It was said a woman had accused the pregnant Helen of slander, and angrily wished 'she might bring forth a monster'. North Ronaldsay folk surely suspected witchcraft, but Brand saw an event that declared 'God's holy and wise Providence'. In contrast, he frowned on the islanders who thought that ships driven onto their shores were 'a favourable providence to them' and who christened the wrecked vessels 'God's Send'.

Wallace was cautiously hopeful about turning the tide of popular superstition. Charms to cure heartburn and rickets were still in evidence, but 'much curbed by the careful industry of our pious ministers'. In 1706, the General Assembly inquired of the conjoined Synod of Caithness and Orkney, 'what paganish customs or superstitions remain among them?' The northern ministers sent an upbeat reply: 'the Presbyteries of Orkney and Caithness have reason to bless God that these are wearing out daily'.

Brand would have considered this unduly optimistic. He rehearsed a woeful litany of charms, rituals and other superstitions in common use among Orcadians: 'tho' their ministers both privately and publicly have spoken to them, yet they cannot get them to forbear and abandon these customs'. Above all, Brand worried about veneration of holy places, and pilgrimages to old kirks and chapels. He feared that if popery were to regain a foothold, 'many of the inhabitants of these isles would readily embrace it, and by retaining some of these old popish rites and customs seem to be in a manner prepared'.

At the end of the seventeenth century, people sought spiritual blessing at the same places that their ancestors had a century and more earlier. Pilgrimages to the Brough of Deerness and Brough of Birsay were, Wallace conceded, not yet 'omitted by the common people'. Also 'much frequented by superstitious people', according to Brand, was the round-towered kirk of St Magnus in Egilsay, where some believed the saint lay buried.

At the time Wallace wrote, it had been more than a hundred years since people in Orkney were taught about saints in their churches, or exhorted to pray to them for intercession. Nonetheless, it was a 'general custom' to observe the feast of the saint to whom the parish church was dedicated, and forgo manual labour on that day. In 1686, Nicol Taylor was hauled before the session in Holm, where the kirk had a dedication to St Nicholas, for cursing a neighbour: 'God and sweet St Nicholas turn thy eyes in thy neck!' Statues of the saints, and books and liturgies in their honour, had long since disappeared from the scene. But an awareness of saintly presence and power was kept alive by remembered associations with times and seasons in the calendar, and familiar places in the landscape.

Brand understood this. From Alexander Grant, staunchly Presbyterian minister of South Ronaldsay, he learned that the island's Ladykirk, home to the miraculous Ladykirk Stone (p. 30), was now in roofless disrepair. Yet the people rejected all entreaties to hear sermons in a rented barn, and forced Grant to officiate in the dilapidated church. The gentlemen of South Ronaldsay were in favour of building a new kirk in a better location, but ordinary parishioners were 'superstitiously wedded to the place of its present situation'.

The worst attitudes attached themselves to the old chapels – these were, in Brand's view, the 'nest egg' of superstition, and he hoped that in Orkney, as in Shetland and Caithness, the government would raze them all to the ground. The desire was reinforced by a visit Brand made with other ministers to St Tredwell's Chapel in Papa Westray, a place people 'have such a veneration for, that they will come from other isles in considerable numbers'.

The chapel itself was a tiny ruinous edifice, but local people laboured to keep the remaining walls from collapsing. In front of the door stood a heap of small stones, 'into which the superstitious people when they come, do cast a small stone or two for their offering'. The waters of the surrounding loch were thought to possess medicinal qualities, 'whereupon many diseased and infirm persons resort to it'. The method was one we have encountered previously (p. 214) – pilgrims would walk around the loch several times in silence, before washing in its waters. The new minister of Westray, William Blaw, told Brand he had recently encountered a group making the circuit – only with difficulty could he get them to talk to him, and admit they came to be cured.

Brand spoke further with the man 'who was minister of the place for many years', James Heart, current minister of Shapinsay. Heart told him about a gentleman's sister in Papa Westray, 'who was not able to go to this loch without help, yet returned without it'. A gentleman from another part of Orkney, 'much distressed with sore eyes', had expended a great deal of money on medical remedies, without success. After washing himself at St Tredwell's Loch, however, he again became 'sound and whole'.

It is a revealing snippet. The (probably mythical) female saint to whom the chapel was dedicated – Tredwell or Triduana – was a Greek nun said to have come to Scotland in the fourth century. According to the Aberdeen Breviary (p. 79), she attracted the lustful attentions of a Pictish king, who admired her fine eyes. Her steely response was to pluck them out, and send them to him skewered on a pin. Inevitably, Triduana (like the better-known St Lucy) became a saint renowned for tackling eye complaints. Before the Reformation, her cult was spread across Scotland, with her relics supposedly preserved in a chapel attached to the parish church of Restalrig outside Edinburgh, re-established as a royal foundation by James III in 1477. Papa Westray's fragment of the shattered devotional culture of Scotland was a site where the specialism was evidently remembered, raising the question of whether other Orkney places of pilgrimage were linked to specific ailments.

Brand was perturbed by the well-attested evidence of cures. He considered the possibility of medicinal virtue in the water, or whether improvements were wrought by 'the force and strength of the imagination' – recovery we might today term psychosomatic. Yet it also seemed likely to the devoutly Protestant Brand that people were healed at such popish places 'by the aid and assistance of Satan, whom God … may permit so to do, for the further judicial blinding and hardening of these who follow such unwarrantable and unlawful courses'.[11]

The participation of gentry families was passed over in silence. It was an article of ministerial faith that superstition held sway among the 'commonalty'. But Orkney's lairds were perhaps not as elevated above the outlook of their illiterate tenants as they generally liked to pretend.

Brand was confident change was coming. Such customs were but 'the sour dregs of pagan and popish superstition'. He expected readers to join with him in marvelling at 'the wonderful mercy and grace of our God, in sending the Gospel to the isles afar off, that these gentiles who have not formerly heard of his fame, nor seen his glory, should have the same declared among them'. Brand, indeed, is yet another witness to the paradoxically pivotal place of the periphery, a man persuaded that the islands on Britain's northern edge represented both the supreme test and the ultimate vindication of godly Protestant Reformation.

Had Brand lived to read George Low's accounts of Orcadian manners and customs, three-quarters of a century later, his heart would surely have sunk. Low, in 1773, could still describe Papa Westray as 'famous for the superstitious regard paid by the inhabitants to St Tredwell's Chapel and Loch'. Other places 'to which superstition has attached an extraordinary veneration' included the chapels on the Broughs of Birsay and Deerness, and another at Cleat on the north side of Sanday, dedicated to the Virgin Mary. Nonetheless, Low reckoned that 'these superstitions are now much wearing out and are practised by only a few of the oldest and most ignorant'.

One custom which people seemed stubbornly determined to retain, however, was that of lighting a large fire in some conspicuous, south-facing place in the parish on 24 June, midsummer by conventional calendar-reckoning, the old feast day of St John the Baptist. For sick or distressed persons, an alternative to promising a pilgrimage to some church or chapel was a vow to carry peats to the Johnsmas Fire. Diseased horses were led around it, 'always taking care to follow the course of the sun', and onlookers joined the procession.

Low was pleased by reports that 'even this is not kept up with that spirit as in former times'. It was probable, he thought, that the Johnsmas Fires, like other old customs, would in due course be forgotten. Yet it is possible that Low's informants were suffering from the strain of middle-aged nostalgia which makes everything from one's youth seem livelier and more vibrant. Certainly, the Johnsmas Fires were not in the 1770s on the cusp of extinction; evidence from across Orkney shows the custom persisting through much of the nineteenth century.[12]

Old customs continued, but not always in exactly the same way. Jo. Ben., in the sixteenth century, described pilgrims depositing stones at the Brough of Deerness chapel – a certified 'magical' method of expelling sickness from the body of a sufferer. Archaeologists examining the site in the 1970s, however, recovered over thirty small-value coins. Around two-thirds of these dated from the seventeenth century, nearly all post-1640; the others were from the eighteenth century, and one was minted in 1806. The coins were corroded, but showed little evidence of wear from handling and transaction – pilgrims, it seems, were selecting for deposit the newest coins available to them.

Habitual possession of coinage as part of a functioning 'money economy' arrived late in Orkney's rural parishes. But it is conceivable the adjustment of votive objects, with coins supplementing or replacing stones, reflected a shift to more 'petitionary' mentalities on the part of pilgrims, the small pieces of money intended as thanksgiving offerings, or tokens of sincere intent. At St Tredwell's Chapel, Brand noted that, in addition to stones, 'some will cast in money'. Here, too, nineteenth-century excavations in the interior of the chapel recovered several dozen copper coins, from the reign of James VI to that of George III.

We have Wallace to thank for knowledge of another tradition about St Tredwell's Loch: 'they say that it will appear like blood before any disaster befall the Royal Family'. Possibly this belief was centuries old when

Wallace recorded it in the early 1680s, but the most conspicuous disasters befalling the royal house were fairly recent, particularly the overthrow of Charles I. There are other cases in Britain of waters reportedly turning red in providential judgement on Charles's execution.[13] If this was a new accretion to knowledge about St Tredwell's Loch, it supplies further evidence of the ability of Orcadian popular belief to adopt and adapt from different explanatory systems. That very flexibility may be an important reason why 'superstition' proved so difficult to eradicate.

Hellie Boot and Ganfers

It was not for want of trying. 'Restraint of superstition' and enforcing 'work upon superstitious days' were long-acknowledged responsibilities, about which ministers like Walter Stewart and Patrick Graham were quizzed at the presbytery visitation of 1643 that led to the destruction of St Peter's statue (p. 120). As we have also seen (p. 223), there was a concerted drive against charmers in the 1640s, and further directives from the episcopal synod after the Restoration.

On 21 June 1708, as Midsummer's Eve approached, the synod at Kirkwall gave thought to 'superstitious practices [that] are yet in many places … notwithstanding of former acts made thereanent'; Johnsmas Fires were 'very common'. Ministers were to denounce them from the pulpit, and meet individually with anyone involved, to 'bring them to a sense of their sin'. Meanwhile, a message went to the stewart, asking him to order parish baillies to punish offenders upon complaint made by a minister or kirk session. How enthusiastically the secular authorities cooperated in a campaign against superstition – patently unsuccessful, in the case of the Johnsmas Fires – is distinctly moot. There is little to suggest it was much of a priority for the stewartry court in the early eighteenth century, at a time of political uncertainty and endemic feuding among ministers and lairds.

There are question marks, too, over how proactive kirk sessions were in stamping out 'paganish' and 'popish' customs – though the disappearance of nearly all Orkney's session records for the first half of the seventeenth century, and most of them for the second, hampers our ability to know. There are a handful of cases involving charmers in the surviving Shapinsay and South Ronaldsay minutes, and one instance of superstitious resort to a chapel in the case of the latter (p. 214) – nearly all linked to suspicions of witchcraft.

The records for the first half of the eighteenth century are fuller, but yield even fewer prosecutions for 'superstition'. A rare example turns up in Orphir session minutes for 1741. As his dead wife was laid in her coffin, Thomas Loutitt 'took corn and put [it] between her fingers and toes, and put some barley corns in her mouth, and laid some in the chest, and threw the rest in the chest about her'. He grudgingly confessed to doing so because of a quarrel between his daughter and his late wife, the girl's stepmother. 'She should not come back again and trouble his daughter'. Denouncing Loutitt's 'gross ignorance', the session fined him and required public penance – largely because of his surly attitude. I have found no comparable cases in other session records, though it seems unlikely that Loutitt invented his own ritual for deterring a vengeful spectre. Low reckoned there was much anxious concern 'among the more ignorant vulgar' with ghosts, 'or as they are here called, ganfers'. The quaint-sounding word was, predictably, of Norn provenance – *gagn ferð*, a return meeting, or haunting.

In the same year, 1741, the Firth Session contemplated an equally unusual case. Five parishioners – Elspeth Bews, Margaret Bews, Christian Bews, Janet Hervey and Nicol Wood – were charged with 'travelling by sea to the Isle of Damsay to the chapel there to pay their vows'. Elspeth confessed that 'a vision came to her in the night to go to the Chapel of Damsay and get her health'. The Damsay chapel was entirely ruinous, but its reputation was intact among the farm-folk of Firth. According to Low, it was known to the 'country people' as 'Hellie Boot' – another phrase from Old Norse, straightforwardly identifying the purpose of the place: *heilagr bót*, holy cure or betterment.

The church on Damsay, consecrated to the Virgin Mary, was, according to Jo. Ben., a place 'to which women frequently go when pregnant'. There is nothing to prove that this was Elspeth's condition, but the fact she went in largely female company makes it a strong possibility – further testimony to the tenacity in Orkney of medically useful beliefs, and to the resourcefulness of Orcadian women in taking their own matters in hand.

The case, however, was highly unusual. Investigations of the extant kirk session and presbytery records have yielded precisely no prosecutions for visiting the Broughs of Deerness and Birsay, or the Chapel of St Tredwell, nor any denunciations for attending Johnsmas Fires, refusing to work on saints' days, participating in birthing rituals, or collecting elf-shot. Indeed, the offence for which Elspeth was actually punished was breach of the Sabbath: she and her companions rowed to Damsay on a Sunday, 'a travelling of no necessity'. They were not charged with superstition, nor – sadly for us – did the session inquire further into the 'vision' Elspeth experienced.[14]

One possibility is that infractions of this kind were much less common than the nagging complaints of ministers like Wallace, Brand and Low made out. But a mass of folkloric evidence on the persistence of such beliefs and practices through much of the nineteenth century makes this intrinsically unlikely. Kirk sessions, as we shall see, had few scruples about imposing strict punishments for actions that today would not be considered remotely criminal. But – at least when there were no suspicions of witchcraft – they were seemingly reluctant to involve themselves in the clandestine cultural economy of protective ritual, votive pilgrimage and magical cure.

With the exception of the minister, the members of kirk sessions were themselves shaped by the culture of the parish community; moreover, policing activities depended to a considerable extent on individuals stepping forward with denunciations. When it came to the dense cultural weave onto which the ministers lazily stitched the label of 'superstition', it would seem that most of the time the sessions didn't ask, and the people didn't tell.

Why the common people continued to do such things was a mystery to the ministers. It may have been a mystery to the people themselves. Low observed wearily in respect of Johnsmas Fires that 'the inhabitants can give no further account, but it was an old custom transmitted to them by their fathers, and which they incline to continue'.

Perhaps, however, these were simply matters that people were reluctant to discuss in front of the minister. In the mid-nineteenth century, it was well known that the purpose of Johnsmas Fires was 'to drive away the trows and fairies'. An old lady from Orphir, interviewed in the early

1900s, long after the tradition had died out, remembered boys running about the fire with burning stems of heather – 'I wadna winder if this wis nae sumthin tae dae aboot keepin aff fairies for the year.'

Orkney's Presbyterian clergy did not face much in the way of open defiance and hostility – though, as we shall see, there were exceptions to that rule. Even the hyper-censorious Brand conceded the people to be 'generally tractable, submissive and respectful to their ministers'. Yet something, they sometimes felt, was not quite right – there were hints of things to which the minister was not privy; codes and customs qualifying the people's full commitment to the values of the Protestant Kirk.

Jo. Ben. had long ago sensed this – the inhabitants of Orkney were 'sly and very subtle' – 'vafri sunt et subtilissimi'. Nearly two centuries later, George Low vented his suspicions about the people of North Ronaldsay: 'they have a great deal of cunning and hidden artifice. They use a great many ambiguous expressions, and there is always something more couched under their language than is expressed.'[15]

Low was fascinated by language, and by the odd way his parishioners spoke: 'to this day there are many sounds in the English language which the Orkney people cannot master, but pronounce according to their old Norn dialect'. These included the digraphs 'qu' and 'th'. In trying to enunciate 'queen', 'question' or 'quarrel', they would say 'wheen', 'whestion', 'wharrel'; 'ting', 'tree', 'tousand', for 'thing', 'three', 'thousand'. For the belles-lettered Lowland minister, such peculiarities went beyond the normal spectrum of regional dialect: to any stranger, 'this corruption of their English is observable at first hearing them talk'.

In making sense of the relationship between Protestant Christianity and Orcadian culture, language looms large as a question. We have noted already the decline of Norn (pp. 179–80), which ceased to function as a written language in Orkney in the fifteenth century, and was less widely spoken in the seventeenth century than in the sixteenth.

Norn's retreat continued inexorably thereafter. Low, in the early 1770s, judged 'it is now so much worn out that I believe there is scare a single man in the country who can express himself on the most ordinary occasions in the language'. The late-middle of the eighteenth century seems to have been the moment of final linguistic extinction. Richard Pococke, travel writer and Church of Ireland bishop, visited Orkney in 1760 and reported 'there is now no Norn or Norwegian spoken, but all English with the Norwegian accent'.

This, if true, was a recent development. Witnesses in the pundlar proceedings swore in 1757 not only that Norn was spoken widely in their youth, but that within the last couple of years they had overheard conversations in the language 'among the country people'. Half a century earlier, John Brand observed that Orcadians 'generally speak English', and found them surprisingly easy to understand, yet Norn was not quite extinct among them, and in Harray 'there are a few yet living who can speak no other thing'.

How, within a bilingual society, one language displaces or absorbs another, is a problem that linguistic specialists have long debated without arriving at a consensus. It is probably enough to say that the disappearance of Norn was exceptionally slow and gradual, and never absolutely complete. In several of the isles, Brand discovered, even people speaking English 'have some words and phrases peculiar to themselves'.

Norn's decline is hard to track precisely because it never openly challenged the dominance of Scots or the written and spoken forms of Scottish-English becoming more prevalent over the course of the eighteenth century. It became, increasingly, a method of private and domestic communication, used within the family, and out of earshot of people of substance – including, and perhaps especially, the minister. Wallace observed in the early 1680s how 'some of the common people among themselves speak Norse'. Writing in the 1670s, Matthew Mackaile, an Aberdeen apothecary who spent time in Orkney a few years earlier, thought there were only three or four parishes where 'Norse or rude Danish' was spoken, 'and chiefly when they are at their own houses'.[16]

An episode from the summer of 1703 illustrates both Norn's social retraction and its potential as an instrument of secrecy and connivance. In the parish of Rendall, soldiers impounded two horses from the tenants of William Craigie of Gairsay for non-payment of the cess or land tax. One of the owners objected, and gathered two dozen men, armed with scythes and other weapons, to confront the troops and release the horses. The ringleader was called Hugh Marwick – pleasingly, a name shared with twentieth-century Orkney's greatest scholar of the Norn language.

At the subsequent trial, witnesses admitted to seeing and hearing the armed conspirators around the smithy in the Rendall township of Isbister. But their testimony proved of little use to the prosecution. John Linay, from St Ola, noticed 'Gairsay's tenants come in and talk to one another upon Norn', but could not comprehend them. Magnus Laughton and William Taylor, from the parish of St Andrews, fared no better. Marwick's associates 'spoke among themselves upon Norn, which the deponent did not understand'. They lived only a few miles apart, but the East Mainlanders could plausibly deny having any idea what the West Mainlanders were saying. There is an irony, however, in the manner of the denial. To describe someone as speaking *upon* a language is itself a characteristically Scandinavian syntax ('in English' is *på engelsk* in Norwegian, *á ensku* in Icelandic) – one of a number of distinctively un-English locutions found in Orcadian-Scots.[17]

Orkney slowly ceased being a bilingual society over the course of the seventeenth and eighteenth centuries. But Norn, in the gradual arc of its descent, set the conditions for a kind of cultural bilingualism, or cultural amphibiousness. People could be upstanding Protestant Christians while also possessing the ability to access alternative imaginative spaces – ones to which most of the ministers had no means of admittance. We have seen already (pp. 198–9) that traces of Norn lingered around the world of witchcraft and magic, and the same is probably true for Orcadian folk culture more broadly. The Orkney landscape was stickered with Old Norse designations of place. And the English language, often literally, lacked the words for things which lay beyond the learned vocabulary and biblical imagination of the ministers – for hogboons and ganfers, Finfolkaheem and Hellie Boot.

From Sanday, a memorable tradition is preserved for us by Walter Traill Dennison, like most of his stories, frustratingly undated. It concerned the 'laird' of Northskaill, a farm on the eastern side of the island's northern spur. Lying on his deathbed, the laird summoned the minister and asked him to say a prayer. But as soon as the clergyman began to speak, the dying man blurted out, 'Awa', awa' wi' your sly Scotch tingue! I want you tae pray i' the Danska tong. Hid s'all never be said that ony een prayed ower me i' the ferry-lupper's jabber!' The minister was forced to confess that he could pray only in his own language. 'Hid's a peur teel' (a poor deal), complained the laird, 'that a college-bred man cinno' pray i' the Danska tong ... deus th'u t'ink that onyt'ing bit a Danska prayer wad apen the yetts [gates] o' paradise for me?'

It seems unlikely to be a reliable account of a genuine incident. The tale's comic juxtaposition of a canny Sanday udaller and a hapless Scots minister radiates the romantic Orcadian patriotism of the later nineteenth century, along with a cool disdain for the carpet-bagging Scots who after 1468, in Traill Dennison's judgement, 'descended like a cloud of locusts on these devoted islands'.[18]

But the suggestion of potential distance and discomfiture between Scots ministers and Orcadian parishioners is not intrinsically implausible. It was a matter of concern in Shapinsay in 1659 that many dying persons 'are careless to send for the minister'. 'Ferry-louper', the derogatory term used by Traill Dennison's disappointed laird, was a well-known expression, unique to Orkney, and still, sad to relate, occasionally heard there today. It was, as the Banff-born sheriff Alexander Peterkin noted in 1822, 'the name by which all persons not natives of Orkney are designated by the vulgar'.

Ferry-loupers were not transient visitors but permanent residents, people who had louped or leaped onto ferries over the Pentland Firth and not louped back. The great majority of Orkney clergymen, through both the seventeenth and eighteenth centuries, fitted the description. Of thirty-five ministers appointed to Orkney parishes in 1640–79, a mere five appear to have been Orcadians, and only seven out of thirty-four were in 1680–1719. Across the whole period from 1600 to 1799, of the 151 ministers instituted to an Orkney parish for the first time, a meagre total of thirty (19.9 per cent) were born in Orkney and, of these, fifteen were themselves the sons of local ministers.

To folk in the pews, the minister's pronunciations probably sounded as dissonant as the speech of Birsay and Harray farmers did to the Reverend George Low. Indeed, the appointment of the Aberdeenshire minister John Reid to Orphir in 1745 was strenuously opposed in the parish on the grounds 'they could not understand his tongue'. Though people were generally courteous and deferential, the minister's outsider status was often an inescapable social fact, and sometimes turned against him. In South Ronaldsay, in 1661, Dean Edward Richardson and his wife Margaret complained to the session that a woman had called them 'thieves and vagabonds and runagates and ferry loopers' – the word's earliest recorded instance, though undoubtedly not its first utterance.[19]

Boniewords

The Sanday laird's desire to be prayed for in the 'Danska tong' may well be Traill Dennison's flight of fancy. Certainly, there was no attempt by the post-Reformation Kirk in Orkney to promote prayer in any language but Scots and Scottish-English. This presents a marked contrast with other parts of what is sometimes called the 'Atlantic Archipelago' where Protestant Reformation encountered bilingual societies: Ireland, Wales, the Scottish Highlands and Western Isles, the Channel Islands, the Isle of Man. With the exception of Cornwall – where timescales for the decline of Cornish broadly mirror those of Orkney Norn – there was everywhere a recognition by Church authorities that translations of the Bible and liturgical texts, and provision of preachers competent in the local tongue, were necessary tools of evangelisation and conversion. In Orkney, the clergy's assumption from the outset seems to have been that there was no need to go to special lengths, as Norse was withering and Scots and English already winning.

But we cannot dismiss entirely the idea that Protestant Orcadians felt moved to speak to God in the language of their Norse Catholic ancestors. We have seen already (p. 85) that the sole surviving post-medieval text in Orkney Norn is a version of the Lord's Prayer, 'Fa vor, i ir i chimrie'. Its origins within Orkney are obscure, and no original manuscript of it survives. It is usually assumed that the Kirkwall minister James Wallace was responsible for transcribing the text, yet it is not in Wallace's 1684 'Account of the Ancient and Present State of Orkney', nor the posthumously printed edition of 1693. The prayer first appears in the second printed edition of 1700, authorship of which was credited to 'James

Wallace, M.D., and Fellow of the Royal Society', the minister's eldest and far-travelled son, who had taken part in the Scottish first expedition to Darien in 1698–9.

Conceivably, Wallace junior inserted the text of the prayer after coming across it in papers sent by his father to Sir Robert Sibbald. It certainly intrigued him on historical and philological grounds – nothing of the sort, he noted, was found in the work of the sixteenth-century polymath and cataloguer of languages Conrad Gesner, nor in a compilation by the former bishop of Chester, John Wilkins, which provided parallel texts of the Lord's Prayer in forty-nine different tongues. This was improved on by a German scholar of Asian languages, Andreas Müller, whose polyglot anthology of 1680, *Oratio Orationum* (The Language of Prayer), boasted a hundred separate versions. A revised edition of Müller's work was printed in London in 1700, promising readers 'The Lord's Prayer in above a hundred languages'. One of the additions was Wallace's Norn text, taking its place in the firmament of global devotion under the exotic label of 'Orcadica'.[20]

For the people who actually recited its words – perhaps at the Brough of Deerness or in the quiet of their own homes – the prayer was no ethnographic curiosity, but a means of access to divine favour. Assuming it had been transcribed accurately, some of the anomalies exhibited by the text strengthen the case for regarding it as part of a living tradition. Dr Wallace thought it surprising that the prayer contained 'but little of the Danish or Norwegian language, to which I thought it should have had more affinity'. Yet Norn, like Faroese or Icelandic, had evolved in its own way from Old Norse roots. There were a couple of intriguing borrowings from Scots – petitioners asked not to be led into *tumtation*, and for God to '*delivera* vos fro olt ilt'.

There were yet more 'loanwords' in a second version, which George Low came across in Foula during his 1774 tour of Shetland. It included the so-called 'doxology' – 'for do i ir konungdum' (for thine is the kingdom ...) – the ending recited devotionally by Protestants, but not by Catholics, and which does not appear in Scandinavian vernacular versions from the middle ages.[21] By the mid-eighteenth century, Shetlanders had apparently updated an earlier formula to mirror the English one recited in the kirk. The absence of the doxology from

Wallace's Orkney version arguably suggests a more direct transmission from pre-Reformation times.

An ability to understand Norn was not – in the eyes of ministers or other authorities – any kind of admirable trait. It was, rather, a sign of ignorance, for which the remedy was education. Provision of schools was the slowest of uphill battles in Orkney, bedevilled by disagreement over schoolmasters' salaries, lack of suitable premises, and parents' reluctance to lose the productive labour of younger family members. At Bishop Graham's visitation of 1627, ministers and parish elders were asked if there was any school in their district, and delegation after delegation replied there was not. 'Without the education of the youth,' lamented Birsay's representatives, 'the travails of the minister upon the elder persons are lost.'

William Hair of Shapinsay, in 1627 a rare Orcadian in the ferry-louping ranks of the ministry, saw the matter clearly: 'na school in the parish, nor never was, because the people are puir labourers of the ground, and therefore are content that their bairns be brocht up to labour with them'. A generation later, it was much the same story. In 1659, the heritors of Shapinsay agreed to make provision for a schoolmaster, 'but to no effect through the unwillingness of many to put their children to school'.

The reluctance was not confined to Shapinsay. In November 1660, the Orkney Presbytery fulminated about parents being 'not careful, yea unwilling, to put their children to school' – a principal cause of 'the great ignorance that did abound among their people'. Things had improved, but not much, by the visitation of 1702. Half a dozen parishes now reported a school of some sort, but outside Kirkwall none possessed a reliable source of funding.

This came after a century and more of orders and exhortations. The *Book of Discipline* called for schools in every parish, and parliamentary acts of 1633, 1646 and 1696 required the establishment of church-supervised parochial schools, funded by a tax on heritors and their tenants. In 1724, the Kirkwall Presbytery plaintively told the county tax officials, the commissioners of supply, that they had ordered heritors to meet for this purpose, 'yet they have failed therein'.[22]

Voluntary action helped to turn things around. The Society in Scotland for Propagating Christian Knowledge (SSPCK), founded in

1709, was a Presbyterian counterpart to the Anglican Society for Promoting Christian Knowledge (SPCK). It sought, through charitable donations, to establish and maintain schools for teaching reading, writing and arithmetic to both boys and girls in 'the Highlands, Islands and remote corners of Scotland'. An avowed aim, intensifying over time, was the civilising and pacifying of the Highlands by advancing English at the expense of Gaelic.

Orkney and Shetland were not, and never were, Gaelic-speaking regions, and the society paid lower stipends to its schoolteachers there than in Highland parishes, where fluency in Gaelic was an ironic prerequisite for contributing to its eradication. But the SSPCK took seriously its responsibility for all 'remote corners', and some of its earliest foundations were in Orkney: between 1712 and 1730, SSPCK schools were established in Harray, Shapinsay, South Ronaldsay, Rendall, Westray, Eday, Graemsay, Stenness, Orphir, Sandwick, Evie, Deerness, Rousay and Walls.

The issue of language was not entirely absent. In 1725, the SSPCK committee received a letter from John Nisbet, minister of Sandwick and Stromness. His large parish was in desperate need of a charity school, as it 'lies in a remote corner, where the old broken Danish language is used among many of the people'. This, he added, 'occasions ignorance in the place'.

Within a generation, SSPCK schools were being credited with reading Norn the last rites. In 1750, Murdoch Mackenzie wrote in his *Orcades* that thirty or forty years earlier it had been the primary language of two Mainland parishes, 'since which, by the means of charity-schools, it is so much wore out, as to be understood by none but old people'. His brother, James, agreed: the Norse language had disappeared 'within this present age, by means of those English schools erected by the Society for Promoting [*sic*] Christian Knowledge'.[23]

The fate of Norn, symbol of a superseded past, evoked few wistful or nostalgic reflections. At the start of the nineteenth century, the Shapinsay minister George Barry could write that the language had long been 'almost entirely forgotten', surviving 'only in a few vulgar and obsolete words, and in the names of men and places'. Among the educated, a metaphor of choice was to describe the language as 'worn out' – the painted façade of an archaic age, peeling away from the fabric of modern society.

An alternative metaphor might help make sense of a complex linguistic and cultural transition: Norn gradually dried out rather than wore out, and in the process left a retentive residue, an indelible tint on ways of speaking and thinking. As late as the first decades of the twentieth century, Hugh Marwick could identify approximately 3,000 Orkney dialect words he judged to be derived from Norn.

One example will suffice. Orcadian children, well into modern times, were asked by parents at bedtime if they had remembered to say their 'boniewords' – their *bœnar-orð*, words of prayer.[24] It is noteworthy that devotions sanctioned by the Scots Presbyterian Kirk were still known by a name bequeathed by its Scandinavian Catholic precursor. No suggestion of Catholic affiliation remained attached to the phrase, and we could choose to see this as a sign of the overall success of the Protestant Reformation – of what, in a rather different context, historians of the Americas have termed 'spiritual conquest'. Yet conquests, spiritual or otherwise, are rarely complete and bereft of complication: in Papa Westray as in Peru, in Birsay as in Bolivia, officially sanctioned Christianity managed to absorb as well as to reject elements drawn from an earlier system of the sacred.

Meeting Judicially

The spiritual imperatives of the Church and the outlook of ordinary Orcadians met in the deliberations of the kirk session. Minutes survive for Kirkwall from 1626, Shapinsay (patchily) from 1645, South Ronaldsay from 1657, Holm from 1673 and Lady Kirk (Sanday) from 1698. Other parish sequences start sporadically through the following decades; only for Westray, Walls and Flotta, and Hoy and Graemsay are no eighteenth-century minutes extant. The records paint a vivid picture of the successes and failures of Protestant Christianity in Orkney, and of attempts by local people to evade its strictures or adapt them to their own needs.

A kirk session comprised the minister and elders. Parishes might have twenty of the latter – all, of course, men – though usually not more than half a dozen attended any given meeting. Sessions were generally held in the kirk itself, after Sunday service, with nervous suspects waiting at the door. By the later part of our period, some parishes were using a purpose-built 'session house' in the form of a small extension to the kirk building. Occasionally, sessions took place in private houses – in part because of the endemic dilapidation of Orkney parish kirks in the seventeenth and eighteenth centuries. The heritors' responsibility for maintaining churches in good repair – like their responsibility to fund parochial schools – was an obligation more regularly shirked than embraced.

Heritors themselves occasionally served on the session, but Orkney's eighteenth-century lairds, whose sympathies were in any case often Episcopalian, largely steered clear of the eldership, with its onerous duties and aura of subservience to the minister. Most elders were of less exalted rank – respectable farmers or, in Kirkwall, merchants and wealthier artisans.

Political changes were weakening kirk session authority in Orkney, just at the point we start to learn most about it. In 1712, parliament restored lay patronage over parish livings, at the expense of the rights of local elders (p. 320). The Toleration Act of the same year effectively removed Episcopalians from the Kirk's disciplinary remit. Magistrates were no longer required to help enforce attendance before the session, or to imprison excommunicates. The great majority of Orcadians, however, remained formal adherents of the Church of Scotland, and submitted voluntarily to what was, in effect, a form of community policing. Elders represented particular parish districts, with responsibility for repressing vice and promoting virtue within them. The districts often corresponded to individual townships, strengthening the connections between kirk discipline and daily community life.

Knox's *First Book of Discipline* of 1560 envisaged congregational election of elders, as well as an annual rotation of membership. But under a revision of 1578, eldership was made into a life appointment, and a 'function spiritual': elders, like ministers, were 'ordained', and took oaths of good conduct. During services, they sat in a special pew beneath the pulpit, scrutinising the congregation. Popular elections, if they ever took place in Orkney, had disappeared without trace by the mid-seventeenth century. When elders died, or became too infirm to continue, the session simply decided on replacements, though invitations were sometimes conveyed to the congregation to offer objections, if they knew of any, to the nominee's 'life and conversation'; that is, his moral behaviour.

Not all were eager to embrace the role. In 1699, at Lady Kirk in Sanday, Andrew Muir 'modestly refused' nomination, but was co-opted when 'he could offer no relevant reason'. Hugh Gibson and Robert Grieve in Egilsay were, in 1736, 'unwilling to engage in that office'. But the minister, James Jamieson, judging they had 'a competent knowledge of the principles of our Holy Religion', managed to talk them round. The clerical voice was decisive, and no elder could expect to be chosen without the minister's consent.[25]

Minister and elders brought to the session meeting what they happened to know of local sins and transgressions. Suspected offenders and relevant witnesses were summoned to attend on a subsequent date. Other cases were initiated by parishioners themselves, in person or in writing, to request favours or make complaint about their neighbours.

Business was not always brisk. John Ballantyne, recently installed minister of South Ronaldsay, met with the elders in the Ladykirk at Burwick on a cold January evening in 1736. He asked if there were matters to discuss, they 'answered in the negative' and, after a quick prayer, everyone went home. Meetings were sometimes fortnightly, or even weekly, but in Orphir minutes from May 1717, the elders felt obliged 'to give the reason of their not meeting judicially these seven months bygone'. From Lammas to November, 'there was little or no business for their consideration', and after that the minister had fallen ill.

Sessions were expected to keep calm and carry on, but absences of the minister sapped their energy and efficiency. The Stronsay elders had to be admonished in 1702 'to double their diligence in taking inspection of the manners of the people … while they wanted a minister'. Orkney's systemic understaffing, with kirks and congregations always outnumbering ministerial remits, hampered the application of discipline. On a visit to North Ronaldsay in September 1740, the Sanday minister Thomas Covingtrie 'sharply reproved the baillie and elders for their being so long in sending for him'.[26] Though not invariably observed, the principle of each kirk and congregation maintaining its own session was a matter of local pride – there were, for example, independent sessions for the conjoined parishes of St Andrews and Deerness, and separate ones for the Ladykirk and the Peterkirk in South Ronaldsay.

Trelapse in Fornication

One transgression massively outperformed the others in the league table of kirk session business: forbidden sex. Those hauled in to answer for it were predominantly – though not exclusively – young, and a majority were women. The Kirk did not, in theory, recognise any gender disparity in the offence, but pregnancy was often the trigger for a prosecution. Pregnant women were ordered to name their baby's father, but men persisting in denial could be let off if they dared to take a solemn oath of purgation, hoping to 'die in desperation and in the great day be cast into hell' if they were not speaking the truth. Many offenders appeared only once, but others were guilty of 'relapse'. A third offence was 'trelapse', a fourth, quadrilapse, and so on.

Adultery, where one or both of the parties was married, was a more serious matter than simple fornication. In June 1704, the elders of Cross and Burness in Sanday contemplated a conundrum in the classification of sin. John Scott had conceived a child with Marion Mowat, whose husband had been absent eight years, not known if he was alive or dead: 'it is therefore to be considered whether it be fornication or adultery'.

Some offenders were dealt with in-house, but difficult cases were referred to the presbytery. In Burray, there was no way to prevent the adultery of William Leith and Janet Smith, the ministers concluded in April 1644, 'except they were put in sundry isles'. It is not known what their successors were feeling in April 1693, as they sentenced Adam Brebner from Papa Westray to his thirty-first public penance. Probably a similar sense of fighting a losing battle to that experienced in Sanday when, in October 1701, Bessie Moodie, 'adulteress', appeared before the Lady Kirk Session for a remarkable forty-second occasion. Meeting after meeting, through the ebb and flow of the 1600s and 1700s, the elders were swimming in a sea of sex.

If the Church could not prevent illicit copulation, it could control its public meanings, and exact symbolic retribution. Primarily, this meant putting culprits on public display, in rituals of shaming designed to advertise the gravity of their sin and underline submission to godly discipline. Most Orkney kirks possessed a 'stool of repentance', a seat or bench for offenders to sit on during or after the sermon, on one or more specified Sundays. The place was usually specified as 'before the pulpit', the kirk's most prominent location. In St Magnus Cathedral, the stool of repentance was a corner of the so-called 'sailors' loft' on the south side of the choir. The Kirkwall Session also sentenced delinquents to stand on the 'white stone', a marble slab close to the pulpit, and made occasional use of physical restraints in the form of the 'cuckstool' and the 'jougs' – a pillory located in the kirkyard, and a hinged iron collar on a chain attached to the kirk door.

The spectacle was enhanced by attire. Penitents were displayed either 'in lineis' – simple linen undergarments – or in gowns of rough sackcloth. The Orkney Synod was insistent in 1663 that 'sacco', rather than linen, was appropriate for adulterers and relapsed fornicators.

One key weapon in the session's armoury was the sacrament of baptism, which parents invariably wanted for their child. It could be conditional on evidence of repentance, or dangled to make unmarried mothers name their baby's father. A central – arguably, the central – doctrinal tenet of the Kirk of Scotland was the concept of predestination formulated by John Calvin: the belief that God, before the creation of the world, had determined the salvation or damnation of every single soul, without consideration of individual merits. One consequence of this austere teaching was that baptism – though an important token of God's covenant with humankind – was not formally necessary for salvation. Infants dying unbaptised were as likely to be saved as those sprinkled by the minister. But ordinary folk never seemed quite to believe this, and usually wanted the rite performed as quickly as possible – in 1659, the presbytery had to insist that baptisms take place only on Sundays.

A string of folktales, gathered in Orkney in the nineteenth and twentieth centuries, featured the wandering souls of unchristened children, sometimes inadvertently laid to rest after a name was casually given to an animal or other shape that the child's spirit had assumed. Naming was critical. Walter Traill Dennison, in 1846, was asked by illiterate parents

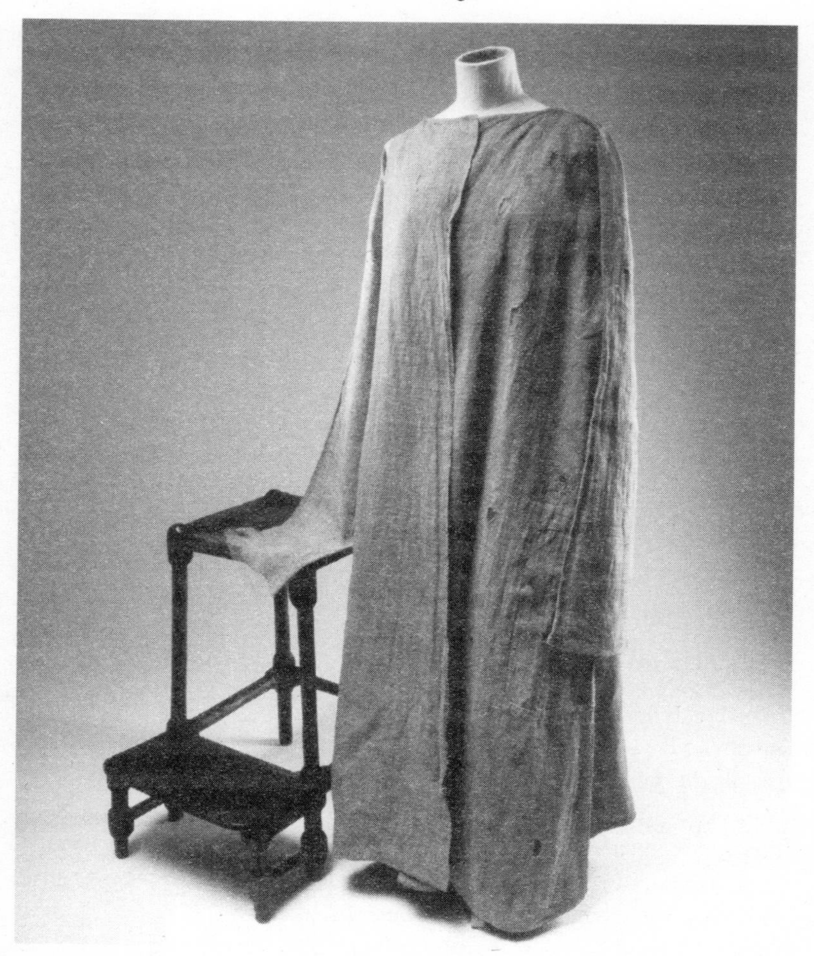

Figure 8.2: A rough sackcloth gown, of the type worn by women making public penance in Orkney churches by order of the kirk session (this eighteenth-century example from West Calder in Midlothian).

to write the name of their unbaptised child on a slip of paper, so they could place it over her breast in the coffin.

Unchristened infants were among the items sailors considered bad luck. In 1678, the Kirkwall Session refused baptism to the illegitimate child of Elspeth Sutherland, conceived while in service in Edinburgh. She was ordered first to return to the city and make satisfaction there for the sin. But the session was forced to relent when Elspeth explained that 'no ship will accept her company unless the child be baptised'.

Other deterrents were pecuniary. Sessions levied fines for fornication, or made demands for a bail payment or 'caution', to be confiscated if there were continuance in sin. By the end of the seventeenth century, most Orkney parishes had instituted a system of 'marriage money'. Sums were posted when couples' banns were called in church, later to be returned to them – or forfeited, if a child's premature arrival implied 'antenuptial fornication'.

Illicit sex was indeed often pre- rather than strictly extra-marital, taking place in the 'handfasting' period after a couple exchanged formal promises of marriage. There was a mismatch between kirk teaching and popular intuition on the degree of sinfulness involved here. Handfasting was believed to possess a solemn and binding character, constituting as much as anticipating a union between the partners. Yet the Kirk saw things differently: the Holm Session decreed in March 1700 that the same punishments be given to fornicators 'having ane intention to marry ... as others not being under promise or contract'.[27]

Handfasting sometimes involved a rather remarkable ritual. Just to the north of the Stones of Stenness was another prehistoric monolith, an eight-foot slab with an artificial hole in its lower portion, perfectly round and described as the size of a hand or a head. Its existence was noted by James Wallace and John Brand, and elaborated on by later commentators. In 1784, the Edinburgh minister Robert Henry, perhaps drawing on information supplied by George Low, described a custom falling into disuse only 'within these twenty or thirty years'. Young couples, having offered prayers to 'the god Wodden', grasped hands through the hole, 'and there swore to be constant and faithful to each other'.

The idea of a dedication to Odin sounds improbable, but is confirmed by a visitor in 1781, the Catholic priest Alexander Gordon. After viewing the stones at Stenness, he was told about an incident that had taken place around twenty years before. A young man deserting his pregnant fiancée was treated with unusual severity by the session. When the minister demanded to know why this was, the elders explained that he had broken 'the promise of Odin'. Around the same time as Gordon, a naval surgeon, James Ker, visited the site in the company of a Kirkwall doctor. He too was informed by his host about this custom, persisting until 'about twenty years ago', and told that, having sworn fidelity, couples 'proceeded to consummation without further ceremony'.

There is no documented reference to the pledging of troths through the stone before the mid-eighteenth century, so it is hard to know if it really was an immemorial custom, and whether the associations with Odin were rooted in ancient folk memory or echoed a more recent scholarly revival of interest in pagan Norse religion. If the practice did come to an end about 1760 – just as Norn went through its death throes – then this was a tactical victory for the Kirk in its attritional war with Eros.

The war was not necessarily unwinnable. Premarital sex was, for example, sometimes spontaneously confessed. In April 1736, the minister of Firth, Andrew Grahame, revealed how John Hervey in Quatquoy 'acquainted him that he was guilty of antenuptial fornication with Katherine Yorston his wife, and therefore craved that they might be allowed to satisfy church discipline'. After paying a fine and receiving 'sharp rebuke', they were absolved of their sin, 'giving evident tokens of their grief and sorrow'.

Figure 8.3: The Reverend Robert Henry's 1784 sketch of the Ring of Brodgar and Standing Stones of Stenness. At the Odin Stone (labelled 'D'), a couple grasp hands through the artificial aperture.

It was a far from unique instance of the internalising of kirk teaching. In March 1760, Jean Sandison 'wept and declared that she had been in bad company' when confronted by one of the Shapinsay elders with rumours about her pregnancy. In 1755, Helen Sinclair was fined by the Birsay Session for conceiving a child with Magnus Linklater. Twenty-three years on, now living in Sandwick, Helen unexpectedly visited George Low, minister of Birsay, 'wounded in her spirit', to confess that she had never formally been absolved for the sin and to ask for it to be put right.

We can only speculate as to the spiritual and psychological impact of accumulated feelings of guilt and shame, amplified by exposure before friends, neighbours and figures of authority. The humiliation fell disproportionately upon the humble. When men in authority themselves transgressed, they were rarely put to public penance, but absolved privately, and allowed to commute the penalty into a fine.[28]

It is undeniable that women – for whom reputation and identity were configured around ideals of sexual 'honesty' – were more affected than men. From a twenty-first-century perspective, the Kirk's policing of sexuality seems intrusive and oppressive. But eighteenth-century people were not twenty-first-century people, and we cannot be certain that penance and repentance were not at times experienced as restorative or cathartic.

Sabbath-Breach

In second place to the regulation of sex, it was what people did on Sunday that absorbed the largest chunk of the sessions' time. The commandment to 'remember the Sabbath Day and keep it holy' meant more than refraining from remunerative labour. It encompassed games, pastimes and recreation, patronising alehouses or making social visits to neighbours, along with all forms of 'unnecessary' travel.

The deliberations on Sundays of the sessions themselves were exempt from the prohibition, but we can enjoy an irony preserved for posterity in the Deerness minutes of April 1704. Before making a trip south, the minister, John Cobb, left strict instructions with the elders 'to inspect and oversee the manners of the people ... and to take particular notice that they transgress not the Sabbath'. But Cobb was himself spotted in Morayshire, travelling on a Sunday, and reported to the Presbytery of Inverness and Forres. It notified the Kirkwall Presbytery, and Cobb was called to account on his return.

The myriad cases of 'Sabbath-breach' found in the session records comprise a mosaic of the activities considered sufficiently important, or enjoyable, to risk prosecution for undertaking. They included varieties of domestic and semi-domestic labour – bleaching clothes, drying washing, brewing ale, thatching houses, threshing corn, baking bread, or, as at Burwick in South Ronaldsay in 1659, 'laying out of fishes to dry'. In Shapinsay, a regular concern was 'trocking with the Inglishes' – that is, bartering with English sailors. In September 1682, Nicol Aim and John Work were fined for having 'received salt fra the Inglishes on the Lord's day', and a young boy who traded eggs was ordered to be whipped.[29]

There are vivid snapshots of Orkney's subsistence economy: a man in Holm reported in 1698 'for going through the hills seeking his horses'; another reproved in Shapinsay in 1660 'for going to the crags to draw [catch] fish'. A band of Shapinsay folk were denounced in 1684 for being 'on the spout sand taking spouts'. The harvesting of razor clams – a delicacy known then and now in Orkney as 'spoots' – was skilled and spine-stretching work. It involved going slowly backwards over a low-tide beach, scanning the sand for little spouts of water betraying the submerged location of a clam.

Another munificence of the receding tide was dulse, edible seaweed. In Holm, at midsummer in 1700, Anna Spence and Elizabeth Crawford were 'alleged to have been in the ebb and carrying dulse out of the same on the Sabbath day'. Accused of the same offence by the Stromness Session in 1702, Anna Hestwall pitifully confessed to gathering seaweed to feed her children, 'not having anything else to give them'.

The shore's bounty proved likewise troublesome along the upper coast of North Ronaldsay. In January 1739, reports emerged from the township of Ancumtoun, Orkney's most northerly place of habitation, of people 'receiving a selkie into their house' – a butchered seal, rather than a shapeshifting spirit. Magnus Swannay confessed to finding and killing it in one of the seal-nets fixed along the rocky shore. He reckoned, with studied vagueness, this was 'off-going Sunday and incoming Monday morning'. The session ordered the Ancumtoun tenants henceforth to pull up their nets on Saturday.[30]

Motion disturbed the sanctity of the Sabbath. The minister of Stronsay spoke from the pulpit in 1758 against 'unnecessary travelling both by sea and land, and people going from house to house'. In October 1706, Nicol Gorn (p. 232) was cited to the Holm Session 'for riding on the Sabbath Day to Deerness'. The case was dismissed when he explained he went there to fetch food, as 'he had neither milk nor fish nor meal in his house'. There was less understanding for James Hepburn, convicted in Shapinsay in 1687 of carrying peats on the Sabbath. Hepburn claimed he needed 'to kindle his fire fra his sister's house'. But the session decided on public penance, to deter others from violations 'upon whatsoever pretence or excuse'.

Travel between islands was a routine matter of sociability and commerce, but on the Sabbath it became a provocation and transgression. We have seen already that timing was the Firth Session's principal objection to Elspeth Bews's visit to Damsay. The session in Shapinsay, an island enticingly close to Kirkwall, was particularly concerned about Sunday sailings. In 1668, it ruled 'no person or persons shall pass the ferry on the Sabbath day except in the case of urgent necessity'. In March 1676, the Kirkwall Session summoned the entire crew of a vessel for carrying letters to Gairsay on the Lord's Day, and as part of their penance ordered them to deliver a boatload of sand to the cathedral.

When ferries went to the Scottish mainland, the embarrassment of Orcadian Sabbath-breach was made manifest to outside observers. In 1714, the Orkney Synod was warned by a committee of the General Assembly 'that the Sabbath Day is greatly profaned by the ferrymen in Walls, who cross Pentland Firth that day very frequently'. Three years later, 'the scandal of the Orkney ferries' was the subject of a complaint to the synod from the Presbytery of Caithness.[31]

Recreation on the Sabbath – the one day when people had time on their hands – was a persistent worry. The 'great profanation' of drinking and drunkenness produced session acts about selling ale on the Lord's

Day in Kirkwall and elsewhere. The Holm Session issued an ordinance against Sabbath drinking in 1682, and again in 1685 and 1686 – the practice was easier to proscribe than prevent.

Equally unrestrainable was the exuberant disorderliness of youth. South Ronaldsay's Ladykirk Session was satisfied with its rebuke to boys playing football on Sunday 26 December 1658: 'since that time not any such abuse committed.' But football was evidently a Christmas tradition in South Ronaldsay, necessitating an annual seasonal warning which did not prevent further prosecutions.

In Kirkwall, names were taken in 1670 of 'idle boys [who] play at football in time of and after sermon' – Sunday services themselves supplied occasions for young people to congregate and commit Sabbath-breach. Earlier complaints involved bowling, and servants and apprentices 'playing at the pennystone' – a game similar to quoits, for which youths were also in trouble in Orphir in 1712. In Lady parish, Sanday, a flurry of concern in 1698–1700 – coinciding with the arrival of a new minister, Patrick Guthrie (formerly of St Andrews and Deerness) – surrounded the Yuletide custom of 'young persons meeting in the night-time … for sporting and dancing in barns'.[32]

Kirkwall had a concentration of young people, in service or apprenticeship, and juvenile socialising there all too easily took alarmingly antisocial turns. A special session meeting was convened in June 1643 to deal with the latest Sunday fad: young men and women 'wading in the sea promiscuously and using lascivious gestures'. This was arguably no worse than the behaviour of a dozen apprentices, rebuked in February 1658 'for profaning the Sabbath in casting of snowballs and abusing the bygoers'. In future, the session decided, masters would be held responsible for such misconduct. In June 1674, a weaver and a tailor were each fined twenty shillings Scots after their apprentices were caught 'climbing houses and taking bird nests in time of sermon'.

A strategy of punishing the parents was adopted by the Holm Session in 1715. The offence here was 'vaging' – wandering around in an aimless manner. Youngsters were prone to 'going abroad in the fields on the Sabbath day'. It was still more of a problem in Kirkwall, people 'standing in multitudes conferring about worldly and sinful things'. In 1710, session elders, accompanied by town officials, were authorised to make sweeps of their districts following afternoon sermon, and arrest 'idle vaging persons' unable to give convincing reasons for being out and about.

To modern sensibilities, the Kirk's Sabbath discipline seems desperately dour and oppressively legalistic. In a poignant case from the summer of 1727, John Welie was punished by the Holm Session for 'sporting himself with a little play ship, or some childish fancy, and his taking down a boat, near the time that divine worship was beginning, in order to recover his play ship'. John admitted launching his boat to rescue the imperilled toy, offering in defence that it belonged to his son and 'he did it out of pity of the weeping child'.

Kirk discipline, however, was never something that was simply imposed. It was a system which, to varying degrees, the people of Orkney embraced, and with which they actively collaborated. Respectable opinion was often in favour of curbing the boisterous habits of youth, and of criminalising liaisons that threatened to burden the community with illegitimate bairns.

Sabbath-violation was an offence which, almost by definition, required witnesses. When John Lennard and Jean Leask were prosecuted by the Rousay Session in September 1734, it was because they 'were seen' travelling in a boat the previous Sunday 'from this isle to the isle of Egilsay'. In Shapinsay in 1664, in an unusually cultured instance of Sabbath-breaking, 'John Work was rebuked for playing on the viol'. Someone must have heard the courtly strains of music and gone off to inform the session.

In some cases, a willingness to report may have stemmed from an aggrieved sense that Sabbath-breakers were gaining an unfair advantage. Witnesses against Hugh Bews of Firth declared in 1736 that 'they saw sheaves of corn lying down in his yard drying upon the Sabbath morning, and at night it was scrowed up' – built into a stack. Harvesting by some Birsay folk till late on a Sunday in 1784 was said to be 'a grief to their neighbours and such an offence against the Law of God that it is wondered at in other parishes'.[33] Kirk sessions, like the baillie courts with which they cooperated, were thermostats of community cohesion and the assumption of 'limited good' (p. 192). They helped to maintain an egalitarian ethos of collective prosperity and shared experience, even at the expense of football, gossip, music and play.

Charitable and Uncharitable Acts

Kirk sessions were in a variety of respects the conscience of the community. They were responsible for poor relief, paid out of the fines of fornicators. Sessions arranged for the fostering of orphans, and covered the costs of pauper funerals. Burial was a largely secular matter, in the view of a Kirk of Scotland allergic to any suggestion of praying for the dead, and no funeral services were held in the church. Attendance at burials was nonetheless a neighbourly duty enforced by the session. In Orkney, individual elders' 'bounds' were generally de facto burial districts, defined by an obligation on residents to take part in funeral proceedings. Next of kin were supposed to inform the elder so he could send round a timely 'burial warning'. The Orphir Session reacted crossly when, in January 1715, James Groundwater failed to start the notification for Janet Gunn in the township of Tuskerbister. Few attended the funeral, 'and those that came were but weak boys, which occasioned great disorder and gave offence'.

A sometimes onerous and time-consuming duty, funeral attendance tested the bonds of social solidarity. In 1658, South Ronaldsay's Ladykirk Session condemned the 'unnatural and unchristian carriage' of neighbours who did not show up. 'Slackness to convene for that end' was in 1671 castigated by the Shapinsay Session, which later instituted fines for such 'unchristianity'. The Holm Session did the same in 1695, concerned particularly about reluctance to turn up 'where the party deceased is of the meaner sort'. In 1741, the South Ronaldsay Session threatened to bring in the civil magistrate 'if the people do not attend the buryings of the poor'.

Solidarity with the living generally fared better. People who, through misfortune or misadventure, lost homes, crops or livestock could petition the session for relief. In Rousay, in 1744, a collection was organised to purchase a share in a boat for John Gibson, one of the parish elders, 'reduced to low circumstances'.

Charity began at home but didn't end there. The session minutes contain many incidental confirmations of Orkney as a place of concourse and encounter. There was a collection at Kirkwall in 1658 for 'ane distressed Dutchman and his wife', and in South Ronaldsay in 1661 for 'an poor Shetland man whom God had wonderfully preserved into a storm at sea into [*sic*] his little boat'. Two years later, the Peterkirk congregation raised thirty shillings for 'some Inglish strangers who had been shipbroken'. In later seventeenth-century Shapinsay, collections were taken up for 'an impotent south country man', 'a distressed south country man', 'the relief of a French Protestant'; in Holm, there were payments to an impoverished Irishman, and 'a distressed stranger named James Dallas whose houses, corns and cattle had been burnt'. It was a public scandal, in the late winter of 1690, when a poor woman from Eday died from cold after being denied shelter in the Shapinsay township of Meoness. The session punished two women 'for such ane unchristian and uncharitable ane act.'[34]

Charity meant more than benevolence to the disadvantaged. In pre-modern society, it was an encompassing social and religious virtue, expected to infuse all aspects of neighbourly relations. Spreading scandalous rumours was a quintessentially uncharitable act, and defamed individuals resorted for redress to the session. A 1681 supplication from James Wishart and his wife, Katherine Robertson, to the Orphir Session demanded that a man they accused of defaming them 'appear personally before the congregation in time of divine service', in order to repent and to restore 'myself and my wife to our good names and reputation'. In October 1758, the Birsay Session received 'The Complaint of James Twatt, Shepherd', outraged at being called a thief by Bess Allen. Twatt grandiloquently required any elder able to confirm such an accusation to speak up, 'for shepherds ought to be honest men'.

Masculine honour, or perhaps just hurt feelings, persuaded John Swannay to contact the Shapinsay Session in January 1668. He was 'nicknamed a sow's cunt'. The session formally ordered that 'no person within the parish shall hereafter upbraid him with such a vile nickname'. Whether this succeeded in stifling the slander, or simply amplified it, must remain a matter of conjecture.

In July 1671, the Kirkwall Session received a bill of complaint from Magnus Spence, parishioner in St Andrews. It accused William Anguson of calling him 'ane buggerer and ane cowlooper dog'. Spence brought to the session a judgement delivered by the JPs in 1659. It stated that Isabel

Mucklenson had slandered Spence 'in saying he committed bestiality with ane cow', a charge she was unable to prove. Anguson, like Mucklenson twelve years previously, was ordered to make public satisfaction. And, at Spence's request, proclamations were made in the cathedral and St Andrews parish kirk the following Sunday threatening the presbytery's wrath against anyone daring to 'reproach the said Magnus, his wife or children with the said unchristian and gross slander'. The case bears witness to the spiteful staying-power of injurious rumours in a small community, and to the importance people placed on the maintenance of their reputations, even at the risk of giving publicity to the original slur.[35]

Kirk sessions were willing to punish slanderers, but their principal concern was often to reconcile bickering neighbours, and to restore parish harmony. In a Birsay case from George Low's time as minister, Janet Sabiston brought a complaint against the weaver Robert Stensgar for saying she and her husband had retained a half-guinea Stensgar accidentally dropped in their house. In his efforts to locate the lost coin, so Sabiston claimed, Stensgar had 'gone to Stromness to wizards and necromancers' – a habit no longer prosecuted by church authorities, but not yet abandoned by aggrieved Orcadians. In June 1777, both were summoned before the session, and Low's 'admonitions and exhortations' persuaded them to set aside their quarrel. Stensgar took Sabiston by the hand to beg forgiveness, and all were dismissed, 'with suitable exhortations to behave for the future as becomes Christians'.

Session discipline was severe when it needed to be, but could sometimes show a surprisingly pragmatic face. In August 1684, the Holm Session agreed to grant ferrymen a licence to transport cattle on Sunday, since 'the multitude of them was so great that there could be no accommodation for them without intolerable skaith [damage] to corns'. The Kirkwall Session hesitated in 1721 over the case of a man who had taken a boat out on Sunday to secure a whale stranded on a sandbank. Such a rich bounty was not easy to forgo, and the session was aware 'others in the country did so on like occasions'. It was decided to pass the buck and refer 'this matter of catching whales' to the judgement of the presbytery.

Quite often, 'rebuke', rather than fines or public penance, was offered to youthful or first-time offenders. In Shapinsay, Magnus Cumming was in 1659 dismissed with a warning after he admitted to being 'overtaken in drink' and made tearful promises of reform. It was a mitigating consideration, too, that 'his scandalous miscarriage was not within our isle'. Alas, Magnus proved a hopeless recidivist, and a year later did public penance after again becoming conspicuously drunk in Kirkwall. In Sanday's 1701 crackdown against 'scandalous night dancing', the Lady Kirk Session decided the youngsters involved 'were of ane unblameable conversation [conduct], except in this particular', and let them off with a warning.[36]

Elders were members of the communities they watched over. That cut both ways, making them informed judges and prosecutors, but creating conflicts of interest and suspicions of partiality. At presbytery and episcopal visitations, ministers were always asked about elders' conduct and capacity. The replies were generally positive, but a persistent refrain was the desire for them to be 'more diligent in delating [reporting] scandals'. In 1680, John Hendrie said of his elders in Orphir that some 'did speak too much for delinquents and some of them did divulge what was acted and spoken in session'. In 1702, John Pitcairn of Firth thought his elders 'too favourable with scandalous persons in the modifying of their penalties' and asked the moderator to issue a 'gentle reproof'.

Ill Kirk Keepers and Wicked Lairds

The authority of the session was acknowledged but not invariably respected, and even bare attendance at services could not always be taken for granted. The 'great abuse of the parishioners in absenting themselves from the church on the Sabbath', reported in Kirkwall in 1657, perhaps reflects a loss of control during the Cromwellian occupation. Yet, in 1670, the bishop and elders were still concerned to identify who in the town 'comes not to the church, they being in health'.

Absenteeism was a serious concern in Shapinsay in 1679, and in Holm in the 1680s, with schedules of fines for those who 'after many exhortations to the contrary absent themselves from the church'. Some young Orphir people were in 1709 noted as 'ill kirk keepers'; and in Deerness, in 1703, Jane Paplay confessed she had been in church only three or four times within the last three or four years.

The motives for skipping services varied. William Swanson of St Margaret's Hope in South Ronaldsay told the Peterkirk Session in 1659 that 'he would come to the kirk or not come to the kirk when he pleased himself'. But shame, not shamelessness, accounts for the absenteeism of James Swannay and his wife in Shapinsay in 1683. They stayed away, they said, 'for want of clothes'.

Orkney's unpredictable and often intemperate weather affected attendance. There was no service at South Ronaldsay's Peterkirk on 27 August 1665, amid 'winds, rain, thunder and fireflaught' (lightning). The minister, Edward Richardson, went to the kirk and waited, but no one turned up. On 11 October, 'the people were not so well convened as that duty required, to the minister his great grief – God be merciful to them'. In December 1668, a day of 'extraordinary tempest' again left Richardson on his own at St Lawrence Kirk in Burray.

Many attended irregularly because there was no weekly service in 'their' kirk. At the 1702 visitation, the minister and elders of Westray admitted that people went to their own churches – Cross Kirk and Lady Kirk on the island itself, St Boniface in Papa Westray – 'but neglected to come to their neighbour churches when sermon fell there'. As John Brand discovered in 1700, there were seventeen clerical postings in Orkney, but thirty-one functioning kirks. Clergymen were lucky if they had but one place of worship to see to: a note of satisfaction graces the statement of the Shapinsay minister and elders at the visitation of 1627: 'the parish is not united to any other parish, but is of itself, having one kirk, which serves the people thereof.' Elsewhere, Brand learned, the pattern was one of people coming to services only every second or third week, as it was 'expensive and dangerous for them to travel from isle to isle'. This, he lamented, 'cannot but obstruct the progress of the Gospel'.[37]

Unsurprisingly, sessional process could create resentment and resistance among those subjected to it. One response was to flee. The Deerness Session heard in July 1706 that John Corner and his servant Isabella Thomson, spotted kissing 'in the face of the peat-stack', had now 'fled to Caithness from the discipline of the Church'. In such cases, sessions notified the presbytery, and presbyteries liaised with each other. In the summer of 1666, two other forbidden couples – Margaret Ballantyne and

Patrick Sandison, Isabel Ballantyne (perhaps Margaret's sister) and Francis Strachan – decamped from Walls. The presbytery ordered ministers to look for them in their parishes 'and charge them to repair to Walls for obedience to the discipline of the church'. In 1678, searches were started for John Mylles of Lady parish in Sanday, suspected of raping Christine Moar. He had been ordered to appear in the church in sackcloth, while Christine was to repeat publicly before the congregation 'what she did confess privately ... before the session'. Mercifully, she was spared the ordeal – Mylles's flight was taken as confirmation of the offence, 'and no guilt to be laid to her charge'.

To stop people evading justice, kirk sessions in Orkney, like those elsewhere in Scotland, operated a testimonial system. Anyone permanently leaving the parish applied for a certificate of good character, to present to the session in their new place of residence. These written attestations of 'honest conversation', or of having 'lived inoffensively', were a confirmation of individual honour by the collective judgement of the community, and for that reason valued by recipients. James Manson of Stromness took offence in 1723 at the sparse testimonial his new wife brought with her from Orphir. He wrote an angry letter to the Orphir Session, saying such efforts 'would be a laughter in other parishes', and that their session clerk deserved to have his right hand removed.

The system was not watertight. Several ministers complained to the Orkney Presbytery in 1680 about people relocating without certificates, and fines were introduced for giving lodging to undocumented incomers. Missing testimonials were a problem in Deerness in 1702 and Holm in 1705.[38]

Some sinners opted for fight over flight. In Kirkwall, the fornicator William Marwick was imprisoned in 1682 for declaring, 'in face of session', 'that hemp or lint was not [in existence] that would be a sack cloth to him!' From 1679 onwards, the weaver Edward Rynd was in serious trouble for a sexual assault on a sleeping relative, undertaken as a drunken bet: 'the shaving the secret parts of ane woman'. His case progressed from the magistrates to the Kirk, and, rather than accept punishment, 'ye did answer in public that ye should rather be hanged or shot before ye would answer any church censure'. The session clapped Rynd in Marwick's Hole (see Fig. 5.2). Upon release, he was lucky to be banished, rather than hanged, after trying to assault the minister James Wallace in his manse at night.

This behaviour was excessively antisocial, but instances of angry contempt for kirk discipline are not hard to multiply. John Peace of Sanday, hauled before Cross and Burness Session in 1704 for beating a boy on the Sabbath, announced 'he cared not what day it was'. In South Ronaldsay in 1745, Archibald Gordon and his wife, Anna Sinclair, committed Sabbath-breach 'in mutually scolding'. When Anna failed to show up before the session, her husband sneered 'it was not worth the appearing for'. In 1735, the Firth Session ordered Alexander Bews to settle his quarrel with Helen Yorston. But Bews, 'a desperate and furious man', swore that even if 'they should hang him, he should never make satisfaction for that which he was not guilty of!'

Such cases played out in intensely 'face-to-face' societies, where people were well acquainted both with their accusers and with the elders sitting in judgement over them. Resentment might well up when there was a perception of punishment as intrinsically hypocritical, since the other parties involved were no better themselves. In 1738, William Swannay was charged before the North Ronaldsay Session with fornication and incest (sleeping with a widow and her niece). Swannay did not deny the charges, but claimed his accusers 'were guilty of as bad'. When the elders pressed him for details, he revealed 'the house of Ancumtoun were guilty of killing selkies on the Lord's Day' – triggering an investigation we have taken note of already (pp. 392–3). Swannay himself escaped consequences from the ongoing case: it was reported in September 1740 he had 'gone off from this isle with a ship bound for England'.[39]

Kirk elders could cope with apprentices and servant girls. But the attitudes and behaviour of social elites posed a persistent and sometimes insoluble problem. In Orphir, in 1681, Elspeth Taylor issued an initially defiant response to the minister's demand that she reveal who had made her pregnant: 'she would give any father to her bairn that she pleased'. It took an appearance before the presbytery to elicit a confession that the baby's father was a laird, David Sinclair of Ryssay.

Bringing him to heel proved exhausting and ultimately fruitless. Sinclair refused to submit to the discipline of the session, and when the Orphir minister John Hendrie began excommunication proceedings,

Sinclair interrupted him with 'opprobrious and irreverent speeches'. A presbytery investigation ruled Sinclair should be censured, despite the reluctance of elders, including a fellow laird, William Stewart of Graemsay, to give evidence against him. In June 1684, three years after the business started, the presbytery threw up its hands, deciding 'to desist from further process against him, until it shall please the Lord to open his heart and to work upon his conscience'.

Across the seventeenth and eighteenth centuries, ministers found themselves locked in unwinnable battles with belligerent lairds. Even with the presbytery's support, it could leave them lonely and beleaguered, particularly in islands where a powerful landowner was used to having things his own way. William Watson, minister of Walls and Flotta, clashed in 1639–41 with James Moodie of Melsetter. Watson was physically assaulted after attempting to bring Moodie to book for 'trelapse in fornication', eventually lamenting that with the laird of Melsetter 'the ordinary execution of the laws cannot prevail'. The presbytery spent fruitless hours on the case. It was, at least, nothing especially personal: in January 1646, Walter Stewart of South Ronaldsay made a formal complaint against Moodie 'for abusing all ministers in general, and himself in particular, by his disgraceful and contumelious speeches'.[40]

Watson's problems pale in comparison with those of his successor John Keith, an impoverished father of ten, who had to deal with the malevolently mercurial Christiana Crawford – widow of James Moodie's son, the murdered Captain James Moodie RN (pp. 342–3). As principal heritor, Lady Melsetter was responsible for the minister's stipend, but Keith's attempts to extract payment ignited a feud of epic proportions, with Keith telling the General Assembly in 1727 that his ministry in Walls had for eighteen years been 'a labyrinth of oppressions'. Lady Melsetter maligned Keith to anyone who would listen, accusing him of drinking, swearing and Sabbath-breach. In Walls, she stopped the miller from grinding his corn and the ferryman from carrying his mail.

Matters came to a head in January 1730, when Crawford burst into Walls Kirk, riding-whip in hand, to denounce Keith as a 'false shepherd' and harangue his children as 'sons and daughters of Belial'. She then stormed off through the kirkyard mouthing imprecations – 'the Devil break your minister's neck, and the Devil break all your necks that comes to hear him!' For good measure, she later sent a message to let Keith know 'she had four charged pistols prepared for me'.

The Orkney Synod secured a shaky reconciliation, but Lady Melsetter remained a scourge of Keith and of other ministers who aggravated her – she confiscated peats from John Pitcairn, minister of Hoy (formerly of Firth), claiming they were cut the wrong side of the parish boundary, and had them thrown in the sea. In 1731, she terrorised the minister of Evie, Hugh Mowat, by sending a fierce highlander to sit in the kirk, glowering and handling his dirk as Mowat attempted to preach.

Orkney folk tradition, enshrined in Walter Traill Dennison's tale of 'The Heuld Horn Rumpus', painted Crawford as a predatory adventuress and Mowat as a hapless innocent paying the price for having once fled her amorous nocturnal advances clad only in his nightshirt. The reality was grimmer. Mowat had started proceedings against William Bellenden, laird of Stenness, and Crawford's son by her first marriage, for raping his servant. The highlander was part of a campaign of murderous intimidation.[41]

Lairdly reluctance to be held accountable was mirrored at the northern end of early eighteenth-century Orkney, in an intractable quarrel between the Westray minister William Blaw and his brother-in-law, Thomas Traill of Holland, heritor of Papa Westray – a story which has been ably told by the Papa Westray historian Jocelyn Rendall. The Traills of Holland in Papay were accomplished sinners in successive generations. Thomas's father, George, agreed in 1702 to build a pulpit and loft in St Boniface Kirk (see Plate 25), to cover a fine imposed by the session for fornication, but years after his death the obligation remained unfulfilled. Blaw complained to the presbytery that, in addition to withholding this debt, Thomas Traill never attended services or paid his teinds, and that he profaned the Sabbath by sending his factor to assign tasks to the tenants from the kirk steps. Blaw's attempts to rebuke Traill generated 'an implacable hatred'.

Traill's son Peter was also prosecuted by the session for fathering an illegitimate child, but he refused to do penance, and in June 1718 came to the kirk with his father and brother to demand the minister absolve him. When Blaw refused, Thomas Traill 'swore with a dreadful oath' that he would close the kirk 'and break the bottom of the boat that should bring the minister to the isle'. He turned Blaw's boat away at his next attempt to visit and threatened to evict any tenants who dared convey the minister over from Westray. For a time, Blaw was forced to arrive in the island outside of Traill's land, on a section of flat stone beach that became known as 'minister's flag'. From there, he struggled several miles along the rocky shore to reach the kirk. Eventually, Blaw gave up coming to Papa Westray altogether.

In later years, Blaw was to attain national notoriety for being such a zealous Sabbatarian that he hanged his cat for catching a mouse on Sunday.[42] Whether the Traills played a role in instigating this seemingly baseless rumour is uncertain, but Blaw's travails illustrate just how inhospitable a place for orderly Presbyterian ministry Orkney could sometimes be, if island topography and lairdly animosity decided to combine forces against it.

Out of the Swine and into the Ministers

A few Orkney ministers were more or less martyrs to their calling, but some of their wounds were self-inflicted. An avoidable incident took place in Harray in July 1730, when parishioners marched out of the kirk in response to a sermon by the minister of Birsay and Harray, George Copland. The ringleader, James Flett, afterwards confronted Copland in 'the most insolent and opprobrious language', saying that he was unworthy of the ministry and telling him 'to go home to Birsay and sell tobacco, for he was more fit to be a packman' – an intriguing hint of how the 'Birsay' minister might seem like an interloper even in neighbouring Harray, heartland of the independent udallers and the fading Norn language.

The Cairston Presbytery, established five years earlier in 1725, launched an investigation and took statements. We don't know what Copland said from the pulpit, but some parishioners described the sermon as so 'weighty and heavy that they could not bear the same'. Copland's own story was that 'he was only denouncing God's judgements against unbelieving and impenitent sinners', which does little to alter an impression that his preaching was at best ill-judged. Flett was rebuked before the congregation and referred to the civil magistrate for corporal punishment. But Copland was quietly advised 'to take a more gaining way with the people by suiting his doctrine to their capacities'.

More common as a clerical misstep was breach of the moral codes that ministers were charged with upholding. In 1710, in Cross and Burness in Sanday, the Reverend Murdoch Mackenzie married his servant woman, Janet Thomson, without the presbytery's permission. They were suspected of intimacy 'both before contract and betwixt contract and marriage'. Mackenzie's denials rang hollow, since he admitted saying to friends in the alehouse, prior to the wedding, 'What if she be with child?' The presbytery subsequently learned Janet 'was delivered of child on 26th January 1711, 6 months 13 days from their marriage'. For good measure, Mackenzie was 'very deficient in his duty to North Ronaldsay'. The process was in the end settled by Mackenzie's decision to resign his charge and emigrate to New England.

Housekeepers and female servants were recurrent stumbling blocks. George Copland's predecessor in Birsay, Andrew Giles, confessed to the presbytery in November 1727 'his fall unto the sin of fornication with his servant woman', Catherine Taylor. In 1729, the same story unfolded in Firth: Jean Paplay, servant to the minister James Weir, named him as the father of her child. Both ministers were deposed by the presbytery, but, exhibiting 'much sorrow', they were later restored to ministerial duties.[43] Some doubtless saw a commendable Christian willingness to forgive; others, an institution's instinct to close ranks and protect its own.

Scandal dogged the tracks of the mid-eighteenth-century ministry. From 1736, the Presbyteries of Cairston and Kirkwall were locked in dispute over the character of Alexander Geddes, a Moray man appointed by the earl of Morton, proprietor of Orkney, as Copland's successor in Birsay. The West Mainland ministers were well satisfied with him, but in the East he was suspected of 'undecent behaviour', and the Kirkwall Presbytery boycotted his ordination. Cairston condemned Kirkwall's attitude as 'contrary to that spirit of meekness and tenderness that should prevail among ministers of the Gospel', but showed limited reserves of that spirit itself in referring to the quarrel as a 'war'. The *casus belli* was a broken heart. In December 1736, Geddes married Copland's widow, Elizabeth Fea, but there had been an earlier offer to Elizabeth's cousin Jean, daughter of the veteran Kirkwall minister Thomas Baikie and his wife, another Elizabeth Fea (p. 323).[44]

Rumours of involvement with a married woman nipped at the heels of a second incoming minister. James Tyrie was recommended to the Cairston Presbytery by a committee of the General Assembly in January 1743, shortly before swearing judicially to innocence of adultery. He was serving in a parish on the southern shores of Loch Ness, but was sent north, officially, for 'want of the Irish [that is, the Gaelic] language'.

Tyrie was a carpet-bagger with more than the usual quantity of baggage. The son of a prominent Aberdeenshire Jacobite family, he had trained in Rome as a Catholic priest, and was working as a missionary in Banffshire when he 'renounced the errors of popery and embraced the true Protestant Religion'. It made him an exotic – and suspect – specimen in Orkney, whose ministers had only a few years earlier assured the General Assembly that 'there are no papists within the bounds of this Synod'.

The first attempt to settle Tyrie in an Orkney parish, by Morton's presentation, was at Cross and Burness in 1744. It was opposed by the North Isles Presbytery, and aroused fierce resistance in Sanday, orchestrated by allies of the disappointed candidate, Mr George Traill of Hobister. He was a kinsman of the troublesome Traills of Holland, and an opponent of Morton in the pundlar controversies. According to Morton's stewart-depute, Andrew Ross, a lairdly cabal 'used all their might to persuade the poor ignorant people of the vacant charge that Mr Tyrie was still a papist in his heart'. Tyrie was finally installed at Cross and Burness in September 1746, but the heritors, elders and other leading parishioners walked out of the ceremony.

Within six months, Tyrie was reappointed by Morton to the conjoined parishes of Sandwick and Stromness. Rumours of popishness and immorality were again circulated by the earl's opponents, and provoked a succession of violent disorders designed to prevent Tyrie's admission – first at Stromness on 4 June 1747, and then at Sandwick, where for successive weeks the doors of the little kirk of St Peter, on the Atlantic-facing Bay of Skaill, were barred to the new minister.

So soon after the suppression of Jacobite rebellion, the earl of Morton and his supporters were not inclined to let matters drift: troops were deployed to install Tyrie at Stromness and arrests were made. Of those charged, no fewer than sixteen were female. The presbytery blamed the trouble on 'a multitude of women', 'a numerous crowd of women'. This was in part a matter of tactics: mass protest from women was intrinsically less threatening than riotous behaviour by men, and less likely to demand an armed or violent response. The women were wives and daughters of Stromness tradesmen and Sandwick farmers, and the presbytery received a formal written protestation from Elizabeth Smith, spouse to the Stromness heritor William Gordon, 'in name of my said husband'.

But there is no reason to suppose these women were mere proxies, and not themselves angered and energised by this ministerial intrusion, which seemed to show the Kirk preaching one thing – often specifically towards women – and practising another. At least one of the protest leaders, Catherine Laing, was the unmarried daughter of a deceased Stromness shipmaster, and seemingly acting on her own account.[45]

Matters in time settled down, and Tyrie served the parish respectably enough through to his death in 1778; his acceptance into the incestuous 'tribe' of Orkney ministers is marked by his marriage to the granddaughter of the Reverend Thomas Baikie, and the marriage of his own daughter to George Low of Birsay. But it was not long before clerical scandal again roiled Orcadian society, setting ministers at odds, and securing unwelcome attention on a national stage.

This time the setting was Lady parish in Sanday, where Thomas Lyell, a native of St Andrews (Fife), was presented by Morton in August 1753. There were soon rumours about a maidservant, Elspeth Smith, sent off to Leith in 1756 to be delivered of a child, which was maintained there at Lyell's expense. In 1762, Lyell's unmarried housekeeper, Margaret Scot, gave birth in his house. This, unavoidably, was a matter for the session, but only two elders convened with the clerk and the minister in Lyell's own manse for a sorry charade of sessional justice. By prior arrangement, Margaret named Lyell's younger brother David as the father of her child. Both were dismissed with a cursory rebuke, and Margaret Scot and her baby were set up in a little house by the side wall of the minister's garden.

All this was soon common knowledge, in Sanday and beyond, producing a series of oddly ineffectual interventions. The Synod of Orkney (dissevered in 1725 from union with the Synod of Caithness and Sutherland) was dissatisfied with inaction by the North Isles Presbytery, and in 1763 established a committee to undertake a parish visitation, which on the basis of its findings suspended Lyell. The synod referred the matter to the General Assembly, but its commission detected irregularities in the suspension and directed the North Isles Presbytery to begin fresh proceedings. The six ministers of the presbytery proved incapable of united action, three siding with Lyell, and the other two advocating prosecution. At separate schismatical meetings, one faction acquitted Lyell while the other drew up a new 'libel' of charges. In 1766, the case came back before the General Assembly, which again decided the libel was flawed and dismissed the case.

In the meantime, the Presbytery of Cairston was in knots over the minister of Firth and Stenness, William Nisbet. In 1763, Margaret Agnew, a sister-in-law of Nisbet's brother James, left her husband in Arbroath and came to live in Kirkwall. Nisbet offered her accommodation in the manse in Firth, but what at the outset seemed a charitable gesture was soon a notorious scandal. Nisbet disregarded first private warnings, from James Tyrie and others, and then formal directives from the presbytery and synod, to bring the arrangement to an end. Presbytery admonitions produced merely token compliance: Nisbet temporarily moved Mrs Agnew out of the manse into a tenant's house. At New Year, 1764, Nisbet married a young kinswoman called Elizabeth Ritch, who later admitted she entered the union 'little suspecting to see or meet with any practices repugnant to her just notions of virtue and modesty'. It was soon brutally clear to her how things stood, and she stayed barely four months in the manse.

The presbytery referred the matter back to the synod, which in January 1764 ordered Nisbet to remove Mrs Agnew from his house within four months – a remarkably generous deadline which Nisbet only narrowly managed to meet, by once more relocating her to a convenient nearby address. In August, the synod again ordered them to separate, but after that the case faltered, as the Presbytery of Cairston, like that of the North Isles in the parallel case of Lyell, collapsed into recrimination. Tyrie of

Stromness, John Reid of Orphir and Edward Irvine of Walls were eager to prosecute. But the presbytery moderator, Thomas Hepburn of Birsay, along with Robert Sands of Hoy and the still unsuspended Nisbet conspired to stymie further disciplinary measures. In a succession of increasingly acrimonious interactions, Tyrie sued Hepburn for defamation, Hepburn suspended Irvine from the presbytery and Irvine appealed against Hepburn to the General Assembly.

In the end, Nisbet's case was resolved not by the Orkney Kirk but by outside intervention. Adultery remained a criminal offence in eighteenth-century Scotland, though one seldom prosecuted by the secular courts. Nisbet, however, had made powerful enemies. In the spring of 1765, he was tried and convicted before the circuit court of Inverness, and sentenced to two months' imprisonment, followed by transportation to the West Indies. In passing sentence, the judges demanded an investigation into 'why no prosecution was properly brought before the ecclesiastical courts'. In July 1766, Nisbet was finally deposed from the ministry, at the initiative of the Presbytery of Inverness.

Lyell's case, meanwhile, had not been allowed to drop, largely due to the determination of George Traill of Hobister. Unusually, Traill was both cleric and laird: minister of Dunnet in Caithness, but principal heritor in Lyell's parish in Sanday. Backed by the Synod of Caithness and Sutherland, Traill ensured the libel against Lyell was drawn up again in the proper form, and brought a case in the court of session to compel the giving of testimony. Several dozen witnesses from Sanday and elsewhere testified to Lyell's multiple offences, including various attempted sexual assaults, and threats of violence against other ministers.

Lyell went down fighting, claiming that there had been nothing in the Church of Scotland 'since its first establishment after the Reformation similar to the persecution he has suffered'. The case against him arose from an irrational antipathy to ferry-loupers on the part of inbred lairds, while the evidence provided by witnesses was 'the tittle-tattle of the very meanest of the rabble of Orkney'. In May 1768, he too was at long last deposed by the General Assembly.

The conjoined cases of Lyell and Nisbet were unedifying public spectacles, wrangled over in the General Assembly and widely reported in newspapers. In his petition of May 1766, Edward Irvine lamented the indignation of 'all ranks of people' in Orkney against the behaviour of the clergy: 'they saw vice and immorality, if not publicly allowed, at least tacitly permitted'. Looking in from the near outside, the ministers of the Synod of Caithness and Sutherland warned that 'reports concerning both Mr Lyell and Mr Nisbet were generally believed all over that country, and were attended with consequences very hurtful to religion'.

Religion aside, the cases had political and economic dimensions, which help to explain, if not excuse, the determination of some Orkney ministers to shield Lyell and Nisbet from the consequences of their folly. The most prominent of their opponents were known critics of the earl of Morton, and the affair was a proxy battle between the pro- and anti-Mortonian forces recently locked in conflict over the pundlar process. Thomas Hepburn of Birsay feared that 'Mr Lyell was to be a victim to Orkney patriotism, to Jacobitism'.[46]

There were also connections to long-running quarrels between ministers and heritors over the financing of the clergy. In 1750, the General Assembly had recommended to parliament a raising of Scottish ministers' stipends. The minimum should correspond to the local value of ten of the large units of grain measurement known as 'chalders'. But since the quality of crops was thought to be lower in Orkney, the recommendation was for a payment there equivalent to the market price of ten chalders from Sutherland. The proposal was vehemently opposed by most of the Orkney lairds, who arranged public meetings and warned darkly of 'the souring and discontenting the minds of the people against their pastors'. In the event, parliament took no action, but some Orkney ministers were doubled in their determination to have whatever the rules allowed them.

At the time of his imbroglio with Mrs Agnew, Nisbet was suing several parishioners in the court of session over sums in lieu of vicarage butter teinds – a matter, he conceded, 'trifling in appearance', but 'of very great importance to him and all the other ministers in the islands of Orkney'. The men refusing to pay at the required rate were tenants of Patrick Honyman of Graemsay, and at his trial in 1765 Nisbet ascribed the whole process to 'the effect of malice and resentment conceived against him by Mr Honyman'.

In 1760, Thomas Hepburn took to print to praise the earl of Morton and defend the Orkney ministers against the heritors. Morton himself perused an advance copy of the work, and considered it, in the wake of the pundlar process, likely to 'prove a means of keeping up the animosity'. Hepburn, a minister arriving in the islands from East Lothian in 1752, condescendingly considered the Orkney gentry both riven by faction and suffocatingly interrelated, 'a family of cousins'. Yet he did not go quite as far as an anonymous pamphleteer who, at the height of the controversy over the increased stipends, sneered at the Orcadian lairds for making their living from pubs and shipwrecks. According to this author, the lairds claimed the clergy had sacrificed their honour by failing to consult with them over the proposed augmentation. But such a loss was something the Orkney ministers could surely live with: 'among the Hottentots it is reckoned an honour to be pissed upon'. The prestige of associating with the lairds of Orkney, the author implied, was of a similarly damp and steamy kind.

Hepburn supported the legal manoeuvres of another disgruntled minister, James Alison of Holm, who complained of undervaluation in stipendiary malt payments from the tacksman of the bishopric. It produced a stinging rebuke in an anonymous pamphlet of 1772. Hepburn had brought 'eternal shame' on himself by covering up the crimes of 'Willie Nisbet and Tom Lyell'. The reputation of the clergy in Orkney was now at such a low ebb that a poor fisherman had been heard to say 'he had read in the Bible that the devils had entered into swine, and now they had come out of the swine, and gone into the ministers'.[47]

The Devil's Clawmarks

The Lady Kirk in Sanday is today a roofless ruin, resting quietly in its kirkyard beside a sand-filled tidal basin on the island's southern shore. My great-grandparents, George Fotheringhame and Elizabeth Angus, lie buried under a memorial here, but visitors come to find and photograph a stone of a different kind, set in the parapet at the top of an exterior staircase once leading to an elevated gallery. It has on its upper face six deep parallel groves: the fingerprints, or clawmarks, of the Devil (see Plate 26).

Tradition tells two stories about how they got there. A new minister was reported to be such a fine preacher that the Devil himself came to listen, and left a token of disapproval on the outside of the building. The other, probably older story tells of a minister who preached hypocritically against adultery in the parish while pursuing secret affairs with married women. The Devil came to carry him off to hell, but the terrified minister turned and fled, managing to escape – just – into the sanctified interior of the church, and leaving the enraged Devil to vent his frustration on the surface of the stone.[48]

The association has now been lost to local memory, but the minister in question can surely only have been Thomas Lyell. The 'clawmarks' – perhaps a result of natural erosion – are a lasting witness to the mindset of the community, at once a judgement on a wicked and unworthy pastor and a confirmation of the enduringly sacred character of the kirk building.

Lyell was not the only Orkney minister to whom stories of the supernatural attached themselves. John Pitcairn – whose peat-digging provoked the displeasure of Lady Melsetter – was later remembered in Hoy in more heroic guise, supposedly bursting into a coven of witches, bible in hand, to confront and vanquish the Devil. South Ronaldsay preserved a tradition about Edward Richardson, the Restoration-era dean, that he had 'no less than three thumbs on each hand' – perhaps a suggestion of the occult powers that in many cultures are associated with the possession of additional digits. And then there was James Morison, the charismatic and secretly promiscuous minister of Evie and Rendall. A curious report about him came to the notice of Alexander Peterkin, an invaluable antiquarian of Orkney as well as its early nineteenth-century sheriff: 'there is still a tradition in the parishes where he ministered that he was a warlock, and that having gone to sea one day in a boat with some of his parishioners he disappeared miraculously from among them'.[49]

The Presbyterian ministers chalked up many victories, but they never succeeded in eradicating mentalities which attributed numinous power to unusual individuals and to selected sites and structures of the landscape. Indeed, as embodiments of sacred power, the clergy themselves could be allotted roles in the old ways of thinking. The ministers of Orkney's synod and three presbyteries were figures of authority and influence whose preaching and moral exhortations worked to embed a deep and conscious commitment to the cause of Protestant Christianity. But, at the same time, their conspicuous failings, individual and collective, helped by the middle of the eighteenth century to stall drives for reform and evangelisation.

Orkney was a place both well and badly suited to projects of social control and moral reformation. Its compact and collectively minded communities responded instinctively to rules and regulation, but its intensely localist identities, and the potential for conflict between rival poles of resident authority, served to complicate the mechanisms of enforcement, and to create opportunities for recalcitrance and resistance.

Programmes of 'social discipline' were attempted across Europe in the centuries immediately following the Reformation. They aimed to instil values of Christian morality and virtuous conduct, and, in the process, to shape obedient, educated and productive citizens and subjects. Studying the phenomenon in Orkney, through the clarity of the island lens, allows us to observe the successes and failures, the fine detail as well as the bigger picture, in ways that are usually harder to bring into focus elsewhere. At the time Thomas Lyell was sent packing from Sanday, by the Devil and the deep designs of his enemies, the people of Orkney were already standing on the cusp of modernity, but it was a modernity they would step into on their own, not the ministers', terms.

9
Arcs of Empire

The Kelp Riots

In the spring of 1741, a wave of anger swept over Orkney. On Monday 11 May, in the parish of Firth, west of Kirkwall, a resentful crowd marched to the shore at Rennibister, just across the water from the little pilgrimage island of Damsay (p. 89). They threw into the sea the ware cut and gathered by servants of the Kirkwall merchant Donald Groat, along with quantities of seaweed which had already been burned to produce the substance to which the name 'kelp' was then usually applied. The rioters proceeded along the coast to Quanterness, and there destroyed further 'great quantity' of Groat's ware and kelp. Disorderly assemblies, and nocturnal destruction of kelp and kelp-making tools, happened all over the islands in 1741–2. Andrew Ross, the exasperated stewart-depute, thought that everybody had gone 'mad upon the matter and desperate to a strange degree'.

The most serious trouble took place in Stronsay, where it was orchestrated by a charismatic farmer called Peter Fea. There, on Sunday 16 May 1742, Fea persuaded the baillie's officer to publicise from the kirk door a gathering of all tenants to address the problem of kelp. After a boisterous meeting at the mill on the following day, a crowd of fifty or more – armed with sticks and confirmed in a mood of valour and violence – set off around their island.

First stop for the insurgents was Clestrain on the north-west side of Stronsay, home to a major kelp-manufacturer, the Jacobite laird James

Fea. He was not currently in Orkney, but his wife, Janet Buchanan, bravely confronted the rioters, who brawled with her servants and destroyed sets of tools. The crowd continued north along the coast to Huip – the farm where, a century earlier, the witch Scota Bess (p. 216) was reputedly beaten to death, and now Stronsay's main centre of kelp-production. A kiln was extinguished, kelp thrown in the sea and more instruments broken. There was angry talk of rounding up the Huip kelp-workers and setting them adrift in a boat, but another resourceful woman, Elizabeth Cockburn, had charge of the operations there and managed to hide the workers within the house. She and her maidservant were handled roughly by the mob before it descended on a third large farm, Strenzie, dousing the kiln and destroying tools there before dispersing across the island in search of further targets.

Such lawlessness demanded a robust response. A week later, Ross's stewart-substitute, John Riddoch, arrived on the island with a posse of armed Kirkwall tradesmen to arrest Peter Fea and the other ringleaders. The inhabitants gathered in numbers to liberate Fea and his lieutenants, and Riddoch was forced to flee from Stronsay with a deep gash to his head and a deeper one to his pride.

Ross appealed for help to the lairds, despite being at odds with many of them in the ongoing pundlar process. Eventually – and possibly on false promises of immunity – Peter Fea and his brother John gave themselves up. Some kelp-producers wanted to press capital charges, but, from the safe distance of London, the earl of Morton advised leniency. The Fea brothers were heavily fined, and, in a pantomime of kirk sessional justice, Peter was sentenced to appear after divine service at church doors in Stronsay, Kirkwall, St Andrews, Deerness, Firth and Orphir – parishes all heavily involved in kelp-manufacture – with a placard around his neck declaring his offence. Fea probably wore it as a badge of pride in his resistance to what many Orcadians regarded as a – literally – noxious business.

Kelp was an improbable treasure. The dried, grey-black lumps of incinerated seaweed were unappealing things. They were, however, rich in concentrations of sodium carbonate and potassium carbonate, and these alkalis were crucial for the manufacture of glass, soap, pottery and dyes. The kelp was shipped to Leith, Dundee, Glasgow and, most of all, to Newcastle – over time, merchant vessels increasingly brought back north the coal with which the English city was proverbially associated, and for which treeless Orkney was beginning to develop an insatiable demand.

With an aggregate coastline of over 500 miles, and many stretches of shallow sea between its neighbouring isles, Orkney possessed an extraordinary abundance of accessible seaweed. Both before and after its wider potential was suspected, nitrogen-rich ware was harvested from the shore for spreading on fields. As we have seen (p. 264), it was a valued and at times contested resource.

By the start of the 1740s, the manufacture and sale of kelp had become a mainstay of the Orkney economy. By 1780 – and for a half century following – it was far and away its most important component, especially in those of the North Isles blessed with many flat beaches, such as Stronsay and Sanday. For lairds who, at little capital outlay, oversaw kelp's production and export, it was a source of previously unimaginable profit. For ordinary Orcadians, the seasonal waged labour collecting and incinerating seaweed – in the doldrums of the year between bere-sowing and peat-cutting in May, and the gathering of the harvest in September – was a welcome supplement to precarious incomes. By the end of the eighteenth century, many humble Orcadians had come to regard their principal occupation as that of 'kelp-maker'; in Sanday, they included John Fotheringhame – my great-great-great-grandfather.[1]

For folk like John, kelp-making was wet, weary and foul-smelling work. The varieties of seaweed most valued for kelp in the eighteenth century were not the long-stalked laminariae, known in Orkney as 'tangles', and cast ashore in abundance by the tide, but rather 'tang' – the weed growing tenaciously on rocks between the high and low tide. To allow for regrowth, it was cut with a serrated hook, and brought up to be burned in one of the countless kelp kilns dotting the Orkney shoreline – simple circular pits, a foot or so in depth and not more than five in diameter. The fires produced a seething liquid, raked to even out the texture. Dried and cooled, it became a hard slag, for breaking into chunks and then carting to a kelp store to await shipment to the south.

The lairds usually organised the sale and export themselves, rather than contracting it to middlemen. James Fea of Whitehall in Stronsay –

Figure 9.1: Women burning kelp in Birsay in the traditional way in around 1904 (the ruins of Earl Robert's palace prominent in the background).

son of the privateer Patrick Fea (p. 323) and a cousin of the pirate-catching James Fea of Clestrain – is thought to have introduced kelp-making to Orkney in about 1720, but others were soon in on the act. In the middle decades of the eighteenth century, over 93 per cent of the nearly 5,000 tons of kelp produced annually in Scotland came either from Orkney or the Western Isles, and before about 1780 most of this was produced in Orkney.

Kelp draws an unexpectedly straight line between the remote beaches of Orkney and the Industrial Revolution transforming Britain in the second half of the eighteenth century. And seaweed is an appropriately slippery starting point for the final chapter of this book, which explores the ways Orcadians became entangled – economically, socially, militarily, culturally and imaginatively – with the British state and empire up to around 1800. It is a story of increasing integration, but also one of how island people resisted, refused or negotiated its terms.

Kelp was a blessing which some regarded as a curse, diverting energies and investment away from a pressing need to effect agricultural reform. As suggested by Willie Thomson, the pre-eminent historian of Orkney's kelp-boom, there was also a divergence of interest between lairds and larger farmers, who were able to organise production and export on a significant scale, and small-to-middling farmers like Peter Fea who were not.

The kelp riots of 1741–2 were a product of unique circumstances. They came on the back of a run of exceptionally bad harvests – for probably the last time, people starved to death in Orkney during the long freezing winters and short chill summers of 1739–41. It was 'the common opinion in Orkney', Peter Fea declared at his trial, that the billows of acrid smoke from the burning of tang were the cause of the poor harvests and of deaths among island cattle. Reduced levels of seaweed were driving fish from Orkney's coasts, and encouraging limpets – 'being sometimes the food of the poor' – to fall away from rocks on the shore.

These were mistaken, but not intrinsically unreasonable suppositions, reflecting an instinctive sense for the interdependence of organisms in a fragile maritime environment. At the same time, they channelled old beliefs and anxieties about the origins of misfortune. In the 1790s, the minister George Barry remarked on how kelp-manufacture faced a long hard struggle against 'the strong and rapid stream of popular prejudice'. People even feared that it would make women infertile – the kind of accusation, along with the blasting of crops and the death of livestock, once levied against witches. In banishing fish from Orkney's shores, kelp-smoke drifted intriguingly across the malign intentions of the fin-men (p. 362).

Natural and supernatural causes of calamity overlapped with each other. The folklorist Walter Traill Dennison considered it well known that people believed 'the burning of seaweed for kelp gave terrible offence to Nuckelavee, and filled him with diabolical rage'. This evil sea-monster (p. 191) was responsible for various misfortunes, but his malice against kelp-makers caused him to unleash on Stronsay a deadly horse-disease called 'Mortasheen', which rapidly spread to other islands where kelp was manufactured.[2] Participation in new practices of production and exchange might energise rather than erode traditional patterns of belief. Orkney's road to modernity was circuitous and rutted.

Improvement

'A treasure buried underground'. This was the acclamatory but ambivalent assessment of Orkney made in 1775 by yet another James Fea, a son of the kelp-pioneer and cousin of the Jacobite rebel. Fea inherited the Clestrain estate after a career as a lieutenant in the army and then as a surgeon in the navy. He struggled to manage his financial affairs, and was hiding from creditors in Edinburgh when he published *The Present State of the Orkney Islands Considered*. Fea's straitened circumstances did not, however, discourage him from prescribing remedies for the ills of his homeland.

Orkney was a place of many God-given advantages: secure and accessible natural harbours, a rich and fertile soil, the absence of predators and an abundance of birdlife, a polite and hospitable population. Yet 'no people in the world can be in a state of greater poverty or dejection of spirits than ourselves'. The situation was one of the Orcadians' own making. By exposing 'the negligence and misconduct of my countrymen', Fea hoped to shame them into effecting improvements.

Agriculture was the root of most of the malaise. An old bête noire – the earldom feu duties (p. 343) – encouraged cultivation of bere at the expense of 'other kinds of more useful grain'. Fea recited a litany of bad farming practices: poor drainage, late sowing and harvests, the shortcomings of seaweed as a fertiliser, lack of enclosures to cultivate grass for making hay, poor quality and badly made butter, the 'rooing' of sheep (pp. 176–7). Innovation was discouraged by short leases and compulsory labour services; Fea advocated abolishing them, while dividing commons, and allocating land to the impoverished and insecure labouring tenants known as cottars.

In its current condition, farming could support only a part of the population. It accounted for a growing tide of emigration, and for Orcadian men in their thousands leaving to join the navy, which made for 'a much greater number of women than men among us'. Like others before him, Fea regarded fishing as a key to prosperity ready to hand, since 'almost all our poor people are bred to the sea'. In particular, he hoped that Orcadians would step up to challenge Dutch domination of the North Atlantic whaling trade. Fea recommended that a peat-powered station for boiling whale blubber and converting it to oil be established at the natural harbour called the Pan Hope off the island of Flotta – the very place where, almost exactly 200 years later, a modern terminal was built to receive and process oil of a different sort.

Fea was frustrated by Orkney's apparent embrace of its own poverty and marginality. He imagined the islands – positioned halfway between Shetland and the Western Isles, and with equal access to Britain's western and eastern seaboards – becoming a hub of busy manufacture and 'the very centre of trade to all the northern kingdoms of Europe'. If the government was willing to settle in the islands families able to instruct Orcadians in different branches of manufacture, then 'there is great reason to hope that from a state of mere insignificancy we might very soon become a useful part of the British Empire'.[3]

Fea was not alone in his eagerness to see Orkney turn into a fully functioning cog in the British national and imperial system. John Campbell, a distinguished Scottish historian and royal agent for the North American colony of Georgia, published in 1774 a two-volume *Political Survey of Britain*, which devoted much thought as to how a set of islands 'of little use even to themselves, and scarce at all known to the inhabitants of the southern parts of Britain', might be transformed into 'a very valuable and profitable province of the British Empire'.

The development of fishing – 'what surely was the design of Providence' – lay at the centre of Campbell's proposals. Observing that Orkney was roughly equivalent in extent to the wealthy region of Zealand, he advocated a Dutch model of combining fishery with boat-building and the production of hemp and flax for making nets, twine and cordage. Instruction in these and other manufactures would enable Orcadians to conduct their own extensive foreign trade and reduce incentives for young men to emigrate.

The Orkney Islands, Campbell declared in a challenge to conventional geographical thinking, enjoyed a 'centrical situation'. They made the ideal location for a 'general magazine' to service North Atlantic fishing. Such a depot would 'put this trade totally and forever into the hands of British subjects', while simultaneously supporting the Greenland and North American whale fisheries. As a bonus, in time of war the islands would be 'an excellent station for a small squadron of his majesty's ships, as well for the protection of our own commerce, as for annoying that of our enemies'.

These were all suggestions aired by Murdoch Mackenzie in 1750, but Campbell went further in suggesting that the East India Company establish a base in Orkney, to receive customs due on East India commodities and to crack down on smuggling – an enforcement vessel, or cutter, should be based there for the purpose. From Orkney, the Company could send its goods to Hamburg, Lübeck and other North European ports, without interference from the Dutch.

Campbell further proposed that Orkney become the site of a new university, funded from the revenues of the bishopric estate. Young people would no longer need to leave the islands for higher education, and in consequence become 'so weaned from their country' that they never returned. The university would be 'a means of cultivating genius amongst the natives', and would draw to Orkney students (and their money) from Norway, Denmark and Germany. In such ways, the naturally 'frugal and diligent' Orcadians would achieve their potential, and 'make these British Islands more and more resemble Britain'.

Campbell's was not a lone voice. Proposals were under discussion in Lord North's government in the summer of 1774 'for rendering the islands of the Orkneys more useful to the Empire of Great Britain' by turning them into the centre of herring and whale fishery; a group of Edinburgh investors pondered a similar plan in 1784.[4] But Campbell's most ambitious schemes were scarcely practicable. Shares in the East India Company helped Sir Lawrence Dundas to the fortune that allowed him to purchase the earldom estates in 1766, but the Company itself had no inclination for an Orkney base. The university would have to wait until 1989, when Edinburgh's Heriot Watt established an International Centre for Island Technology in Stromness, followed by the arrival of another higher educational institute in 2011, the year Kirkwall's admirable Orkney College became an affiliated campus of the newly accredited University of the Highlands and Islands.

The eighteenth century nonetheless saw sporadic but serious interest in the commercial potential of the islands. From the 1730s, proposals to make Orkney a centre of cod or whale fishing were repeatedly laid before parliament. They achieved partial fruition in 1750, when a charter was granted to the Free British Fishery Society to raise capital and equip vessels for the patriotic purpose of challenging Dutch hegemony. The society made a base in Orkney, and at its height employed 300 or more Orcadians in its fleet of broad-decked 'busses'. The lapse of its charter in 1771 reduced the scale of commercial fishing, and was regarded by James Fea as a disaster for the islands, though kelp, he grudgingly conceded, provided alternative employment for poor men and women.[5]

Alongside kelp, linen was an important source of supplementary income – a 'proto-industry' compatible with existing patterns of agricultural and household labour. The elaborate manufacturing process involved harvesting flax, which had to be threshed or 'rippled' to remove the seeds, soaked, and then 'heckled' or combed to separate the fibres. After that, the material was spun into thread on spinning-wheels – a time-hallowed form of female labour which made 'spinster' into the conventional designation for an unmarried woman. Male weavers turned the thread into cloth, bleached by kelp-alkali and exposure to the air: large bleachfields were laid out in Birsay and on the outskirts of Kirkwall.

The industry took off with the establishment of the British Linen Company in 1746, and its decision, influenced by the earl of Morton, to choose Orkney, where labour was cheap, as a principal base of operations. As with the Free British Fishery, there were political motives in play. Developing the commerce of Highland and northern Scotland, and providing gainful employment for the poor there, was seen as a means of reducing support for Jacobitism, and of binding outlying parts more closely into 'Great Britain'. For a decade or so, a monopoly of Orkney linen production was held by the company's agents, Thomas and William Lindsay, nephews of the Mortonian stewart-depute, Andrew Ross. After 1760, the company relinquished direct oversight of manufacturing, allowing enterprising merchant-lairds such as Benjamin Moodie of Melsetter (p. 348) and Patrick Graeme of Holm to take a significant role in linen production and export.[6]

Kelp, fish and linen provided ordinary Orcadians with some protection from the sheer precariousness of existence. They needed it. The late summer of 1778 brought storms of extraordinary ferocity, which destroyed the entire harvest. This was followed, in the early 1780s, by a run of poor weather and bad harvests, affecting Scotland as a whole but hitting the Northern Isles particularly hard. English newspapers in 1785 reported 'very melancholy representations of the distressed state of the inhabitants of those parts'. Relief shipments were organised by the government, by landowners like Patrick Graeme, and by Sir Thomas Dundas, who in 1781 succeeded to his father's baronetcy and Orkney estates.

The crisis persisted for much of the decade. A letter from Birsay, printed in the *Caledonian Mercury* in June 1787, documented the efforts to relieve an ongoing crisis: 'a ship arrived lately from the South, with meal and potatoes, of which this miserable country has too much need'. But it does not seem as if anyone in Orkney actually starved. Many tenants were able to pay for portions of famine-relief shipments with cash earned from Greenland whaling, or sent back by emigrant kin and sons in naval service, or from their own home-produced linen. Writing in 1786 to the Board of Trustees for the Improvement in Fisheries, Manufactures and Arts, Patrick Graeme said of the linen production of his Holm tenants, 'in these past years of scarcity I cannot conceive how they could have existed without dependence on this manufacture'.

A falling rate of mortality due to dearth was accompanied by increased protection against disease. Professional medical services of any sort were patchy in Orkney before the start of the eighteenth century. But in the 1720s, the former schoolmaster John Watt established a general practice in Kirkwall, and was succeeded in the following decade by Dr Hugh Sutherland, an Orcadian trained in the famous medical faculty of the University of Leiden. Over the course of a thirty-year career, Sutherland consulted with patients throughout Orkney. His clients were principally the better-off, receiving treatment in return for an annual retainer. But he also administered medicine to the poor, for which kirk sessions organised regular collections.

Sutherland was an impassioned advocate for the long-established but still controversial technique of inoculation against smallpox, which involved the insertion of matter from the pustules of a sufferer into the skin of a healthy individual. Orkney, indeed, became an unlikely role model for the public-health messaging of the mid-eighteenth century. Lamenting the 'prejudices against inoculation which do yet influence the vulgar', a correspondent to the *Scots Magazine* in February 1758 pointed to a recent outbreak of the disease in Kirkwall, in supposedly remote and backward Orkney. About sixty children there inoculated by Dr Sutherland had recovered, but of 273 inhabitants who contracted the disease 'in the natural way', seventy had died.

Sutherland's successor, Dr Andrew Munro, was still more of an evangelist for inoculation; press reports marvelled at his 'uncommon success' in protecting children from smallpox. A serious outbreak in December 1769, Munro declared, 'induced me to use all the persuasive arguments I was master of with some of the common people to submit their children to that operation'. Between January and April 1770, he inoculated 332 children in the North Isles alone, nearly all 'of the lowest class', and reckoned it a great success that only four of these died.

James Fea, a medical man himself, rejoiced that the people of Orkney had 'happily surmounted their prejudices against inoculation': hundreds of lives – he later claimed thousands – had been saved. To a self-declared 'improver', it was encouraging to learn that Orcadians were not averse to all innovations, though Fea noted the persistent influence throughout the islands of 'old women quacks'. These, he wryly remarked, seemed 'rather to be in league with the Grim Tyrant [Death], and to have agreed to execute his purpose, than to do anything contrary to his rule'.[7]

The end of the eighteenth century witnessed measures to tackle another of the Grim Tyrant's rackets: the profusion of shipwrecks on the Orkney coast. If people in eighteenth-century Britain knew anything about Orkney, they knew that it was a graveyard of maritime transportation: shipwrecks were by far the most common reason for the islands to appear in the periodical press. Indeed, the *Caledonian Mercury* considered it newsworthy in March 1773 that, with exception of the *Betsey* of Liverpool, 'there have been no wrecks ... on the coast of Orkney for these twelve months past'.

Shipwrecks, the newspapers repeatedly pronounced, could have been prevented by lighthouses to warn vessels of impending peril. Orkney's 'north-eastern extremity' – North Ronaldsay, and the adjacent, low-lying, and so almost invisible island of Sanday – were the places where passing vessels most commonly foundered.

Decades of agitation on the issue culminated in a 1786 act of parliament establishing the Commissioners of the Northern Lighthouses, later the Northern Lighthouse Board, with a remit to construct beacons at key locations around the Scottish coast: Kinnaird Head in Aberdeenshire, Scalpay in the Outer Hebrides, the Mull of Kintyre and North Ronaldsay. In 1789, under supervision of the engineer Thomas Smith, a manned seventy-foot lighthouse was built at Dennis Head, North Ronaldsay's easternmost point. It was conceived as part of a pair with an unlit tower on Start Point, the eastern tip of Sanday, constructed in 1802. But the arrangement produced limited results, and the North Ronaldsay tower was superseded in 1806 by the refurbishment of Start Point, where Robert Stevenson, Smith's stepson and successor as engineer to the Lighthouse Board, installed a literally revolutionary innovation: the first Scottish lighthouse equipped with a rotating lamp. In the meantime, Smith oversaw the 1794 construction of another lighthouse on the Pentland Skerries.

The case for the erection of lighthouses seemed self-evident to all who wrote on the matter. But not everyone in Orkney welcomed their arrival. In North Ronaldsay, it was believed in the 1790s that the new lighthouse was driving away the seals the inhabitants were accustomed to catching in nets. Like pods of beached whales, wrecks were a providence of the sea from which ordinary islanders derived fortuitous profit. Safety of sailors aside, the construction of lighthouses in Orkney symbolised a subordination of community customs to the commercial interests of wider British society. A lighthouse, inevitably, is a metaphor as well as a pillar of stone and glass.

The advocates of improvement were eager to play down suggestions of conflict. James Fea mocked the 'ridiculous charges' levelled against the inhabitants of North Ronaldsay; that 'they used spells and other delusory methods to procure those shipwrecks'. The islanders accounted fairly for goods cast onto their shore, and shipwrecked mariners often attested to receiving from the inhabitants 'the most singular proofs of humanity and attention'. Complacent contrasts were regularly drawn between the civility of Orcadians and the barbarity of wreckers in places like Cornwall; for example, by Murdoch Mackenzie, and by William Clouston, late eighteenth-century minister of Cross and Burness, who cited one captain as saying that if he had to be wrecked anywhere 'he would wish it to be in the isle of Sanday'.[8]

It was, perhaps, a sanitised view. The late eighteenth-century Orcadian writer George Eunson thanked God for the charity displayed to shipwrecked seamen by some inhabitants of Sanday and North Ronaldsay, though he conceded that 'many others are of more hardened dispositions ... not caring even for the distress of their own countrymen more than that of the most remote foreigners'.

Accounts of actual shipwrecks bear out the generalised contrasts. The crew of the *Peggy*, driven onto the west side of Papa Westray in a gale in November 1771, were all saved, and managed, 'with the assistance of the people of the island', to get most of their sails and rigging ashore. But the crew of the *Elizabeth* of Banff, and its captain, James Ord, had a different experience when they were wrecked off Mull Head in January 1775: 'never were any poor men used with such barbarity as we were by these people at Deerness ... who stripped us of our very clothes, and plundered everything about the ship'.

A still more notorious case, from the early years of the nineteenth century, involved the *Albion* of Blyth in Northumberland, driven onto the coast of Hoy between the Old Man and Rora Head, after nearly all its crew were washed overboard. Some Rackwick fishermen, plundering the wreck, were believed to have overlooked one unconscious survivor strapped to the rigging, and to have moved another onto a shelf in the cliff-face, where he perished in the course of a cold November night. There were no witnesses and no prosecutions after the wreck of the *Albion*, though tradition maintained that the minister of Hoy, Gavin Hamilton, witnessed the whole episode in a dream, and berated the astonished fishermen at the funeral of the two sailors – another instance of unearthly powers ascribed to charismatic parish ministers (p. 413).

In another case, justice followed more conventional courses. The *Jupiter* of Wick, and its cargo of dressed flax, leather, cotton, woollen cloth and sugar, came to grief in the Bay of Skaill in October 1790. Nine men from Sandwick were subsequently indicted in the Edinburgh high court of admiralty, accused of endangering the lives of the crew by cutting a cable anchoring the ship during a storm. This drove it onto the rocks, where they robbed the wreck of its stores, cargo, timbers and rigging. The alleged wreckers were men of substance in the parish – they included Isaac Kirkness, one of the 'Knights of Stove' claiming an ancient association with King James V (p. 14). Another was Thomas Tyrie, a younger son of the controversial clergyman James Tyrie (p. 407), and brother-in-law of the distinguished Birsay minister George Low. Also indicted was the factor of the Skaill estate, Robert Graham, brother-in-law to the laird, and a cousin of the namesake who brought Low to Orkney as tutor to his children.

Before Graham's complicity emerged, the ship's master begged him to take protective charge of the sloop and its freight. Graham replied he could do nothing, as 'it was the cant [old custom] of the country that vessels wrecked on that particular spot became the property of the first finders'. Writing about the incident in 1794, the minister of Sandwick claimed an exonerating circumstance: parishioners who helped salvage a vessel on the same spot in 1788 were cheated of a promised share of the cargo.

Graham was eventually convicted for his involvement; Tyrie's guilt was 'not proven'. Kirkness, with five others, had evaded arrest in the first place by fleeing to Scotland in an open boat. Newspapers advertised a reward for their apprehension and circulated a description: 'These men have the provincial accent, having never been before out of Orkney, and the boat in which they escaped is of the shape peculiar to the country, resembling a Norway skiff.'[9] People hailing from Orkney, even in the last decade of the eighteenth century, and even in Caithness or Sutherland, still stood out as strangers.

Land's End to the Orkneys

The 'otherness' of Orcadians was an impression that influential members of the community were eager to contradict. George Eunson's *Ancient and Present State of Orkney*, published in 1788, radiated a concern for the islands' rightful place within the common bounds of a wider national polity: 'For *Orkneymen*, may surely claim/British rights, and British fame.'

The patriotic Eunson was an eccentric and quixotic character, a former cooper and naval deserter, who during the dearth years of the early 1780s was briefly imprisoned for breaking into the girnel (grain store) of Kirkwall – either for his own profit or, depending on whom you believed, to make supplies available to the town poor. Eunson, moreover, was an accomplished smuggler, who in 1784 turned his back on that illicit trade with a surprise appointment as 'Extraordinary Officer of Customs'.

Orkney, in the second half of the eighteenth century, was notorious throughout Britain not just as a location for shipwrecks but as a destination for contraband, its international maritime connections and profusion of quiet bays and harbours making it a handy point of entry for tea, tobacco, brandy, Dutch gin and French wine. 'Smuggling', pronounced the *Oxford Journal* in March 1773, 'was never more brisk in the Orkney Islands than at present'.

Reformers condemned the impact of the traffic on societal morals and public health. Thomas Hepburn, former minister of Birsay, considered the 'pernicious trade' injurious to all aspects of Orkney life, but admitted it was carried on 'with the general consent and concurrence of the country'. The eighteenth century saw periodic attempts by the heritors to clamp down on smuggling, but many lairds were deeply complicit in the practice, and some of the clergy, too. At the turn of the nineteenth century, a visitor claimed that Orkney's kirks were often used to store smuggled goods, foreign liquor in particular, and waggishly remarked that the people of the islands seemed to have misunderstood St Paul's injunction for believers to be filled with the spirit.

Eunson's role as a special customs officer came about not through any disinterested desire for reform but was an appointment thoroughly mired in local politics. Having chafed under the fiscal and social dominance of the earl of Morton, some Orkney lairds found the absentee lordship of Sir Lawrence and then Sir Thomas Dundas to be just as irksome. Prominent among opponents of the Dundas 'interest' was Robert Baikie of Tankerness, who twice in the early 1780s stood for election as MP for Orkney and Shetland, but failed to oust the sitting representative, Sir Thomas's cousin.

In 1784, with the aim of discrediting Dundas's supporters, Baikie and his allies established an association 'for preventing the progress and increase of the pernicious practice of smuggling'. Eunson was their agent, and he pursued contraband with a tactless fervour that twice resulted in him being arrested by the magistrates of Kirkwall. Critics pointed to Eunson's lack of official accreditation from the Board of Customs in Edinburgh, though his efforts enjoyed some cautious backing from the sheriff, Patrick Graeme.

Eunson's *Ancient and Present State of Orkney* was laced with patriotic effusions about Anglo-Scottish union and the gloriousness of 'Great Britain', but these had a clear strategic purpose. His goal, in contrasting Orkney's sad plight with 'happy England's state', was to highlight misgovernment at the hands of Dundas and tyrannical lairds, and to arouse indignation at the spectacle of such behaviour taking place within a united kingdom whose influence was now spread across the world. The inhabitants of Orkney, Eunson implausibly declared, laboured under oppressions 'in many respects more intolerable than what the slaves experience in the sugar islands of the West Indies'.

> Can Britons then poor Orkneymen forsake,
> And let a *few* despotic measures take?[10]

The implication was that the value the nation placed on the welfare of its most northerly parts was no less than the moral measure of Britain itself. Orcadians as well as outsiders understood the islands' symbolic utility.

The commotion over customs and excise was also tied up with the politics of a moment when Orkney found itself briefly catapulted into the national headlines, with a part to play in one of the great rivalries of British parliamentary history. The general election of 1784, fiercely contested in the aftermath of Britain's defeat at the hands of the American colonists, pitched the supporters of Prime Minister William Pitt the Younger – called by their enemies 'Tories', and favoured by George III – against a faction of the old Whig party led by the radical liberal Charles James Fox – the political colossus of his age, friend of the dissolute Prince of Wales and self-styled 'man of the people'.

Fox was one of two sitting members for the borough of Westminster, a constituency notable for the unique size of its electorate (12,000) and the prestige attaching to victory there. Campaigning in the borough in 1784 was enlivened by the celebrity involvement of Georgiana, Duchess of Devonshire, Fox's distant cousin and reputed lover, who offered kisses to tradesmen in return for their votes. This inducement notwithstanding, Fox finished only marginally ahead of the third-placed Pittite candidate, who demanded a recount. Pitt, victorious in the election nationally, used his new majority to prevent Fox from taking his seat, and not until almost a year after the election was the matter finally resolved in Fox's favour.

In the meantime, a contingency plan had been set in motion. Through the good offices of his friend and supporter Thomas Dundas, Fox was hurriedly enrolled as a burgess of Kirkwall. He thus became an eligible candidate in 1784 for a second contest, in the 'Northern Burghs' (p. 328). Its representative was elected by a single delegate from each of the five little towns north of Inverness, the stark antithesis of democratic Westminster.

Yet it proved no shoo-in. Kirkwall's delegate was chosen by the town council, where allies of Robert Baikie of Tankerness strongly opposed Fox's candidature. Provost John Riddoch – scarred veteran of the Stronsay kelp riot – was eventually chosen on a majority vote, but only after three *ex officio* councillors who were deacons (presidents) of Kirkwall's 'Incorporated Trades' (hammermen, shoemakers, tailors and weavers) were excluded from a fractious electoral meeting, on the questionable grounds that they were not freemen of the town. Riddoch joined with delegates from Dingwall and Tain to elect Fox, against those of Wick and Dornoch, supporters of the Pittite candidate, the Caithness landowner Sir John Sinclair, who loudly alleged procedural irregularities. Fox himself, far from the fray, graciously sent his portrait as a gift to the burgh of Kirkwall.

The episode elicited from Fox's opponents howls of hostility and derision. A satirical cartoon printed in London (see Plate 28) depicted Fox, dressed half in the blue and buff colours of the Whigs and half in a tartan plaid, sitting in a communal latrine with his left leg thrust down one of two adjoining holes – a reworking of earlier vicious caricatures suggesting Scots were too ignorant to know how to use a 'bog-house'. 'I have a right to two seats in the House', Fox is shown as lamenting, 'but damn me if I know how I shall get into the other', as the Duchess of Devonshire advances to offer him a purse. In the House of Commons, William Pitt remarked sarcastically that his great antagonist 'had undergone a sort of banishment, being driven by the impulse of patriotic indignation, as an exile from his native clime, to seek refuge in the stormy and desolate shore of the Ultima Thule'.

One Pitt-supporting newspaper informed its readers that the region in which the borough of Kirkwall was situated 'lies almost beyond the habitable earth … wrapped in frost and impenetrable darkness for near six months in the year'. The author imagined its 'benumbed electors' had not even known of the election until a summons arrived commanding them into 'a concurrence with the will of Sir Thomas Dundas'. Another report, noting Fox's reputation as a great gambler, always declaring for 'heads or tails', observed he had evidently decided that if he could not represent Westminster he would be returned for 'a district of boroughs in Orkney, the remotest corner of the British Isles'.

Some anti-Fox commentators struggled to remember what they knew about the place represented by the leader of the opposition in parliament. One rather lame satirical effort declared that Orkney, 'being a famous place for geese, and Kirkwall, for which the Man of the People is elected, being in Orkney, it is very surprising that the Fox should be solicited to protect the Geese!' A 1785 spat over the authorship of an anonymous pamphlet inspired another anti-Foxite writer to ask how long his opponent had been endowed with the 'second sight' – 'is it before or since Charles Fox cast a wishful eye towards the "hospitable shores of the Orkneys?"' Prophetic vision was principally a cliché of Highland culture, a sphere to which the author, quoting the reported ironic words of Prime Minister Pitt, carelessly assumed Orkney belonged.

Similar misapprehensions surfaced in a letter, printed in the *Derby Mercury* in October 1784, which purported to be an account of a cruise to Orkney and Shetland. There, the author claimed, the people speak 'the Erse of the Highlands, mixed with bad English' – only the most oblivious of visitors could have supposed Gaelic to be the language of Orkney. This supposedly eyewitness account was remarkable for its tone of sneering antagonism to the islands: 'I assure you, I never desire to see them again.' From a distance, Kirkwall 'might well be taken for a collection of dunghills, with a church much like a pigeon-house, and yet' – this perhaps the real point of the diatribe – 'a candidate for the great and opulent City of Westminster did not think it beneath him to represent this place'.[11]

Knowledge of Orkney in mid- and later eighteenth-century Britain, from the evidence presented by passing references in various published works, was patchy and often hand-me-down. The unique genesis of the Orcadian barnacle goose (p. 21) – a cause of scepticism and wonder in the middle ages – continued in the era of the Enlightenment to attract speculative comment and complaisant witticism. A social-climber in a London stage comedy of 1785 was described as having 'come into life through as many shapes as an Orkney Barnacle; he was first a block, then a worm, and is now a goose'.

To some, Orkney remained a byword for poverty and remoteness, an idea of a place, which was sometimes useful for settling arguments as a test case or *reductio ad absurdum*. As proposals reached parliament in 1778 to relax the penal laws against Catholics, the Church of Scotland minister at Paisley declared Britain's entire constitution to be grounded on the exclusion of papists, 'from the king himself ... to the lowest judge or magistrate in the Orkneys'. In the aftermath of the anti-Catholic Gordon Riots of 1780, which erupted in response to this limited step towards religious toleration, the Welsh philosopher David Williams stressed the importance of government operating under moral as well as legal constraint, observing that it would be perfectly constitutional for parliament to pass a law 'by which every Englishman may be banished to the Orkneys or put to death'. It was not quite clear which fate was thought worse. The anonymous author of the gothic novel *Ranspach, or, Mysteries of a Castle* (1797), expected readers to experience a frisson of pleasurable horror at the heroine's imprisonment in a fortress on 'the most northerly of the Orkney Islands'. Meanwhile, one of the earliest British visitors to the faraway Falkland Islands in the South Atlantic described finding there 'a large cluster of them like our Orkneys or Orcades'. In 1823, a similar resemblance would suggest itself to the explorer James Weddell, who christened an archipelago still closer to the Antarctic landmass 'the South Orkney Islands'.

The islands might be outlandish, but were nonetheless 'our Orkneys'. There was a growing tendency to acknowledge them as an integral part of the national polity, subsumed into a confident and expansive concept of Britishness. In 1762, for example, a pamphlet defended the divisive government of the earl of Bute – the first British prime minister to come from Scotland. It argued, using what seems intended as an ascending scale of exoticism, that 'it is absurd to condemn a minister because he is a Scotchman, an Irishman, a native of Orkney, or a native of the woods of America, if he is a Briton'.

Whether Americans, native woodsmen or otherwise, were in fact Britons was an increasingly moot point, as the grievances of the Thirteen Colonies propelled them towards rebellion in 1775. Their actions were denounced as the result of 'enthusiastic spirit and overgrown wealth' in a humble address from the provost and magistrates of Kirkwall, presented at court by Thomas Dundas in January 1776, and received 'very graciously' by George III. On the eve of the conflict, an essayist calling himself 'Britannicus' dismissed the colonists' demands for representation at Westminster, though on the basis of fundamental incompatibility of trade and manufacturing interests, rather than because of difficulties in sending MPs to London from a great distance. This was 'a circumstance in which the inhabitants of Rhode Island do not essentially differ from those of the Orkneys'. Conversely, during debates leading to the Act of Union with Ireland in 1800, several commentators argued, with specific reference to Orkney, that physical distance and an intervening sea were no sound arguments against incorporation into the British state.[12]

The meaning of 'Great Britain', meanwhile, had gradually morphed – it was no longer so much the name of a particular island as the accepted designation for a wider geographical and political entity, with 'great' a measure of quality as well as extent. From the mid-eighteenth century – in literature, travel writing, sermons and political pamphlets – descriptions of the scope of 'Great Britain' often portrayed it as extending from 'Land's End to Orkney', occasionally substituting Penzance or Dover as the companion national pole. 'We are all Britons', declared the Sussex clergyman John Lettice in 1794, 'from the Land's End to the Orkneys.'

Noticeably absent from descriptively geographical or floridly metaphorical statements about Britain was the modern habit of invoking the Caithness village of John o' Groats – there was as yet no settlement to speak of at the site from which a fifteenth-century Dutchman, Jan de Groot, supposedly operated a ferry service across to South Ronaldsay. More surprisingly, only very rarely did writers point to Shetland as

Britain's defining northern boundary. It is hard to be certain why this was. Orkney, as we have seen, enjoyed greater political prominence than Shetland, which was sometimes silently subsumed under the label 'Orcades'. Yet, as this book has already argued, it was the very closeness of Orkney to mainland Scotland that often underlined its challenging otherness. In thinking about Britain, Orkney had once been the place just beyond; now, perhaps, it was the place just within.[13]

Where's the North?

The relative, relational nature of 'peripheries' was something understood by people at the time. A much-quoted passage, in Alexander Pope's 1733 poem, *An Essay on Man*, positioned Orkney in an ever-receding sequence of beckoning locations:

> Ask where's the north? at York, 'tis on the Tweed;
> In Scotland, at the Orcades; and there,
> At Greenland, Zembla, or the Lord knows where.[14]

There were norths beyond Britain's north, and, for these, Orkney was an important point of access. No interactions seem to be recorded between Orkney and Zembla – Novaya Zemlya, a pair of desolate islands in the Barents Sea north of Russia. But Orkney's role as a way station for the Icelandic fishing fleet, and as a home base for Greenland whaling ships, was already long established when Pope wrote his poem.

In another region of the north, Orcadians were at the forefront of British exploration, commercial exploitation and colonial control. In 1670, Charles II issued a royal charter to 'The Governor and Company of Adventurers of England, trading into Hudson's Bay'. It granted the new joint-stock company a monopoly of trade along with sweeping political powers in the region known as Rupert's Land, after the Company's chief investor, the king's cousin, Prince Rupert. The territory was defined as comprising all lands drained by rivers running into Hudson Bay – a vast area, nearly a million and a half square miles, or some two-fifths of the entire extent of modern-day Canada.

The principal value of the land lay in the colonies of beavers with which its forests and waterways teemed. Waterproof beaver pelts were the ideal material for making into smart gentlemen's hats, and across Europe there was an insatiable demand for them. The employees of the Hudson's Bay Company, dispersed in outposts known variously as houses, forts and factories, traded with indigenous peoples who did the actual trapping – Cree, Assiniboine, Ojibwe. Over time, the Company's operations moved further inland, west and north, as beaver stocks dwindled in the initial zones of trade near the shores of Hudson Bay. Individual agents embedded themselves with the tribes, and learned to speak native languages.

The London-based committee running the Hudson's Bay Company decided from the outset to recruit its workforce from among Scots, and Orcadians in particular. The first dozen were signed up in Stromness in 1702. They were regarded, in a condescendingly stereotyped manner, as hardy, obedient, sober and industrious – in contrast to equally stereotyped views of the London populace as disorderly, and the Irish as drunken and lazy. Crucially, Orcadians were willing to work for low wages.

Figure 9.2: The harbour at Stromness, with the hills of Hoy behind, from an aquatint in William Daniell's 1821 Voyage Round Great Britain.

For ships looking to start the Atlantic crossing without venturing too close to an often hostile France, Stromness was a convenient final stop for water and supplies. Through the eighteenth century and beyond, Company vessels invariably called there. It helped turn an insignificant hamlet into a thriving little port, matching Kirkwall for population, and in the 1740s and 50s, in a bitterly fought law case, successfully challenging Kirkwall's right as a royal burgh to tax the Stromness merchants for independent trading.

Over time, the scale of recruitment in Orkney to the Hudson's Bay Company became such as to make the early development of Canada almost an Orcadian colonial enterprise. By 1789, at least 70 per cent of the entire workforce of the Company were Orcadians, and it had risen to around 80 per cent by the turn of the century. Longstanding arrangements for hiring labour were formalised in 1791 with the appointment of a Stromness merchant, David Geddes, as a resident local agent of the Company.

Life in the Company's forts and factories was hard, and often dangerous. Canadian winters were gruelling beyond anything experienced in Orkney. There was periodic trouble with people of the First Nations, and agents of the Hudson's Bay Company were embroiled in what amounted to low-level warfare with employees of the rival North West Company, a conflict lasting until the merger of the companies in 1821.

A few Orcadians were taken on as craftsmen – carpenters, blacksmiths or tailors – but most were hired as simple labourers. The Company's officers were usually English, and low literacy rates among Orcadian recruits acted as a barrier to promotion, though a talented or fortunate handful did rise through the ranks. Even poorly educated Orcadians often displayed an impressive facility for native languages – perhaps a reflection of Orkney's own historic bilingualism and its location as a crossroads of travel and commerce. 'The inhabitants of these islands', wrote one traveller in 1788, 'have long been visited from almost every nation in Europe.' A willingness to engage with native customs and languages advanced the career of William Tomison from South Ronaldsay, taken on as a twenty-year-old in 1760. Tomison spent half a century in the Company's employ, and by 1778 had reached the rank of 'inland master', overseeing westward expansion along the Saskatchewan River.[15]

Tomison was an untypical recruit, yet more exceptional still was John Fubbister from Orphir, who enlisted as a labourer in 1806, just as Tomison was returning to Canada from Orcadian retirement to lead a final expedition. Fubbister went in the autumn of 1807 with a work party to the distant trading post at Pembina, near the Red River in present-day North Dakota. There, on 29 December, Fubbister, evidently unwell, called on the fort commandant, who permitted him to rest for a while in his (the commandant's) quarters, and shortly afterwards found him 'extended out upon the hearth uttering most dreadful lamentations'. It transpired that 'Fubbister' was in fact 'an unfortunate Orkney girl, pregnant and actually in labour'; her real name was Isobel Gunn.

It is an extraordinary story, though there are parallel cases of women enlisting in the army or navy in the eighteenth and nineteenth centuries and maintaining the deception over long periods of time. Whether the Orcadians labouring alongside Isobel knew her real identity is uncertain, though one at least must have done. She named as the father of her baby John Scarth, a veteran of Company service, who sailed with her from Stromness in 1806. Shortly after the delivery of her son, Isobel was ordered to return to the headquarters at Fort Albany. She worked there as a washerwoman until September 1809, when she was sent back to Orkney, to pass her remaining years in penury and obscurity.

Perhaps, as many assumed at the time, an infatuated Isobel felt compelled to follow her lover to Canada, though it is equally possible she had rational reasons of her own for signing on as a Company labourer: a wish to make money, a yearning for adventure, or a desire to escape the poverty and monotony of life in Orphir, where opportunities for choice and satisfaction in marriage were limited by the imbalance of women and men. The prospect of labouring was unlikely to have been a deterrent; lower-class women in Orkney undertook many forms of hard physical work. One cynical observer, a few years earlier, attributed low mortality rates among Orcadians in the Hudson Bay forts not to hardy male constitutions but to intrinsic idleness, 'for the women do all the hard work in Orkney'.[16]

Isobel Gunn's motivations may not have differed that much from those of young Orcadian men signing on with the 'HBC', lured by lively tales from old hands, and by the printed bills which appeared on kirk doors and other public spaces to advertise the wages on offer: a meagre but nonetheless enticing £6 per annum for labourers, rising to £15 for writers or clerks, and potentially double that for skilled craftsmen.

It was no accident that peak recruitment coincided with the crop failures and economic downturn of the 1780s. Emigration from Orkney had already begun on a noticeable scale, with ships sailing out of Stromness for Georgia in 1770, 1773, 1774 and 1775, after leaflets circulated containing 'a very flattering description of the colony'. Some of the voyages were organised by an Orcadian sea-captain, William Manson, others by his former employer, the Whitby trader Jonas Brown. With his son Thomas, and the Stromness merchant James Gordon, Brown purchased land for a settlement in Georgia, and offered plots to migrants. Those unable to afford the passage were given the opportunity to sign on for three years as indentured servants. The reasons they gave to customs officers for doing so were a demoralising catalogue of Orkney's social and economic woes. William Bews from Evie went because 'his farm [was] too high rented'; Magnus Halcro, St Ola, 'cannot get bread for his family owing to bad crops and high prices'; Peter Petrie, a young farm servant from the East Mainland, went simply 'in expectation of doing better'.

The mid-1770s, as it turned out, were an inauspicious time for emigration to the American colonies. One passenger was a sixteen-year-old from Kirkwall, Baikie Harvey, leaving the islands 'to seek a better way of living'. He found conditions as an indentured servant to Thomas Brown intolerable, and soon ran away. As he did so, America was igniting into revolution, the back-country of Georgia and South Carolina a site of bitter civil war between 'Loyalist' Tories and 'Patriot' rebels. Harvey joined a rebel expedition against South Carolina Tories, and at the end of 1775 described his experiences in a letter to his patron, Thomas Baikie in Firth: 'dear godfather, tell all my country people not to come here, for the Americans will kill them like deer in the woods'. Harvey himself lost his life during the British assault on Savannah in October 1779.[17]

Baikie Harvey's letter contains several references to 'our people'. Orcadian identity became congealed rather than diluted by distance from home. In the Hudson's Bay Company, Orkneymen were recognised by their largely English officers as a body distinct from the 'Scotch', and generally regarded as close-knit if not secretive. The explorer and fort commandant Samuel Hearne complained in the 1770s of 'their clannish attachment to each other'.

This criticism was in part an affronted response to several unexpected bouts of concerted action by Orcadian employees in pursuit of better pay and conditions: there was a wage strike in 1777, and in 1788 men at the York Factory refused to leave for their winter postings until officers promised to lobby the committee to require returning vessels to stop off in Stromness – having to journey home to Orkney from London was an expensive aggravation. There was another refusal to renew contracts in 1797, and in 1805 the committee received news of a 'conspiracy' in Orkney, impeding the enlistment of labourers until generous bounties were provided for both new recruits and those re-engaging for a second term of employment

Over time, many Orcadians put down permanent roots, particularly in a settlement on the Red River, established in 1812, which would later become the city of Winnipeg. But most viewed service in the Company – for one or two five-year contracts – as a stage in the life cycle before coming back to set up in a trade or settle down on a farm. It was common for employees to take 'country wives' from among the tribes with whom they did business. Sons, though not daughters, of these marriages were sometimes sent back to Orkney for education, but native wives hardly ever accompanied homebound husbands. They were usually abandoned, returning to their tribe, or starting a new relationship with another trader.

Company servants lived frugally and typically saved as much as they could from their pay. Once an agent was in place in Stromness, arrangements were often made for wages to be paid locally to the employees' families.[18] Only a few became truly wealthy in Hudson's Bay service, but much of what was earned in Canada made its way into the islands' economy. The arc of empire bent back towards Orkney.

Napoleon's Greatcoat

Something else drew Orcadians into the expanding networks of British imperial activity, while encouraging them to reflect on Orkney's place in the world. In the reigns of George II and George III, Britain was almost continually at war: in Europe, in the colonies and – crucially for Orcadians – at sea. France was the perennial enemy: in the War of the Austrian Succession (1740–8); the French-Indian and Seven Years Wars (1754–63); the American War of Independence (1775–83; entered by France in 1778); the French Revolutionary Wars (1792–1802); and Napoleonic Wars (1803–15; including, in 1812–15, a second war with America).

Orcadians, declared the poet, patriot and ex-smuggler George Eunson, 'are great lovers of war'. It seems unlikely. The general view among officers of the Hudson's Bay Company was that Orkneymen had little stomach for fighting. Indeed, their reluctance to enlist in the army became almost proverbial. George Low, writing shortly before his death in 1795, believed not a single man from Birsay joined the army in the entire course of the eighteenth century. William Clouston, minister of Sandwick and Stromness (and formerly of Cross and Burness), thought Orcadians 'have no turn to the military line', and the Reverend James Watson of South Ronaldsay remarked on their 'aversion' to soldiering. Even Eunson, in fact, admitted that 'not all have an inclination to list in the army'.

One islander who did enlist, Magnus Flett from Harray, almost instantly regretted doing so. He complained to the sheriff in March 1778 about a persuasive sergeant in Lord Seaforth's newly raised Highland regiment: 'every unfairness were used against him in order to seduce him unto that service'. Triggering the complaint was Flett's dawning realisation that the recruiting party 'intends to carry him out of the country contrary to his inclination'.[19] By 'country', Flett meant Orkney rather than Britain; it is quite likely that – a landsman from Harray – he had never travelled outside the islands.

An aversion to army enrolment, it was generally agreed, did not carry across to service in the navy. It was a commonplace of eighteenth-century commentary that Orcadians (Harraymen perhaps excluded) were natural-born sailors, and that the islands were a prolific 'nursery' of crews for the Royal Navy. James Fea's claim that 12,000 men from Orkney and Shetland served in the navy of the Seven Years War is undoubtedly inflated. A more plausible estimate has 1,200 Orkneymen employed on Royal Navy vessels during the War of American Independence. This represented 1.2 per cent of the Navy's total manpower at the time, and suggests that Orcadians, proportional to population, were four or five times more likely to be serving as sailors than people from other parts of Britain. As many as 2,000 may have served during the Napoleonic Wars.

By no means all were volunteers. In time of war, men were pressed into service for both the army and navy – either by local constables acting on orders from the justices and tax officials, the county commissioners of supply, themselves in receipt of demands from the War Office, or by press gangs commanded by a naval 'regulating officer'. These were empowered to conscript any men lacking legal exemptions, on land or at sea – the impressment of British-born sailors from intercepted American ships was a principal cause of war with the United States in 1812. For young Orcadian men, an attraction of Hudson's Bay service was as a means of avoiding the press gang. The crews of Greenland whaling vessels, and ships servicing the coal trade, were also supposed to be exempt, but the niceties were not always observed.

Conscription on a significant scale began during the Seven Years War. In 1757–8, Orkney's JPs and commissioners of supply laid down quotas for island and Mainland parishes, and in 1759 Edinburgh newspapers reported 'a very hot press' in Orkney. Impressment reached a national peak in 1795, when the threat from Revolutionary France persuaded William Pitt to introduce the Quota System, requiring from every British county a stipulated number of men for the navy.

Quotas delegated a degree of control to local authorities, but impressment was without doubt the most direct and demanding intrusion of central government into the rites and rhythms of island society. It was keenly resented, notwithstanding George Eunson's confidence about Orcadians' eagerness to fight 'for their king, their country, and their glory'. Not surprisingly, impressment was often resisted – most commonly by the simple expedient of keeping clear of constables and naval shore parties. Virtually every parish in Orkney had secluded coastal caves, where young men reputedly concealed themselves. Countless traditions have been passed down about these places, and about other expedients to frustrate the press gangs' intentions: feigning illness or simplicity, and dressing in women's clothing – a reversal of the strategy of Isobel Gunn – were some of the most commonly reported stratagems.

On occasion, resistance went further. In 1758, two constables were beaten and driven off by a mob of thirty or forty after they 'seized on a young fellow' in Hoy. Attempts to impress men in Sandwick during the harvest of 1812 provoked similarly violent confrontation: the constables and soldiers were attacked with stones and the 'ware picks' used to rake seaweed from the sea. The culprits, a local laird complained, were 'a mob of women' – an echo of the parish's female activism from a couple of generations earlier (p. 408), when Sandwick women assembled to frustrate the installation of the Reverend James Tyrie.

Local traditions tell, too, of informers and betrayal, the acting out of grudges and settling of scores.[20] The demands of war in distant places placed new pressures on social relations, between people of equivalent status, and between rulers and ruled. When details of the quota arrived in Orkney, the magistrates divided it among the parishes. Within each parish, heritors then met to draw up a list of who was to be taken, and this was handed to the constables. But as the constables began their work, no one knew whose names were in the document. The involvement of locals in the selection of candidates made impressment an inescapably personal business. A revealing comment appears in an 1803 letter from the factor of the Graemeshall estate in Holm: 'the lads are much more averse to be taken by their own people than by a regular gang authorised by government'.

The expectation of officialdom, as expressed in a 1778 War Office circular to sheriffs, was that only 'able-bodied, idle and disorderly persons' were to be impressed. No man unfit for service was to be put forward by heritors, 'however obnoxious he may be'. The fear was that the parliamentary recruitment acts might become a convenient cover for landlords to purge their parishes of dependent cases or troublesome tenants.

More commonly, however, Orkney heritors and parish constables reported to the JPs that there were no men in their districts fitting the required description. Lairds sometimes sought to protect individual workers from unscrupulous press gangs. John Traill of Elsness petitioned the justices in February 1758 on behalf of Edward Sinclair, a Stronsay farmer who came over to Kirkwall as one of Traill's boatmen, and was seized 'in a most inhuman manner'. A year earlier, the Reverend James Tyrie complained to the JPs about the impressment of the sole ploughman on land which he rented at Hurkisgarth in Sandwick – adding, in the moralising tones of the kirk session, that there were plenty of 'stout young fellows who profess no other caring than that of the bagpipe, or stroll about the country as chapmen'. Throughout the century, lairds worried about the impact of impressment on the labour supply for summer kelp-work.[21]

Britain's eighteenth-century wars were fought far away from the islands, but not always as far as islanders would have liked. In the years after the Jacobite Rebellion, French privateers remained active in the waters around Orkney – a Newcastle-bound ship loaded with kelp was seized in September 1746, and ransomed for £100. The pattern was replicated during the Seven Years War. In 1761, Spanish privateers, allies of the French, landed in Orkney to seize livestock and other provisions.

Conflict with the American colonies made Orkney a still more tempting target: three French warships were reported lurking off Orkney in September 1778, 'with an intention to intercept the homeward-bound fleet from Hudson's Bay'. An English merchant naval officer, whose vessel was taken in July 1777 by Captain John Grimes of the *American Tartar*, learned that Grimes's ship was one of ten privateers come over from Boston, and that he 'intended to land or do mischief at the islands of Orkney and Zetland'. One prominent Massachusetts privateer, indeed, was an Orcadian by birth: Captain John Clouston of the *Freedom*, who took a dozen and more prizes in and around the English Channel before his capture in September 1777.

The Franco-American War reignited antagonism with another old foe of England. Attempts to stifle Dutch trade with America and France led to a fourth Anglo-Dutch War (1780–4). It ended with British victory, and the loss of Dutch possessions in both the West and East Indies. In the meantime, however, Netherlandish privateers launched a minor reign of terror against Orkney. Between 1781 and 1783, newspapers carried multiple reports of armed incursions by Dutch sailors, of cattle and other plunder seized, and money extorted with threats of house-burning. On at least one occasion, it involved – in the euphemistic discourse of the eighteenth-century press – 'treating some young women in a very indecent manner'.

July of 1785 brought some welcome reassurance that Orkney had not been forgotten. The Royal Navy frigate *Hebe*, captured from the French in 1782, dropped anchor at Kirkwall, where its captain dined on board with assorted lairds. His officers included the nineteen-year-old third lieutenant HRH Prince William Henry, younger son of the king, and the future William IV – a first royal visit to Orkney since that of James V in 1540. William came ashore with his fellow officers and, 'in compliment to the city of Kirkwall', paraded the full length of the street. 'On this glorious occasion', according to a local account printed in the press, 'nothing was to be heard but ringing of bells and shouting of people, as demonstrations of their joy, on seeing a prince of the blood royal in the *Ultima Thule* of his royal father's dominions.' It was not an unpolitical occasion. With memories of the contested election still fresh, Prince William pointedly snubbed the town council's offer of the freedom of the burgh, saying 'they were all foxes', but accepted it from the Pittite tradesmen's association.[22]

Orkney's vulnerability to external maritime attack – and its potential to become an attic door into Scotland or Britain – was an old tune played in quick time after the outbreak of war with Revolutionary France. In 1793, it prompted the formation of a regiment of Orkney and Shetland Fencibles – full-time soldiers, but limited to duties of home defence. The regiment was raised and commanded by Thomas Balfour of Elwick in Shapinsay. In 1746, his father had been an impoverished Jacobite fugitive, but the family pulled off a remarkable social and political rehabilitation. William Balfour rebuilt his position through close alliance with Sir Lawrence Dundas. His eldest son, John, made a fortune in Madras, and on his return in 1790 was elected MP for Orkney and Shetland; Thomas made a sparkling marriage to the sister of Viscount, later Earl, Ligonier. One of the company officers of the Fencibles was

James Moodie of Melsetter. James's father, Benjamin, had burned down Thomas's father's house (p. 350), but old feuds faded in the shadow of the guillotine.

The Orkney and Shetland Fencibles – never, in truth, a very formidable force – were dissolved at the end of 1797, leading to renewed fears that the islands lay open to French attack. In Birsay, at the start of 1798, the minister, Andrew Anderson, convened a parish meeting to discuss ways to preserve 'the safety of our religion and liberty, which we cannot expect to enjoy should the French Republic prevail'.

French plans for an attack on Britain in 1798 came to nought, though an expeditionary force was sent in support of rebellion in Ireland, where Thomas Balfour had earlier been dispatched at the head of a new, more mobile regiment, the North Lowland Fencibles. In 1803–4, however, fears of invasion resurfaced, as Napoleon assembled a vast Armée d'Angleterre in camps at Boulogne and other Channel ports.

Orkney seemed unlikely to be its intended destination. In Rousay, Thomas Balfour's sister Mary Craigie reflected ruefully in September 1803 that 'our poverty will preserve us from French visitors', a view shared by Balfour's aristocratic wife, Frances: 'this poor country is so far from being an object of plunder, it would not even afford existence to an enemy while they passed through'.

Others, though, were less sanguine. There was real concern in the early part of 1804 about a Franco-Dutch squadron operating in North Sea waters under the command of Jean-Jacques de Saint-Faust, a charismatic and unscrupulous French naval officer. His ships carried a complement of fierce French grenadiers, capable of landing in Shetland, Orkney or the north of Scotland. Saint-Faust's squadron was based in Bergen, in Danish-administered Norway – Denmark was still officially neutral, but hostility to Britain had increased following Admiral Nelson's 1801 attack on Copenhagen, designed to destroy the Danish fleet and prevent it falling into French hands. In February 1804, a group of non-resident lairds, the Committee of Orkney Heritors, wrote from Edinburgh to the commissioners of supply and the town clerk of Kirkwall urging them to exercise vigilance. From Kirkwall, circular letters travelled across Orkney to the ministers, requiring them to impress on their congregations the gravity of the situation and the need to report all sightings of suspicious vessels. Patrick Neill, an Edinburgh printer and naturalist who toured the Northern Isles in 1804, considered it extremely unwise that no troops were quartered in Orkney and the islands were left 'totally defenceless'.

There was no French invasion of Orkney in the early nineteenth century, and it seems unlikely one was ever seriously contemplated. But it was not entirely delusional self-importance that fed dark visions of Gallic occupation among the islands' landowners and ministers. The emperor himself knew something of Orkney, and its political as well as strategic value. In the summer of 1804, Napoleon went on an inspection of the camps at Boulogne, and in a speech to the troops made supportive reference to Denmark's long-dormant claim to the islands. During his years of victory, Napoleon imposed numerous border and boundary changes on continental Europe. Had he triumphed over Britain, it is far from inconceivable the emperor might have decided to reward allied Denmark by reversing the impignoration of 1468 (p. 18) and returning the islands to dimly remembered Danish dominion.

After the invasion crisis of 1803–5 passed, there were some who still thought French designs on Orkney needed to be taken seriously. In June 1812, as Napoleon began the invasion of Russia that would ultimately lead to his downfall, Graeme Spence, an Orcadian with almost forty years' experience as a maritime surveyor for the Admiralty, sent a set of maps and a memorandum to the Lords Commissioner of the Admiralty Board. The communication contained a plan ahead of its time: to establish Scapa Flow as the principal place of rendezvous for the country's wartime fleet – a strategy later to be adopted in both of Britain's twentieth-century wars against Germany.

Spence prefixed his scheme with a sweeping historical account of the islands, emphasising the importance of 'the Kingdom (as it was then called) of the Orcades' to the emperors of Rome, and stressing how Danes, Swedes and Norwegians regarded the conquest of Orkney 'as an object of the first importance towards their obtaining that footing and possession which they ultimately got in Britain and Ireland'. By illustrating the esteem in which Orkney was held by 'the warlike nations of antiquity', Spence hoped to 'dispel the cloud of neglect which has hung over this Thule of the Ancients'.

Scapa Flow's advantages, natural and strategic, were evident – a fleet based here could disrupt enemy trade with Russia and Scandinavia, and prevent the French going 'northabout' to interfere in Ireland. It would also be able to operate throughout the year, as despite lying so far north the seas here never froze. Although the waters around Orkney were often considered hazardous to shipping, a combination of reliable maps and new lighthouses made navigation here as safe as anywhere in Britain. Conversely, the dangers of neglecting Orkney were considerable. Certainly, 'the dark and deep designs of the present ruler of France seem only known to himself', but it was quite likely that Bonaparte would consider a northern attack easier than a southern one. With ten or twelve ships, and equivalent thousands of men, Napoleon could conquer the islands and turn them into a base for strikes against the rest of Britain. Orkney's harbours 'would swarm with his privateers'.

Spence concluded his proposal with a roll-call of well-known Orcadian seafarers, and other prominent individuals, whom the Lords Commissioner could turn to for independent confirmation of the scheme's viability. They included Admiral Alexander Graeme, current laird of Graemeshall, no fewer than four Royal Navy captains (Baikie, Balfour, Honyman, Richan) and a former RN master (Traill), as well as Malcolm Laing MP (a renowned historian and friend of Charles James Fox) and the distinguished judge, William Honyman, Lord Armadale. Spence's meticulous cataloguing of honourable islanders in the naval and political establishment testifies to his proud provincial patriotism, to a strong sense of Orcadian identity perfected through participatory Britishness. The document itself, one presumes, was read and then filed.[23]

There is an intriguing postscript to Orcadian involvement with the Napoleonic Wars, which involves the emperor himself. After his defeat at Waterloo in 1815, Napoleon planned to escape to America, but British ships were blocking the French Atlantic ports, and on 15 July the emperor surrendered to Captain Frederick Maitland of HMS *Bellerophon*, which conveyed him to England prior to his final voyage into exile on the Atlantic island of St Helena.

On board the *Bellerophon*, Napoleon made a new friend: James Tait from Deerness, a sailor thought to have been pressed into the navy on his way back from Hudson Bay years before. The Royal Navy had a habit of pitching Orcadians into proximity with key individuals and defining historical episodes. Alexander Graeme, who headed Spence's list of eminent Orcadian sailors, was a friend of Horatio Nelson, and like him had lost his right arm on active service. After they dined together in July 1801, Nelson wrote jokingly to Lady Hamilton, 'I expect we shall be caricatured as the lame defenders of England.' At least one Orcadian, William Mainland from Rousay, served as an able seaman at Trafalgar on Nelson's flagship, the *Victory*, and a dozen others were certainly present at the battle. There was a tradition in Kirkwall, recorded by Ernest Marwick, that an Orkneyman called Cooper helped carry the dying Nelson below decks.

A generation earlier, and on the other side of the world, an Orcadian, Forbes Sutherland, had become the first European to be buried in Australia. Sutherland was a participant in Captain James Cook's first expedition to the South Pacific, and died at Botany Bay in May 1770. News of Cook's own death on his third mission arrived in Stromness in 1780 with the expedition's returning vessels.

One of the visiting officers was William Bligh, and seven years later he agreed to take the son of his Stromness hosts as a midshipman on the *Bounty*. George Stewart, a great-great-grandson of the Reverend Walter Stewart of South Ronaldsay, played a key role in the fateful events on the ship in April 1789. He became second-in-command to the mutineer Fletcher Christian, took a wife in Tahiti and drowned in 1791 on his way back to England to face trial. Lord Byron, in his 1823 poem *The Island*, painted a picture of a tragic and romantic figure, 'blue-eyed northern child/Of isles more known to man, but scarce less wild' – though Byron displayed a less than certain grasp of the geography of northern Scotland in portraying Stewart as

> The fair-haired offspring of the Hebrides,
> Where roars the Pentland with its whirling seas.

Byron's account of Stewart's appearance was also distinctly fanciful. Bligh's own description of the mutineers notes him as having 'dark hair ... and black eyes, tattooed on the left breast with a star, and on the left arm with a heart and darts; is also tattooed on the backside'.[24]

The story of James Tait's encounter with Napoleon was preserved by a much less eminent poet than Byron: William Delday of Quoybelloch, Deerness (b. 1855).In the later nineteenth century, Delday recorded what he remembered being told by two of Tait's neighbours, Robert Stove and James Hourie: a tale of Tait going ashore for supplies at Rochefort, and being approached by Napoleon, who offered him his sword – though Tait very properly insisted he keep it to surrender to Captain Maitland. The two conversed amiably together on the boat, and often afterwards on the voyage to England, Napoleon complimenting Tait on his excellent French, which he had learned from a great-uncle in Deerness. On parting, the emperor presented his Orcadian confidant with a personal gift, his military greatcoat – 'James Tait accepted of the coat gratefully and it kept him warm for about twenty years and was very highly prized by him'.

Sadly, it seems unlikely to be true. Ernest Marwick, who in 1969 typed up a copy of a now lost manuscript, then in the possession of Isabelle Stove, imagined Quartermaster Tait to have been the commander of the

Figure 9.3: A Victorian depiction of Napoleon as a prisoner in 1815 on board HMS Bellerophon, *wearing the greatcoat which Orcadian tradition claimed was given by the emperor to James Tait and taken to a farm in Deerness.*

barge sent to collect Napoleon, but we know that this was the first lieutenant, Andrew Mott. There is no mention of Tait in Captain Maitland's published reminiscences, and more to the point he does not seem to have been part of the ship's crew at all. A couple of other Orcadians, John Wood and James Towers, appear in the *Bellerophon*'s muster book for July 1815, but Tait himself does not.[25]

Napoleon's greatcoat, then, may have come to Orkney only as a tall tale told by returning sailors. Even so, it is an irresistible metaphor for how Orcadians involved themselves in the affairs of the world, yet gave instinctive priority to the concerns of home. An entire philosophy of life is enacted in the pageant of its improbable journey from imperial shoulders commanding the fate of thousands on the battlefield of Waterloo to a retired sailor's mud-caked back on Braebuster Farm, the fabric worn out through years of digging drains and raising dykes, with pauses to rest and remember, and glance west towards Tankerness over the sheltered waters of Deer Sound.

A Statistical Account

A distillation of key themes of this chapter, and of this book, appears in a set of volumes published in Edinburgh between 1791 and 1799. They were billed as a *Statistical Account of Scotland*, and were the brainchild of Sir John Sinclair of Ulbster – the Caithness landowner who unsuccessfully challenged Charles James Fox in the election for the Northern Burghs in 1784. This hiccup notwithstanding (the government obligingly found him a rotten borough in Cornwall), Sinclair sat in parliament almost continuously between 1780 and 1806, usually for his birth county of Caithness. Trained as a lawyer, he was a lay member of the General Assembly, and a committed reformer and 'improver', acquiring the nickname 'Agricultural Sir John'.

'Statistical' and 'statistics' were relatively new concepts in the later eighteenth century. Borrowed from terms then current in Germany, they signified the compilation of useful information about the capacities of a state. Sinclair's vision was of a comprehensive historical, political, economic and social survey of Scotland – to serve as a 'means of future improvement', and help increase the country's 'quantum of happiness'. Societal advancement through the accumulation of knowledge was a characteristic instinct of the Enlightenment, and of Scotland's own contribution to that wider European project. Sinclair's objectives had affinities with those of Adam Smith's *Wealth of Nations* (1776), and of the *Encyclopaedia Britannica*, first published in Edinburgh in 1768–71.

The entry for 'Orkney Islands', in the *Encyclopaedia Britannica*'s 1797 edition, bristled with facts of history, geography, agriculture, fauna and flora. It exuded a rational, scientific outlook, dismissing 'absurd and unphilosophical' speculations about the reproductive patterns of the barnacle goose. There was discussion of 'ancient monuments and curiosities of art', in which – with some understatement – Orkney was said to be 'not altogether destitute'. The article concluded by noting that 'the Orkneys contain 30,000 inhabitants, and are equal in extent to the county of Huntingdon'. The author was aware of the islands' Scandinavian past, and its legacies in the present. He nonetheless judged that 'the inhabitants of the Orkneys may be now justly deemed a Scotch colony. They speak the language, profess the religion, follow the fashions, and are subject to the laws, of that people.'[26] The better informed contributors to Sinclair's *Statistical Account* showed all of that to be true – up to a point.

Those contributors were ministers of the Church of Scotland. Sinclair's inspired idea was to invite clergymen to write entries on the parishes in which they served. They were men on the ground, with fingers on the pulse of local affairs, but at the same time educated observers, with the detached and contextualising perspective of the outsider. Starting in 1790, Sinclair sent a detailed questionnaire to every minister in Scotland. If they failed to respond, he wrote again, and again, falling back as a last resort on information supplied by paid surveyors. The Orkney ministers, whose replies were published in volumes between 1793 and 1798, rose admirably to the challenge. Only in St Andrews and Deerness did the account have to be compiled by 'a friend to statistical enquiries'.

Taken together, the entries are an illuminating companion piece to the source with which this book started, Jo. Ben.'s *Descriptio Insularum Orchadiarum*, perhaps compiled by another clergyman exactly 200 years earlier. In place of random myths, anecdotes and snippets of parish hearsay, the ministers of the 1790s provided an abundance of detailed and verifiable facts, laced with informed interpretation and reasonable conjecture. Scapa Flow, readers were to understand, was 'a small Mediterranean'.

The minister of Hoy, Robert Sands, rehearsed none of the fabulous stories about the Dwarfie Stane (p. 29), but he did supply its exact dimensions, '32 feet in length, 16½ broad, 7 feet 5 inches in height'. 'Antiquities' were a topic of Sinclair's questionnaire, and the ministers scrupulously documented the appearance of marvels like Maeshowe, the Ring of Brodgar and the Odin Stone. George Barry, the Kirkwall minister transferred to Shapinsay in 1793, projected onto the prehistoric past the values of his own culture in speculating that a standing stone on the island might be 'a monument of some signal battle or victory, or to preserve the memory of some celebrated hero'. But James Watson's discussion of Pictish brochs in South Ronaldsay was more cautious: 'the inquisitive mind everywhere meets with darkness visible'.

In St Andrews, where the Reverend Charles Alison ignored Sinclair's entreaties to contribute to the project, the substitute author exacted some sardonic revenge. There were ruins, probably of a broch, on a neck of land jutting into the Loch of Tankerness, just behind the parish manse. But 'the minister, who considers modern enclosures as more ornamental and useful in a country than ancient ruins, has taken a great number of the stones of this building for inclosing his glebe'.[27]

Jo. Ben.'s suspicions about the motives and mindset of the ordinary people of Orkney had in large measure been replaced by a warm, occasionally condescending, regard. In the 1590s, ordinary people still spoke in a language strange to the minister's ear; by the 1790s, even in Harray, 'the ancient Norse language ... is now worn out'. Indeed, observed Francis Liddell of Orphir, visitors would often remark on how the people spoke 'with less of a provincial accent' than in the southern parts of Scotland.

The *Statistical Account*'s 'Friend' in Deerness perceived the people there as 'sober, regular, industrious'. William Clouston of Sandwick and Stromness thought his parishioners 'decent in their behaviour, respectful to their superiors, and modest in their carriage and conversation' – the latter word's modern meaning of verbal communication now starting to supplant its older one of moral conduct. Reflecting that Stromness was barely a century old, Clouston conceded 'there has not as yet, perhaps, been time enough to prove whether the genius of the people of this village may lead them to excel in literature, and the higher attainments of science' – he might have been pleased to learn, in a little over another century, of the birth there of the locally beloved poet George Mackay Brown and the acclaimed landscape painter Sylvia Wishart.

A discordant note was struck by George Barry, who detected in the people of Kirkwall 'a sort of low cunning', brought on by their complicity in smuggling. There was a tendency to murmur and complain, and 'a constant endeavour to throw the thick veil of mystery over every transaction they engage in' – traits, Barry emphasised, only to be found among the lower class of people.

While most ministers considered their parishioners to be fundamentally honest, there was general frustration with what John Anderson of Stronsay termed 'pertinacious adherence to old customs'. The people's reluctance to embrace necessary agricultural changes and accept the need for comprehensive 'improvement' was a continual refrain of the parish entries, echoing much that reformers had been saying for the best part of a generation. Obstacles to improvement included popular attachment to the single-stilted Orkney plough, the prevalence of short leases, payment of rents in kind, labour services, intermingling of fields, aversion to enclosure, ignorance of crop rotation, neglect of fishing, and the poor or non-existent state of the islands' roads.

In the view of some ministers, the traditional township and hill-dyke system – the bedrock of Orkney's cultural identity and an ancient source of social solidarity – was itself a wellspring of malaise. It meant that as soon as one tenant harvested his crops, all of them had to do so, and the whole farmland became a common for beasts. Anderson of Stronsay ventured that agriculture was retarded by 'traces of the manners, customs, language and law of the Norwegians ... still to be found in these islands'. There was a recognition that the old ways had been coming under increasing strain since the Heritable Jurisdictions Act of 1746, when, in the wake of the Jacobite Rebellion, the remaining judicial rights of landowners were abolished. Matters that were previously regulated in parish baillie courts – communal maintenance of the dyke, and arrangements around the grazing, marking and shearing of sheep – were at risk of becoming a messy free-for-all.

On some matters, the ministers were at odds. James Izat of Westray believed that agricultural improvement would forever be an afterthought 'while kelp continues to sell at any tolerable price', but Barry recognised kelp's importance in raising living standards, and for encouraging workers, aware of their value, to throw off a former demeanour of 'servile subjection' and embrace 'more and more of the spirit of liberty'.

Views differed on whether entanglement with the Hudson's Bay Company was a boon or a bane for Orkney. In South Ronaldsay, there was delight at the recent news that William Tomison had decided to establish a school in the parish, with a salary of £20 a year for the teacher. Clouston of Stromness suggested that, if work was slack for the youth of Orkney at home, it was better for them to 'go to hunt the harmless and civilised beaver than, like the Swiss, to fight the wars of other nations for hire'.

John Malcom of Firth, however, disliked the idea of servants and sons leaving 'to spend the prime of life in cold and drudgery', and reckoned that when they returned to take up farming, they found 'their skill in that line not improved by their absence'. The most hostile voice belonged to Liddell of Orphir. His instinct was to castigate the young men who

> instead of offering an honourable service to their King and country, or staying at home to cultivate their lands and protect their wives, their children and their parents, for the sum of £6 per annum hire themselves out for slaves in a savage land … Fie be on the man who would rather be the slave of a company of private merchants than enter into the fleets and armies of Great Britain, and fight bravely for his King and country, our religion, our liberties and our laws.[28]

With all their woes and worries, the ministers found matters to celebrate: the success of inoculation, or the adoption of the potato – unknown in Evie as late as 1784, but 'now universally cultivated with success'. Carts were a literal vehicle of progress, non-existent in some parishes fifty years ago, but 'getting into use'. William Clouston, leaving Cross and Burness for Sandwick and Stromness in 1793, was pleased to report there were now no fewer than thirty-seven carts in Sanday. There was praise for the occasional heritor, like Thomas Balfour in Shapinsay, willing to push through agricultural reform in the face of his tenants' scepticism.

Comparisons with even the recent past marked an advance towards 'civility'. In Birsay and Harray, George Low reflected that 'when I came first to the parish, there was not a piece of English cloth to be seen on a man's back, no figured waistcoats or velvet breeches, their stocking made of their own wool, their shoes of their own leather.' Now, however, 'the young fellows, instead of bonnets, almost all wear hats; upon Sunday, a suit of decent south country clothes, with cotton waistcoats and corduroy breeches'.

William Clouston produced a meticulous tabulation of social and economic differences between Stromness in 1700 and Stromness in 1794. At the start of the century, the place comprised five houses and a few scattered huts; now there were 130 houses with slate roofs, and a total of 222 inhabited dwellings. There were then only two small sloops belonging to the village, of thirty tons each; in 1794, two brigs and four sloops, at a combined tonnage of 500. In 1700, even the wives and daughters of lairds spun and sewed their own clothes; 'in 1794, the wives and daughters of tradesmen and mechanics dress in cottons and printed muslins'. Orkney cheese, oatcakes, and ale brewed without hops were once the fare at christenings. The guests now expected to see English cheese, white bread, cinnamon waters and wine. Tea, unknown in the houses of leading gentlemen in 1700, was today drunk by people of all qualities: 860 pounds of this quintessentially imperial commodity were imported to Stromness in 1792. And – the ultimate test of manners and sophistication – a dancing-master had just opened a salon in Stromness, teaching forty or fifty students.[29]

Orkney's ministers, at the close of the eighteenth century, liked to think that they embodied civility and progress. George Eunson had little time for them, 'men studious for their own interest … more greedy than godly'. That, perhaps, was unfair. The detailed and thoughtful articles produced for the *Statistical Account* paint a picture of a cohort with broad intellectual interests, and a conscientious sense of oversight for the parishes in their care. New sensibilities were starting to emerge – evident in petitions from the Kirkwall Presbytery in 1788, and North Isles Presbytery in 1792, which, like others from churches across Scotland, castigated the Atlantic slave trade as 'cruel and unjustifiable'. This was a trade from which Orcadian emigrants in the Caribbean – as well as the disgraced and transported minister William Nisbet (pp. 409–10) – derived direct profit.

By the time the *Statistical Account* was published, the moral scandals of the mid-eighteenth century were largely a thing of the past – though Francis Liddell of Orphir, scourge of the Hudson's Bay Company, was a lingering thorn in the side of his brethren. In 1788, he was rebuked by the General Assembly for slandering George Barry, and his relations with the presbytery continued downhill thereafter. In 1802, by his own admission, Liddell gave his enemies a handle against him: 'having drunk too large a quantity of champagne, mistaking it for cider', at an election-night dinner in Kirkwall, he became 'uncommonly witty' at his colleagues' expense. Liddell went too far when, before a notary, he married his housekeeper – the ministers of Kirkwall, who assumed she was pregnant, refused to perform the ceremony. His deposition by the presbytery was upheld by the synod, reversed by the General Assembly on appeal, and reimposed in 1808.

Liddell went down fighting, sending offensively satirical letters to other Orkney ministers, and publishing extravagant printed justifications of his own character and conduct. He made much of his Orcadian birth and descent, 'in the midst of the storm and tempest', and mocked the generality of the ministers, 'imported like lumber into this country'. They were, he said, 'a troop of poor ragged students from the College of Aberdeen, who, after being shin-burned and smoke-dried for twelve or fourteen years in a country school, are at length introduced into an Orkney living, dressed in the garb of humility, and with all the appearance of gratitude and sanctimonious mortification'.

It is a revealing glimpse into the prejudices against ferry-loupers which might lurk not far beneath the skins of some native Orcadians, but scarcely a fair characterisation of men like Low and Barry, accomplished scholars who contributed much to knowledge of the history of Orkney

– a place Barry earnestly hoped 'would emerge from obscurity, and assume the character of a respectable province, and add not only to the strength, but the splendour of the British Empire'.[30]

In respect of their own calling, the ministers contributing to the *Statistical Account* expressed relative satisfaction. The physical condition of churches was an enduring complaint, but the spiritual condition of parishioners was heartening. 'There are no dissenters or sectaries of any kind in these parishes,' wrote Watson of Burray and South Ronaldsay. 'No mischief is dreaded here, either from the flame of fanaticism, or the fire of sedition.' It was the same story from Sandwick and Stromness: 'all are of the established Church … neither enthusiastic nor superstitious'. There was a revealing judgement from John Malcolm on his Firth congregation, 'happily ignorant of the controversies on speculative points' causing animosity among Christians elsewhere. James Bremner in Walls and Flotta likewise attributed the contented orthodoxy of his parishioners to their being 'strangers to all the … vain disputations, violent dissentions, and strifes about words so frequently to be met with in other places'.

There was some truth to these assessments. Catholicism failed to make any headway in eighteenth-century Orkney, and there was equally little sign of the Cameronian strand of radical Covenanting (p. 312). By the 1790s, a once politically powerful Episcopalianism had declined dramatically in numbers and influence. But the complacent confidence of some ministers that ignorance and isolation would keep their people loyal and obedient to the Church of Scotland was spectacularly misplaced. 'They are', declared Hugh Ross of Evie and Rendall, 'free from profaneness on the one hand, and, hitherto, from fanaticism on the other.'

Ross was writing in September 1797, and his 'hitherto' indicates a new nervousness. A few weeks earlier, a powerfully charismatic lay preacher, James Haldane, arrived in Orkney with two companions and stayed for eighteen days, preaching in the open air right across the islands. It was part of a missionary tour around northern Scotland, its Orkney leg timed to coincide with the Lammas Fair in Kirkwall. On 12 August, Haldane addressed an audience of 800 in the courtyard of the Palace of the Yards; in the evenings of the following week, with the fair underway, he claimed to have drawn sermon congregations of 3,000 and more.

Evangelical revivalist fervour was rising in eighteenth-century Scotland, driven by a widely shared belief that the established Church of Scotland had squandered the zeal and spirit of the national Covenants, and – after the restoration of lay patronage over ministerial appointments in 1712 – was dangerously prey to corrupting political influence. In 1733, a number of ministers and ordinary members departed from the Kirk, and in 1747 this 'secession' itself split into 'Burgher' and 'Anti-Burgher' congregations. At issue was an oath requiring newly enrolled town burgesses to uphold 'the true religion presently professed within this realm'. This was interpreted by the Burghers as an uncontroversial commitment to the principles of Protestantism, and by the Anti-Burghers as an unacceptable endorsement of the Church of Scotland in its current institutional form.

These controversies had been present in Orkney even before Haldane arrived, carried along the arteries of kelp and coal. John Rusland, a Kirkwall tradesman transacting business in Newcastle in 1791–2, was enraptured by his experiences at an Anti-Burgher church there, and on his return set up a weekly meeting for prayer and religious fellowship. In 1795, work was started on the construction of a church in Kirkwall; Haldane attended Sabbath service there, but found it 'unfortunately too small'. Appeals by the modest but growing congregation to the Anti-Burgher Synod produced two temporary ministers, and in 1798 a permanent one, William Broadfoot.

A demand for the kind of religion the Church of Scotland ministers disparaged as 'enthusiastic' was not restricted to Kirkwall. Anti-Burgher congregations were founded in Stronsay in 1800, in Birsay the same year and in Stromness in 1807. Francis Liddell counted seven such places of worship in 1808, attributing their appearance, with no discernible sense of irony, to 'the dissolute lives of the Orkney clergy, in some of whose parishes it is a lamentable fact that the sacrament of the Lord's Supper has not been administered for half a century'. In a Parthian shot at his erstwhile brethren, Liddell predicted that in a dozen years, 'out of 26,000 inhabitants, there will not be found 500 to attend the established Church'.

Such a startling collapse did not ensue, but the Kirk's de facto spiritual monopoly was beginning to unravel in Orkney at the end of the eighteenth century. Haldane thought it the ministers' own fault. He discovered a fact that had disconcerted earlier ecclesiastical visitors, too: many parishes comprised two or three different islands, and due to disrepair of kirks, and difficulties of crossing firths, 'to say nothing of the want of zeal, many of the people see their pastor but seldom in the course of the year'. In his journal of his mission, Haldane offered a bleak assessment of a landscape of spiritual desolation:

> The Islands of Orkney, according to our information, which is rendered strongly credible by what we actually witnessed, have been for a period beyond the memory of any man living (excepting in one or two solitary instances) as much in need of the true Gospel of Jesus Christ, so far as respects the preaching of it, as any of the islands of the Pacific Ocean.

It was a provocative comparison, negating Orkney's claims to be an ordinary constituent part of the British state, and likening it to Tahiti and other Polynesian societies known about from the colonial naval expeditions of Cook and Bligh. On their travels across Orkney, Haldane and his assistants found countless individuals to be 'grossly ignorant', even as 'ignorant of the gospel as heathens' – though at the same time they were usually receptive to the message of true Christianity as the missioners chose to define it, with an emphasis on human sinfulness and the unmerited grace of Jesus Christ. They made a point of seeking out the sick, elderly and dying. A ninety-two-year-old man in Shapinsay knew nothing of Christ, but said he 'sometimes prayed to God'. In Flotta, an old lady thought she would go to heaven, 'as she said she had done nothing bad in this world, excepting once that she had had an illegitimate child'.[31]

The 'ignorance' that Haldane and his colleagues encountered was a mirror of their own attitudes and values. Still, the charge carried weight as an indictment of the long and laborious efforts of the Church of Scotland to fashion an Orcadian worldview that conformed to the outlook of educated Protestantism. An embrace of the modernity hailed by the *Statistical Account* was hampered by the habits of the past. As the *Encyclopaedia Britannica* fastidiously put it, Orkney's inhabitants 'are much addicted to superstitious rites; in particular, interpreting dreams and omens, and believing in the force of idle charms'.

'Superstition' was an ongoing problem the ministers themselves recognised, and which Barry attributed to 'the remote and secluded situation of this place'. He reported the reluctance to turn boats against the sun's course, or to marry during the moon's waning. Unless cows were killed during the waxing of the moon, the meat would spoil – a belief about the importance of the Martinmas slaughter-season noted by Jo. Ben. two centuries earlier (p. 28), and undoubtedly then already ancient.

There were many days in the year, Barry complained, when people would not fish at sea, nor labour in the fields. Clouston noted that people in Sandwick would do no work on 3 March, the consecration day of the parish kirk, and would work for wages but not for their own husbandry on 29 June, the feast day of St Peter to whom the church was dedicated.

The saints lingered in the Orkney landscape, shadowy figures of ancestral allegiance, to be propitiated or petitioned. Barry's account of Kirkwall and St Ola, the least 'remote' of Orkney parishes, addressed the persistence of pilgrimage:

> In the time of sickness or danger, they often make vows to this or the other favourite saint, at whose church or chapel in the place they lodge a piece of money as a reward for their protection; and they imagine that if any person steals or carries off that money he will instantly fall into the same danger from which they, by their pious offering, had been so lately delivered.

His comments confirm what we have already had occasion to observe (p. 215): that the instinct to undertake pilgrimage in Orkney was closely linked to practices of magical healing, and that maladies were understood to transfer themselves to the objects – stones, coins – employed as part of a ritual cure.

The practice, perhaps, was on its way out. George Low reckoned pilgrimage to the Brough of Birsay 'now much neglected'. But other, more domestic, forms of 'superstition' seemed frustratingly resilient. A catalogue of them was supplied by James Watson of South Ronaldsay, an Aberdeen graduate presented to the parish by Sir Thomas Dundas in 1786, after disappointed hopes of a Presbyterian posting in England. Like

Barry, Watson noted an aversion to marriage unless during a waxing moon, and preferably a rising tide. Within the last few years, he had twice been interrupted by parishioners for baptising a female baby before a male one. It was explained to Watson afterwards that the girl would undoubtedly later grow a beard, while the boy would be unable to – further evidence of the magically retentive properties of water.

Some believed seriously in the existence of witches and fairies, drawing imaginary circles and placing knives – the iron to which fairies were notoriously allergic (p. 364) – in the walls of houses. The worst was that when anyone in the parish lost a horse or cow, 'it sometimes happens that a poor woman in the neighbourhood is blamed, and knocked in some part of the head, above the breath, until the blood appears'. It was nearly a hundred years since people could hope to bring the force of law to bear on suspected witches, but the vigilante counter-magic of drawing blood 'above the breath' (p. 235) was still available to them. Watson, however, was confident that in South Ronaldsay and Burray 'there are many decent, honest and sensible people, who laugh at such absurdities'.

John Anderson, minister of Stronsay and Eday, another Aberdeen graduate, and former teacher at Kirkwall Grammar School, offered a slightly different perspective. He conceded that the common people of his islands 'remain to this day so credulous as to think that fairies do exist, that an inferior species of witchcraft is still practised, and that houses have been haunted'. There had indeed been a case in Stronsay in 1791, affecting the family of the boat-builder John Spence. Strange and persistent knockings were detected coming from a boat under construction on the stocks, and later heard in the house itself. They eventually fell away after Spence issued a solemn challenge, 'in the name of the Holy Trinity', for the suspected spirit to speak.

Anderson was aware that such stories would be ridiculed by 'those who glory in being superior to vulgar prejudices'. But he challenged readers to consider whether remnants of superstition, which affected character in small ways, were more or less harmful to the human mind, and the tranquillity of Church and state, than religious scepticism, 'which gives a new and totally different direction to the understanding and the will'. In the wake of the French Revolution, one might well understand 'why the minister and people of this district are not very hasty to exchange old prejudices for new and strange doctrines'.[32]

Anderson's earnest and sympathetic questioning of John Spence and his wife invites comparison to a minister's involvement with another haunted house in Stronsay, two centuries earlier, when Jo. Ben. offered counsel to a farmer's wife troubled with a sea-spirit (p. 191). There were unfathomable constants in Orcadian culture, held in place by age-old patterns of communal island living, and the invisible shaping hand of the elements.

Yet things changed, and sometimes changed utterly. Orkney in the 1590s stood at one historical crossroads, adjusting from a communitarian Scandinavian past to its uncertain future in a socially and politically unstable kingdom of Scotland. By the 1790s, it had arrived at a second crossroads, pointed in one set of directions by the commercial and cultural imperatives of imperial Protestant Britain, and in another by attachment to immemorial custom and the impulse to prioritise its own domestic affairs.

The islands of Orkney, positioned on the geographical edge of recurrent rounds of conflict and consolidation, are a seldom-heard witness to the storm-tossed birth of modern Britain. Orkney's quiet but compelling testimony nonetheless offers rewarding possibilities for thinking about what times of tempestuous historical change must have meant to the people who lived through them. It reminds us, not least, that there can never be any single authoritative narrative of national destiny and development. There are a multitude of stories still to be told about 'British' history and identity, and islands are good places to pause and start to listen to them.

Epilogue
The Lighthouse Inspector

The sloop was known simply as the *Lighthouse Yacht*, and was built to carry the Commissioners of the Northern Lighthouses on an annual tour of inspection around the Scottish coast. For the tour of 1814, the vessel weighed anchor at Leith on Friday 29 July, following the route north along the east coast of Scotland taken by James V and his pilot Alexander Lindsay in June 1540 (p. 15).

It was a summer of momentous realignments, political, military and literary. Delegates were preparing to meet in Vienna, to discuss the shape of Europe's borders in the wake of what was believed to be Bonaparte's final defeat; Napoleon himself brooded in restless Mediterranean exile on the island of Elba. In America, British forces were campaigning in the Chesapeake, and would soon launch an attack on Washington and set fire to the White House. On the day of the *Lighthouse Yacht*'s departure, Swedish troops crossed into Norway, to crush the nation's brief hopes of independence after the dissolution of the country's long union with Napoleon's ally, Denmark.

In London, Lord Byron had just completed another lengthy narrative poem, *Lara*, following the phenomenal success earlier in the year of his tale of Mediterranean piracy, *The Corsair.* A day before the commissioners sailed from Leith, the radical poet Percy Bysshe Shelley eloped with the sixteen-year-old Mary Wollstonecraft Godwin. They were heading for France, but the young woman's extraordinary imagination would before long swing round to the north. In Mary Shelley's pioneering gothic novel of 1818, the reckless scientist Victor Frankenstein removes himself to 'one of the remotest of the Orkneys'. There, in a 'desolate and appalling landscape', he begins work on a mate for his embittered and vengeful 'creature'.

The voyage of the *Lighthouse Yacht* generated important literary sequels of its own. With the commissioners, and Chief Surveyor Robert Stevenson, a number of passengers were on board. They included the sheriff of Orkney, William Erskine; the minister of Tingwall in Shetland, John Turnbull; and the acclaimed poet Walter Scott. Three weeks earlier, Scott had published his first novel. It was an instant popular hit, an initial print run selling out within days. Formally, the work remained anonymous, but the true identity of 'the author of *Waverley*' was no great secret. Set during the Jacobite Rebellion of 1745, *Waverley* was the first of twenty-six historical tales that would go on to make Scott quite possibly the most influential and internationally famous novelist Britain has ever produced. Scott has been called 'the man who invented Scotland'; arguably he invented Orkney, too, or at least he tried to.[1]

We know about the 1814 trip from an account in Scott's journal, headed 'Voyage in the Lighthouse Yacht to Nova Zembla, and the Lord knows where' – the author of *Waverley* knew his Alexander Pope (p. 435). He regarded it as a bracing adventure, though when Byron heard from a correspondent that Scott was 'gone to the Orkneys in a gale of wind', his instinct was to scoff. 'Lord, Lord! If these home-keeping minstrels had crossed your Atlantic or my Mediterranean, and tasted a little open boating in a white squall … how it would enliven and introduce them to a few of the sensations!' The problem with these 'Scotch and Lake troubadours', Byron confided, was they were 'spoilt by living in little circles and petty coteries. London and the world is the only place to take the conceit out of a man!' Peripheries, as we have learned, are always relative and subjective.

From the east coast of Scotland, the sloop sailed first to Shetland, to drop off the Reverend Mr Turnbull and scout locations for new lighthouses. On 11 August, it arrived in Orkney to inspect the installation at Start Point: 'all in excellent order'. While in Sanday, Stevenson made a tactless comment about the sails on a local farmer's boat, and received the tart retort that if 'you hadna built sae many lighthouses hereabout, I would have had new sails last winter'. There was a price to pay for progress.

Not all was well in the island of Sanday. From Sheriff Erskine, Scott learned of a recent dastardly crime: over the course of three or four years, a farmer had lost twenty-five cows, stabbed by an unknown assailant. No certain motive could be ascribed, but the farmer, by order of the justices, had been given the duty of 'taking up names for the militia', and resentment was suspected. The pressures of war were felt far from the battlefields of Germany and America. Scott witnessed evidence of another slaughter as the ship rounded the headland of Tofts Ness: the carcasses of 265 beached whales. The new lighthouse had stinted one source of providential beneficence to the inhabitants of Sanday, but it left another undisturbed.

A day later, the *Lighthouse Yacht* dropped anchor at Kirkwall. It was the eve of the annual Lammas Fair. A Scots clergyman visiting the town at this season just a few years earlier found much to admire: 'here were to be seen as fashionable and as well-dressed people as any in the capital of the kingdom'.[2] Scott, however, was not greatly impressed. 'The town looks well from the sea, but is chiefly indebted to the huge old cathedral that rises out of the centre. Upon landing we find it but a poor and dirty place.' Kirkwall's antiquities – the cathedral, remains of the castle, and the crumbling earl's and bishop's residences – were 'all situated within a stone's cast of each other'. Earl Patrick's palace was, admittedly, once a fine structure, but its ruins were now 'so disgustingly nasty' that Scott's antiquarian zeal was seriously challenged by them. Scott was pleased that St Magnus Cathedral had 'escaped the blind fury of Reformation', and he complimented the kirk session for keeping it in a tidy condition, but it was a building he found 'rather massive and gloomy than elegant'.

At the end of a second day in Orkney's capital, Scott penned a snide letter in verse to his friend the duke of Buccleugh:

> We have now got to Kirkwall, and needs I must stare
> When I think that in verse I have once called it fair;
> 'Tis a base little borough, both dirty and mean –
> There is nothing to hear, and there's nought to be seen,
> Save a church, where, of old times, a prelate harangued,
> And a palace that's built by an earl that was hanged.

As an epitome of the three momentous centuries of struggle and endurance that have supplied the subject matter for this book, it was at least admirably condensed.

The land around Kirkwall Scott judged 'tolerably cultivated', but in general he was critical of the poor state of Orcadian husbandry. His views on 'improvement' coincided closely with opinions expressed by the *Statistical Account* of the 1790s, by the naturalist Patrick Neill in 1804 and by the East Lothian estate manager John Shirreff, whose comprehensively critical survey of agriculture in Orkney was published in the year of Scott's visit.[3]

Scott ran through a familiar litany of agrarian errors. The neglect of fishing by Orcadians was inexplicable, and though their soil was intrinsically fertile, its pastoral and arable potential was ruined by excessive digging of peats and failures of crop rotation. Scott reserved similar strictures for Shetland – including the inadequacies of the single-stilted plough favoured in both sets of islands – but he nonetheless felt rather more warmly about Shetlanders. Their 'general tone of urbanity and intelligence' exceeded that of the Orcadians, 'by no means an alert or active race'. Orkney's landscape, moreover, 'is far less interesting, and possesses none of the wild and peculiar character of the more northern archipelago'.

Like other 'improvers', Scott regarded ordinary Orcadians as the architects of their own misery. 'The prejudices of the people' frustrated all efforts at modernisation beyond some willingness to experiment with potatoes. But Scott, writing in the era of the 'clearances' that depopulated swathes of the Highlands (but had not, yet, directly affected the Northern Isles), was aware of a human cost to the rationalising of estate management. Scott's guide around the West Mainland was John Rae, factor to William Honyman, Lord Armadale, and father to another John Rae, then a babe-in-arms, but later to become a Hudson's Bay Company surgeon, intrepid Arctic explorer and discoverer of the fate of the lost Franklin expedition. Rae senior explained there were no fewer than 300 tenants on Armadale's Graemsay estate, each paying about £7 in rent. 'How', Scott mused, 'is the necessary restriction to take place, without the greatest immediate distress and hardship to these poor creatures? It is the hardest chapter in economics, and if I were an Orcadian laird, I feel I should shuffle on with the old useless creatures, in contradiction to my better judgment.'

The author of *Waverley* had his head in the future, but his heart was tugged at by the past. In both Shetland and Orkney, Scott was fascinated, and at times horrified, by traditions and folk beliefs, such as the persistent 'horrid opinion' that 'he who saves a drowning man will receive at his hands some deep wrong or injury' (p. 359). He learned from Mr Turnbull, 'who is not credulous upon these subjects', that one of his parishioners had recently seen a sea-monster, swimming along the top of the waves, lifting and bending its serpentine head. Scott considered it 'must have been of the species of sea-snake, driven ashore on one of the Orkneys two or three years ago'. There had been great excitement among Edinburgh naturalists in 1808, when the corpse of a strange creature, fifty-five feet in length, and with six protruding limbs, was washed ashore in Stronsay. Some experts confidently declared the 'Stronsay Beast' an animal previously unknown to science, but a leading London surgeon and anatomist claimed it was only the decaying body of a large basking shark.

As locales of mystery and superstition, the Northern Isles both appealed and appalled. Scott wrote down what he had learned of trows, their similarities to Lowland fairies and Highland *sighean*, and the enduring belief that melancholy or low-spirited persons had in fact been stolen away by them, leaving a wraith-like substitute in their place. There was comic potential in a story Scott heard from one of his travelling companions about North Ronaldsay, an island he was disappointed not to have been able to see. A year or two earlier, a missionary preacher had visited the place, and, as a short, dark-bearded man, with noticeably small feet, was suspected by the inhabitants of being an ancient Pict. The local schoolmaster entreated Robert Stevenson, then staying on the island, to come and confirm whether the visitor was an ancient Pict or not. The islanders were mollified only when Stevenson crept into the exhausted preacher's bed-chamber, and recognised in the sleeping figure an acquaintance from youthful days in Edinburgh.

Scott was truly intrigued, however, by another North Ronaldsay story, related to him by the veteran political laird Robert Baikie of Tankerness (p. 429). Sometime in the later eighteenth century, an Orkney clergyman took to the island a copy of Thomas Gray's 1768 poem 'The Fatal Sisters', a loose translation of some Old Norse verses about fate-weaving Valkyries and the destiny of Orkney warriors, appearing originally in the thirteenth-century Icelandic *Njal's Saga*. The minister thought that the islanders might enjoy hearing this recovered tale of Viking ancestors. But after he began reading it aloud, some elderly inhabitants interrupted him to say they knew the poem well; they called it 'The Enchantresses', and had often sung it to the minister himself in the Norn original when he requested from them a song in that language.[4]

The long reach in Orkney of the Scandinavian and pre-Christian past was a source of fascination for Scott, who himself had recently produced an abridgement of the Icelandic *Eyrbyggja Saga*. Scott arranged to be rowed over to Hoy to clamber inside the Dwarfie Stone, and he took a horseback excursion from Rae's house at the Hall of Clestrain to visit the Stones of Stenness and Ring of Brodgar, lying in the heart of the West Mainland.

The quality of guidance at the sites was not, perhaps, what it is today. 'Mr Rae seems to think the common people have no tradition of the purpose of these stones,' Scott complained, adding, 'but probably he has not enquired particularly.' By the start of the nineteenth century, folk tradition and intellectual inquiry were starting to seem like allies not enemies in the task of making sense of the distant past. Like earlier visitors to the Stenness site, Scott took particular notice of the Odin Stone (p. 389), documenting the 'odd superstitions' of the lovers who used it to plight their troth. He speculated that the hole was originally intended to bind sacrificial victims, though he regarded as 'justly exploded' the theory that stone circles were 'exclusively Druidical'. To the constitutionalist mind of Regency Britain, it seemed evident these must have been places of assembly, 'whether for religious purposes or civil policy'.

As it turned out, Scott was the last visitor to document their impressions of the Odin Stone. Four months later, it was pulled down by a local tenant farmer, Captain William Mackay, along with two megaliths from the Stenness circle. It is uncertain whether Mackay meant to plough up the land or wanted to use the stone for building a cattle byre, or whether he was simply irritated by a stream of inquisitive visitors onto his land.

But the destruction would undoubtedly have gone further had an outraged interdict not been obtained from the sheriff court by three Orcadian gentlemen, among them the influential former MP Malcolm Laing.

Mackay maintained he had Factor Rae's permission to remove some stones, and grovelingly sought to reassure the lairds 'I was not in the smallest degree aware of giving them, or the meanest individual in the county, offence by doing so'. The apology did him little good. Laing regarded Mackay's action as an affront to 'the community at large', and expressed regret that so little reverence for Orkney's ancient monuments had been inspired by Rae's 'residence in the country' – a revealing contrast to Mackay's 'county' of Orkney.

Alexander Peterkin, the recently appointed sheriff-substitute who issued the injunction, wrote in 1822 that Mackay – like Rae, a ferry-louper – had 'suffered a sort of mean persecution ever since'. There was no formal prosecution, but the local peasantry formed various conspiracies against him, and twice tried to set fire to his property. Peterkin's own view was that Mackay was more sinned against than sinning, but the defacing of any ancient monument was nonetheless an injury: 'a link in the chain of our associations is broken; the landmarks between different generations of men are thrown down'.[5]

It was a sentiment Walter Scott would have whole-heartedly endorsed. His poignant, and unintentionally ironic, description of the stones at Stenness was 'that those which have fallen down (about half the original number) have been wasted by time, and not demolished'. Scott's assessment of the remarkable monoliths populating the narrow strip of land between the Lochs of Harray and Stenness was that 'Stonehenge excels these monuments, but I fancy they are otherwise unparalleled in Britain'. This silver medal of a judgement conferred praise on Orkney, but also kept it, in more senses than one, in its proper place – within the geographical boundaries of Britain, and as a junior contributor to a shared national heritage.

Before Scott left Orkney, there was one more business to attend to: he went to visit a witch. She was to be found in Stromness, a place which, despite the praise lavished on it in the *Statistical Account*, scarcely commended itself to the illustrious visitor. Scott found it 'a little dirty straggling town, which cannot be traversed by a cart, or even by a horse, for there are stairs up and down, even in the principal streets. We paraded its whole length like turkeys in a string, I suppose to satisfy ourselves that there was a worse town in the Orkneys than the metropolis, Kirkwall.'

The witch, whose name was Bessie Millie, lived just outside the settlement, near the top of a hill called Brinkie's Brae, with unrivalled views of Stromness harbour, and the unfolding expanse of Scapa Flow. Bessie's interest in the panorama of anchorage was a professional one, for she was in the business of selling winds to sailors. Scott described her as 'an old hag', 'a miserable figure; upwards of ninety, she told us, and dried up like a mummy', though her face bore 'a ghastly expression of cunning'. It was the practice for every captain of a merchantman to visit her hovel, and there, 'between jest and earnest', to offer the old woman sixpence. A favourable wind was then secured by Bessie's boiling of her kettle.

Scott observed with amusement the old lady's pre-prepared 'rigmarole', and her delight at the duplication of the usual donation from her party of important visitors. The stiff breeze bearing the *Lighthouse Yacht* out of Stromness on 17 August 1814 proved Bessie to be 'a woman of her word', though a headwind encountered the following day off the north coast of Sutherland suggested to Scott that 'Bessie Millie's charm has failed us'.

We are both a long and a short distance from the frenzy of the witch trials, and the horrors of the Loan Head. Educated sentiment now looked on the supernatural with a superior smile rather than a murderous frown, but the extent to which popular beliefs had altered is harder to judge. Scott saw them as poised between 'jest and earnest', but the former may have been a brave face on the latter. The ritual of purchasing a fair wind, and of pacifying persons with the power to turn it to a foul one, was not lightly set aside.

Orkney's reputation as a haunt of witches, notorious to James VI in the later years of the sixteenth century (p. 185), was apparently robust in the early decades of the nineteenth. Not very long after Scott's visit to Stromness, Walter Traill Dennison was taken as a child to see the docks at Leith. There, he recalled, he got talking with a friendly old sailor, who regaled him with seamen's tales. But when the mariner learned the boy came from Orkney, he visibly recoiled: 'O, my lad, you hail from that lubber land where so many cursed witches dwell.'[6]

The witch did not only sell Walter Scott a wind, she also told him a story. She could remember, she said, the pirate John Gow (p. 341). It is just possible that she could, though even if Bessie was indeed over ninety, she would have been extremely young when Gow sailed from Stromness to his fate. Bessie was, nonetheless, a well-stocked repository of local beliefs and traditions. Scott learned that in Stromness Gow courted a Miss Gordon, and that the couple were pledged to each other by the clasping of hands (through the Odin Stone?). After Gow was arrested, the lady believed the engagement could only be dissolved by their taking up hands again. Miss Gordon travelled to London to request an audience with the prisoner, but she arrived too late and had to content herself with the hand of Gow's swinging corpse. 'Without going through this ceremony according to the superstition of the country', Scott later remarked, any future betrothal would have brought a visit from the ghost of her departed lover.[7]

Scott's encounter with a witch, and his listening to the eerie tale of an Orkney pirate, were memories which lingered. Between them, they provided the inspiration for a novel he published at the end of 1821. By then, the now Sir Walter Scott was at the height of his powers and fame, having written several of the books on which his reputation still rests: *Old Mortality*, *Rob Roy*, *The Heart of Midlothian*, *Ivanhoe*. Scott's novel set in Shetland and Orkney, *The Pirate*, is today less well known than these other works, but in its time it was a commercial success, and would put Orkney on the map as a desirable destination for adventurous Victorian tourists.

To say that Scott made a mark on the landscape of Orkney is a little more than a metaphor. When the acclaimed English painter William Daniell travelled to the islands in 1818, he arrived armed with a letter from Scott advising him on subjects for his brush: the cathedral and palace ruins in Kirkwall; 'a noble view' from Stromness towards Graemsay and Hoy; the Dwarfie Stone in its 'lonely valley'; and, of course, the 'most noble assemblage of huge upright stones about eight or ten miles from Kirkwall … Mr Daniell must not omit this subject.'

Daniell's depictions of these and other scenes were reproduced in the fifth (1821) volume of his *Voyage Round Great Britain*, an ambitious, decade-long project to produce a comprehensive pictorial record of the British coastline, its dramatic vistas and the social condition of its people. Daniell agreed that the Stenness Stones were 'very singular and interesting monuments', though 'by no means comparable with Stonehenge'. He regarded the recent throwing down of the Odin Stone as regrettable, 'but farmers are not bound to be antiquaries'. Daniell also recorded how, viewed across the water from the manse in Stromness, a high headland known as the 'Kame of Hoy' was considered 'to present a profile likeness of the most distinguished living poet of Scotland, Sir Walter Scott'. This, Daniell thought, 'evinces the warmth of affection with which his name is cherished in the remotest recesses of Scotland, and which would assign to him a monument formed by nature herself, in the country which he honours and adorns'.[8] One struggles somewhat to see it, but the claim was reproduced by numerous nineteenth-century writers.

The action of *The Pirate* takes places in the spring and summer of 1689. Its convoluted and melodramatic plot involves Basil Mertoun, a reclusive widower and incomer to Shetland, where he lives with his son, Mordaunt, the hero of the novel. The Mertouns are tenants of a big-hearted udaller laird, Magnus Troil, who has two beautiful daughters: the intense and mystical Minna, and the cheerfully practical Brenda. All are watched over by Magnus's relative, Norna of Fitful Head, a mysterious prophetess with power over spirits and winds – a figure inspired by, but scarcely modelled on, Bessie Millie of Stromness.

Relationships are unsettled by the rescue of a shipwrecked sea-captain, the Byronic antihero, Clement Cleveland – a heavily creative reworking of John Gow. Unbeknown to his hosts, Cleveland is a pirate with a Caribbean past, whose second ship has survived the storm and sailed to await rendezvous in Orkney. Having become Mordaunt's rival for the affections of Minna, Cleveland wounds him in a struggle and flees to Orkney to rejoin his pirate companions, leaving behind an infatuated Minna. The injured Mordaunt is spirited away by Norna to Hoy, where she reveals herself to be his mother, and Mertoun the lover who long ago deserted her. The Troils, too, travel to Orkney for the great Lammas Fair at Kirkwall, and are caught up in the intrigues as the pirates negotiate with the town magistrates to be allowed to proceed to Stromness to take on provisions and depart from the islands.

The arrival of a Royal Navy frigate precipitates a dramatic showdown at the Stones of Stenness, after which Cleveland and his accomplices are arrested. Norna, it transpires, was mistaken; her long-lost son is not Mordaunt, but Cleveland. Mertoun – his real name Vaughan – had once himself been a pirate captain. Clement served an apprenticeship under him, before father and son lost contact with each other, and Mertoun, with his child from a failed marriage to a lady of Hispaniola, fled to Shetland to atone for his sins. Conveniently, it turns out that both pirates have already been pardoned, as a result of heroically selfless past deeds. Cleveland and Minna part sorrowfully, and we learn that he later loses his life in government employment. Mordaunt and Brenda are married. Norna gives up her practice of the occult to devote herself to works of Protestant piety.

This brisk summary does scant justice to Scott's meticulous depictions of Orkney and Shetland locales and society, or to the compelling themes he contrives to weave into the work. *The Pirate* differs from Scott's other historical novels with a seventeenth- or eighteenth-century Scottish setting in its exclusive focus on the Northern Isles, which he endeavours to portray as culturally and politically distinctive – a place of glamorous otherness, and a frontier zone resisting full assimilation into the emergent British nation.

It required some research. By the early 1820s, Scott knew the history of Scotland like he knew the grounds of his grand home at Abbotsford in the Borders, but the Northern Isles were another matter. He read the histories of Orkney by James Wallace and George Barry, along with other works about Shetland and the antiquarian researches of Sir Robert Sibbald. On request, Alexander Peterkin sent him an account of the downfall of John Gow. Scott also consulted Olaus Magnus's *Description of the Northern Peoples* (p. 21), a book which Magnus Troil finds lying around in Norna's secluded magical lair.

The Pirate is packed with motifs and incidents we have already come across in this book. Its characters well remember the cruelty of 'Pate Stuart' (Earl Patrick). Magnus Troil, then living in Orkney, 'was pressed to serve under Montrose, when he came here about the sixteen hundred and fifty'. Troil's bardic neighbour, Claud Halcro, makes reference to the 'mickle bicker of Scapa', offered to new bishops to drink from on their arrival in Orkney, and Norna compares a stream in Shetland to the Well of Kildinguie in Stronsay, able to 'cure all maladies save Black Death'. There is depiction and discussion of shipwrecks – 'Godsends come on our coast' – and of the morality of unauthorised salvage. The old prejudice against rescuing a drowning person becomes a crucial lever of the plot: the secret half-brothers Mordaunt and Clement save each other from this fate, and go on to inflict mutual hurt. Gow's story is, of course, a central point of reference, though much sanitised and even inverted – protecting female captives from the threat of rape is a measure of Cleveland's character in both the West Indies and Orkney.

'Superstition' is a recurrent motif in the novel, at times wistfully celebrated as a mark of a more innocent age: 'it had charms which we fail not to regret, even in those stages of society from which her influence is well-nigh banished by the light of reason'. We hear of petitionary visits to ruined chapels, of trows, witches and mermaids, of sea-monsters still to be glimpsed 'on the deep and dangerous seas of the north', of refusals to work on St Magnus's Day, of reluctance to turn fishing boats against the course of the sun, and the necessity of equal sharing of catches.

All of this was mentioned in the works that Scott consulted, but he also drew on his own memories of the *Lighthouse Yacht*'s visit. Cleveland and his pirate lieutenant climb Wideford Hill to take in the magnificent views over Kirkwall and the promontories of Inganess and Quanterness, as Scott himself did in 1814 (and I found myself doing, not for the first time, in 2023). Critical encounters take place in the 'solemn old edifice' of St Magnus Cathedral. There are descriptions, poetic and precise, of 'the Orcadian Stonehenge'. Norna's career as a white witch and visionary begins after a mystical encounter with a Norse-speaking spirit at the Dwarfie Stone in Hoy. Her union with Vaughan is sealed with clasped hands at the Stone of Odin, 'wedded after the ancient manner of the Norse … with such deep vows of eternal fidelity, as even the laws of these usurping Scots would have sanctioned as equivalent to a blessing before the altar'.

The hostility of island characters towards 'usurping Scots' is a recurring theme of the novel. Scott carefully selected a period setting when he considered this would be historically plausible. The udaller Magnus Troil is descended from 'an old and noble Norwegian family', and he loses few opportunities to assert his ethnicity, swearing colourful oaths by the blood or the bones of St Magnus. 'I like you the better for being no Scot', he says to the Englishman Mertoun, comparing the Scots to a flock of barnacle geese, descending on Shetland to roost and gobble up resources. Indeed, Troil is critical of the gentry of Orkney, whom he disdainfully regards as 'always in a hurry to draw the Scotch collar tighter round their own neck'.

Minna is a chip off the old block, and takes her anti-Scottish patriotism to dangerous lengths, which Triptolemus Yellowley, a Scots agent of the chamberlain of Orkney and Shetland, thinks 'can end in naething but trees and tows' (ropes). She resents 'the tyrannical laws of our proud neighbours of Scotland', and sees in the revolutionary events of 1688–9 a golden opportunity 'to shake off an allegiance which is not justly due from us, and to return to the protection of Denmark, our parent country'. Prospects of emancipation from the Scottish yoke are, however, hindered by 'the tame spirit of the Orcadians', whose gentry 'have mixed families and friendship so much with our invaders, that they have become dead to the throb of the heroic Norse blood which they derived from their ancestors!'

We can see what Scott is up to. Several of his 'Scottish' novels are deeply infused with the spirit of Jacobitism – noble, romantic, alluring and long past its sell-by date in Regency Britain. Magnus and Minna's Scandi-separatism is a twist on the Jacobite theme, allowing readers the pleasurable contemplation of a fiercely tribal and troubled past – one now politically tamed and defused precisely by being celebrated and romanticised.

Minna and Brenda are not, in fact, pure-bred Shetland Norse: their mother was the daughter of a Sutherland Highland chieftain, driven from his homeland 'during the feuds of the seventeenth century'. There is thus a 'Celtic' tinge to their exotic provincialism, tying the Northern Isles to Scott's wider fascination with the untamed Highlands, and their perhaps regretful but nonetheless inevitable assimilation to nationhood and modernity.

Orkney, it seems, is closer than Shetland to recognising the new political realities – this is a device of characterisation for the Troils, but probably also a reflection of real cultural and economic differences that Scott thought he detected on his tour of 1814. It falls to the pirate Cleveland, of all people, to disabuse Minna of her sentimental and anachronistic notions: 'Think not of such visions. Denmark has been cut down into a second-rate kingdom, incapable of exchanging a single broadside with England.' Centuries of subjection have reduced the islanders' desires for independence to 'a few muttered growls over the bowl and bottle'; and, in any case, were Minna's neighbours to prove as willing warriors as their ancestors, 'what could the unarmed crews of a few fishing-boats do against the British navy?'

Minna's father, Magnus, proves susceptible to patriotic persuasions of a different sort: Cleveland's talk of fighting the Spanish in the Americas with colours nailed to the mast – and of the Armada's destructive descent on Fair Isle in 1588 – fills the old udaller with martial verve: 'That is the way, the old British [Union] jack should never down! When I think of the wooden walls, I almost think myself an Englishman, only it would be becoming too like my Scots neighbours …'

The ageing udaller's curmudgeonliness can be generously indulged, since it belongs to an old order in the process of being superseded. The pattern is reflected in Magnus Troil's debates with Triptolemus Yellowley, an evangelist for agricultural improvement, whose striking name is borrowed from the inventor of the plough in Greek mythology. Of the two, Troil is the more appealing character, but can offer little by way of rebuttal to Yellowley's strictures about inefficient farming, or the woeful inadequacies of the single-handled plough, save for bursts of offended honour: 'You come to us from a strange land, understanding neither our laws, nor our manners, nor our language, and you propose to become governor of the country, and that we should all be your slaves!' 'My pupils, worthy sir, my pupils!' Yellowley condescendingly replies.

For a book set in the spring and summer of 1689, the Glorious Revolution – that categorical moment in the creation of modern Britain – is a surprisingly muted and peripheral presence in *The Pirate*. But it nonetheless frames and enfolds the novel's key theme. Society is transitioning painfully but inexorably towards the modern – no matter the attractions of tradition, superstition, magic or a glorious warrior past.

Scott's purpose in *The Pirate* is not dissimilar to the task I have set myself in writing this book: to tell a story about history, identity and nationhood rooted in the peculiarities and priorities of a particular place. Scott's version culminates in a succession of sometimes sad but nonetheless necessary resolutions. Cleveland gives up piracy and is later reported to have died nobly in the service of the British crown; Norna abandons her claims to supernatural control over the elements, turning erstwhile clients away with pious reflections that 'the winds are in the hollow of His hand'. Mordaunt – like many of Scott's heroes – in the end makes the sensible romantic choice, preferring the down-to-earth Brenda to her complex and poetic sister, Minna. He finds Brenda 'neither superstitious nor enthusiastic, and I love her the better for it' – an uncanny echo of language used by the complacent ministers of the *Statistical Account* about their supposedly docile Orkney parishioners.

Walter Scott's journey to Orkney, and his factual and fictional renderings of it, was always intended to be the ending for *Storm's Edge*; a conveniently congruent chronological bookend with James V's sojourn of summer 1540. Between them, the two visits frame this book's appraisals of what changed in Orkney, and in its relationship with Scotland and the rest of Britain, over the transformative course of nearly three centuries.

Scott's portrayal of the islands in the north was an illustration in miniature of his panoramic vision of Britain. He acknowledged, and even celebrated, the distinctive character of eccentric individuals and out-of-the-way places, but in the end he wanted to colour their history and experience with the brushstrokes of an emergent bigger picture, one that depicted rational progress, political order and national unity.

Not just for Scott and his readers, but for visitors and outsiders of many stripes, the Orkney Islands were a territory always fastened to the fringe, the margin, the periphery; a cluster of small communities clinging to the storm-tossed rim of Britain. The islands, literally as well as metaphorically, were places on the edge of the map.

It did not always make them appear unimportant. Possession of Orkney was a trophy and token of power, for rulers of Denmark, Scotland and Britain, as well as for a succession of externally anointed maintainers of local sway and swagger. Orkney's allotted position on the outer boundaries of the realm made the place a decorative adornment – and, at times, a structural component – of larger social and political edifices. The islands, according to William Daniell, were 'the coronet of Great Britain'.[9]

A coherent history of the making of Britain can certainly be told through a focus on Orkney's absorption into the Scottish and British nations; the conversion of its people from Catholicism to Presbyterian Protestantism, and the transformation of their ethnic and social identities from that of colonial Scandinavians to compliant Scots to British citizens of empire. But it is not, in the end, the history this book has wanted to tell.

For the islands and people of Orkney, as I have tried to show, all of these absorptions, conversions and transformations were gradual, tortuous and never definitively completed. Orkney changed in very significant ways between the early sixteenth century and the early nineteenth. But it remained a resolutely 'lumpy' place, with priorities, preferences and problems unmistakably its own. Writing about Orkney has confirmed my conviction that the first instinct of historians should be to regard local communities not as 'case-studies' for assessing wider processes of change, but as sites of social alchemy, where different admixtures of external and internal pressure produced unpredictably transformative results. The distinctive character, the 'uniqueness', of islands is often fairly evident, but it should encourage us to acknowledge that nothing and nowhere is ever really 'typical'.

Shining a spotlight on place poses questions about perspective. Orkney was at the centre of historic networks of trade, religion and politics, as well as of its own people's aspirations and experience. A position on the periphery, we need to recognise, is something allocated and assigned – a cultural and political assertion, not a geographical 'fact'.

It raises issues about identity, too. Nations, and the nationalisms they generate, demand a hierarchy of affiliation, where local, regional or ethnic senses of belonging are subordinated to an identification with the wider community. Over time, some Orcadians perhaps did come to see their principal 'country' as Britain, rather than as Orkney, but it was seldom, if ever, an either-or. In the ethnically, linguistically and culturally layered landscape of Orkney, there was little that required multiple identities to be ranked and ordered in this way, or even to make being an Orcadian take priority over belonging to Birsay, South Ronaldsay, or Lady parish in Sanday.

Orkney, in the twenty-first century, regularly takes the top spot in social surveys of the 'best place to live' in Scotland, and sometimes Britain, too. It is extremely unlikely that, had such surveys existed in the sixteenth, seventeenth or eighteenth centuries, the islands would have fared anything like as well. By modern standards, life for the majority of people was almost unimaginably hard – a round of unremitting labour, laced with hazards of social oppression, economic immiseration and physical or supernatural violence.

Yet, under intense external and internal pressures, island society endured. In an era of profound political, social and religious upheaval, Orcadians proved capable of adapting to change while retaining much of what was distinctive about their communal existence, inherited traditions and instinctual beliefs. Orkney also demonstrated an abiding ability to accept and absorb incomers and immigrants, and acclimatise them to local patterns of living and thinking. There are lessons to learn here, if we are willing to look: a myriad of small social miracles, unfolding quietly between the hill, the sea and the shore.

Acknowledgements

This book has been both a journey home and an unfamiliar departure, and many friends, old and new, have helped with the coming and going. I want to thank my agent, Doug Young, who was from the first an enthusiast for a book about Orkney, and encouraged me to imagine what it might look like. At HarperCollins, Arabella Pike has been a reassuringly supportive editor, while Katy Archer and Sam Harding have helped to make the process of turning my manuscript into a book remarkably painless and pleasant. The final version of the text has benefited immeasurably from the superb copy-editing of Kit Shepherd.

There are some required institutional acknowledgements, though these mainly involve debts to individuals. I am grateful to the Leverhulme Trust for the award of a Research Fellowship which supplied me with time to think and write, as well as funding for travel to Orkney and other necessary places. Alec Ryrie and Alex Walsham wrote in support of the application, thus adding to an already long list of deeds of kindness and friendship, compiled over very many years. Additional financial provision was provided by the Montreal-based collaborative project on 'Early Modern Conversions', though more valuable than the money was the companionship and stimulus provided by a team of outstanding researchers, under the inspiring leadership of Paul Yachnin. Closer to home, I have received much practical and moral support from colleagues at the University of Warwick; among them, Beat Kümin, Bernard Capp, Claudia Gray, James Green, Martha McGill, Nyasha Dandara, Rebecca Earle and Tim Lockley. I also need to thank Robin Urquhart at the

National Records of Scotland in Edinburgh, and Ulrike Hogg at the National Library of Scotland, for welcome advice and assistance. Further north, Lucy Gibbon, Sarah Maclean and Andrea Massey at the Orkney Archive in Kirkwall have all, over several years, been unfailingly helpful and hospitable. I can't quite say that the Orkney Archive is the most agreeable record office in the country to work in, only that it is the most agreeable one I have ever worked in.

Numerous well-wishers have shown an (I think more than polite) interest in the progress of this book, and gone out of their way to provide references and suggestions: my thanks to Angela McShane, Arnold Hunt, Bruce Gordon, Charlotte Methuen, Chris Langley, David Trim, Evan Jones, Jane Dawson, Jonathan Westaway, Judith Jesch, Laura Sangha, Michael Graham, Stephen Conway, Terry Gunnell and Tim Harris. Several Norwegian friends and colleagues have assisted with queries and translations, and graciously indulged my faltering efforts to master their beguiling language: Ian Peter Grohse and Sigrun Høgetveit Berg in Tromsø, Henning Laugerud in Bergen, Jo Rune Ugulen at the Norwegian National Archives in Oslo, and, in Orkney, Ragnhild Ljosland. A planned visit to Lerwick was cancelled by Covid, but I have greatly appreciated advice and encouragement from Shetland's renowned archivist-historian, Brian Smith.

My indebtedness to earlier historians of Orkney will be evident from the preceding pages. I have for a long time now looked up to two giants in the field, Ray Fereday and the late Willie Thomson. These remarkable scholars were, during the years I was a pupil there, the Principal Teacher of History and the Rector at Kirkwall Grammar School, an outstanding institute of comprehensive education, founded two centuries before Eton. I learned a great deal from them, in person and in print. Other able custodians and interpreters of Orkney's past are carrying forward the tradition: I have benefited immensely from correspondence and conversations with Brian Tulloch, Fran Flett Hollinrake, James Irvine, Jocelyn Rendall, Sarah Jane Gibbon, Spencer Rosie and Tom Muir.

Special thanks are due to those who read and commented on my manuscript, in whole or in part. Particularly helpful were Eric Carlson's extensive notes on a draft set of early chapters – a perfect combination of the candour of an old friend and the enthusiasm of a Californian convert to all things Orcadian. I was fortunate indeed to have my chapter on witchcraft read by two of the world's leading experts on the subject, Julian Goodare and Ronald Hutton. Julian also provided helpful advice on aspects of Scottish administration, politics and language, while Ronald's approval for the book stems from an instinct for generous affirmation of which I have long been a grateful beneficiary. Laura Stewart brought her immense expertise on seventeenth-century politics and religion to bear on Chapter 6, as did Andrew Lind. Andrew also sent me scans of important documents and checked some references for me in the Orkney Archive – true acts of scholarly benevolence. Alan Ford and Salvador Ryan, loyal friends as well as superlative scholars, were kind enough to read the manuscript in its entirety, and made numerous helpful suggestions. So did Leslie Burgher, Orkney's leading architectural historian. Leslie has been giving me sensible advice since about 1970, and I was greatly reassured by his assessment of the book as (more or less) error-free.

Visits to Orkney in all weathers have been brightened by the hospitality, conviviality and cheery encouragement of, among others, Angus Konstam, Clare Hammond, Ishbel and Ian Fraser, John Rendall, Judith Symes, Maria Eunson, and Roy and Isla Flett. I owe a particular debt of gratitude to my Kirkwall cousins, Christine and Michael Sinclair. Their boundless warmth and generosity has taught me that home is a state of the mind and an inclination of the heart. Between the inception of this book and its completion, my daughters, Bella, Maria and Kit, all finally left to make their way in the world; they continue to amaze and inspire me in a joyous diversity of ways. I don't have words to express what I owe their mother, Ali. I think, I hope, she knows.

Peter Marshall
Leamington Spa
Lammastide, 2023

Glossary

assize: the jury in a criminal trial, usually fifteen in number.

baillie: a term for office-holders of various kinds. Kirkwall's baillies were magistrates sitting on the town council; parish baillies, presiding over baillie courts, were local judges delegated by the sheriff to hear cases involving agricultural matters and petty crimes, and to cooperate with *kirk sessions* (q.v.).

band: a bond or contractual obligation to perform an act or pay a sum; bands could be voluntary alliances, or imposed mechanisms for keeping the peace.

bannock: a flatbread made of beremeal.

bere: a type of ancient barley, low-yielding but fast-growing, found predominantly in Orkney.

bismar: a distinctive type of Orkney weighing-beam, used for calculating payment of rents in lighter commodities such as butter and cheese. See also *pundlar*.

breviary: a book containing the daily services for Catholic worship.

broken men: lawless characters without an immediate superior; outlaws.

bu: a large farm, operated as a single unit rather than worked in *runrig* (q.v.).

cess: a tax on the rental value of land, originating in payments imposed by Oliver Cromwell to support the occupying English army in Scotland.

charmer: a practitioner of healing and other forms of beneficent magic (charming); usually, but not always, female.

commissary: a deputy or delegate, either of the bishop or the sheriff.
dittay: the formal indictment, or list of charges, in a criminal trial.
elder: in the Presbyterian Church of Scotland, a member of the community chosen to sit on the *kirk session* (q.v.) and exercise a general oversight over the conduct and welfare of the rest of the congregation. See also *ruling elder*.
escheat: the forfeiture of land or property upon conviction of a crime.
ferry-louper: a disparaging Orcadian term for an incomer.
feu or *feu-farm*: a form of land tenure introduced to Orkney from Scotland, involving annual payment of a specified fee, and under which estates needed to be passed on undivided to a single heir of the tenant.
feuing: the process of setting up a feu or feudal tenure, sometimes by converting an existing *udal* (q.v.) arrangement.
glebe: a portion of land assigned to a minister to supplement his salary.
head courts: Orkney's most important judicial occasions, convened by the *lawman* (q.v.) or sheriff: the *lawthing* (q.v.) at midsummer; the All Hallows court in November; and the Hirdmanstein or Hird court at New Year year.
heritor: a landowner or proprietor of a heritable property, with a responsibility to contribute to the upkeep of the parish church.
hill-dyke: a wall, usually built of turf, separating the agricultural land of the *township* (q.v.) from the common grazing land of the 'hill'.
holm: a small, uninhabited, grass-covered isle, off the coast of a larger island; pronounced 'home' and not to be confused with the Orkney Mainland parish of Holm (pronounced 'Ham').
kelp: a word employed today to designate various types of seaweed, but in pre-modern times predominantly applied to the substance created by the burning and cooling of seaweed, rich in alkalis and used in a variety of manufacturing processes.
kirk session: a parish-based religious court, staffed by the minister and *elders* (q.v.), and responsible for matters of social welfare and personal morality.
ky, kyne: cattle.
lawman: the chief judicial officer in Orkney during the period of Norse rule.

lawrightman: a substantial member of the community acting as a lesser magistrate; a member of the lawthing in Norse times and of local *baillie* courts (q.v.) in later centuries.
lawthing: the principal court of the Orkney lawman, functioning also as a kind of popular assembly.
mark: a unit of weight peculiar to Orkney and Shetland, used principally for butter, oil and tallow.
merk: a monetary unit, worth two-thirds of a pound Scots (i.e. thirteen shillings and fourpence).
moderator: the minister (on a rotating basis) chairing meetings of a *presbytery* (q.v.). *Synods* (q.v.), and the General Assembly of the Church of Scotland, also had moderators.
Norn: distinctive form of Old Norse language spoken in Orkney and Shetland.
notary: a person licensed to witness, record or certify legal transactions.
ogang: the judicial practice of judges travelling on circuit, and hold court in the Orkney parishes.
pennyland: a unit of land for rental, sale and valuation purposes; of varying size, but usually equivalent to a small farm.
prebend: see *stouk*.
precentor: the cathedral clergyman responsible for the music.
presbytery: an administrative and judicial body of the Church of Scotland, comprising the ministers and one *ruling elder* (q.v.) from each parish in a designated locality. Orkney was a single presbytery till 1707, when the Presbytery of the North Isles was formed. A Presbytery of Cairston (Stromness) was added in 1725.
profit: a word, often encountered in witchcraft cases, meaning the yield or increase of agrarian substances like grain, milk and butter, as well as the essential health or goodness of the product.
provost: the clergyman in charge of the running of a cathedral.
pundlar: like the *bismar* (q.v.), an Orcadian weighing-beam, used principally for *bere* (q.v.) and malt.
rental: a document listing the rents owed by tenants to the proprietor or landowner.

Riksråd: the Norwegian council of state, comprising nobles and bishops.

roithman: a Norse term meaning 'adviser' or 'councillor'; a leading inhabitant of Orkney entitled to participate in the *head courts* (q.v.).

ruling elder: a member of the *kirk session* (q.v.) chosen to represent the parish at meetings of the *presbytery* (q.v.).

runrig: a system of agriculture in which land in large fields was held in individual strips (rigs), which were periodically reallocated among the farmers.

selkie: a Scots and Orcadian dialect word for a seal; also, a magical aquatic creature able to take on human form.

skat: a Norse word meaning 'treasure' or 'tribute'; here, specifying the land tax, usually payable in malt, meat or butter, due from *udal* (q.v.) landowners and from tenants of the earldom and bishopric estates in Orkney and Shetland.

stouk: a cathedral-based benefice or source of income in Orkney, of a type known elsewhere as a prebend; used also to designate the lands endowing the office.

synod: an ecclesiastical term meaning either an intermediate court between a *presbytery* (q.v.) and the General Assembly of the Church of Scotland, or, during periods of episcopal dominance, a meeting of the clergy of the diocese presided over by the bishop.

tack: a Scots word meaning a 'lease' or 'tenancy'; in this book, usually referring to the leasing out by the crown of rents and other dues from land forming part of the earldom or bishopric estates in Orkney and Shetland.

tacksman: an individual paying the monarch a fixed sum for a tack.

teinds: the Scots word for 'tithes', a 10 per cent tax on produce owed to the Church.

township, town: an area containing a number of farms, characterised by the sharing of subdivided fields (see *runrig*) and participation in collective practices of agriculture.

trow, trowie: a term, derived from the Norse word *troll*, for a variety of potentially dangerous supernatural creatures, found both on land and at sea.

udal (or *odal*) *law*: the Norse and Orcadian legal system regulating udal landownership and its obligation for estates to be divided among heirs; some aspects of it remained in place after Orkney's transfer to Scotland in 1468.

udaller (or *odaller*): a landowner holding an estate in outright ownership, paying *skat* (q.v.) but no rents to any feudal superior.

ware: seaweed, gathered for use as a manure, or in the manufacture of *kelp* (q.v.).

wrack: anything washed up on shore, including stranded whales and goods from wrecked ships.

wrack and waith: the legal right to possession of wrack.

Abbreviations

Annals: John Smith, *Annals of the Church of Scotland in Orkney from 1560* (Kirkwall, 1907)

Black: G. F. Black and Northcote W. Thomas (eds), *County Folklore, Vol. III: Orkney and Shetland Islands* (London, 1903)

BUK: Thomas Thomson (ed.), *Booke of the Universal Kirk: Acts and Proceedings of the General Assembly of the Church of Scotland* (3 vols, Edinburgh, 1839–45)

Canmore: *National Record of the Historic Environment, Scotland* (https://canmore.org.uk/)

Craven: J. B. Craven, *History of the Church in Orkney* (3 vols, Kirkwall, 1893–1901)

CSP: Calendar of State Papers

DN: *Diplomatarium Norvegicum* (https://www.dokpro.uio.no/)

Fasti: Hew Scott (ed.), *Fasti Ecclesiae Scoticanae: The Succession of Ministers in the Church of Scotland from the Reformation, Volume VII: Synods of Ross, Sutherland and Caithness, Glenelg, Orkney and of Shetland* (Edinburgh, 1928)

Folklore: Ernest W. Marwick, *The Folklore of Orkney and Shetland* (London, 1975)

Hossack: B. H. Hossack, *Kirkwall in the Orkneys* (Kirkwall, 1900)

LP: J. S. Brewer, J. Gairdner and R. H. Brodie (eds), *Letters and Papers, Foreign and Domestic, of the Reign of Henry VIII* (21 vols in 33 parts, 1862–1910)

Macfarlane: Arthur Mitchell and James Toshach Clark (eds), *Geographical Collections Relating to Scotland Made by Walter Macfarlane* (3 vols, Edinburgh, 1906–8)

Mooney: John Mooney (ed.), *Charters and Other Records of the City and Royal Burgh of Kirkwall* (Kirkwall, 1950)

NHO: William P. L. Thomson, *The New History of Orkney* (3rd edn, Edinburgh, 2008)

NLS: National Library of Scotland

Notes: Alexander Peterkin, *Notes on Orkney and Zetland Illustrative of the History, Antiquities, Scenery, and Customs of Those Islands* (Edinburgh, 1822)

NRS: National Records of Scotland

OCR: Orkney Library and Archive, Orkney Church Records

ODNB: *Oxford Dictionary of National Biography* (https://www.oxforddnb.com/)

OLA: Orkney Library and Archive

OP: J. Storer Clouston (ed.), *The Orkney Parishes, Containing the Statistical Account of Orkney, 1795–1798* (Kirkwall, 1927)

OSR: Alfred W. Johnston and Amy Johnston (eds), *Orkney and Shetland Records* (3 vols, London, 1907–13)

Rentals: Alexander Peterkin, *Rentals of the Ancient Earldom and Bishoprick of Orkney* (Edinburgh, 1820)

REO: J. Storer Clouston (ed.), *Records of the Earldom of Orkney, 1299–1614* (Edinburgh, 1914)

RPC: John Hill Burton et al. (eds), *The Register of the Privy Council of Scotland* (3 series, 36 vols, Edinburgh, 1877–1933)

RPS: *Records of the Parliaments of Scotland* (https://www.rps.ac.uk/)

SD: John H. Ballantyne and Brian Smith (eds), *Shetland Documents, 1195–1579* (Lerwick, 1999)

SEO: Peter Anderson, *The Stewart Earls of Orkney* (Edinburgh, 2012)

TNA: The National Archives

Tudor: John R. Tudor, *The Orkneys and Shetland: Their Past and Present State* (London, 1883)

Notes

Preface

1. 'Trials for Witchcraft, Sorcery, and Superstition, in Orkney', in William B. D. D. Turnbull (ed.), *Miscellany of the Abbotsford Club, Volume First* (Edinburgh, 1837), 143–9.
2. J. Storer Clouston, *A History of Orkney* (Kirkwall, 1932), 329.
3. Henrietta E. Marshall, *Our Island Story: A Child's History of England* (London, 1905); Pete Hay, 'A Phenomenology of Islands', *Island Studies Journal*, 1 (2006), 19–42, and many subsequent contributions by various authors in that journal; Roland Greene, 'Island Logic', in Peter Hulme and William H. Sherman (eds), *'The Tempest' and Its Travels* (London, 2000), 138–45; Jodie Matthews and Daniel Travers (eds), *Islands and Britishness: A Global Perspective* (Newcastle, 2012).

Prologue: A Journey and Two Maps

1. *LP*, XV, 632, 634; *State Papers Published under the Authority of His Majesty's Commission, King Henry VIII* (11 vols, London, 1830–52), V, 177–9; Jamie Cameron, *James V: The Personal Rule, 1528–1542* (Edinburgh, 2011), 245; *Historical Manuscripts Commission: Report on the Manuscripts of the Earl of Mar and Kellie* (London, 1904), 15; *CSP, Venice*, V, no. 194.
2. *LP*, XV, 697, 736; *State Papers, Henry VIII*, V, 178.
3. J. B. Kaulek (ed.), *Correspondance politique de MM. de Castillon et de Marillac* (Paris, 1885), 186; Jane Dawson, *Scotland Reformed, 1488–1587* (Edinburgh, 2007), 43–7, 71–4, 145; Cameron, *James V*, 228–38.
4. T. Dickson and Sir J. Balfour Paul (eds), *Accounts of the Lord High Treasurer of Scotland* (12 vols, Edinburgh, 1877–1916), VII, 309–14; VIII, 159.
5. James undertook an earlier trip around the north coast in summer 1536. The Holy Roman Emperor Charles V's English ambassador reported the king's intention to visit Orkney, but there is no evidence he did. The voyage, clouded in secrecy and interrupted by storms, may have been a failed attempt to reach France by the western route: *LP*, XI,

358; Andrea Thomas, 'Renaissance Culture at the Court of James V, 1528–1542', University of Edinburgh PhD thesis (1997), 221–2.

6. M. Livingstone et al. (eds), *Register of the Privy Seal of Scotland* (8 vols, Edinburgh, 1908–82), I, 74; John Major, *A History of Greater Britain*, tr. A. Constable (Edinburgh, 1892), 36; Hector Boece, *The History and Chronicles of Scotland*, tr. John Bellenden (2 vols, Edinburgh, 1821), I, l–li; *Treasurer*, VII, 307, 449; *LP*, IX, 1049.
7. John Lesley, *The History of Scotland from the Death of King James I in the Year MCCCCXXXVI to the Year MDLXI*, ed. Thomas Thomson (Edinburgh, 1830), 156; Mooney, 17; John Stuart (ed.), *Extracts from the Council Register of the Burgh of Aberdeen* (2 vols, Aberdeen 1844–8), I, 173–4; *Treasurer*, VII, 328; George Buchanan, *History of Scotland* (4 vols, Glasgow, 1827), II, 311.
8. *State Papers, Henry VIII*, V, 182; Craven, I, 146; *LP*, XV, 962, 963.
9. Craven, I, 144; Hossack, 64; *Folklore*, 150–2; David Stevenson, '"The Gudeman of Ballangeich": Rambles in the Afterlife of James V', *Folklore*, 115 (2004), 187–200.
10. For what follows, D. G. Moir, *The Early Maps of Scotland to 1850* (3rd edn, 2 vols, Edinburgh, 1973–83), I, 19–23; Alexander Lindsay, *A Rutter of the Scottish Seas, c. 1540*, ed. A. B. Taylor, I. H. Adams and G. Fortune (London, 1980).
11. Lesley, *History*, 289.
12. Mooney, 107.
13. *NHO*, chs 12–14; Gordon Donaldson, 'Problems of Sovereignty and Law in Orkney and Shetland', in David Sellar (ed.), *The Stair Society: Miscellany Two* (Edinburgh, 1984), 13–40; Brian Smith, 'When Did Orkney and Shetland Become Part of Scotland? A Contribution to the Debate', *New Orkney Antiquarian Journal*, 5 (2010), 1–18.
14. Olaus Magnus, *Ain kurze Auslegung und Verklerung der neuuen Mappen von den alten Goettenreich und andern Nordlenden* (Venice, 1539), A4v.
15. *NHO*, 143; P. Munch and D. Laing, 'Why Is the Mainland of Orkney Called Pomona?', *Proceedings of the Society of Antiquaries of Scotland*, 1 (1852), 15–18; Mooney, 1, 16. The earliest known user of 'Pomona' for the Orkney Mainland may be the fourteenth-century John of Fordun: *Johannis de Fordun Chronica Gentis Scotorum*, ed. William F. Skene (Edinburgh, 1871), 44.
16. Magnus, *Auslegung*, A4v; Olaus Magnus, *A Description of the Northern Peoples, 1555*, ed. Peter Foote, tr. Peter Fisher (3 vols, London, 1996–8), III, 957, 1013; Michael Allaby, *Animals: From Mythology to Zoology* (New York, 2010), 75–7; Aeneas Sylvius, *Opera Geographica et Historica* (Helmstadt, 1699), 318.

1: Between Norway and Scotland

1. George Barry, *History of the Orkney Islands*, ed. James Headrick (2nd edn, London, 1808), 450; Hamish Haswell-Smith, *The Scottish Islands* (rev. edn, Edinburgh, 2015), 353–4; Canmore ID 9465, 9461; 9469, 9472, 9473; James M. Irvine (ed.), *The Orkneys and Schetland in Blaeu's Atlas Novus of 1654* (Ashtead, 2006), 15.

2. Quotations in the following discussion are my translations from the Latin texts printed in Barry, *History*, 437–51, and Macfarlane, III, 302–13, drawing on the English versions in Macfarlane, III, 313–23, and Margaret Hunter, 'Jo: Ben's Description of Orkney', *New Orkney Antiquarian Journal*, 6 (2012), 34–47. For dating and authorship, see James M. Irvine, 'Jo: Ben Revisited', *New Orkney Antiquarian Journal*, 6 (2012), 48–58.
3. James G. Kyd (ed.), *Scottish Population Statistics* (Edinburgh, 1952), 82; Michael Anderson, 'Population Patterns', in Michael Lynch (ed.), *The Oxford Companion to Scottish History* (Oxford, 2001), 487–8.
4. *NHO*, 206–18; David Griffiths, 'Status and Identity in Norse Settlements: A Case Study from Orkney', in James H. Barrett and Sarah-Jane Gibbon (eds), *Maritime Societies of the Viking and Medieval World* (Leeds, 2015), 219–36.
5. *Orkneyinga Saga: The History of the Earls of Orkney*, tr. Herman Pálsson and Paul Edwards (Harmondsworth, 1981), 155; William P. L. Thomson, 'The Ladykirk Stone', *New Orkney Antiquarian Journal*, 2 (2001), 31–46.
6. The Latin is *utuntur idiomate proprio* (Macfarlane, III, 310), rather than *lingua* (language), though something stronger than accent or dialect is undoubtedly implied.
7. Michael P. Barnes, *The Norn Language of Orkney and Shetland* (Lerwick, 1998), 19.
8. Ian Peter Grohse, 'From Asset in War to Asset in Diplomacy: Orkney in the Medieval Realm of Norway', *Island Studies Journal*, 8 (2013), 255–68; Ian Peter Grohse, *Frontiers for Peace in the Medieval North: The Norwegian–Scottish Frontier, c. 1260–1470* (Leiden, 2017).
9. Randi Bjørshol Wærdahl, *The Incorporation and Integration of the King's Tributary Lands into the Norwegian Realm, c. 1195–1397*, tr. Alan Crozier (Leiden, 2011), 207–10; *NHO*, 183–5; Barbara E. Crawford, *The Northern Earldoms: Orkney and Caithness from AD 870 to 1470* (Edinburgh, 2013), 77–9; Crawford, 'Two Seals from Orkney: The 15th Century Community Seal and a Seal Matrix Dating to c. AD 1300', in Irene Baug, Janicke Larsen and Sigrid Samset Mygland (eds), *Nordic Middle Ages – Artefacts, Landscapes and Society* (Bergen, 2015), 105–17.
10. Crawford, *Earldoms*, 332. For thoughtful qualification, Grohse, *Frontiers*, ch. 2.
11. Steven G. Ellis and Raingard Eßer, 'Introduction', in Ellis and Eßer (eds), *Frontiers and the Writing of History, 1500–1850* (Hannover-Laatzen, 2006), 9–20.
12. Steinar Imsen, 'Public Life in Shetland and Orkney, *c.* 1300–1550', *New Orkney Antiquarian Journal*, 1 (1999), 53–65; Grohse, *Frontiers*, 196–202; Gregor Lamb, *Orkney Family Names* (Kirkwall, 2003).
13. Steinar Imsen, 'The Country of Orkney and the Complaints against David Menzies', *New Orkney Antiquarian Journal*, 6 (2012), 9–33 (original text and translation at 28–33); *REO*, 45–8, 51–5; James Kirk (ed.), *The Books of Assumption of the Thirds of Benefices* (Oxford,

1995), 666–7; Hunter, 'Jo: Ben's Description', 38.

14. Imsen, 'Country', 17–20, 30; *REO*, 34, 51; *NHO*, 191; *OSR*, I, lvi–lviii; Grohse, *Frontiers*, chs 6–7.
15. *OSR*, I, 246; *REO*, 31–3; *NHO*, 175; *DN*, VIII, no. 276.
16. *RPS*, 1358/1/2; *NHO*, 190–1; Mooney, 16; *REO*, 194.
17. *REO*, 99–100, 217–18; James Irvine, 'The Irvines of Orkney', *Sib Folk News* (Feb. 1997), 8–9; David Balfour, *Oppressions of the Sixteenth Century in the Islands of Orkney and Zetland* (Edinburgh, 1859), 95.
18. 'The Islands' Names', *Orkneyjar*, http://www.orkneyjar.com/placenames/pl-isle.htm; Macfarlane III, 205, 307; Hugh Marwick, *Orkney* (London, 1951), 203.
19. William P. L. Thomson (ed.), *Lord Henry Sinclair's 1492 Rental of Orkney* (Kirkwall, 1996), 61; *SD*, 21; N. Nicolaysen (ed.), *Bergens Borgerbog, 1550–1751* (Christiania, 1878), 1–54; Knut Helle, *Orknøyene i Norsk Historie* (Bergen, 1988), 23–7; Jon Leirfall, *Vest i Havet* (Oslo, 1976), 130; Hugh Marwick, *Merchant Lairds of Long Ago* (2 vols, Kirkwall, 1939), II, 43–8; *REO*, 299; *Bergen Rådstueprotokoll, 1592–94*, fols 205a–213a, transcript at https://xml.arkivverket.no/diverse/rd12011592.htm.
20. Nina Østby Pedersen, 'Scottish Immigration to Bergen in the Sixteenth and Seventeenth Centuries', in Alexia Grosjean and Steve Murdoch (eds), *Scottish Communities Abroad in the Early Modern Period* (Leiden, 2005), 135–67; Christian Lange (ed.), *Norske Samlinger* (2 vols, Christiania, 1849–60), II, 481–92; *DN*, XXII, no. 147; Knut Helle, '"Lille Jon" Jon Thomessøn', *Norsk Biografisk Leksikon* (nbl.snl.no/"Lille_Jon"_Jon_Thomessøn).
21. Andreas Hanssen, 'Hollandske Kedeførere', in Villads Christensen (ed.), *Samlinger til Jydsk Historie og Topografi*, 3rd ser., vol. 5 (Copenhagen, 1906–8), 5; N. Nicolaysen (ed.), *Norske Magasin: Skrifter og Optegnelser Angaaende Norge og Forfattede efter Reformationen* (3 vols, Christiania, 1858–70), I, 103–7, 129; II, 19; Helle, *Orknøyene*, 26; *DN*, XXII, no. 129.
22. Mooney, 107; *NHO*, 198–204; Ian Peter Grohse, 'The Lost Cause: Kings, the Council, and the Question of Orkney and Shetland, 1468–1536', *Scandinavian Journal of History*, 45 (2020), 290; *RPS*, 1472/14.
23. Grohse, 'Cause', 291–3; Mooney, 115; 'Valghåndfestning', in Steinar Imsen and Harald Winge (eds), *Norsk Historisk Leksikon* (https://lokalhistoriewiki.no/wiki/Leksikon:Valghåndfestning); Torjus Rolid Hagen 'Konge og Kirke, 1507–1523: Christian IIs Forhold til den Norske Kirkeprovinsen', NTNU Trondheim Master's thesis (2016), 36–7.
24. Brian Smith, 'When Did Orkney and Shetland Become Part of Scotland? A Contribution to the Debate', *New Orkney Antiquarian Journal*, 5 (2010), 5–8; *RPS*, 1526/6/14; *DN*, VI, no. 609; I, no. 1042; Marguerite Wood (ed.), *Flodden Papers: Diplomatic Correspondence between the Courts of France and England, 1507–1517* (Edinburgh, 1933), 113–14; *LP*, II, 588; Caspar Frederik Wegener

(ed.), *Aarsberetninger fra det Kongelige Geheimearchiv Kilder* (7 vols, Copenhagen, 1852–3), III, 113, 118.

25. Grohse, 'Cause', 297; Smith, 'Orkney', 8–9; Wegener, *Aarsberetninger*, II, 86; Henning Laugerud, *Reformasjon uten Folk: Det Katolske Norge i Før-og Etterreformatorisk Tid* (Oslo, 2018), 111; *SD*, 44; *DN*, XXII, no. 429.
26. *NHO*, 227–32; Barbara E. Crawford, 'Sinclair Family (per. 1280–*c.* 1500)', *ODNB*; Sally Mapstone, 'Older Scots Literature and the Court', in Thomas Owen Clancy and Murray Pittock (eds), *The Edinburgh History of Scottish Literature, Volume 1: From Columba to the Union (until 1707)* (Edinburgh, 2007), 280.
27. *REO*, 85, 87, 90, 92, 95; J. Storer Clouston, 'The Lawthing and Early Officials of Orkney', *Scottish Historical Review*, 21 (1924), 101–14.
28. *NHO*, 233–4; George P. McNeil (ed.), *The Exchequer Rolls of Scotland, Vol. XV: A.D. 1523–1529* (Edinburgh, 1895), lxviii–lxvix, 151–2.
29. *REO*, 57–60.
30. John Lesley, *The History of Scotland from the Death of King James I in the Year MCCCCXXXVI to the Year MDLXI*, ed. Thomas Thomson (Edinburgh, 1830), 141; Macfarlane, III, 309; George Low, *A History of the Orkneys*, ed. Olaf D. Cuthbertson (Kirkwall, 2001), 324; *OP*, 182; Barry, *History*, 245.
31. James T. Calder, *Sketch of the Civil and Traditional History of Caithness, from the Tenth Century* (Glasgow, 1861), 95–7.
32. Balfour, *Oppressions*, xvi–xvii; J. Storer Clouston, *A History of Orkney* (Kirkwall, 1932), 287–92; John Gunn, *The Fight at Summerdale* (London, 1913).
33. Jamie Cameron, *James V: The Personal Rule, 1528–1542* (Edinburgh, 2011), 242–4; *REO*, 216, 219; *NHO*, 246; TNA, SP 1/82, 203r.
34. *REO*, 60–1, 219–20, 235; *OSR*, I, 109–13; *NHO*, 241–2. For the claim about James Sinclair's duplicity, see James Wallace, *An Account of the Islands of Orkney* (London, 1700), 95; Barry, *History*, 246.
35. Stephanie Malone Thorson, 'Adam Abell's *The Roit or Quheill of Tyme*: An Edition', University of St Andrews PhD thesis (1998), ii–iii, 250–1; Wallace, *Account*, 96. See Alasdair M. Stewart, 'The Final Folios of Adam Abell's "Roit or Quheil of Tyme": An Observantine Friar's Reflections on the 1520s and 30s', in Janet Hadley Williams (ed.), *Stewart Style, 1513–1542: Essays on the Court of James V* (East Linton, 1996), 227–53.
36. *NHO*, 242–3; *OSR*, I, 115–16; Cameron, *James V*, 244–5; *SD*, 40, 46–7; David Caldwell, 'Sinclair, Oliver (d. in or before 1576)', *ODNB*.
37. *REO*, lxxiv–lxxv, 61–3; *SD*, 47; *RPS*, 1540/12/26.
38. *OSR*, I, xxxiii, 251; Hector L. MacQueen, '"Regiam Majestatem", Scots Law, and National Identity', *The Scottish Historical Review*, 74 (1995), 1–25; *RPS*, 1504/3/45; Gordon Donaldson, 'Problems of Sovereignty and Law in Orkney and Shetland', in David Sellar (ed.), *The Stair Society: Miscellany Two* (Edinburgh 1984), 26.
39. *REO*, lxiii, 101–3 (Agatha Sinclair), 109–10, 119–21, 134–5, 140–3, 201–2,

211–12, 208 (Margaret Sinclair), 211–12, 290–1 (Janet Irving); *OSR*, I, 56.

40. *REO*, 82, 205, 214, 222, 237; *OSR*, I, 53.
41. *OSR*, I, 61–4, 251–3.
42. Peter Marshall, *Heretics and Believers: A History of the English Reformation* (New Haven, 2017), 75–6, 184–5; John D. Shafer, 'Where Is Orkney? The Conceptual Position of Orkney in Middle English Arthurian Literature', *Journal of the North Atlantic*, Special Volume 4 (2013), 189–98; *LP*, VIII, 1160.
43. Joseph Bain (ed.), *The Hamilton Papers: Letters and Papers Illustrating the Political Relations to England and Scotland in the XVIth Century* (2 vols, Edinburgh, 1890–2), I, 202, 240, 246–7.
44. *Hamilton Papers*, I, 337–8; *LP*, XVII, 1194, 1230; Caldwell, 'Sinclair'; Cameron, *James V*, 325.
45. *SEO*, 12–14; *REO*, 109; *The Scottish Correspondence of Mary of Lorraine*, ed. Anne I. Cameron (Edinburgh, 1927), 46–7, 85–6; *SD*, 28.
46. Frederik Krarup, 'Chr. de Treschow: Contributions to the History of Queen Elizabeth', *Historisk Tidsskrift*, 4th ser., 2 (1870–2), 901; Thormodus Torfaeus, *Orcades seu Rerum Orcadensium Historiae* (Copenhagen, 1697), 207–8; Amy Blakeway, *Regency in Sixteenth-Century Scotland* (Woodbridge, 2015), 200–3; Thorkyld Lyby Christensen, 'The Earl of Rothes in Denmark', in Ian B. Cowan and Duncan Shaw (eds), *The Renaissance and Reformation in Scotland* (Edinburgh, 1983), 68–73.

2: The Islands of St Magnus

1. William P. L. Thomson (ed.), *Lord Henry Sinclair's 1492 Rental of Orkney* (Kirkwall, 1996), 27; *REO*, 87, 90, 99, 101, 103, 202, 204, 244; *OSR*, I, 52, 102, 104, 106, 108, 119–20, 253, 260.
2. Craven, I, 149; *Sixth Report of the Royal Commission on Historical Manuscripts, Part I* (1877), 670; Olaf D. Cuthbert, *A Flame in the Shadows: Robert Reid Bishop of Orkney, 1541–1558* (Kirkwall, 1998), 179–83.
3. Steinar Imsen, 'Nidarosprovinsen', in Imsen (ed.), *Ecclesia Nidrosiensis, 1153–1537* (Trondheim, 2003), 15–45; Augustin Theiner (ed.), *Vetera Monumenta Hibernorum et Scotorum Historiam Illustrantia* (Rome, 1864), 464–8.
4. *OSR*, I, 56–7; M. Livingstone et al. (eds), *Register of the Privy Seal of Scotland* (8 vols, Edinburgh, 1908–82), I, 111, 151–2. See Barbara Crawford, 'Sir David Sinclair of Sumburgh: "Foud" of Shetland and Governor of Bergen Castle', in J. R. Baldwin (ed.), *Scandinavian Shetland: An Ongoing Tradition?* (Edinburgh, 1978), 1–12.
5. Halvard Bjørkvik, 'Gaute Ivarsson', *Norsk Biografisk Leksikon* (https://nbl.snl.no/Gaute_Ivarsson); *DN*, I, no. 1041; VI, no. 675; VII, nos. 602–3; LP, II, 588; Konrad Eubel (ed.), *Hierarchia Catholica Medii Aevi, Vol. III* (Regensberg, 1923), 263; Øystein Rian, 'Olav Engelbrektsson', *Norsk Biografisk Leksikon* (https://nbl.snl.no/Olav_Engelbrektsson); T. M. Y. Manson, 'Shetland in the Sixteenth Century', in Ian B. Cowan and Duncan Shaw (eds),

The Renaissance and Reformation in Scotland (Edinburgh, 1983), 202; Nelson H. Minnich, 'The Participants at the Fifth Lateran Council', *Archivum Historiae Pontificae*, 12 (1974), 164.

6. *Rentals*, No. 1, 14–18; *NHO*, 227–30; A. D. M. Barrell, *Medieval Scotland* (Cambridge, 2000), 48; D. E. R. Watt and A. L. Murray (eds), *Fasti Ecclesiae Scoticanae Medii Aevi ad Annum 1638* (rev. edn, Edinburgh, 2003), 343–4.
7. Brian Smith, 'In the Tracks of Bishop Andrew Pictoris of Orkney and Henry Phankouth, Archdeacon of Shetland', *Innes Review*, 40 (1989), 91–105; Alan Macquarrie, Roland J. Tanner and Annie I. Dunlop (eds), *Calendar of Scottish Supplications to Rome, Volume VI, 1471–1492* (Edinburgh, 2017), 235; Watt and Murray, *Fasti*, 338, 341; Anton Wilhelm Brøgger, *Ancient Emigrants: A History of the Norse Settlements of Scotland* (Oxford, 1929), 184.
8. *OSR*, I, 121; *REO*, 190, 208, 197–8, 419; Watt and Murray, *Fasti*, 331–2.
9. *REO*, 83–4, 225–30, 274; Alfred W. Johnston and Amy Johnston (eds), *Orkney and Shetland Miscellany: Old-Lore Series, Vol. I* (London, 1907–8), 169; John M. Thomson et al. (eds), *Registrum Magni Sigilli Regum Scotorum/The Register of the Great Seal of Scotland* (11 vols, Edinburgh, 1882–1914), II, 726–7; Roland Saint-Clair, *The Saint-Clairs of the Isles* (Auckland, 1898), 424, 428; *Privy Seal*, I, 17.
10. Smith, 'Tracks', 101–2; *REO*, 215, 229 (see also 78, 193–4, 195, 196, 205, 206, 207, 208, 209); Iain MacDonald, *Clerics and Clansmen: The Diocese of Argyll between the Twelfth and Sixteenth Centuries* (Leiden, 2013), ch. 4.
11. James Kirk, 'Reid, Robert (d. 1558), *ODNB*; J. H. S. Burleigh, 'The Scottish Reforming Councils, 1549 to 1559', *Records of the Scottish Church History Society*, 11 (1953), 199–200; David Patrick (ed.), *Statutes of the Scottish Church, 1225–1559* (Edinburgh, 1907), 115–16.
12. Mooney, 17–18; *REO*, 105, 217, 338, 340, 343, 363–71, 374, 436; *SD*, 26; Hossack, 39; J. Storer Clouston, 'The Old Prebends of Orkney', *Proceedings of the Orkney Antiquarian Society*, 4 (1925–6), 30–6; Margareth Buer Søiland, 'Orkney Pilgrimage: Perspectives of the Cult of St. Magnus', University of Glasgow PhD thesis (2004), 165–6; George Burnett (ed.), *The Exchequer Rolls of Scotland, Vol. X: A.D. 1488–1496* (Edinburgh, 1887), 585; Bryce Wilson, 'Why St Magnus Cathedral Was Built, and How It Changed', in Steve Callaghan and Bryce Wilson (eds), *The Unknown Cathedral: Lesser Known Aspects of St Magnus Cathedral in Orkney* (Kirkwall, 2001), 22; Tom Turpie, The Many Lives of St Duthac of Tain', *Northern Studies*, 44 (2013), 3–4.
13. *REO*, 340, 363–71; Stephen Mark Holmes, 'Catalogue of Liturgical Books and Fragments in Scotland before 1560', *Innes Review*, 62 (2011), 183, 185; Ludwig Eisendorfer, *The Liturgy of the Roman Rite* (New York, 1961), 445.
14. *REO*, 243; L. N. Nowosilski, 'Robert Reid and His Time', Pontifical

University of St Thomas Aquinas PhD thesis (1965), NLS, typescript Acc. 3987, 555; *NHO*, 248–9, Peter D. Anderson, *Robert Stewart: Earl of Orkney, Lord of Shetland, 1533–1593* (Edinburgh, 1982), 32–3; Cuthbert, *Flame*, 120.

15. *REO*, 364, 370; Gordon Donaldson, *Reformed by Bishops: Galloway, Orkney and Caithness* (Edinburgh, 1987), 24–5; *NHO*, 252; Robert S. Barclay (ed.), *Orkney Testaments and Inventories, 1573–1615* (Edinburgh, 1977), 172.
16. *SD*, 44; *REO*, 107, 110, 364–6, 370–1, 417; James Wallace, *An Account of the Islands of Orkney* (London, 1700), 68–9.
17. A. N. Galpern, 'The Legacy of Late Medieval Religion in Sixteenth-Century Champagne', in C. Trinkaus and H. O. Oberman (eds), *The Pursuit of Holiness in Late Medieval and Renaissance Religion* (Leiden, 1974), 149; *REO*, 91, 95, 104, 201, 211, 212, 216, 231, 240, 245, 248, 250, 267, 332, 334, 335, 336; Spencer J. Rosie, *Saints and Sinners: Memorials of St Magnus Cathedral* (Kirkwall, 2015), 18; *REO*, 208, 338, 340; *OSR*, I, 106.
18. Ernest Marwick, *An Orkney Anthology* (Edinburgh, 1991), 301–6; 'The Well o' Kildinguie' (www.orkneyjar.com/tradition/sacredwater/kildinguie.htm); Canmore, ID 3312; *OP*, 318.
19. Canmore, ID 3250, 2882. On chapels, J. Storer Clouston, 'The Old Chapels of Orkney', *Scottish Historical Review*, 15 (1918), 89–105, 223–32, and 'Old Kirks and Chapels in Orkney', *Scottish Historical Review*, 15 (1918), 233–40, have been superseded by the meticulous researches of Sarah Jane Gibbon, 'The Origins and Early Development of the Parochial System in the Orkney Earldom', University of the Highlands and Islands PhD thesis (2006), and 'Medieval Parish Formation in Orkney', in Beverley Ballin Smith, Simon Taylor and Gareth William (eds), *West over Sea: Studies in Scandinavian Sea-Borne Expansion and Settlement Before 1300* (Leiden, 2007).
20. Macfarlane, III, 307, 309; Canmore, ID 2072, 2149.
21. OLA, D24/1/127 (printed, with mistranscription of 'Fann[en]' as 'Forinen', in *REO*, 209–10); Martin Blindheim, 'The Cult of Medieval Wooden Sculptures in Post-Reformation Norway', in Søren Kaspersen (ed.), *Images of Cult and Devotion: Function and Reception of Christian Images in Medieval and Post-Medieval Europe* (Copenhagen, 2004), 47–8.
22. *Orkneyinga Saga: The History of the Earls of Orkney*, tr. Herman Pálsson and Paul Edwards (Harmondsworth, 1981), 91–7, 103–8, 130. See William P. L. Thomson, 'St Magnus: An Exploration of his Sainthood', in Doreen Waugh (ed.), *The Faces of Orkney* (Edinburgh, 2003), 46–64; Haki Antonsson, *St Magnus of Orkney: A Scandinavian Martyr-Cult in Context* (Leiden, 2007); Maria-Claudia Tomany, 'Sacred Non-Violence, Cowardice Profaned: St Magnus of Orkney in Nordic Hagiography and Historiography', in Thomas A. DuBois (ed.), *Sanctity in the North: Saints, Lives, and Cults in Medieval Scandinavia* (Toronto, 2008), 128–53.

23. Stewart Crudden, 'The Cathedral and Relics of St Magnus, Kirkwall', in M. R. Apted, R. Gilyard-Beer and A. D. Saunders (eds), *Ancient Monuments and Their Interpretation* (London, 1977); Peter Yeoman, *Pilgrimage in Medieval Scotland* (London, 1979), 96–7; *Orkneyinga Saga*, 106; Øystein Ekroll, 'St Olavs Skrin i Nidaros', in Imsen, *Nidrosiensis*, 325–50; *DN*, XVII, no. 552 (abbreviated translation in Ian B. Cowan, 'Appendix', in H. W. M. Cant and H. N. Firth (eds), *Light in the North: St Magnus Cathedral through the Centuries* (Kirkwall, 1989), 121–2).
24. Yeoman, *Pilgrimage*, 97; *St Magnus Graffiti Project, Stage 1*, Orkney Research Centre for Archaeology (Kirkwall, 2021); *REO*, 370, 388; Craven, I, 141–2; Tom Muir, 'The Stone Carvings', in Callaghan and Wilson, *Cathedral*, 59–60.
25. Alexander Carmichael (ed.), *Carmina Gadelica* (6 vols, Edinburgh, 1900–71), I, 179–81; Barbara E. Crawford, *The Northern Earldoms: Orkney and Caithness from AD 870 to 1470* (Edinburgh, 2013), 224; Cosmo Innes, *Origines Parochiales Scotiae: The Antiquities Ecclesiastical and Territorial of the Parishes of Scotland* (2 vols, Edinburgh, 1851–5), II (ii), 757–8; *Privy Seal*, I, no. 787 (pp. 115–16); Alan Macquarrie (ed.), *Legends of Scottish Saints: Readings, Hymns and Prayers for the Commemorations of Scottish Saints in the Aberdeen Breviary* (Dublin, 2012), 100–14; William James Anderson, 'Nidaros and Aberdeen', *Innes Review*, 16 (1965), 130–4.
26. Hector Boece, *The History and Chronicles of Scotland*, tr. John Bellenden (2 vols, Edinburgh, 1821), II, 394; Stephanie Malone Thorson, 'Adam Abell's *The Roit or Quheill of Tyme*: An Edition', University of St Andrews PhD thesis (1998), 194; Thomson, 'St Magnus', 58; Macfarlane, I, 49; Macquarrie *Legends*, 104–5.
27. Barclay, *Testaments*; *REO*, 231, 242; John Lesley, *The History of Scotland from the Death of King James I in the Year MCCCCXXXVI to the Year MDLXI*, ed. Thomas Thomson (Edinburgh, 1830), 141; Raphael Holinshed, *Chronicles of England, Scotland and Ireland* (London, 1577), II, 440; (London, 1587), V, 300.
28. *Magnus' Saga: The Life of St Magnus, Earl of Orkney, 1075–1116*, ed. and tr. Herman Pálsson and Paul Edwards (Kirkwall, 1996), 40, 45–7; Marwick, *Anthology*, 301; Hugh Marwick, *The Place-Names of Rousay* (Kirkwall, 1947), 50, 78; Macfarlane, III, 306; Søiland, 'Pilgrimage', 26–8; Sarah Jane Gibbon and James Moore, 'Storyways: Visualising Saintly Impact in a North Atlantic Maritime Landscape', *Open Archaeology*, 5 (2019), 235–62.
29. *REO*, 225–7.

3: Mutation of Religion

1. John Foxe, *Germaniae ad Angliam de Restituta Evangelii Luce Gratulatio* (Basel, 1559), 46–7.
2. P. L. Hughes and J. F. Larkin (eds), *Tudor Royal Proclamations* (3 vols, New Haven, 1964–9), II, 103; Gilbert Goudie, *The Celtic and Scandinavian Antiquities of Shetland* (Edinburgh, 1904), 217–18; Thomas Hughes, 'The

Misfortunes of Arthur', in *Certaine Devises and Shews Presented to Her Majestie by the Gentlemen of Grayes-Inn* (London, 1587), 44.

3. Peter Marshall, *Heretics and Believers: A History of the English Reformation* (New Haven, 2017), 388; John Foxe, *Actes and Monuments* (London, 1583), Book 8, p. 1296.
4. *Rentals*, 20; Hugh Marwick, *The Place-Names of Rousay* (Kirkwall, 1947), 41; John Strype, *Ecclesiastical Memorials* (3 vols, Oxford, 1822), III (ii), 86–7; *CSP, Spain*, XIII, no. 345; Macfarlane, III, 308; Robert Lindsay of Pitscottie, *The Historie and Cronicles of Scotland*, ed. Aeneas J. Mackay (3 vols, Edinburgh, 1899–1911), II, 118.
5. *OSR*, I, 104; R. K. Hannay (ed.), *Acts of the Lords of Council in Public Affairs, 1501–1554* (Edinburgh, 1932), 549; TNA, SP 50/3, 81; John Roche Dasent (ed.), *Acts of the Privy Council* (46 vols, London, 1890–1964), II, 114, 289; M. Livingstone et al. (eds), *Register of the Privy Seal of Scotland* (8 vols, Edinburgh, 1908–82), IV, 153; *Forty-Sixth Annual Report of the Deputy Keeper of the Public Records* (London, 1886), App. 2, No. 1, 63.
6. Margaret H. B. Sanderson, *Biographical List of Early Scottish Protestants* (Edinburgh, 2010), 106; Heinz Scheible (ed.), *Melanchthons Briefwechsel: Band 12, Personen F–K* (Stuttgart, 2005), 197.
7. *SD*, 44; J. Maitland Anderson (ed.), *Early Records of the University of St Andrews* (Edinburgh, 1926), 207, 252, 263.
8. Charlotte Methuen, 'Orkney, Shetland and the Networks of the Northern Reformation', *Nordlit*, 43 (2019), 32–8; Charlotte Methuen, '"Islands Not Far from Norway, Denmark and Germany": Shetland, Orkney and the Spread of the Reformation in the North', and Henrik von Achen, '"Another Age Will Damage and Destroy": The Radicalised Reformation in Denmark-Norway in the Later Part of the Sixteenth Century', in James Kelly, Henning Laugerud and Salvador Ryan (eds), *Northern European Reformations: Transnational Perspectives* (London, 2020), 77–111, 191–211.
9. John Finlay, 'Bellenden, Sir John, of Auchnoul [of Auchinoul] (d. 1576)', *ODNB*; Duncan Shaw, 'Bothwell, Adam (1529?–1593)', *ODNB*; Oluf Kolsrud, 'Den Norske Kirkes Erkebiskoper og Biskoper', in *Diplomatarium Norvegicum: Oldbreve til Kundskab om Norges Indre og Ydre Forhold* (Christiania, 1913), 307.
10. Gladys Dickinson (ed.), *Two Missions of Jacques de la Brosse* (Edinburgh, 1942), 75; *CSP, Scotland*, I, no. 695; Gordon Donaldson, *Reformed by Bishops: Galloway, Orkney and Caithness* (Edinburgh, 1987), 25–6; W. C. Fairweather and R. Meldau, 'The Men of Norroway Contra the Men of Orkney', *Old-Lore Miscellany of Orkney, Shetland, Caithness and Sutherland, Vol. X* (London, 1935), 36–42.
11. John Knox, *Works*, ed. David Laing (6 vols, Edinburgh, 1846–64), I, 206; 'Noltland Castle, Westray', *Orkneyjar*, http://www.orkneyjar.com/history/noltlandcastle.htm; Mark Napier, *Memoirs of John*

Napier of Merchiston (London, 1834), 65; *REO*, 263; NRS, GD96/78, 80; *SEO*, 16–17.

12. Napier, *Memoirs*, 63, 67, 68, 73; *RPS*, 1560/8/6; Knox, *Works*, II, 276; *SD*, 79; *OSR*, I, 121.
13. Napier, *Memoirs*, 67–70. Bothwell's letter says he approached 'the sheriff' – conceivably, Oliver Sinclair, but context makes Edward seem likelier.
14. Robert Pitcairn (ed.), *Criminal Trials in Scotland, 1488–1624* (3 vols, Edinburgh, 1833), I (ii), 413; *CSP, Scotland*, I, no. 967; Napier, *Memoirs*, 65–6.
15. Napier, *Memoirs*, 72; *SEO*, 33; Alan MacDonald, *The Jacobean Kirk, 1567–1625* (Aldershot, 1998), 7–8, 191; Donaldson, *Bishops*, 30–1; *BUK*, I, 32.
16. George Mackay Brown, *An Orkney Tapestry* (London, 1969), 41–3; Ninian Winzet, *The Buke of Fourscoir-Thre Questions Tueching Doctrine, Ordour, and Maneris* (Antwerp, 1563), A5v.
17. Ian B. Cowan, *The Scottish Reformation* (London, 1982), ch. 7; Gordon Donaldson, 'Reformation to Covenant', in Duncan Forrester and Douglas Murray (eds), *Studies in the History of Worship in Scotland* (2nd edn, Edinburgh, 1996), 37–57; *The Book of Common Order of the Church of Scotland*, ed. George W. Sprott and Thomas Leishman (Edinburgh, 1868), 79, 90, 125; Hugh Marwick, *The Place Names of Rousay* (Kirkwall, 1947), 73; Hugh Marwick, *The Place Names of Birsay* (Aberdeen, 1970), 37, 62.
18. Craven, II, 114–15; *REO*, 118, 122, 132, 267, 272, 274, 281, 285; *OSR*, I, 235; Robert S. Barclay (ed.), *The Court Book of Orkney and Shetland, 1612–1613* (Kirkwall, 1962), 18, 32, 65, 77; Donaldson, 'Reformation', 44–5; Mooney, 26; Ronald Hutton, *The Stations of the Sun: A History of the Ritual Year in Britain* (Oxford, 1996), 330; *The Forme of Prayers and Ministration of the Sacraments &c. Vsed in the English Church at Geneua* (Edinburgh, 1564), 'A Kallender'.
19. OCR/4/1, 249; Macfarlane, III, 302; *OP*, 341; *The New Statistical Account of Scotland* (15 vols, Edinburgh, 1845), XV, 117–18.
20. Craven, II, 29; *RPS*, A1560/8/4.
21. Knox, *Works*, III, 179; Henry Dryden, *Description of the Church Dedicated to Saint Magnus and the Bishop's Palace at Kirkwall* (Kirkwall, 1878), 66.
22. Barbara E. Crawford, 'An Unrecognised Statue of Earl Rognvald?', in Crawford (ed.), *Northern Isles Connections* (Kirkwall, 1995). The nickname may derive from the designation 'Mense' (cf. 'mensal', referring to revenues or 'table' of a bishop), from 'Mass Tower', or as a corruption of 'Massy More', meaning 'dungeon': George Barry, *History of the Orkney Islands*, ed. James Headrick (2nd edn, London, 1808), 237; Hossack, 70.
23. Knox, *Works*, II, 183; Craven, II, 145–6; Walter Scott, *The Pirate*, ed. Mark Weinstein and Alison Lumsden (Edinburgh, 2001), 353; *Description of the Islands of Orkney and Zetland by Robert Monteith*, ed. Robert Sibbald (Edinburgh, 1845), 3, 9.
24. George Low, *A Tour through the Islands of Orkney and Schetland*, ed.

Joseph Anderson (Kirkwall, 1879), 61; John Mooney, *The Cathedral and Royal Burgh of Kirkwall* (Kirkwall, 1943), 88–9; Judith Jesch and Theya Molleson, 'The Death of Earl Magnus Erlendsson and the Relics of St. Magnus', in Olwyn Owen (ed.), *The World of the Orkneyinga Saga: 'The Broad Cloth Viking Trip'* (Kirkwall, 2005), 127–42.

25. I have found helpful the suggestions of Peter Yeoman, *Pilgrimage in Medieval Scotland* (London, 1999), 97.
26. *REO*, 342; Frank D. Bardgett, *Two Millennia of Church and Community in Orkney* (Edinburgh, 2000), 63–4; *BUK*, I, 112.
27. *BUK*, I, 112, 114; Rosalind K. Marshall, 'Hepburn, James, Fourth Earl of Bothwell and Duke of Orkney (1534/5–1578)', *ODNB*.
28. John Guy, *'My Heart Is My Own': The Life of Mary Queen of Scots* (London, 2004), 370–84; *Letters of Mary Queen of Scots*, ed. Agnes Strickland (2 vols, London, 1844), I, 331–2; *RPC*, 1st ser., XIV, 198; *RPS*, 1567/12/60.
29. *BUK*, I, 131, 162–3, 165–8; 'Introduction' in James Kirk (ed.), *The Books of Assumption of the Thirds of Benefices* (Oxford, 1995), lxxxii; Donaldson, *Bishops*, 28–35; Duncan Shaw, 'The Sixteenth Century and the Movement for Reform', in H. W. M. Cant and H. N. Firth (eds), *Light in the North: St Magnus Cathedral through the Centuries* (Kirkwall, 1989), 51–6.
30. Knox, *Works*, II, 310; Gordon Donaldson (ed.), *Accounts of the Collectors of Thirds of Benefices, 1561–72* (Edinburgh, 1949), 144; *Register of Ministers, Exhorters and Readers, and of Their Stipends* (Edinburgh, 1830); Bardgett, *Church*, 60–5; *Books of Assumption*, 559–60.
31. Craven, II, 40–1; Hossack, 228–9; Spencer J. Rosie, *Saints and Sinners: Memorials of St Magnus Cathedral* (Kirkwall, 2015), 29; Marwick, *Rousay*, 29; *The Protocol Book of Mr Gilbert Grote, 1552–1573*, ed. William Angus (Edinburgh, 1914), 53–4; MacDonald, *Kirk*, 9–10; *BUK*, I, 286–7.
32. Craven, II, 20–1; *REO*, 271, 280; Charles H. Haws (ed.), *Scottish Parish Clergy at the Reformation, 1540–1574* (Edinburgh, 1972), 108, 222, 244.

4: The Return of the Earls

1. 'The Earl's Palace, Birsay', *Orkneyjar*, http://www.orkneyjar.com/history/earlspalace.htm; *Description of the Islands of Orkney and Zetland by Robert Monteith*, ed. Robert Sibbald (Edinburgh, 1845), 4; John Knox, *Works*, ed. David Laing (6 vols, Edinburgh, 1846–64), II, 88; British Library, Ms. Cotton Caligula B/X, 287v.
2. Peter D. Anderson, 'Stewart, Robert, First Earl of Orkney (1533–1593)', *ODNB*; *RPS*, 1581/10/90; *NHO*, 263–4; TNA, SP 52/10, 115r; *REO*, 118–19, 287; *OSR*, I, 141–5; British Library, Ms. Cotton Caligula B/X, 389r; Thomas Miller, 'The Palace of Birsay in Orkney', *Scottish Historical Review*, 15 (1917), 50–1.
3. *REO*, 123–4; Peter D. Anderson, *Robert Stewart: Earl of Orkney, Lord of Shetland, 1533–1593* (Edinburgh, 1982), 55–65, 70; *BUK*, I, 162, 165–7.

4. Mooney, 80; James Mackenzie et al., *The General Grievances and Oppression of the Isles of Orkney and Shetland* (Edinburgh, 1836), 36; *SEO*, 99; David Balfour, *Oppressions of the Sixteenth Century in the Islands of Orkney and Zetland* (Edinburgh, 1859), xlvii–li.
5. *SD*, 162–7, 169, 183–224; *SEO*, 87, 90–105; TNA, SP 52/23/1, 380.
6. Dionyse Settle, *A True Reporte of the Laste Voyage into the West and Northwest Regions* (London, 1577), B2r–v; George Best, *A True Discourse of the Late Voyages of Discouerie* (London, 1578), 3–4; Kenneth E. Andrews, *Trade, Plunder and Settlement: Maritime Enterprise and the Genesis of the British Empire* (Cambridge, 1984), 175–6. For 'St Magnus Sound', see Evan T. Jones (ed.), 'The Journal of the Voyage of the *Marigold* to Iceland, 1654', in Susan Rose (ed.), *The Naval Miscellany, Volume VII* (Aldershot, 2008), 100–2, 108.
7. Robert Sempill, *Ane New Ballet Set out be ane Fugitiue Scottisman* (St Andrews, 1572); Ernest Marwick (ed.), *Orkney Poems* (Kirkwall, 1949), 154, 192; Robert Bruce, *Sermons vpon the Sacrament of the Lords Supper* (Edinburgh, 1591), H1v–2r.
8. David Moysie, *Memoirs of the Affairs of Scotland*, ed. James Dennistoun (Edinburgh, 1830), 34; *RPS*, 1581/10/90.
9. *SEO*, 118, Spencer J. Rosie, *Saints and Sinners: Memorials of St Magnus Cathedral* (Kirkwall, 2015), 31–2; *REO*, 448 (incorrectly designating Barbara a daughter of Robert); Peter Anderson, 'The Reformation and the Stewart Earls', in Donald Omand (ed.), *The Orkney Book* (Edinburgh, 2003), 86; Anderson, *Robert Stewart*, 130.
10. *NHO*, 272–3; *REO*, 304, 311–13.
11. *CSP, Scotland*, III, no. 732; Gilbert Goudie, *The Celtic and Scandinavian Antiquities of Shetland* (Edinburgh, 1904), 219–28; Moysie, *Memoirs*, 53; Brian Smith, 'When Did Orkney and Shetland Become Part of Scotland? A Contribution to the Debate', *New Orkney Antiquarian Journal*, 5 (2010), 11–13.
12. *RPS*, 1587/7/114; Balfour, *Oppressions*, 93–8; *SEO*, 128; *CSP, Scotland*, IX, no. 396.
13. A. Teulet (ed.), *Papiers d'Etat, pièces et documents inédits ou peu connus relatifs à l'histoire de l'Ecosse au XVIe siécle* (3 vols, Edinburgh, 1851–60), III, 525; Anderson, *Robert Stewart*, 116; Walter Traill Dennison, 'Armada Notes', *Northern Notes and Queries*, 4 (1890), 120–6.
14. Jim Wilson, 'Following the Fletts', talk at Orkney International Science Festival, 11 Sept. 2023 (https://oisf.org/fest-event/following-the-fletts/); *CSP, Scotland*, X, App. no 5; Concepción Sáenz-Cambra, 'Scotland and Philip II, 1580–1598', University of Edinburgh PhD thesis (2003), 159–60, 214–15.
15. *CSP, Scotland*, X, nos. 434, 440, 443, 449, 454, 458, 464, 468, 472, 477, 479, 482; Anderson, *Robert Stewart*, 118–28; *NHO*, 285–6.
16. *RPS*, 1592/4/116; *Calendar of the Manuscripts of the Most Honourable the Marquess of Salisbury* (London, 24 vols, 1883–1976), V, 111, 167; *SEO*, 155–6.
17. John H. Ballantyne and Brian Smith (eds), *Shetland Documents, 1580–1611*

(Lerwick, 1994), 86–8; *SEO*, 163–4; Liv Kjørsvikk Schei, *The Islands of Orkney* (Grantown-on-Spey, 2000), 218–19; *Rentals*, No. II, 50, 94.

18. Robert Pitcairn (ed.), *Criminal Trials in Scotland, 1488–1624* (3 vols, Edinburgh, 1833), I (i), 373–7, 386–8, 392–7; *SEO*, 177–81; *NHO*, 277–9.
19. *Shetland Documents*, 113–14; *RPC*, 1st ser., V, 436–7, 535–7.
20. Peter Anderson, 'The Stewart Earls of Orkney and the History of Orkney and Shetland', *Northern Studies*, 29 (1992), 43–52; J. Storer Clouston, *A History of Orkney* (Kirkwall, 1932), 304–29.
21. *SEO*, 186–8; *RPC*, 1st ser., V, 551; *CSP, Scotland*, XIII (1), no. 359; John M. Thomson et al. (eds), *Registrum Magni Sigilli Regum Scotorum/The Register of the Great Seal of Scotland* (11 vols, Edinburgh, 1882–1914), VI, nos. 1022, 1038; *NHO*, 285–7.
22. NRS, GD1/236/2; *SEO*, 164; Canmore, ID 2496; Thomas Thomson (ed.), *The Historie and Life of King James the Sext*, (Edinburgh, 1825), 386–7.
23. *CSP, Venice*, X, no. 739; James F. Larkin and Paul L. Hughes (eds), *Stuart Royal Proclamations* (2 vols, Oxford, 1973–83), I, 95–7; *Notes*, App. 58–9; Thomas Craig, *De Unione Regnorum Britaniae Tractatus*, ed. and tr. C. Sanford Terry (Edinburgh, 1909), 299–300.
24. Robert Pont, *De Unione Britanniae* (Edinburgh, 1604). An English translation (quoted from here) was produced for a non-appearing London printing: 'Of the Union of Britayne', in Bruce R. Galloway and Brian P. Levack (eds), *The Jacobean Union: Six Tracts of 1604* (Edinburgh, 1985), 18, 28.
25. Peter Marshall, *Reformation England, 1480–1642* (3rd edn, London, 2022), 139–40; Alan R. MacDonald, *The Jacobean Kirk, 1567–1625* (Aldershot, 1998), 21, 26, 31; *Fasti*, 230, 238; Craven II, 74–5.
26. *BUK*, II, 724; Henry Fitzsimon, *Words of Comfort to Persecuted Catholics*, ed. Edmund Hogan (Dublin, 1881), 121; Thomas McCoog, *The Society of Jesus in Ireland, Scotland, and England, 1598–1606* (Leiden, 2017), 304.
27. Craven, II, 65–6, 68–9: the passage in question is Johannes Brenz, *In Acta Apostolica Homiliae Centum Viginti Duae* (Frankfurt, 1546), 195; Thomas Francis Knox (ed.), *The First and Second Diaries of the English College, Douay* (London, 1878), 199, 204; *The Diary of Mr. James Melvill, 1556–1601* (Edinburgh, 1829), 291–2; *Annals*, 101.
28. James Kirk, 'Pont, Robert (1524–1606)', *ODNB*; Robert Wodrow, *Collections upon the Lives of the Reformers and Most Eminent Ministers of the Church of Scotland*, ed. William J. Duncan (2 vols, Glasgow, 1834–48), I, 197, 514.
29. A. S. Wayne Pearce, 'Law, James (d. 1632)', *ODNB*; University of Glasgow, Sp Coll Bk4-i.15. flyleaf; David Calderwood, *The History of the Kirk of Scotland*, ed. Thomas Thomson (8 vols, Edinburgh, 1842–9), VI, 210; MacDonald, *Kirk*, 104–5, 110–11, 125–6; *RPC*, 1st ser., VII, 26; William M. MacPherson, *Materials for a History of the Church and Priory of Monymusk* (Aberdeen, 1895), 238–9; *Diary of Walter Yonge,*

Esq., 1604–1628, ed. George Roberts (London, 1848), 11.

30. *James the Sext*, 385; *NHO*, 288; *SEO*, 218–19; *Rentals*, App. 88–92.
31. *SEO*, 188–210 (quote at 188); *Notes*, App. 54–5, 58; Pont, 'Union', 22; *CSP, Scotland*, XIII (2), no. 714; *RPC*, 1st ser., V, 523; VII, 737–7.
32. *Notes*, App. 59; *RPS*, 1607/3/41; *SEO*, 220–2.
33. *BUK*, III, 1078–82, 1085, 1095–6; MacDonald, *Kirk*, 144–5; Craven, II, 90.
34. David Laing (ed.), *Original Letters Relating to the Ecclesiastical Affairs of Scotland* (2 vols, Edinburgh, 1851), I, 267–8; *Notes*, App. 61–2, 64–5, 72; Craven, II, 108–9.
35. *Notes*, App. 63; *RPC*, 1st ser., IX, 181–2; Jane Ryder, 'Udal Law: An Introduction', *Northern Studies*, 25 (1988), 1–20.
36. *Notes*, App. 78–83, 85–9; *NHO*, 283–5.
37. Hossack, 24n; Calderwood, *History*, VI, 194.
38. *SEO*, 157, 235–6; *Original Letters*, I, 289–91; *RPC*, 1st ser., IX, 533.
39. *RPS*, 1612/10/22; *RPC*, 1st ser., IX, 479–81; *NHO*, 296, *Criminal Trials*, III, 89.
40. *SEO*, 241; *Trials*, III, 304, 306–7, 324–5.
41. *Trials*, III, 295, 303, 305–6, 322, 325; Calvin F. Senning, 'The Visit of Christian IV to England in 1614', *The Historian*, 31 (1969), 555–72; Edward J. Cowan, 'Clanship, Kinship and the Campbell Acquisition of Islay', *Scottish Historical Review*, 58 (1979), 132–57.
42. *Trials*, III, 275–6, 289, 293–4, 295, 300; *SEO*, 250–1.
43. *SEO*, 257–63; Robert Gordon, *A Genealogical History of the Earldom of Sutherland*, ed. Henry Weber (Edinburgh, 1813), 299; *Trials*, III, 287, 291; *Original Letters*, II, 369; *State Papers and Miscellaneous Correspondence of Thomas, Earl of Melros*, ed. James Maidment (2 vols, Edinburgh, 1837), I, 147–50, 155.
44. *Original Letters*, II, 379, 381; *Trials*, III, 292; *NHO*, 298–9; Robert S. Barclay (ed.), *The Court Books of Orkney and Shetland, 1614–1615* (Edinburgh, 1967), xxii–xxiii.
45. *Trials*, III, 292; *RPC*, 1st ser., X, 319–23; Hossack, 27.
46. Evan MacGillivray (ed.), 'Richard James, 1592–1638, Description of Shetland, Orkney and The Highlands of Scotland', *Orkney Miscellany*, 1 (1953), 50–1; Calderwood, *History*, VI, 194.
47. *Trials*, III, 301.
48. John H. Ballantyne (ed.), *Shetland Documents 1612–1637* (Lerwick, 2016), 143–5.
49. *Court Books*, 25–6; Craven, II, 112; *Original Letters*, II, 412–13.
50. James M. Irvine (ed.), *The Orkneys and Schetland in Blaeu's Atlas Novus of 1654* (Ashtead, 2006), 11–27, with a convincing argument for Stewart's authorship at 48–52.
51. Craig, *De Unione*, 288–9; *SEO*, 213.

5: Devilry and Witchcraft

1. Leslie Burgher, *Orkney: An Illustrated Architectural Guide* (Edinburgh, 1991), 22; Hossack, 342–3, 402.
2. The figure comprises seventy-one names in *The Survey of Scottish Witchcraft* database (https://witches.hca.ed.ac.uk/), seven cases in *Rentals*, three in Ernest Marwick, *An Orkney Anthology* (Edinburgh,

1991), 376–9, and sixteen others from sheriff court, presbytery and session records, omitting individuals against whom accusations were judged libellous. Relative frequency from recalculation of figures provided by Lauren Martin, 'Scottish Witchcraft Panics Re-examined', in Julian Goodare, Lauren Martin and Joyce Miller (eds), *Witchcraft and Belief in Early Modern Scotland* (Basingstoke, 2008), 123–4. I am indebted to Julian Goodare for advice on this point.

3. OCR/4/2, 19r.
4. *Rentals*, No. II, 28, 38, 43, 64, 94, 103; *Notes*, App. 90; Peter Marshall, 'The Reformation and the Idea of the North', *Nordlit*, 43 (2019), 4–24 (quote at 15).
5. Robert S. Barclay (ed.), *The Court Books of Orkney and Shetland, 1614–1615* (Edinburgh, 1967), 25, 33.
6. Marwick, *Anthology*, 338–9; Knud Krogh, *Viking Greenland* (Copenhagen, 1967), 97.
7. OCR/4/1, 292.
8. Julian Goodare, 'The Framework for Scottish Witch-Hunting in the 1590s', *Scottish Historical Review*, 81 (2002), 240–50; Michael Wasser, 'The Privy Council and the Witches: The Curtailment of Witchcraft Prosecutions in Scotland, 1597–1628', *Scottish Historical Review*, 82 (2003), 20–46. The appearance of a handful of Orkney trial papers in the records of the privy council suggests some kind of at least fitful liaison.
9. Barclay, *Court Books, 1614–1615*, 18; Julian Goodare, 'The Scottish Witchcraft Act', *Church History*, 74 (2005), 39–67.
10. Margaret Hunter, 'Jo: Ben's Description of Orkney', *New Orkney Antiquarian Journal*, 6 (2012), 37; Barclay, *Court Books, 1614–1615*, 18–20; NRS, SC10/1/6, 99r. See Julian Goodare, 'Away with the Fairies: The Psychopathology of Visionary Encounters in Early Modern Scotland', *History of Psychiatry*, 31 (2020), 37–54.
11. Hunter, 'Jo: Ben's Description', 37; *Folklore*, 22–3; George M. Foster, 'Peasant Society and the Image of Limited Good', *American Anthropologist*, 67 (1965), 293–315.
12. For what follows, NRS, GD217/570; SC10/1/3, 65r–67r.
13. NRS, SC10/1/3, 75v.
14. 'Acts and Statues of the Lawting, Sheriff, and Justice Courts, within Orkney and Zetland, M.DC. II–M.DC.XLIV', *Miscellany of the Maitland Club, Volume II* (Edinburgh, 1840), 187–191, *SEO* 249, 256, 269.
15. NRS, SC10/1/3, 60v–61v; Michael F. Graham, 'Kirk in Danger: Presbyterian Political Divinity in Two Eras', in Bridget Heal and Ole Peter Grell (eds), *The Impact of the European Reformation: Princes, Clergy and People* (Aldershot, 2008), 171.
16. Hossack, 30; Laura Paterson, 'Executing Scottish Witches', in Julian Goodare (ed.), *Scottish Witches and Witch-Hunters* (Basingstoke, 2013), 196–214.
17. Berit Sandnes, *From Starafjall to Starling Hill: An Investigation of the Formation and Development of Old Norse Place-Names in Orkney* (http://www.spns.org.uk/Starafjall.pdf, 2010), 237; *Dictionaries of the Scots Language* (https://dsl.ac.uk/), 'Riggin'; Hugh Marwick, *The*

Orkney Norn (Oxford, 1929), 187; Black, 117, 126 and, for the following, 103–11.

18. NRS, SC10/1/5, 48v.
19. 'Trials for Witchcraft, Sorcery, and Superstition, in Orkney', in William B. D. D. Turnbull (ed.), *Miscellany of the Abbotsford Club, Volume First* (Edinburgh, 1837), 139; NRS, SC10/1/6, 50v, 261r; *RPC*, 2nd ser., VIII, 71.
20. For the following, NRS, SC10/1/6, 97r–98r, and two separate drafts of the dittay: OLA, D20/2/16; D23/14/8.
21. J. G. Dalyell, *The Darker Superstitions of Scotland* (Edinburgh, 1834), 455; OCR/4/1, 45; Alexander Fenton, *The Northern Isles: Orkney and Shetland* (Edinburgh, 1997), 314–15.
22. OCR/4/1, 109, 131, 231, 232, 250, 278, 292; NRS, SC10/1/6, 68v; OCR/23/1, 140, 166; J. B. Craven, *Church Life in South Ronaldsay and Burray in the Seventeenth Century* (Kirkwall, 1911), 30.
23. NRS, SC10/1/5, 86v–87r.
24. *RPC*, 2nd ser., VIII, 71–5; NRS, SC10/1/3, 73v.
25. For the following, NRS, SC10/1/5, 85r–v, 88v–89v.
26. NRS, SC10/1/3, 94r–95v; SC10/1/5, 87r; 'Trials', 164–9, 171–80; Owen Davies, 'The Material Culture of Post-Medieval Domestic Magic in Europe: Evidence, Comparisons and Interpretations', in Dietrich Boschung and Jan Bremmer (eds), *The Materiality of Magic* (Paderborn, 2015), 396–400.
27. NRS, SC10/1/6, 51r; SC10/1/5, 87r; Black, 133–4.
28. Duncan J. Robertson, 'Orkney Folk-Lore', *Orkney Herald*, 20 Feb. 1924; NRS, SC10/1/5, 87r; *RPC*, 2nd ser., VIII, 72.
29. NRS, SC10/1/3, 94r; SC10/1/6, 68v; Black, 140–7.
30. NRS, SC10/1/5, 48v; 'Trials', 167; Craven, *Life*, 31.
31. Barclay, *Court Books, 1614–1615*, 26 ('walks' here might have the meaning of vigil, rather than perambulation); OCR/14/74, 8; OCR/4/3, 57; Craven, *Life*, 53.
32. OCR/13/1, 105; Craven, *Life*, 45, 69.
33. Walter Traill Dennison, *The Orcadian Sketchbook* (Kirkwall, 1880), 162; Tom Muir, *Orkney Folk Tales* (Stroud, 2014), 133.
34. NRS, SC10/1/3, 74r; SC10/1/6, 264r; SC10/1/5, 89r.
35. NRS, SC10/1/6, 256v, 272v; SC10/1/3, 234r; OCR/23/1, 165; Craven, *Life*, 37.
36. 'Trials', 143; OLA, D23/14/8; NRS, SC10/1/3, 73r, 74r.
37. 'Trials', 144–5, 159; NRS, SC10/1/5, 86v–87r.
38. NRS, SC10/1/3, 177v–178r; 10/1/6, 50r; Craven, *Life*, 30.
39. OCR/4/1, 282; Black, 63.
40. OCR/4/1, 282; OCR/14/74, 170, 226 (brought to my attention by Sarah Jane Gibbon). See Ane Ohrvik, *Medicine, Magic and Art in Early Modern Norway* (London 2018).
41. OCR/4/1, 223–4, 235.
42. The suggestion of Liv Helene Willumsen, *Witches of the North: Scotland and Finnmark* (Leiden, 2013), 162–3. Willumsen's figures (p. 161) on the distribution of cases differ slightly from mine, but demonstrate the same overall pattern.
43. David George Mullan (ed.), *Religious Controversy in Scotland, 1625–1639* (Edinburgh, 1998), 35; James

Wallace, *An Account of the Islands of Orkney* (London, 1700), 64; *OP*, 286; *RPC*, 2nd ser., V, 284–5, 320, 531–2, 659–60.

44. *Records of the Kirk of Scotland*, ed. Alexander Peterkin (Edinburgh, 1838), 279, 327, 354; OCR/4/1, 248. For total accusations by year, see graphs at *The Survey of Scottish Witchcraft* database.
45. For what follows, see Peter Marshall, 'The Ministers, the Merchant and His Mother: Politics and Protest in a 17th Century Witchcraft Complaint', *New Orkney Antiquarian Journal*, 9 (2020), 56–70. Interrogation of Boundie: OCR/4/1, 253–7.
46. George Low, *A Tour through the Islands of Orkney and Schetland*, ed. Joseph Anderson (Kirkwall, 1879), 6n; OCR/14/74, 104–5, 265; OCR/4/3, 40.
47. OCR/23/1, 196, 207; OCR/14/74 (Kirkwall), 257; *Annals*, 213.
48. Craven, *Life*, 54.
49. OCR/4/2, 47v; OCR/23/1, 254; OCR/23/13, 7v–8r, 29v,41r–v, 42v.
50. OLA, CO1/1/1, 118; OCR/14/75, 508; James T. Calder, *Sketch of the Civil and Traditional History of Caithness* (2nd edn, Wick, 1887), 278; OLA, D66/1/19; OCR/4/4, 49.
51. OCR/14/75, 392–3.
52. *RPC*, 2nd ser., V, 558; OCR/14/74, 158–9; OCR/4/1, 50; Craven, *Life*, 58.
53. OCR/23/1, 218; OCR/7/8, 3r; OCR/23/13, 14r, 18v, 28v, 19v. Stanley T. Williams, *The Life of Washington Irving* (2 vols, New York, 1935), II, 247–50.
54. OCR/13/1, 156–7; OCR/16/1, 5.
55. OCR/23/1, 300; 23/13, 22r, 27r.
56. OCR/23/13, 41v, 43r, 61r; OCR/14/75, 205–6.
57. Ronald Hutton, *The Witch: A History of Fear* (New Haven and London, 2017), 16–18.
58. NRS, GD106/203; Julian Goodare, *The European Witch-Hunt* (London, 2016), 113.
59. William Mackenzie, *Gaelic Incantations, Charms, and Blessings of the Hebrides* (Edinburgh, 1895), 4; Marwick, *Anthology*, 343–5.
60. Thomas Edmondston, *An Etymological Glossary of the Shetland and Orkney Dialect* (Edinburgh, 1866), 141.
61. Margaret A. Murray, *The God of the Witches* (1931; reprint, Oxford, 1970), 130.

6: Revolution

1. For the following, Gilbert Gordon of Sallagh, 'Continuation of the History of the Earls of Sutherland', in Robert Gordon, *A Genealogical History of the Earldom of Sutherland*, ed. Henry Weber (Edinburgh, 1813), 552–6; *The Historical Works of Sir James Balfour*, ed. James Haig (4 vols, Edinburgh, 1824–5), IV, 8–12; *A True Relation of the Happy Victory Obtained by the Blessing of God upon April 27, 1650* (Edinburgh, 1650), based in part on a previously unnoticed letter by the chaplain of Strachan's regiment, George Hall: NLS, Wodrow Mss., Fol. XXV (1), 126r; S. R. Gardiner (ed.), *Letters and Papers Illustrating the Relations between Charles the Second and Scotland in 1650* (Edinburgh, 1894), 5–6, 54.
2. Mooney, 89; Balfour, *Works*, IV, 18; *Notes*, App. 106; *RPS*, A1650/5/19; OCR/4/2, 2r; OCR/14/74, 267; Hossack, 432. A comparison with

the Great War is suggested by Ronald Miller, *Orkney* (London, 1976), 93–4.

3. Maurice Lee, *The Road to Revolution: Scotland under Charles I* (Urbana and Chicago, 1985), 9; Richard James, *The Muses Dirge Consecrated* (London, 1625), 2–4, 12.
4. *SEO*, 285–6; *RPC*, 2nd ser., X, 315; Henry Erskine, *The Erskine Halcro Genealogy* (London, 1890), 11, 29–30.
5. *NHO*, 302–3; *SEO*, 286–7; John M. Thomson et al. (eds), *Registrum Magni Sigilli Regum Scotorum/The Register of the Great Seal of Scotland* (11 vols, Edinburgh, 1882–1914), VIII, 643–4.
6. David Marshall, 'Notes of the Connection of the Earls of Morton and Dick of Braid and Craighouse, with the Earldom of Orkney', *Proceedings of the Society of Antiquaries of Scotland*, 23 (1889), 275–313; Steve Murdoch and Alexia Grosjean, *Alexander Leslie and the Scottish Generals of the Thirty Years' War, 1618–1648* (London, 2014), 44–6; *NHO*, 304; *RPC*, 2nd ser., V, 310; 85–7; *RPC*, 2nd ser., III, 171–2.
7. Steve Murdoch, *The Terror of the Seas? Scottish Maritime Warfare, 1513–1713* (Leiden, 2010), 169–70; *RPC*, 2nd ser., II, 281–2, 605; David Worthington, *Scots in Habsburg Service, 1618–1648* (Leiden, 2004), 61–2, 93–4; Hugh Marwick, *Orkney* (London, 1951), 175.
8. *RPC*, 2nd ser., V, xlv, 30; *Naval Tracts of Sir William Monson*, ed. Michael Oppenheim (5 vols, London, 1902–14), V, 258; William Lithgow, *Scotlands Welcome to Her Native Sonne, and Soveraigne* (Edinburgh, 1633), E2r; *The Totall Discourse of the Rare Adventures and Painefull Peregrinations* (London, 1632), 106; Martin Garrett, 'Lithgow, William (b. 1582, d. in or after 1645)', *ODNB*.
9. *RPC*, 2nd ser., IV, 120, 536; V, 122–3; VI, 307.
10. 'Acts and Statutes of the Lawting, Sheriff and Justice Courts within Orkney and Zetland, M.DC. II–M.DC.XLIV', in *Miscellany of the Maitland Club, Volume II* (Edinburgh 1840), 203, 207, 209–10; Mooney, 17, 21–2; *RPC*, 2nd ser., VI, 506–7.
11. Robert Scott Fittis, *Sketches of the Olden Times in Perthshire* (Perth, 1878), 444–5; Alan R. MacDonald, *The Jacobean Kirk, 1567–1625* (Aldershot, 1998), 90, 103, 106; William Scot, *An Apologetical Narration of the State and Government of the Kirk of Scotland*, ed. D. Laing (Edinburgh, 1846), 130; L. G. Graeme (ed.), 'Some Letters and Correspondence of George Graeme Bishop of Dunblane and of Orkney, 1602–1638', *Miscellany of the Scottish History Society (Second Volume)* (Edinburgh, 1904), 233–4, 236; David Calderwood, *The History of the Kirk of Scotland*, ed. Thomas Thomson (8 vols, Edinburgh, 1842–9), VI, 155, 203.
12. *Fasti*, 351; Graeme, 'Letters', 253; *OSR*, III, 14–15; Hossack, 78–9; James M. Irvine, *The Breckness Estate* (Ashtead, 2009), 27–33; Craven, II, 163–6.
13. Craven, II, 161–2; 168–9, 187; *Fasti*, 243, 251, 269, 273; *Annals*, 50–1; Macfarlane, I, 151.
14. Craven, II, 176, 185; *Rentals*, App. 54–5; D. E. R. Watt and A. L. Murray

(eds), *Fasti Ecclesiae Scoticanae Medii Aevi Ad Annum 1638* (rev. edn, Edinburgh, 2003), 332; *Fasti*, 211, 267, 273; OCR/14/74, 103, 111; NRS, GD/190/3/212/2.

15. Lithgow, *Totall Discourse*, 505–6; *Rentals*, App. 52; Craven, II, 125, 145,159–60; Hossack, 264; OCR/14/74, 60, 63, 89, 93, 95.
16. Laura A. M. Stewart, *Rethinking the Scottish Revolution: Covenanted Scotland, 1637–1651* (Oxford, 2016), 38–43; Lee, *Revolution*, 64, 129, 137–8, 195–7; Anthony Milton, *Catholic and Reformed: The Roman and Protestant Churches in English Protestant Thought, 1600–1640* (Cambridge, 1995), 203–4.
17. NLS, Wodrow Mss., Fol. LXII, 182r–183r; Craven, II, 177–81; NRS, RH9/15/20.
18. NLS, Wodrow Mss., Fol. LXII, 183r–187v; Craven, II, 183–90; James Gordon, *History of Scots Affairs, 1637–1641*, ed. Joseph Robertson and George Grub (3 vols, Aberdeen, 1841), I, 156; II, 138; *Letters and Journals of Robert Baillie*, ed. David Laing (3 vols, Edinburgh, 1841–3), I, 163.
19. OCR/14/74, 115–16, 117–18; OCR/4/1, 2, 11, 20.
20. *Acts of the General Assembly of the Church of Scotland, 1638–1842* (Edinburgh, 1843), 35; Craven, II, 191; William Prynne, *The Antipathie of the English Lordly Prelacie* (London, 1641), 359; *The Recantation, and Humble Submission of Two Ancient Prelates* (n.p., 1641); Gordon, *Affairs*, III, 44.
21. *Works of the Most Reverend Father in God, William Laud*, ed. W. Scott and J. Bliss (7 vols, Oxford, 1847–60), VI, 572–3; Joseph Hall, *Episcopacie by Divine Right Asserted* (London, 1640), 'Epistle Dedicatorie', 1.
22. *Fasti*, 353; David Stevenson, 'The Financing of the Cause of the Covenants, 1638–51', *Scottish Historical Review*, 51 (1972), 89–90, 96; *Rentals*, No. III, 12–32 (quote at 20); Hugh Marwick, 'The Baikies of Tankerness', *Orkney Miscellany*, 4 (1957), 35.
23. *Kong Christian den Fjerdes Egenhaendige Breve*, ed. C. F. Bricka and J. A. Fridericia (7 vols, Copenhagen, 1878–91), IV, 378–9; Steve Murdoch, *Britain, Denmark-Norway and the House of Stuart, 1603–1660* (East Linton, 2000), 93–7, 124–5; *RPS*, 1644/6/628; Marshall, 'Morton', 286–7.
24. OCR/4/1, 4, 8–9, 28, 33, 37, 44, 48, 73, 95, 106, 188; *RPC*, 2nd ser., VII, 271; *Rentals*, No. III, 19; *Fifth Report of the Royal Commission on Historical Manuscripts, Part I* (London, 1876), 61.
25. NRS, GD190/3/206/1; 200/1; OCR/4/1, 7, 10, 12, 14, 22, 51–2, 57, 61; *RPC*, 2nd ser., VII, 140.
26. OCR/4/1, 44, 50, 54, 62–7, 80–4; NRS, GD190/1/12; GD190/3/206/3; GD190/3/206/4; GD190/3/206/1; GD190/3/212/4.
27. OCR/4/1, 101, 167–9, 250, 253–4; *Fasti*, 347; Wallace Douglas Kornahrens, 'Eucharistic Doctrine in Scottish Episcopacy, 1620–1875', University of St Andrews PhD thesis (2008), 195–6; Alexander F. Mitchell and James Christie (eds), *The Records of the Commissions of the General Assemblies of the Church of Scotland Holden in Edinburgh in the Years 1646 and 1647* (Scottish

History Society, 1892), 213; Mitchell and Christie (eds), *The Records of the Commissions of the General Assemblies of the Church of Scotland Holden in Edinburgh the Years 1648 and 1649* (Scottish History Society, 1896), 274.

28. The following draws on my 'The Ministers, the Merchant and His Mother: Politics and Protest in a 17th Century Witchcraft Complaint', *New Orkney Antiquarian Journal*, 9 (2020), 56–70.
29. *RPC*, 2nd ser., VI, 307.
30. NRS, GD/106/6; *Acts of the General Assembly*, 140, 191; OCR/4/1, 14, 101, 150, 232, 243, 260, 262–3, 329, 360, 363, 364, 371; Natalie Mears et al. (eds), *National Prayers: Special Worship since the Reformation, Volume 1: Special Prayers, Fasts and Thanksgivings in the British Isles, 1533–1688* (Woodbridge, 2013), 401–2; OCR/14/74, 168–9, 203; Karin Bowie, *Public Opinion in Early Modern Scotland, c.1560–1707* (Cambridge, 2020), 115; Stewart, *Revolution*, 234; Mooney, 87, 88.
31. *Military Memoirs of the Great Civil War, Being the Military Memoirs of John Gwynne*, ed. Walter Scott (Edinburgh, 1822), 83–5; 'A Declaration of the Right Honourable James, Marquis and Earl of Montrose', *The Harleian Miscellany, Vol. XI* (London, 1810), 472; Steve Murdoch and Tim Wales, 'King, James, Lord Eythin (1589–1652)', *ODNB*; Mary Elizabeth Ailes, *Military Migration and State Formation: The British Military Community in Seventeenth-Century Sweden* (Lincoln, NE, 2002), 39, 63–4; 'The King's Bodyguard of the Yeomen of the Guard', *BCW Project*, http://wiki.bcw-project.org/royalist/foot-regiments/yeomen-of-the-guard; David Stevenson (ed.), *The Government of Scotland under the Covenanters, 1637–1651* (Edinburgh, 1982), 65; NLS, Wodrow Mss., Fol. XXV (1), 94r–v.
32. NLS, Wodrow Mss., Fol. XXV (1), 94r–v; Mark Napier, *Memoirs of the Marquis of Montrose* (2 vols, Edinburgh, 1856), II, 724–6; NRS, GD220/6/2068; GD190/3/208; David Stevenson, *The Scottish Revolution, 1637–1644: The Triumph of the Covenanters* (Edinburgh, 2003), 133; Gwynne, *Memoirs*, 86; *Commissions of the General Assemblies, 1648–9*, 310, 322–4, 378.
33. Napier, *Memoirs of Montrose*, II, 735–6; Edward J. Cowan, *Montrose: For Covenant and King* (London, 1977), 277–9; NRS, GD190/3/484; GD18/3112; Mooney, 87, 88–9; Murdoch, *Terror*, 224; Stevenson, *Revolution*, 132, 136; Balfour, *Works*, IV, 24–5.
34. NRS, RH9/15/221/1–4; Gwynne, *Memoirs*, 91–3; George Wishart, *The Memoirs of James, Marquis of Montrose, 1639–1650*, ed. Alexander D. Murdoch and H. F. Morland Simpson (London, 1893), 496–501; Hossack, 252.
35. Gordon, *Sutherland*, 556–7; *Notes*, App. 106; Mooney, 89; James T. Calder, *Sketch of the Civil and Traditional History of Caithness* (Glasgow, 1861), 176; C. J. Lyon, *Personal History of King Charles the Second* (Edinburgh, 1851), 45: Balfour, *Works*, IV, 57–8, 168; Edward M. Furgol, *A Regimental History of the Covenanting Armies*,

1639–1651 (new edn, Edinburgh, 2003), 68; OLA, CO1/1, 21r–v.

36. Calder, *Caithness*, 116–17; Mooney, 89; Hossack, 253.
37. C. H. Firth (ed.), *Scotland and the Commonwealth* (Edinburgh, 1895), 34, 36; C. Sanford Terry (ed.), *The Cromwellian Union* (Edinburgh, 1902), 27; TNA, SP 25/66, 599.
38. Nottinghamshire Archives, CA/4686(a); *Mercurius Politicus*, 11 Mar. 1652/3, 1470.
39. NLS, Adv.MS.19.3.4, 33v–44r; *Poetical Descriptions of Orkney, M.DC.LII*, ed. James Maidment (?) (Edinburgh, 1835); *Prose Works of John Milton* (New York, 1885), 344. Another copy of the poems survives in a seventeenth-century compilation: West Yorkshire Archives, Bradford, 32D86/17. For Emerson, Charles Firth, *The Regimental History of Cromwell's Army* (2 vols, Oxford, 1940), II, 463–6, 469–71. I've found no confirmation of companies of Alured's regiment in Kirkwall before 1653, though troops for the Mull expedition departed from Orkney. Emerson perhaps transferred into Cooper's regiment.
40. Mooney, 90; Hossack, 414–15; *REO*, 448–51; NRS, GD190/3/209/4.
41. Hossack, 42, 50, 167; James Fea, *The Present State of the Orkney Islands Considered* (1775, reprint, Edinburgh, 1884), 14; Tudor, 256.
42. George Barry, *History of the Orkney Islands*, ed. James Headrick (2nd edn, London, 1808), 261; Hossack, 414, 432.
43. Terry, *Union*, xlix, 122–6; *Journal of the House of Commons, Volume 7: 1651–1660* (London, 1802), 202; Allan I. Macinnes, *Union and Empire: The Making of the United Kingdom in 1707* (Cambridge, 2007), 75–7; F. D. Dowd, *Cromwellian Scotland, 1651–1660* (Edinburgh, 1979), 42–7; *CSP, Domestic, 1656–7*, 129.
44. *RPS*, A1650/5/19; A1650/5/20; A1650/5/49; A1650/5/5; M1650/5/18; M1650/5/17; Murdoch, *Terror*, 224; NLS, Wodrow Mss., Fol. XXV (1), 128r; *The Diary of Mr John Lamont of Newton, 1649–1671*, ed. G. R. Kinloch (Edinburgh, 1830), 21.
45. *Commissions of the General Assemblies, 1648–9*, 406–7, 414; III, 189–91, 312, 355; NLS, Wodrow Mss., Fol. XXV (2), 162r–v; Fol. XXVI, 64r; *Fasti*, 116, 222, 243; Craven, II, 218–19; OCR/4/2, 19v.
46. OCR/4/2, 20r, 25v–26r; R. Scott Spurlock, *Cromwell and Scotland: Conquest and Religion, 1650–1660* (Edinburgh, 2007), 102; Bowie, *Opinion*, 80; *Narrative Papers of George Fox*, ed. Henry J. Cadbury (Dublin, IN, 1972), 39; William Stockdale, *The Doctrines and Principles: The Persecution … of the Saints of God by the Priests and Magistrates of Scotland* (London, 1659), 5; J. B. Craven, *Church Life in South Ronaldsay and Burray in the Seventeenth Century* (Kirkwall, 1911), 27; George Fox, *The Great Mistery of the Great Whore Unfolded* (London, 1659), 262; Vincent de Paul, *Correspondence*, ed. and tr. Marie Poole et al. (8 vols, New York, 1985–99), IV, 478; VI, 546.
47. *CSP, Domestic, 1651–2*, 255; *CSP, Domestic, 1652–3*, 143; *CSP, Domestic, 1653–4*, 561; Terry, *Union*, 230; W. Dunn Macray (ed.), *Calendar of the Clarendon State Papers Preserved in*

the Bodleian Library, Vol. II (Oxford, 1869), 141, 249; *CSP, Venice, 1653–4*, 59; Murdoch, *Denmark-Norway*, 176–8.

48. Marguerite Wood (ed.), *Extracts from the Records of the Burgh of Edinburgh* (Edinburgh, 1940), 55; Terry, *Union*, 124; John Shirreff, *General View of the Agriculture of the Orkney Islands* (Edinburgh, 1814), App. 1–8; Marshall, 'Morton', 293–9.
49. Hossack, 251; William R. Mackintosh, *Glimpses of Kirkwall and Its People in the Olden Time* (Kirkwall, 1887), 311.

7: The Earldom and the Kingdom

1. Lieuwe van Aitzema, *Historie of Verhael van Saken van Staet en Oorlogh* (The Hague, 1669), 409–11; *Letters and Negotiations of the Count d'Estrades* (3 vols, London, 1711), II, 538, III, 193, 208, 288; Steve Murdoch, *The Terror of the Seas? Scottish Maritime Warfare, 1513–1713* (Leiden, 2010), 243–6; Gijs Rommelse, *The Second Anglo-Dutch War (1665–1667)* (Hilversum, 2006), 152, 184–5; TNA, SP 29/209, 160r; Peter Willemoes Becker (ed.), *Samlinger til Danmarks Historie under Kong Frederik den Tredies Regiering* (2 vols, Copenhagen, 1847–57), II, 196.
2. Ernst Ekman, 'The Danish Royal Law of 1665', *Journal of Modern History*, 29 (1957), 102–7.
3. Robert Bell (ed.), *Memorials of the Civil War* (2 vols, London, 1849), II, 185–9; Richard Baker, *A Chronicle of the Kings of England* (London, 1674), 714.
4. J. B. Craven, *Church Life in South Ronaldsay and Burray in the Seventeenth Century* (Kirkwall, 1911), 32; Hew Scott (ed.), *Fasti Ecclesiae Scoticanae: The Succession of Ministers in the Church of Scotland from the Reformation, Volume IV: Synods of Argyll, and of Perth and Stirling* (Edinburgh, 1923), 211; *Notes*, 185–6.
5. OCR/4/2, 64v; *RPS*, 1662/5/61; Hossack, 244; *NHO*, 311–14; Francis J. Shaw, *The Northern and Western Islands of Scotland: Their Economy and Society in the Seventeenth Century* (Edinburgh, 1980), 39–42.
6. Hossack, 84, 339–40; Mooney, xiii, 43–51, 86–90; *RPC*, 3rd ser., I, 116–17; *RPS*, 1662/5/18; Spencer J. Rosie, 'Kirkjuvagr to Kirkwall: An Analysis of Kirkwall's Royal Charters', University of the Highlands and Islands MA thesis (2021), 4–5, 61–2.
7. Clare Jackson, *Restoration Scotland, 1660–1690* (Woodbridge, 2003), 104–8; Godfrey Davies and Paul H. Hardacre (eds), 'The Restoration of the Scottish Episcopate, 1660–61', *Journal of British Studies*, 1 (1962), 32–52; Samuel Pepys, *Diary*, ed. Lord Braybrooke (2 vols, London, 1906), I, 170–1; David George Mullan, 'Sharp, James (1618–1679)', *ODNB*.
8. Craven, III, 9–10; *Fasti*, 263; *RPC*, 3rd ser., I, 50, 182; II, 266; III, 603; NRS, PA7/9/1/102; OCR/4/2, 140; *RPS*, 1661/1/38; 1661/1/262; A1661/1/7; Hossack, 236; *RPC*, 3rd ser., VIII, 467.
9. OCR/4/3, 14, 23, 28–9, 35, 56–7, 59, 60; OCR/4/1, 220; *RPC*, 3rd ser., II, 435; III, 118; *Fasti*, 215; James Morison, *The Everlasting Gospel of the Everlasting Covenant Discussed* (Edinburgh, 1668), A2r–v.

10. Mooney, 62; Rosie, 'Charters', 122–3; Hossack, 96; *RPS*, 1670/7/49; 1669/10/33; *NHO*, 306–7.
11. Gordon Donaldson, 'Covenant to Revolution', in Duncan Forrester and Douglas Murray (eds), *Studies in the History of Worship in Scotland* (Edinburgh, 1984), 63–70; J. B, Craven, *The Blazon of Episcopacy in Orkney, 1421–1688* (Kirkwall, 1901), 9–10; Hossack, 32–3, 84.
12. J. B. Craven (ed.), *Descriptive Catalogue of the Bibliotheck of Kirkwall* (Kirkwall, 1897), v–x; Paul Kaufman, 'Discovering the Oldest Publick Bibliotheck of the Northern Isles', *Library Review*, 23 (1972), 285–7; Craven, III, 65–7; Hossack, 86–7; George Buchanan, *History of Scotland* (4 vols, Glasgow, 1827), I, 57–8; OLA, DO88, 28; James M. Irvine (ed.), *The Orkneys and Schetland in Blaeu's Atlas Novus of 1654* (Ashtead, 2006), 26; *RPC*, 3rd ser., IX, 76.
13. *RPC*, 3rd ser., I, 536, 640; II, 315–16, 394, 435; III, 130, 376–8, 459, 575–6, 599–600, 612, 618; VII, 205–7, 252; X, 118–21.
14. *RPC*, 3rd ser., II, 479–80, 495–6, 671; Craven, III, 30–3, 37–41; *Letters and Journals of Robert Baillie*, ed. David Laing (3 vols, Edinburgh, 1841–3), III, 459; Osmund Airy (ed.), *The Lauderdale Papers, 1639–79* (3 vols, London, 1884–5), II, Ap. xxvi; Tudor, 623–6.
15. Robert Wodrow, *The History of the Sufferings of the Church of Scotland* (2 vols, Edinburgh, 1721–2), II, 83; OLA, D1/359/1–2.
16. Claire Jowitt, 'The Last Voyage of the Gloucester (1682): The Politics of a Royal Shipwreck', *English Historical Review*, 137 (2022), 728–62; *RPS*, 1681/7/29; OCR/4/4, 59–60; *RPC*, 3rd ser., VIII, 132, 168, 266; IX, 73.
17. John Willcock, *A Scots Earl in Covenanting Times: Being Life and Times of Archibald 9th Earl of Argyll (1629–1685)* (Edinburgh, 1907), 349–52; *Diary of Thomas Brown, Writer in Kirkwall, 1675–1693*, ed. A. Francis Steuart (Kirkwall, 1898), 34–5; *Journal of the Hon. John Erskine of Carnock*, ed. Walter MacLeod (Edinburgh, 1893), 115; D. Laing (ed.), *Historical Selections from the Manuscripts of Sir John Lauder of Fountainhall, Volume First: Historical Observations, 1680–1686* (Edinburgh, 1837), 164, 167; *RPC*, 3rd ser., XI, 46, 305.
18. TNA, SP 29/62, 159; SP 44/9, 271; Louis Moreri, *Le Grand Dictionnaire Historique, Vol. III* (Paris, 1689), 1089; Louis Moreri, *The Great Historical, Geographical, Genealogical and Poetical Dictionary* (London, 1694), s.v. 'Monmouth'.
19. Brown, *Diary*, 50, 53; *London Gazette*, 1 Oct. 1688; Rosie, 'Charters', 43; John Brand, *A Brief Description of Orkney, Zetland, Pightland-Firth and Caithness* (1701; Edinburgh, 1883), 124; *RPS*, 1689/3/189; A1689/6/15; OCR/23/13, 31v.
20. *CSP, William and Mary, 1689–1702*, I, 295; 'Illustrative Notes' in James Wallace, *A Description of the Isles of Orkney*, ed. John Small (Edinburgh, 1883), 240; Utrecht Archives, Church Records, 1675–92, fol. 33, online at https://www.openarch.nl/hua:093E1C00-93F9-D7BE-E053-4701000AAEAB; W.

Cobbett and. T. B. Howell (eds), *A Complete Collection of State Trials* (34 vols, London, 1809–28), XIII, 790; Hossack, 88–9; Brown, *Diary*, 62–3; *RPC*, 3rd ser., XV, 246, 624–5; Craven, III, 126–8; *A Specimen of Dr Burnet's Behaviour in Private Cases* (London, 1724), 11–17.

21. *NHO*, 309–11; James M. Irvine, *The Breckness Estate* (Ashtead, 2009), 57–8; Karen J. Cullen, *Famine in Scotland: The 'Ill Years' of the 1690s* (Edinburgh, 2010), 31–53; Joseph Redington (ed.), *Calendar of Treasury Papers, 1720–1728* (London, 1889), 276.
22. *NHO*, 310; James M. Irvine, *The Orkney Poll Taxes of the 1690s* (Kirkwall, 2003); *RPS*, A1700/10/35.
23. OCR/4/5, 1, 27–33; J. B. Craven, *History of the Episcopal Church in Orkney, 1688–1882* (Kirkwall, 1883), 22–3; *Fasti*, 235, 252; *Annals*, 116; *Acts of the General Assembly of the Church of Scotland, 1638–1842* (Edinburgh, 1843), 278.
24. OCR/4/5, 24–5, 35–8, 39–42, 51, 54–5, 72–3, 98–111, 113–31, 137–9, 143; James Lyon, *A Short Account of the Divine Original of Episcopacy* (Edinburgh, 1710), 'Preface'; *RPC*, 3rd ser., XV, 352; OCR/23/13, 32v, 33r, 33v, 36v.
25. Brand, *Orkney*, 10, 39–40; OCR/4/5, 59–60, 293–300.
26. Hugh Marwick, 'The Feas of Clestrain', *Proceedings of the Orkney Antiquarian Society*, 11 (1933); 31–43; OCR/4/5, 166–8; Lyon, *Episcopacy*, 'Preface'; OCR/4/6, 55–6, 60, 65, 68; Craven, *Episcopal*, 33–9.
27. Laura Stewart and Janay Nugent, *Union and Revolution: Scotland and Beyond, 1625–1745* (Edinburgh, 2020), 76–8; Jane N. Ross, *Orkney and the Earls of Morton, 1643–1707* (Kirkwall, 1977), 9–15; *RPS*, 1689/3/58; 1693/4/125; M1700/10/27; 1702/643; 1705/6/143.
28. *RPS*, 1706/10/305; 1706/10/257; Stewart and Nugent, *Union*, 99–100; Hugh Marwick, *Merchant Lairds of Long Ago* (2 vols, Kirkwall, 1939), I, 103; John Bates, *Two (United) Are Better than One Alone* (London, 1707), 4; John Philips, *Cyder: A Poem* (London, 1708), 87; Charles Darby, *Union* (London, 1707), A2v; Elkanah Settle, *Carmen Irenicum: The Union of the Imperial Crowns of Great Britain* (London, 1707), 37.
29. Thomas W. Traill, *The Frotoft Branch of the Orkney Traills* (privately printed, 1902), 41–2; RPS, 1705/6/2; David Wilkinson, 'Orkney and Shetland, 1690–1715', *The History of Parliament*, https://www.historyofparliamentonline.org/volume/1690-1715/constituencies/orkney-and-shetland; OCR/4/6, 66–7; *To the Queen's Most Excellent Majestie, the Humble Address and Supplication of the Suffering Episcopal Clergy in the Kingdom of Scotland* (Edinburgh, 1703); James Ramsay, *Remarks upon the Case of the Episcopal Clergy* (Edinburgh, 1703), 30; Jeffrey Stephen, *Defending the Revolution: The Church of Scotland, 1689–1716* (London, 2013), 197–9; Anthony Aufrere (ed.), *The Lockhart Papers* (2 vols, London, 1817), I, 66; Craven, *Episcopal*, 47–51, 54; OCR/1/1, 49; Robert Wodrow, *Analecta, or, Materials for a History of Remarkable Providences*, ed. Matthew Leishman (4 vols, Edinburgh, 1842–3), I, 368–9.

30. *Fasti*, 240; Tudor, 584–90; Lyon, *Episcopacy*, 139–41; OCR/4/6, 251–3; Craven, *Episcopal*, 57–71.
31. William R. Mackintosh, *Glimpses of Kirkwall and Its People in the Olden Time* (Kirkwall, 1887), 54–6; Tudor, 585–95; OCR/4/6, 314, 347, 366, 373; Craven, *Episcopal*, 72–9; Alasdair Raffe, 'Presbyterians and Episcopalians: The Formation of Confessional Cultures in Scotland, 1660–1715', *English Historical Review*, 125 (2010), 595; *A Letter from a Gentleman in Orkney* (Glasgow, 1710); *The Post Boy*, 29 Nov.–1 Dec. 1711; *Newcastle Courant*, 3 Dec. 1711; *A Further, More Full, True and Impartial Account of the Notorious Mutton-Covenant* (Dublin, 1711); Laurence A. B. Whitley, *A Great Grievance: Ecclesiastical Lay Patronage in Scotland until 1750* (Eugene, OR, 2013), 106.
32. James Sands, *A Letter Directed Thus, for Mr. James Lyon in Kirkwall* (Edinburgh,1710), 28; Lyon, *Episcopacy*, 'Preface', 30; Wilkinson, 'Orkney'; James Law (ed.), *The Ecclesiastical Statutes at Large* (5 vols, London, 1847), I, 594–8; OCR/4/6, 389, 394; Wodrow, *Analecta*, II, 72–3; OLA, D46/2/1/1–3.
33. Wilkinson, 'Orkney'; David M. Ferguson, *Shipwrecks of Orkney, Shetland and the Pentland Firth* (Newton Abbot, 1988), 34–5; R. P. Fereday, *Orkney Feuds and the '45* (Kirkwall, 1980), 1–4; *The Arrival of the King* (London, 1714), 7; Marwick, *Lairds*, I, 48.
34. OCR/4/7, 5–7; Lyon, *Episcopacy*, 'Dedication'; TNA, SP 54/11, 315; 'Declaration of King James VIII, October 25, 1715', *The Jacobite Heritage*, http://www.jacobite.ca/documents/17151025.htm; David Laing and Thomas Macknight (eds), *Memoirs of the Insurrection in Scotland in 1715* (Edinburgh 1858), 352–5, 365, 367; Marwick, *Lairds*, I, 57–60; Irvine, *Breckness*, 63–4.
35. OCR/4/7, 8, 11, 330; *Rentals*, App. 105; Craven, *Episcopal*, 95–100.
36. Fereday, *Feuds*, 1–7, 9–10; *Caledonian Mercury*, 6 Apr. 1725; (Daniel Defoe?), *An Account of the Conduct and Proceedings of the Late John Gow* (London, 1725); Marwick, *Lairds*, I, 110–12; (Daniel Defoe?), *The History and Lives of All the Most Notorious Pirates, and Their Crews* (3rd edn, London, 1729), 112–20; *Notes*, 213–24.
37. Fereday, *Feuds*, 7–24; Marwick, *Lairds*, I, 110.
38. Fereday, *Feuds*, 36–52; *Rentals*, App. 125; Horace Walpole, *Journal of the Reign of King George the Third* (2 vols, London, 1849), I, 233; Magnus Troil, *A Second Letter to the Earl of Zetland* (Kirkwall, 1847), App. 12–13.
39. *Derby Mercury*, 27 July 1744; Alistair and Henrietta Tayler, *The House of Forbes* (Edinburgh, 1937), 257; OCR/4/8, 179; Fereday, *Feuds*, 59–65.
40. Fereday, *Feuds*, chs 8–11; *Ascanius, or, The Young Adventurer: A True History* (London, 1746), 178; George Low, 'Tour through the North Isles and Part of the Mainland of Orkney in the Year 1778', in Alfred W. Johnston and Amy Johnston (eds), *Old-Lore Miscellany of Orkney, Shetland, Caithness and Sutherland, Vol. VIII* (London, 1920), 142–3.
41. James Mackenzie, *The General Grievances and Oppression of the Isles of Orkney* (1750; Edinburgh,

1836), quotes at 2, 100; court submissions collated in *Earl of Galloway and Udallers of Orkney v Earl of Morton, 1745–52* and *Pundlar Process ... 1757–59*, OLA, 333Y.

42. OLA, D1/375; *NHO*, 341, 397; Mackenzie, *Grievances*, 21; Murdoch Mackenzie, *Orcades, or, A Geographic and Hydrographic Survey of the Orkney and Lewis Islands, in Eight Maps* (London, 1750), 1–2; Christopher Fleet, Margaret Wilkes and Charles Withers, *Scotland: Mapping the Islands* (Edinburgh, 2016), 88–9.

8: Sin and Selkies

1. NRS, SC10/1/3, 226v.
2. NRS, SC10/1/5, 56v; *Folklore*, 19–23.
3. J. B. Craven, *Church Life in South Ronaldsay and Burray in the Seventeenth Century* (Kirkwall, 1911), 44; OCR/4/4, 76; *Caledonian Mercury*, 31 Dec. 1739.
4. Craven, *Life*, 34; *Poetical Descriptions of Orkney, M.DC.LII*, ed. James Maidment (?) (Edinburgh, 1835), xiii; 'An American's Impression of the Orkneys', *Orkney Herald*, 8 Feb. 1911; Jeanie M. Laing, *Notes on Superstition and Folk Lore* (Brechin, 1885), 72; Duncan J. Robertson, 'Orkney Folk-Lore', *Orkney Herald* 5 Mar. 1924; *George Marwick: The Collected Works of Yesnaby's Master Storyteller*, ed. Tom Muir and James Irvine (Kirkwall, 2014), 352–4; Alexander Fenton, *The Northern Isles: Orkney and Shetland* (2nd edn, East Linton, 1997), 618–22; Black, 167–70; John Brand, *A Brief Description of Orkney, Zetland, Pightland-Firth and Caithness* (1701; Edinburgh, 1883), 94, 105; J. M. McPherson, *Primitive Beliefs in the North-East of Scotland* (London, 1929), 70–1; Christina Hole, 'Superstitions and Beliefs of the Sea', *Folklore*, 78 (1967), 188–9.
5. Brand, *Orkney*, 172–3; Walter Traill Dennison, 'Orkney Folk-Lore', *The Scottish Antiquary, or, Northern Notes and Queries*, 7 (1893), 171–7; J. A. Pottinger, 'The Selkie Wife', *Orkney and Shetland Miscellany*, 1 (1907), 173–5.
6. Traill Dennison, 'Folk-Lore', 173; *Folklore*, 25–9; Muir and Irvine, *Marwick*, 4–5; Andrew Jennings, 'The Finnfolk', *University of the Highlands and Islands*, https://www.uhi.ac.uk/en/research-enterprise/cultural/institute-for-northern-studies/research/conferences/previous-conferences-/the-finnfolk/.
7. OLA, DO88, 26–7; James Wallace, *A Description of the Isles of Orkney* (Edinburgh, 1693), 28; James Wallace, *An Account of the Islands of Orkney* (London, 1700), 61; Brand, *Orkney*, 76–7; Jonathan Westaway, 'The Inuit Discovery of Europe? The Orkney Finnmen, Preternatural Objects and the Re-enchantment of Early-Modern Science', *Atlantic Studies*, 19 (2022), 200–23; Hans Sloane, 'An Account of Four Sorts of Strange Beans, Frequently Cast on Shoar on the Orkney Isles', *Philosophical Transactions*, 19 (1695–7), 298–300; Guinevere Barlow, 'A Northern Charm: Some Popular Uses of Sea-Beans', *Northern Studies*, 46 (2014), 1–14; Muir and Irvine, *Marwick*, 6.
8. Alan Bruford, 'Trolls, Hillfolk, Finns and Picts: The Identity of the Good Neighbors in Orkney and Shetland',

in Peter Narvaez (ed.), *The Good People: New Fairylore Essays* (New York and London, 1991), 120; Duncan J. Robertson, 'Orkney Folk-Lore', *Orkney Herald* 20 Feb. 1924; OLA, DO88, 26.

9. Brand, *Orkney*, 57, 96, 170; George Low, *A History of the Orkneys*, ed. Olaf D. Cuthbertson (Kirkwall, 2001), 50–2; George Low, *A Tour through the Islands of Orkney and Schetland*, ed. Joseph Anderson (Kirkwall, 1879), xli–xlii, 17, 57. Fairies' 'perpetual recession': Lizanne Henderson and Edward Cowan, *Scottish Fairy Belief: A History* (East Linton, 2001), 24–6.
10. Brand, *Orkney*, 122; OLA, DO88, 19, 25–6, 33–5.
11. OLA, DO88, 30; Alfred W. Johnston (ed.), *The Church in Orkney: Miscellaneous Records of the Sixteenth, Seventeenth and Eighteenth Centuries* (Kendal, 1940), 38; OCR/1/1, 13, 21; Brand, *Orkney*, 37, 79–82, 86–94, 232, 243; Alasdair A. MacDonald, 'James III: Kingship and Contested Reputation', in Steve Boardman and Julian Goodare (eds), *Kings, Lords and Men in Scotland and Britain, 1300–1625* (Edinburgh, 2014), 251–3; Carole M. Cusack, 'The Cult of St Triduana in Scotland', *Journal of the Sydney Society for Scottish History*, 17 (2018), 31–45.
12. Brand, *Orkney*, 82, 170–1, 241, 243, 244; Low, *History*, 53, 55; McPherson, *Primitive Beliefs*, 17–18; M. Macleod Banks, *British Calendar Customs: Orkney and Shetland* (London, 1946), 55–62.
13. Christopher D. Morris and Norman Emery, 'The Chapel and Enclosure on the Brough of Deerness, Orkney: Survey and Excavations, 1975–1977', *Proceedings of the Society of Antiquaries of Scotland*, 116 (1986), 304–5, 341–5; Brand, *Orkney*, 87; 'Remnants of Popery in Orkney', *The Orcadian*, 26 July 1856; Hugh Marwick, 'The Antiquities of Papa Westray', *Orkney Herald*, 10 Dec. 1924; OLA, DO88, 19; James M. Mackinlay, *Folklore of Scottish Lochs and Springs* (Glasgow, 1893), 145–6).
14. OCR/4/1, 148–9; OCR/16/1, 140–2; OCR/1/1, 70; Low, *History*, 42, 53; *Dictionaries of the Scots Language* (https://dsl.ac.uk/), 'Ganfer'; OCR/11/, 57–9; R. G. Lamb, 'The Cathedral of Christchurch and the Monastery of Birsay', *Proceedings of the Society of Antiquaries of Scotland*, 105 (1973), 204; Margaret Hunter, 'Jo: Ben's Description of Orkney', *New Orkney Antiquarian Journal*, 6 (2012), 40.
15. Low, *History*, 39, 53; James Omond, *Orkney Eighty Years Ago* (Kirkwall, 1911), 10; Banks, *Calendar*, 58.
16. Low, *History*, 45–6; Richard Pococke, *Tours in Scotland, 1747, 1750, 1760*, ed. Daniel W. Kemp (Edinburgh, 1887), 152; Hugh Marwick, *The Orkney Norn* (Oxford, 1929), 226; Brand, *Orkney*, 25; OLA, DO88, 30; Matthew Mackaile, 'A Short Relation of the Most Considerable Things in the Orkney Islands', in Macfarlane, III, 1.
17. OLA, S1/11/6/40; John Geipel, *The Viking Legacy: The Scandinavian Influence on the English and Gaelic Languages* (Newton Abbot, 1971), 106–7.
18. Walter Traill Dennison, *The Orcadian Sketchbook* (Kirkwall, 1880), ix–xi.

19. OCR/23/1, 164; *Notes*, 21; OCR/2/1, 404; Craven, *Life*, 37. Figures from *Fasti*.
20. Wallace, *Account*, 68–9; John Considine, *Small Dictionaries and Curiosity: Lexicography and Fieldwork in Post-Medieval Europe* (Oxford, 2017), 149–50; *Oratio Dominica Polyglottos* (London, 1700), 65.
21. Wallace, *Account*, 68; Low, *Tour*, 105; Íslensk Hómilíubók (heimskringla.no/wiki/Íslensk_hómilíubók); 'Old Norwegian Homily Book', *Medieval Nordic Text Archive*, https://clarino.uib.no/menota/catalogue.
22. *Rentals*, No. III, 33–98 (quotes at 97, 47); OCR/23/1, 175; OCR/4/2, 60v, 1–28 (1702 visitation, separately paginated); OCR/2/1, 21; *RPS*, 1633/6/20; 1645/11/185; 1696/9/144.
23. NRS, GD215/1501; John Mason, 'Scottish Charity Schools of the Eighteenth Century', *Scottish Historical Review*, 33 (1954), 8; Nathan Gray, '"A Publick Benefite to the Nation": The Charitable and Religious Origins of the SSPCK, 1690–1715', University of Glasgow PhD thesis (2011), 177–8; J. T. L. Campbell, 'The Norse Language in Orkney in 1725', *Scottish Historical Review*, 33 (1954), 175; Murdoch Mackenzie, *Orcades, or, A Geographic and Hydrographic Survey of the Orkney and Lewis Islands, in Eight Maps* (London, 1750), 2; James Mackenzie, *The General Grievances and Oppressions of the Islands of Orkney and Shetland* (1750; Edinburgh, 1836), 12.
24. George Barry, *History of the Orkney Islands*, ed. James Headrick (2nd edn, London, 1808), 230; Marwick, *Norn*; Geipel, *Legacy*, 106–7.
25. Margo Todd, *The Culture of Protestantism in Early Modern Scotland* (New Haven and London, 2002), 370–2; OCR/15/1, 6; OCR/19/1, 52. Disrepair of churches: Jocelyn Rendall, *Steering the Stone Ships: The Story of Orkney Kirks and People* (Edinburgh, 2009), 103–15.
26. OCR/24/1, 12; OCR/16/1, 10; OCR/4/2 (1702 visitation), 14; OCR/18/10, 14.
27. V. C. Pogue, 'Church Life in Orphir Two Hundred Years Ago', *Orkney Miscellany*, 2 (1954), 27; Paul J. Sutherland, 'Church Discipline and Its Enforcement in Early 18th Century Kirkwall', in William P. L. Thomson (ed.), *The Making of Modern Orkney* (Aberdeen, 1995), 9–10; Hossack, 417, 429–30; OCR/4/2, 25r; *Folklore*, 94–5; Bo Almqvist, 'Some Orkney Traditions about Unbaptised Children', *Béaloideas*, 66 (1998), 1–7; Hossack, 431–2; OCR/13/1, 203.
28. Ernest Marwick, *An Orkney Anthology* (Edinburgh, 1991), 309–19; OCR/11/1, 19; OCR/23/13, 69v; OCR/5/1, 38, 213; Hossack, 424.
29. John Mooney (ed.), 'Deerness: The Kirk Session (1702–1706)', *Proceedings of the Orkney Antiquarian Society*, 9 (1931), 53; *Annals*, 118; OCR/14/74, 259 (brewing); OCR/15/1, 4 (thatching); OCR/13/1, 63, 70 (threshing), 69 (bleaching), 189 (washing), 225 (baking); Craven, *Life*, 27; OCR/23/1, 285; OCR/23/13, 19r.
30. OCR/23/1, 179; OCR/23/13, 21r; OCR/13/1, 101, 189; Rendall, *Stone*, 143; OCR/18/19, 5–6.

31. OCR/27/1.1, 17; OCR/13/1, 257–8; OCR/23/13, 28r; OCR/14/75, 127, 196–8, 417; OCR/1/1, 124, 146.
32. OCR/14/75, 29, 62; Mooney, 'Deerness', 54; OCR/13/1, 67, 86, 102; Craven, *Life*, 25, 28, 48; OCR/14/74, 159, 265; Johnston, *Church*, 100; OCR/15/1, 1v, 8v–9r, 11v–13r.
33. OCR/14/74, 159–60, 263; OCR/14/75, 139; OCR/13/2, 29, 66; Hossack, 419; OCR/19/1, 33; OCR/23/1, 211; OCR/11/1, 25; Rendall, *Stone*, 142.
34. J. Storer Clouston, 'The Old Chapels of Orkney', *Scottish Historical Review*, 15 (1918), 93–4; OCR/16/1, 9r–v; OCR/13/1, 164; OCR/24/1, 62; OCR/19/1, 126; OCR/14/74, 266; Craven, *Life*, 25, 36, 44; OCR/23/1, 232, 251; OCR/23/13, 13v, 30v, 32r.
35. NRS, GD106/21; OCR/5/1, 47; OCR/23/1, 229; OCR/14/75, 46–7.
36. OCR/5/1, 209–10; OCR/13/1, 82; William P. L. Thomson, 'The Eighteenth-Century Church in Orkney', in H. W. M. Cant and H. N. Firth (eds), *Light in the North: St Magnus Cathedral through the Centuries* (Kirkwall, 1989), 76–7; OCR/23/1, 170, 183; OCR/15/1, 13r.
37. OCR/14/74, 195; OCR/14/75, 258; OCR/23/13, 4r; OCR/13/1, 81, 94; Johnston, *Church*, 99; Mooney, 'Deerness', 52; Craven, *Life*, 26, 51; OCR/23/13, 18v; OCR/4/2 (1702 visitation), 6–7; *Rentals*, No. III, 47; Brand, *Orkney*, 60–1.
38. Mooney, 'Deerness', 53, 55; OCR/4/3, 57; OCR/4/4, 19–20, 49; OCR/23/13, 5v; OCR/23/1, 221; OCR/16/1, 26v; OCR/13/1, 247.
39. OCR/14/75, 328; OCR/4/4, 32, 60–2; NRS, GD106/252; Hossack, 172–3, 357; OCR/7/8, 3r; OCR/24/1, 90, 92; OCR/11/1, 13–14, 17; OCR/18/10, 2, 9. Other cases of defying session authority: Craven, *Life*, 28–9, 33, 45–6; OCR/4/1, 230; OCR/4/3, 61; OCR/23/1, 116.
40. OCR/4/4, 56–7, 68, 70, 81; OCR/4/1, 5, 20, 29, 38, 42–3, 45, 49, 52 (quote), 57, 74, 77, 147, 353 (quote), 367.
41. OCR/2/1, 18–19, 38, 79–82, 143–4; Melville H. Massue, *The Moodie Book* (London, 1906), 43–4; Rendall, *Stone*, 96–7; OCR/4/7, 315–16, 327–8; R. P. Fereday, *Orkney Feuds and the '45* (Kirkwall, 1980), 29–30.
42. Jocelyn Rendall, *St Boniface Kirk* (4th edn, Papa Westray, 2017), 14–17; Rendall, *Stone*, 95–6; *Fasti*, 419.
43. OCR/2/1, 74, 129, 157, 160–3; OCR/3/1, 35, 37–9, 42–7, 100; *Annals*, 168, 188, 214.
44. OCR/2/1, 236–7, 239, 352, 355; Fereday, *Feuds*, 36–7; *Fasti*, 240.
45. OCR/2/1, 336, 437–44; Clotilde Prunier, 'Representations of the "State of Popery" in Scotland in the 1720s and 1730s', *Innes Review*, 64 (2013), 192; OCR/3/2, 152; *Annals*, 254–6; Fereday, *Feuds*, 141–3.
46. Nathaniel Morren (ed.), *Annals of the General Assembly of the Church of Scotland from the Origin of the Relief in 1752 to the Rejection of the Overture of Schism in 1766*, (Edinburgh, 1840), 347–53, 362–70 (quotes at 352, 364); OCR/2/1, 541; *Annals*, 189–91, 272–3; *Unto the Right Honourable the Lords of Council and Session, the Petition of Mr George Trail of Hobbister, Minister of Dunnet* (Edinburgh, 1766), 2–3; *Aberdeen Journal*, 18 June 1764; *Caledonian Mercury*, 9 June 1764; *Scots Magazine*, 1 June 1767, 333; *Lyell vs Traill and Others*, OLA, 285Y.

47. *Scots Magazine*, 6 July 1750, 346; *Collection of All the Papers Published in Relation to the Scheme for Augmenting the Stipends of the Established Clergy in Scotland* (Edinburgh, 1751), 10, 68, 70; *Memorial for Mr. William Nisbet, Minister of the United Parishes of Firth and Stenness* (Edinburgh, 1762), 20; Hossack, 292; OLA, D1/375; [Thomas Hepburn], *A Letter to a Gentleman from His Friend in Orkney*, ed. William Brown (1760; reprint, Edinburgh, 1885), 25, 28–9, 39, 43–4, 56; *A Detection of Some Real and Artificial Errors* (Edinburgh, 1751), 16; 'A. Rediviv.', *A Familiar Epistle from His Excellency the Lord Lieutenant of Orkney to His Mightiness the Prolocutor of the Athelnstonford Congregation in East Lothian* ([Edinburgh?], 1772), 9, 10–11.
48. Tom Muir, *Orkney Folk Tales* (Stroud, 2014), 145–6; Charles Tait, *The Orkney Guide Book* (4th edn, St Ola, 2012), 464.
49. *Folklore*, 50–1; Craven, *Life*, 59; Craven, III, 35.

9: Arcs of Empire

1. William P. L. Thomson, *Kelp-Making in Orkney* (Stromness, 1983), ch. 10 (at p. 79 giving the name of James Fea's wife as Barbara, though this was in fact his mother). For John Fotheringhame, I am indebted to *North Isles Family History*, https://www.bayanne.info/Shetland/getperson.php?personID=I383387&tree=ID1.
2. Thomson, *Kelp-Making*, 12–19, 26–36, 39–41; *OP*, 37; Walter Traill Dennison, 'Orkney Folklore: Sea Myths', *The Scottish Antiquary*, 19 (1891), 131–2.
3. James Fea, *The Present State of the Orkney Islands Considered* (1775; reprint, Edinburgh, 1884), quotes at 51, 54, 86, 93, 100, 128; 'Dr James Fea (James Fea VIII of Clestrain)', *Fea: A Genealogy with Connections to Orkney, Scotland*, http://genealogy.northern-skies.net/genealogy.php?number=13.
4. John Campbell, *A Political Survey of Britain* (2 vols, London, 1774), I, 639–76; Francis Espinasse, revised by M. J. Mercer, 'Campbell, John (1708–1775)', *ODNB*; *Hampshire Chronicle*, 15 Aug. 1774; *Northampton Mercury*, 6 Dec. 1784.
5. *Derby Mercury*, 3 May 1733; *Stamford Mercury*, 25 Dec. 1746, 21 May 1747; *Newcastle Courant*, 8 July 1749; Bob Harris, 'Patriotic Commerce and National Revival: The Free British Fishery Society and British Politics, c.1749–58', *English Historical Review*, 114 (1999), 285–313; *NHO*, 368; Fea, *State*, 87–8.
6. *NHO*, 363–6; Sheena Wenham, *A More Enterprising Spirit: The Parish and People of Holm in 18th Century Orkney* (Kirkwall, 2001), 164–73.
7. *Northampton Mercury*, 1 Aug. 1785; *Leeds Intelligencer*, 2 Aug. 1785; *Caledonian Mercury*, 23 June 1787; *NHO*, 377; Wenham, *Holm*, 179; P. N. S. Graeme, 'Health and Healing in 17th and 18th Century Orkney', *Orkney Miscellany*, 2 (1954), 34–46; OCR/11/1, 62, 76; OCR/24/1, 520; *Scots Magazine*, 6 Feb. 1758, 106, 1 July 1770, 378; *Leeds Intelligencer*, 4 Mar. 1768; Fea, *State*, 44–5; James Fea, *Considerations on the Fisheries*

in the Scotch Islands (London, 1787), 3, 34.

8. *Caledonian Mercury*, 10 Mar. 1773, 30 July 1774; *Newcastle Chronicle*, 12 Oct. 1771; *Leeds Intelligencer*, 21 Dec. 1773; Murdoch Mackenzie, *Orcades, or, A Geographic and Hydrographic Survey of the Orkney and Lewis Islands, in Eight Maps* (London, 1750), 1–2; *OP*, 255, 284; Paul A. Lynn, *Scottish Lighthouse Pioneers: Travels with the Stevensons in Orkney and Shetland* (Dunbeath, 2017), 6–18; Canmore, ID 3644.
9. George Eunson, *The Ancient and Present State of Orkney* (Newcastle, 1788), 49–50; *Caledonian Mercury*, 28 Nov. 1771, 25 Mar. 1775, 17 Oct. 1791, 27 Feb. 1792; Tudor, 322–3; James M. Irvine, *The Breckness Estate* (Ashtead, 2009), 123–5; *Scots Magazine*, 1 Apr. 1792, 200–1; *OP*, 136.
10. Eunson, *Ancient*, 75, 112–26; *Oxford Journal*, 6 Mar. 1773; [Thomas Hepburn], *A Letter to a Gentleman from His Friend in Orkney, Containing the True Causes of the Poverty of That Country*, ed. William Brown (1760; reprint, Edinburgh, 1885), 23–4; *Caledonian Mercury*, 4 Mar. 1767; James Hall, *Travels in Scotland by an Unusual Route, with a Trip to the Orkneys and Hebrides* (2 vols, London, 1807), II, 57; Hossack, 387–9; Wenham, *Holm*, 215–20; P. N. S. Graeme, 'George Eunson: Orkney's 18th Century Firebrand', *Orkney Miscellany*, 4 (1957), 1–16.
11. John Brooke, 'Westminster, 1754–1790', *The History of Parliament*, https://www.historyofparliamentonline.org/volume/1754-1790/constituencies/westminster; W. R. Mackintosh, *Curious Incidents from the Ancient Records of Kirkwall* (Kirkwall, 1892), 244–8; N. Chapman (ed.), *Select Speeches, Forensic and Parliamentary* (5 vols, Philadelphia, 1807), III, 296; James Hartley (ed.), *History of the Westminster Election* (London, 1785), 266, 270, 274; *Kentish Gazette*, 12 Aug. 1785; *Derby Mercury*, 21 Oct. 1784.
12. Leonard Macnally, *Fashionable Levities, A Comedy* (London, 1785), 13; Daniel Defoe, *A Tour thro' That Part of Great-Britain Called Scotland* (Dublin, 1746), 288; William Smith, *A Natural History of Nevis* (London, 1745), 293; Charles Smith, *The Antient and Present State of the County and City of Waterford* (Dublin, 1746), 347; James Morrison, *An Address to the Protestant Interest in Scotland* (Glasgow, 1778), 18; David Williams, *A Plan of Association, on Constitutional Principles* (London, 1780), 42; *Ranspach, or, Mysteries of a Castle* (2 vols, Uttoxeter, 1797), II, 9; *Leeds Intelligencer*, 2 Oct. 1770; Elizabeth Baigent, 'Weddell, James (1787–1834)', *ODNB*; *Scots Magazine*, 5 July 1762, 373; 1 Jan. 1775, 16; *London Gazette*, 30 Jan. 1776; John Gray, *Practical Observations on the Proposed Treaty of Union* (London, 1800), 47; *Speech of the Right Honourable Sylvester Douglas in the House of Commons* (London, 1800), 35–6.
13. John Lettice, *Letters on a Tour through Various Parts of Scotland* (London, 1794), 36. See also *The Welcome: Two Congratulatory*

Poems (London, 1714), 4; *True Briton*, 23 July 1723; Samuel Foote, *The Englishman Returned from Paris* (Dublin, 1756), 17; Jonas Hanway, *A Journal of Eight Days Journey from Portsmouth to Kingston upon Thames* (London, 1756), 306; John Burgoyne, *The Lord of the Manor, A Comic Opera* (Dublin, 1781), 27; Thomas Lewis O'Beirne, *A Gleam of Comfort to This Distracted Empire* (London, 1785), 53; Theobald Wolfe Tone, *Spanish War* (Dublin, 1790), 1; *Four Sermons, Preached in London, at the Second General Meeting of the Missionary Society* (London, 1796), xx; Andrew Fuller, *Memoirs of the late Rev. Samuel Pearce* (London, 1800), 165.

14. Alexander Pope, *Essay on Man*, ed. Mark Pattison (6th edn, Oxford, 1881), 44.
15. Suzanne Rigg, *Men of Spirit and Enterprise: Scots and Orkneymen in the Hudson's Bay Company, 1780–1821* (Edinburgh, 2011), 3–4, 16, 19, 24–5, 45–6, 56–7; Edith I. Burley, *Servants of the Honourable Company: Work, Discipline, and Conflict in the Hudson's Bay Company, 1770–1879* (Oxford, 1997), 3, 67–8; Ian MacInnes, 'The Alexander Graham Case: The Royal Burgh of Kirkwall and the Unfree Traders of Stromness', *Orkney Heritage*, 1 (1981), 99–130; John Willock, *The Voyages and Adventures of John Willock, Mariner* (Philadelphia, 1798), 243; John Nicks, 'Tomison, William', *Dictionary of Canadian Biography* (http://www.biographi.ca/en/bio/tomison_william_6E.html).
16. *The Journal of Alexander Henry the Younger, 1799–1814*, ed. Barry M. Gough (2 vols, Toronto, 1988), I, 299; Jocelyn Rendall, *Farstraers: Voyages and Homecomings* (Papa Westray, 2021), 228–60; *London Magazine*, 28 (1759), 707.
17. Rigg, *Enterprise*, 41–2; *Chester Courant*, 2 Oct. 1770; *Aberdeen Journal*, 1 Nov. 1773; *Newcastle Courant*, 1 Oct. 1774; Rendall, *Farstraers*, 62–7; Howie Firth, *Orkney* (London, 2013), 409; OLA, D3/385; Barbara DeWolfe (ed.), *Discoveries of America: Personal Accounts of British Emigrants to North America during the Revolutionary Era* (Cambridge, 1997), 210.
18. E. E. Rich, *Hudson's Bay Company, 1670–1870* (3 vols, New York, 1961), II, 128, 269; Burley, *Servants*, 73; Rendall, *Farstraers*, 262–7; Rigg, *Enterprise*, 16, 122–32.
19. Eunson, *Ancient*, 57; Rich, *Hudson's Bay*, II, 268; *OP*, 138, 159, 215; OLA, SC11/5/1778/46.
20. Fea, *State*, 85; Alexander Fenton, *The Northern Isles: Orkney and Shetland* (East Linton, 1997), 596; Ernest Marwick, *An Orkney Anthology* (Edinburgh, 1991), 413; Eunson, *Ancient*, 57; J. D. M. Robertson, *The Press Gang in Orkney and Shetland* (Kirkwall, 2011), 3–4, 25, 29, 71–117, 137–58, 182–3; Rendall, *Farstraers*, 118; W. R. Mackintosh, *Around the Orkney Peat-Fires* (8th edn, Kirkwall, 1975), 91–171.
21. OLA, D31/24/3, Ernest W. Marwick, 'Journey from Serfdom', unpublished manuscript, ch. 3; Wenham, *Holm*, 205–6; Rendall,

Farstraers, 120–1; Robertson, *Press Gang*, 30–6, 278–9, 285.

22. *Newcastle Courant*, 2 Aug. 1746, 28 May 1748, 26 Sept. 1778; *Caledonian Mercury*, 25 Sept. 1746, 17 May 1777, 24 June 1782; *Stamford Mercury*, 23 Oct. 1746, 28 May 1782; *Leeds Intelligencer*, 22 Sept. 1761, 19 Aug. 1777; Gardner W. Allen, *A Naval History of the American Revolution* (2 vols, Boston and New York, 1913), I, 234, 238; *Hereford Journal*, 26 July 1781; *Oxford Journal*, 29 June 1782, 5 Oct. 1782; *Saunders's News-Letter*, 25 Nov. 1782; OLA, D1/660/25 [H1]; *Dublin Evening Post*, 18 Aug. 1785; R. P. Fereday, *The Orkney Balfours, 1747–99* (Oxford, 1990), 103.
23. Fereday, *Balfours*, ch. 6; OCR/5/2, 51–2; Rendall, *Farstraers*, 220–1; R. P. Fereday, *Saint-Faust in the North, 1803–4: Orkney and Shetland in Danger* (Oxford, 1995), 1–3, 51; *Scots Magazine and Edinburgh Literary Miscellany*, 67 (1805), 268; Samuel Laing, *Journal of a Residence in Norway* (London, 1837), 353; P. N. Sutherland Graeme (ed.), 'Scapa Flow in 1812: A Memorial to the Lords Commissioners of Admiralty', *Orkney Miscellany*, 4 (1957), 57–64.
24. Louis G. Graeme, *Or and Sable: A Book of the Graemes* (Edinburgh, 1903), 508; *Victory* muster roll, *Memorials and Monuments in Portsmouth*, www.memorialsinportsmouth.co.uk/dockyard/victory-muster.htm; 'Nelson, Trafalgar, and Those Who Served', TNA, www.nationalarchives.gov.uk/trafalgarancestors/; Marwick, *Anthology*, 413–15; Firth, *Orkney*, 105, 409; Lord Byron, *The Island, or, Christian and His Comrades* (London, 1823), 26; William Bligh, *An Answer to Certain Assertions* (London, 1794), 9.
25. OLA, D31/2/1/9; Marwick, *Anthology*, 415–16; David H. Edwards, *Modern Scottish Poets, with Biographical and Critical Notices* (Brechin, 1889), 39–40; British Library, Add. Ms. 34710 L; F. L. Maitland, *Narrative of the Surrender of Buonaparte* (London, 1826); TNA, ADM 37/5032.
26. Rosemary Mitchison, 'Sinclair, John (1754–1835)', *The History of Parliament*, https://www.historyofparliamentonline.org/volume/1754-1790/member/sinclair-john-1754-1835; *The Statistical Accounts of Scotland, 1791–1845*, https://stataccscot.edina.ac.uk/static/statacc/dist/home; C. W. J. Withers, 'How Scotland Came to Know Itself: Geography, National Identity and the Making of a Nation, 1680–1790', *Journal of Historical Geography* 21 (1995), 371–97; *Encyclopaedia Britannica* (3rd edn, 18 vols, Edinburgh, 1797), XIII, 496–501.
27. *OP*, 10, 66, 181, 210, 222, 361.
28. *OP*, 15, 16, 21, 38, 45, 52–5, 69, 75, 78, 79, 121, 135, 137, 165, 194–5, 215, 298, 340, 354–7.
29. *OP*, 53, 66, 140, 159–60, 192, 270.
30. Eunson, *Ancient*, 59–60; Rendall, *Farstraers*, 95–107; *Caledonian Mercury*, 31 May 1788; *The Melancholy Case of Francis Liddell Minister of Orphir in the Orcades* (Edinburgh, 1808), 16, 20–1, 24; *OP*, 61.
31. *OP*, 137, 177–8, 196, 211, 235; James Haldane, *Journal of a Tour through the Northern Counties of Scotland*

and the Orkney Isles in Autumn 1797 (2nd edn, Edinburgh, 1798), 50–66; Archibald Macwhirter, 'Separations and Unions in the Church of Scotland', *Orkney Miscellany*, 3 (1956), 26–33; Adam Hood, *John Oman* (Milton Keynes, 2012), 8–9; Hossack, 445–6; Liddell, *Melancholy*, 78.

32. *Britannica*, 499; *OP*, 53–4, 135, 158, 214–15, 328–30; *Annals*, 133–4.

Epilogue: The Lighthouse Inspector

1. Mary Shelley, *Frankenstein, or, The Modern Prometheus* (3 vols, London, 1818), III, 35, 37. For what follows, J. G. Lockhart, *Memoirs of the Life of Sir Walter Scott, Bart.*, (10 vols, Edinburgh, 1862), IV, 182–277. The voyage is often described as on the *Pharos*, but that was a separate vessel, fitted as a lightship: 'Our Ships', *Northern Lighthouse Board*, www.nlb.org.uk/history/ships/.
2. Lockhart, *Memoirs*, IV, 387 (Byron's letter); James Hall, *Travels in Scotland by an Unusual Route, with a trip to the Orkneys and Hebrides* (2 vols, London, 1807), II, 516.
3. Scott referred to 'Fair Kirkwall' in Canto 6 of *The Lay of the Last Minstrel* (London, 1805), 179; Patrick Neil, *A Tour through Some of the Islands of Orkney and Shetland* (Edinburgh, 1806); John Shirreff, *General View of the Agriculture of the Orkney Islands* (Edinburgh, 1814).
4. Bill Jenkins, 'The "Stronsay Beast": Testimony, Evidence and Authority in Early Nineteenth-Century Natural History', *Notes and Records: The Royal Society Journal of the History of Science*, 77 (2023), 471–94; Alison Finlay, 'Thomas Gray's Translations of Old Norse Poetry', in David Clark and Carl Phelpstead (eds), *Old Norse Made New: Essays on the Post-Medieval Reception of Old Norse Literature and Culture* (London, 2007), 1–20.
5. OLA, D2/17/4; Hossack, 397–400; *Notes*, 20–2.
6. *Folklore*, 47.
7. 'Advertisement' in Walter Scott, *The Pirate*, ed. Mark Weinstein and Alison Lumsden (Edinburgh, 2001), 4. All quotations in the following paragraphs are taken from this edition.
8. NLS, MS.6140(v), William Daniell, *A Voyage Round Great Britain, Vol. V* (London, 1821), manuscript insert, 8, 12.
9. Daniell, *Voyage*, 1.

List of Illustrations

Integrated images

Chapter opener image, map of Orkney on dark slate. Panther Media GmbH/Alamy Stock Photo.

Plates

27 Start Point Lighthouse, by William Daniell. Heritage Image Partnership Ltd/Alamy Stock Photo.
28 Satirical Print of Charles James Fox, 1784. © The British Museum.
29 Kirkwall and the Cathedral of St Magnus, by William Daniell. © Bridgeman Images.
30 Sir Walter Scott, by Sir Thomas Lawrence, 1820–27. Royal Collection, Public domain.

Index